Machine Learning for Tomographic Imaging

About the Series

Series in Physics and Engineering in Medicine and Biology will allow IPEM to enhance its mission to 'advance physics and engineering applied to medicine and biology for the public good.'

Focusing on key areas including, but not limited to:

- clinical engineering
- diagnostic radiology
- informatics and computing
- magnetic resonance imaging
- nuclear medicine
- physiological measurement
- radiation protection
- radiotherapy
- rehabilitation engineering
- ultrasound and non-ionising radiation.

A number of IPEM–IOP titles are published as part of the EUTEMPE Network Series for Medical Physics Experts.

Machine Learning for Tomographic Imaging

Ge Wang
Rensselaer Polytechnic Institute

Yi Zhang
Sichuan University

Xiaojing Ye
Georgia State University

Xuanqin Mou
Xi'an Jiaotong University

IOP Publishing, Bristol, UK

ISBN   978-0-7503-2216-4 (ebook)
ISBN   978-0-7503-2214-0 (print)
ISBN   978-0-7503-2217-1 (myPrint)
ISBN   978-0-7503-2215-7 (mobi)

DOI   10.1088/978-0-7503-2216-4

Version: 20191201

IOP ebooks

British Library Cataloguing-in-Publication Data: A catalogue record for this book is available from the British Library.

Published by IOP Publishing, wholly owned by The Institute of Physics, London

IOP Publishing, Temple Circus, Temple Way, Bristol, BS1 6HG, UK

US Office: IOP Publishing, Inc., 190 North Independence Mall West, Suite 601, Philadelphia, PA 19106, USA

For my wife Ying Liu, whose support has been extraordinary, my academic mentors Michael Vannier et al, and all other family members and collaborators.

— Ge Wang

For my wife and daughter, who have been always fully supporting my academic career development.

— Yi Zhang

For my family.

— Xiaojing Ye

For my family.

— Xuanqin Mou

Contents

Part II   X-ray computed tomography

4   X-ray computed tomography   4-1

Foreword

We are currently witnessing a revolution in science and engineering: in only a few years, machine learning has become the basis for almost every single algorithm development. I still remember a college course on non-linear systems in the early 1990s: except for a few zealots, my peers and I used to think of neural networks as an exotic and entirely impractical field of science. Little did we realize how wrong we were. Increases in computing power, access to large amounts of data, and the creativity of many scientists have entirely changed that view. Today, neural networks have become more pervasive than the Fourier transform.

Some of the most popular applications of machine learning are in image analysis. You probably have some powerful deep learning networks in your pocket: just type 'dog' or 'hat' on your smart phone and it will instantaneously find all pictures containing your targets of interest. The application of machine learning may be less obvious in other areas, such as image generation. By now, machine learning has been at least considered for almost every imaginable algorithmic challenge. In the field of medical imaging, machine learning techniques can triage patients, determine the best scan parameters, perform image reconstruction, enhance images, analyze the quality of images, perform a diagnosis, compute a treatment plan, etc.

Dr Ge Wang was one of the earliest innovators who has been creatively exploring various ways of using deep learning in tomographic imaging since 2016, which is a long time ago given how fast the field has evolved. His team's main impacts range across the topics of CT, MRI, and optical imaging, but they have also made fundamental contributions to the science of neural networks, such as through the investigation of quadratic deep neural networks.

Dr Wang teamed up with Dr Zhang, Dr Ye and Dr Mou to write this book on machine learning for tomographic imaging. The authors are uniquely qualified to undertake this major project, given their combined expertise in mathematics, computer science, image processing, and tomographic imaging, as well as their pioneering research in machine learning for medical imaging. At a high level, this book teaches medical imaging from the perspective of machine learning, and hence covers two important fields: machine learning and tomographic imaging.

The book's first part gives a colorful and inspiring introduction to machine learning and tomographic imaging. Using the Human Vision System as reference, the authors take us on a journey from sparse representations, through dictionary learning, to neural networks and deep learning. Many 'textbook' deep learning architectures are covered at an introductory level. Parts two and three provide an in-depth tutorial of CT and MR image reconstruction, followed by a wide range of machine learning techniques that were developed in recent years. The fourth part further enriches the content of this book by elaborating on other imaging modalities, image quality evaluation, and quantum computing.

Overall, this book provides an amazingly comprehensive overview of neural networks and tomographic reconstruction methods. It is written in an engaging and accessible style, without lengthy mathematical derivations and proofs. This makes it ideal for introducing machine learning and tomographic imaging in the more applied disciplines (physics and engineering), and also for bringing application contexts into the more theoretical disciplines (mathematics and computer sciences).

Every medical imaging scientist who graduated before machine learning was taught in college should probably learn about this area in order to remain competitive. To my knowledge, this book is the first and only publication capturing all important aspects of machine learning and tomographic imaging in one place.

I highly recommend this book for any medical imaging students/professionals with a STEM background. Start with chapters 1–3. Then, depending on whether you are a CT, PET, MRI, ultrasound, or optical imaging aficionado, you may select one or more of the other chapters for further study. Before you know it, you will *deeply learn* this exciting new science, be able to talk intelligently about it, and perform state-of-the-art research in a world that can no longer be imagined without neural networks.

Bruno De Man, October 2019

Preface

This book arose from discussions among four colleagues with a long-standing collaboration and interest in advanced medical image reconstruction methods and applications. Beginning in 2018, our group realized the gap in the literature and in particular among technical books on the emerging technologies that develop and apply artificial intelligence/machine learning (AI/ML) techniques to tomographic reconstruction or tomographic imaging.

As early as 2012, we recognized the opportunity presented by machine learning in formulating plans for doctoral dissertation research where dictionary learning can be used to recover images from projection measurements. By connecting several contemporary image recovery and signal processing methods, in particular compressed sensing, neural network, and deep learning techniques, our discussions and projects converge to develop and apply ML methods to advance the frontier of image reconstruction, with an emphasis on medical imaging.

The interested reader entering this field may have a background in artificial intelligence or mathematical knowledge of tomographic reconstruction, but few will have all of the knowledge needed to understand the field of ML for tomographic image reconstruction. Hence, we have now created this book to cover what we believe to be a comprehensive collection of key topics in a logical and consistent manner.

The prerequisites for reading this book include calculus, matrix algebra, Fourier analysis, medical physics, and basic programming skills. We believe that PhD candidates in the imaging field are generally well prepared to understand all of the content through serious effort, while advanced undergraduate students can also learn essential ideas and capture selected materials (you can skip the chapters/sections/subsections marked with an '*'). To facilitate teaching and learning, most relevant numerical methods are described in appendix A, and hands-on projects are suggested in appendix B, with sample codes and working datasets.

The logical dependence between the key components of this book is illustrated in the diagram in the introduction below. We strongly recommend that you read this introduction first to obtain an overall perspective. It is also recommended to read the first three chapters sequentially so that you are well prepared with both the imaging context and network basics. However, chapter 3 alone is a good introduction to general knowledge on artificial neural networks. Then, we can proceed in parallel to CT, MRI, or other tomographic modalities, which are covered in parts II, III, and IV of the book, respectively. It would be the best to read part IV after reading parts II and III as deep reconstruction networks are clearly explained for CT and MRI in these two parts. Appendix A can be read as needed, but appendix B is strongly recommended, and should be at least consulted to run the basic networks explained in chapter 3. The network examples for CT and MRI can be adapted for independent class projects.

We hope that this book will be useful for a review course at the graduate level, but it has not been tested yet. As teaching experience is accumulated using this book, homework problems and solutions will become available, along with example class project reports and codes. A book-related website is maintained on the Fully3D community website: http://www.fully3d.org/rpi/.

The materials contained in this book are presented in their first version. As a result, a number of topics are not treated in detail or in depth. Nevertheless, after reading this book you should have state-of-the-art knowledge of a broad spectrum of methods. We welcome your critiques and suggestions so we can make future versions better, with key references cited in a more balanced way.

July 2019

Acknowledgments

The four parts of this book were initially drafted by Professors Mou, Zhang, Ye, and Wang respectively, based on a collectively developed overall layout. The appendices were drafted by Professor Ye. Hands-on examples were developed by Professors Zhang, Ye, and Mou collectively, and integrated by Professor Zhang. All parts were internally reviewed and revised by the four co-authors, and editorially refined by the staff of IOP Publishing.

We would like to express our sincere gratitude to all individuals, publishers, and companies for permission to reproduce some of the images and figures in this book, IOP Publishing staff for guidance during the development of the book, and importantly our students, other lab members, and collaborators for their significant contributions, including but not limited to Hongming Shan, Qing Lyu, Christopher Wiedeman, Harshank Shrotriya, Huidong Xie, Fenglei Fan, Mengzhou Li, and Varun Ravichandran. Drs Michael Vannier and Hengyong Yu offered insightful advice on the strengths and weaknesses of this book for improvements. Last but not least, the following leading companies have graciously given permission to reproduce some of the best figures/images in this book: Cannon, General Electric, Siemens, and Phillips (in alphabetical order). Without these, this book would not have been created in its current form. We are happy that the first version of this book is now complete, and look forward to producing future versions and more excitement in the years to come.

Author biographies

Ge Wang

Ge Wang, PhD wangg6@rpi.edu; https://www.linkedin.com/in/ge-wang-axis

Ge Wang is the Clark and Crossan Endowed Chair Professor and the Director of the Biomedical Imaging Center, Rensselaer Polytechnic Institute, Troy, NY, USA. He published the first spiral/helical cone-beam/multi-slice CT paper in 1991 and has since then systematically contributed over 100 papers on theory, algorithms, artifact reduction, and biomedical applications in this area of CT research. Currently, over 100 million medical CT scans are carried out yearly with a majority in the spiral/helical cone-beam/multi-slice mode. Dr Wang's group developed interior tomography theory and algorithms to solve the long-standing 'interior problem' for high-fidelity local reconstruction, establishing the feasibility of omni-tomography ('all-in-one') with simultaneous CT-MRI as an example. He initiated the area of bioluminescence tomography. He has produced over 480 journal publications, covering diverse imaging-related topics and receiving a high number of citations and academic awards. His results have been featured in *Nature*, *Science*, *PNAS*, and news media. In 2016, he published the first perspective on deep-learning-based tomographic imaging as a new direction of machine learning and a *Nature Machine Intelligence* paper on the superiority of deep learning over iterative reconstruction in collaboration with his colleagues. His team has been continuously well-funded by federal agencies and imaging companies. His interests include CT, MRI, optical tomography, multi-modality fusion, artificial intelligence, and machine learning. He is the lead Guest Editor for five *IEEE Transactions on Medical Imaging* special issues, the founding Editor-in-Chief of the *International Journal of Biomedical Imaging*, a Board Member of *IEEE Access*, and an Associate Editor of *IEEE Transactions on Medical Imaging* (TMI) (recognized as an 'Outstanding Associate Editor'), *Medical Physics*, and *Machine Learning Science and Technology*. He is Fellow of IEEE, SPIE, OSA, AIMBE, AAPM, AAAS and NAI.

Yi Zhang

Yi Zhang, PhD yzhang@scu.edu.cn

Yi Zhang received his bachelor's, master's, and PhD degrees from the College of Computer Science, Sichuan University, Chengdu, China, in 2005, 2008, and 2012, respectively. From 2014 to 2015, he was with the Department of Biomedical Engineering, Rensselaer Polytechnic Institute, Troy, NY, USA, as a Postdoctoral Researcher. He is currently an Associate Professor with the College of Computer Science, Sichuan University, and is the Dean of the Software Engineering Department. His research interests include medical imaging, compressive sensing, and deep learning. He authored more than 60 papers in the field of medical imaging. His group published the first peer-reviewed journal paper on deep learning based low-dose CT and subsequently published more than 20 papers in this rapidly expanding area. These papers were published in several leading journals, including *IEEE Transactions on Medical Imaging*, *IEEE Transactions on Computational Imaging*, *Medical Image Analysis*, *European Radiology*, *Optics Express*, etc, and reported by the Institute of Physics (IOP) and during the Lindau Nobel Laureate Meeting. He received major funding from the National Key R&D Program of China, the National Natural Science Foundation of China, and the Science and Technology Support Project of Sichuan Province, China. He is a Guest Editor of the *International Journal of Biomedical Imaging, Sensing and Imaging*, and an Associate Editor of *IEEE Access*. He is a Senior Member of IEEE.

Xiaojing Ye

Xiaojing Ye, PhD xye@gsu.edu

Dr Xiaojing Ye is an Associate Professor with the Department of Mathematics and Statistics at Georgia State University, Atlanta, USA. Prior to joining Georgia State University in 2013, he was a Visiting Assistant Professor with the School of Mathematics at the Georgia Institute of Technology, USA. He received his bachelor's degree in mathematics from Peking University in 2006, and a master's degree in statistics and a PhD degree in mathematics from the University of Florida, USA, in 2009 and 2011, respectively. His research focuses on applied and computational mathematics, in particular variational methods for imaging problems, numerical optimization and analysis, and computational problems in machine learning.

Xuanqin Mou

Xuanqin Mou, PhD xqmou@xjtu.edu.cn

Dr Xuanqin Mou received his bachelor's, master's, and PhD degrees from Xi'an Jiaotong University, China. Since 1987, he has been a faculty member with Xi'an Jiaotong University and was prompted to Full Professor in 2002. Currently, he is the Director of the National Data Broadcasting Engineering and Technology Research Center, and the Director of the Institute of Image Processing and Pattern Recognition. He served on the Twelfth Expert Evaluation Committee for the National Natural Science Foundation of China. He is currently on the Executive Committee of the China Society of Image and Graphics, the Executive Committee of the Chinese Society for Stereology, also serves as the Deputy Director of the CT Committee of the Chinese Society for Stereology. He was on the editorial boards of several academic journals and technical program committee for many international conferences. In 2017, Dr Mou co-chaired the Fourteenth Fully Three-Dimensional Image Reconstruction in Radiology and

Nuclear Medicine Conference at Xi'an Jiaotong University, which is the prime conference in the field of CT/PET/SPECT image reconstruction research and development. His interests include x-ray medical imaging, CT reconstruction, observer models, perceptual quality assessment, and video coding. He published over 200 peer-reviewed journal and conference papers, and holds over ten Chinese patents as the principle inventor. He received a Second-class Award for Invention issued by the Ministry of Education of China as the principal investigator, and several other awards. As the principal investigator, he received 16 governmental grants and 20 industrial funds.

Art rendering by Ge Wang, July 2019. Panda symbolizes digital and biological technologies, being binary and adorable. As is well known, it is racially representative, being black and white, as well as oriental. The Yin–Yang pattern suggests entanglement of information, and hope for the future.

Introduction

Artificial intelligence/machine learning (AI/ML) is one of the largest diamonds ever discovered in the evolution of science, and has many facets, one of which is AI/ML-based tomography—the central theme of this, first-of-its-kind, book (figure 0.1). First, we hope to explain the big picture behind our book, help you assess if it is valuable to you, and, if so, suggest guidelines for a pleasant and rewarding reading or learning experience. As the book is in its first version, your feedback is most welcome for us to produce the next edition in the future so that representative networks and key references can be covered as completely as possible. We strongly suggest that you read both the preface and this introduction carefully to obtain a general perspective. In the following, we will do our best to present key points in easy language.

0.1 Artificial intelligence/machine learning/deep learning

Currently, deep learning is the mainstream approach of machine learning (ML), which is arguably the hottest research area of artificial intelligence (AI). AI/ML means allowing a computer to think like a human and even outperform humans in certain (if not most) important tasks. While classic science and technology are really about the magnification of humans' physical power (such as steam engines and assembly lines) and the enhancement of non-intelligent functions (such as cars and planes as our legs, and microscopes and telescopes extending our eyes), AI/ML targets the understanding and prototyping of human intelligence so that we not only demystify the ultimate secret of life but also let machines work for us intelligently (figure 0.2).

Over only the past few years, AI/ML techniques have achieved impressive successes in computer vision, image analysis, speech recognition, language processing, and many other areas. A major feature behind these successes is that they use deep artificial neural networks trained with big data. An artificial neural network consists of many artificial neurons. Such neurons are basic data processing units performing a linear (weighted sum) operation followed by a simple nonlinear (thresholding) operation. This was inspired by how a biological neuron works. A biological neuron accumulates multiple stimuli, and when the overall stimulation is over a threshold, the neuron will become excited and respond by sending an electrical signal to other neurons or cells. There are a huge number of biological neurons in our brain, and it is the biological neural network that gives us intelligence, and the unmistaken example

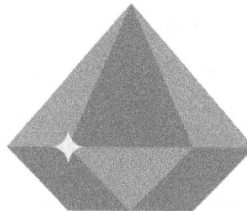

Figure 0.1. AI/ML enables superior tomography.

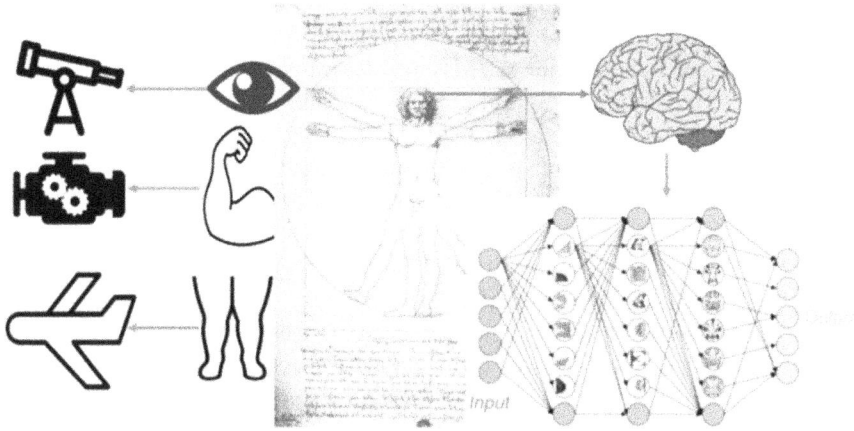

Figure 0.2. Curiosity and needs drive scientific pursuits through the industrial, information, and intelligent revolutions.

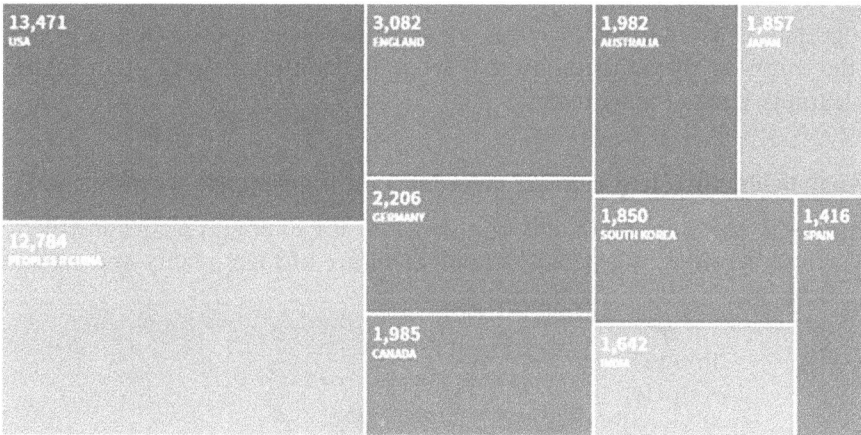

Figure 0.3. Web of Knowledge results with 'deep learning' as the topic term (data collected on 11 July 2019).

showing that intelligence is feasible. Similar to this biological/neurological system, an artificial neural network can behave, to a good degree, like a brain, if the number of artificial neurons is high enough, organized deeply (i.e. with many layers of artificial neurons), and trained well with big data. This resembles the learning process in our childhoods, where our neurological connections are formed adaptively, and we become increasingly intelligent.

The importance and potential of AI/ML has now been well recognized. The AI Executive Order was issued by the White House in February 2019 (https://www.whitehouse.gov/articles/accelerating-americas-leadership-in-artificial-intelligence/), and the response from NIST is also inspiring to read (https://www.nist.gov/topics/artificial-intelligence). International competition is remarkable in advancing AI/ML theory and technologies (figure 0.3).

0.2 Image analysis versus image reconstruction

As the most famous examples of AI/ML applications, computer vision and image analysis deal with existing images and produce features of these images thanks to the great efforts of many talented researchers. We are researchers in the field of tomographic imaging, and our products are tomographic images reconstructed from externally measured, complicated data that look totally different from the underlying images, and are actually various features (attenuated/non-attenuated line integrals, Fourier/harmonic components, and so on) of the underlying images. Currently, machine learning (especially deep learning) techniques are being actively developed worldwide for tomographic image reconstruction, which is a new area of research, with low hanging fruit in terms of data-driven post-processing and high hanging fruit in terms of end-to-end mapping via transfer, adversarial, ensemble, and other forms of machine learning.

This first-of-its-kind book is dedicated to machine learning for tomographic image reconstruction, or tomographic imaging, primarily targeting image reconstruction (from data to images) with some mentions of relevant image analysis (from images to images/features) and end-to-end mapping (from data to features). Tomography is a Greek word, meaning reconstruction of cross-sectional images. It is the emphasis on tomography that sets our book apart from other AI/ML or deep learning books (figure 0.4).

0.3 Analytic/iterative/deep learning algorithms for tomographic reconstruction

Traditionally, there are two kinds of algorithms for tomographic reconstruction—analytic and iterative. When tomographic data are of high quality and sufficiently

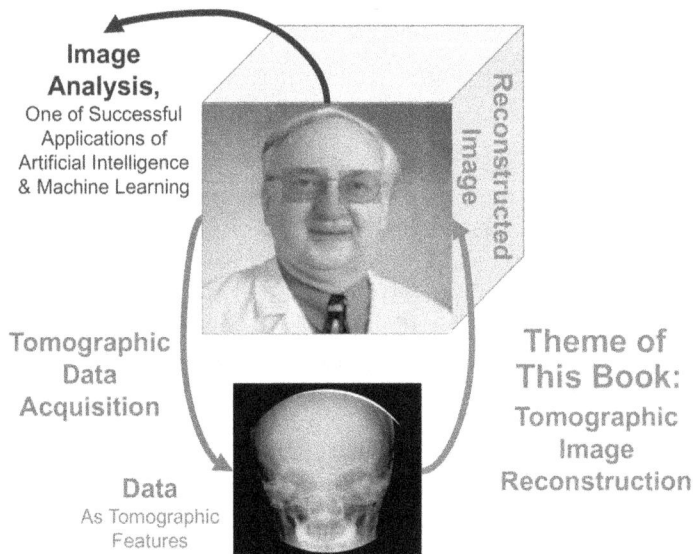

Figure 0.4. Uniqueness of this book, dedicated to tomographic image reconstruction in the AI/ML framework, in contrast to deep learning for image analysis or computer vision that takes images as the input.

collected, the relationship from an underlying image to the tomographic measures can be expressed as a forward model, which can be mathematically and computationally inverted. The inverse transform will transform the data back to the underlying image, which is what image reconstruction means. When such an inverse transform is in a closed form (for example, an inverse Fourier transform), a reconstruction algorithm can be directly obtained, and implemented on a computer.

When tomographic data are compromised, incomplete, or the forward model is too complicated to be analytically inverted, an iterative algorithm can be used for image reconstruction. An iterative algorithm does not solve the problem in one shot. From an initial estimate of the underlying image, which can be as simple as an all zero or all one image or another form when specific prior knowledge is available, the algorithm refines an intermediate solution iteratively (the first one is the initially guessed image). The refinement is guided by two preferences. First, the discrepancy should be small between the measured data and the data computed according to the forward model based on an intermediate image. Second, the characteristics of a reconstructed image should look reasonable or consistent with our prior knowledge, such as non-negative pixel values and no severely oscillating features. These requirements are summarized into an overall objective function, and the reconstruction problem becomes an optimization task. In other words, the iterative algorithm is just for this optimization. In most cases of tomographic imaging, measured data are not enough to determine the underlying image uniquely, and prior knowledge is instrumental for satisfactory image reconstruction. The stronger the prior knowledge is, the better the reconstructed image quality will be. Various kinds of prior knowledge are in use for iterative reconstruction, including non-negativity, maximum entropy, roughness penalty, total variation minimization, low rank, dictionary, and low-dimensional manifold learning. Normally, an iterative algorithm is very time-consuming.

It is clear now that deep learning networks form the third category of image reconstruction algorithms. In contrast to the aforementioned kinds of prior knowledge, each of which can be formulated as one mathematical term in one or two lines, an unprecedented source of prior knowledge is big data. For example, millions of CT scans contain overwhelming information on underlying anatomical and pathological information, and a new scan should be very much correlated to or consistent with the existing scans. If these data can be utilized for image reconstruction, superior image quality is expected. Fortunately, deep neural networks can be trained with big data on a high-performance computing platform so that prior knowledge can be represented by the trained neural network that serves as a mapping from data to images. Because the prior knowledge used by the neural network is task-specific and yet comprehensive, in principle the network may produce better image quality than an iterative algorithm when it falls short of clinical satisfaction. Although training the network is still time-consuming, the trained network only involves forward operations and is computationally efficient.

The analytic, iterative, and deep learning algorithms for tomographic image reconstruction can be compared and contrasted (table 0.1). Briefly speaking, an analytic algorithm can be formulated in the following form, $f(x, y) = O[p(\theta, t)]$,

Table 0.1. Three types of tomographic image reconstruction algorithms in an over-simplified comparison (the penalization of image reconstruction and topology of network architecture can be complicated).

Category	Form	Knowledge	Input	Quality	Speed
Analytic reconstruction	$f = O[p]$	Idealized model, without noise	High SNR, complete	High	High
Iterative reconstruction	$f^{(k)} = O[p, f^{(k-1)}]$	Physical model, image prior	Low in various ways	Decent	Low
Deep reconstruction	$f = O_{\theta_N} \cdots O_{\theta_1}[p]$	Model, prior, big training data	Poor, incomplete	Superior, task-specific	High

where f represents an image in a 2D case without loss of generality, p denotes data as a function of projection viewing angle and detector position, and O is an analytic operation in the closed form, such as an inverse Fourier transform. An iterative algorithm, on the other hand, is expressed as $f^{(k)}(x, y) = O[p(\theta, t), f^{(k-1)}(x, y)]$, where the index k goes from 0, 1, 2, to a sufficiently large number K for the iterative process to converge. The image for $k = 0$ is the initial guess as the starting point of the iterative process. Different from either analytic or iterative algorithms, a deep learning based tomographic reconstruction algorithm is written as $f(x, y) = O_{\theta_N} \dots O_{\theta_1}[p(\theta, t)]$, where each operator corresponds to a layer of artificial neurons whose parameters $\theta_1, \dots, \theta_N$ need to be optimized in the training process with big data. Although a deep algorithm may still be slightly slower than an analytic algorithm, it is much faster than an iterative algorithm, since normally $N \ll K$.

0.4 The field of deep reconstruction and the need for this book

The industrial revolution from the eighteenth century onwards has greatly accelerated civilization, and now we are in the intelligence revolution, synergizing big data, exploding information, instantaneous communication, sophisticated algorithms, high-performance computation, and AI/ML. Over only the past few years, as AI/ML methods have become mainstream, deep learning has affected many practical applications and generated overwhelming excitement (figure 0.5). As a result, more and more students and researchers are motivated to learn and apply AI/ML.

Our field is tomographic image reconstruction, which is experiencing a paradigm shift towards deep-learning-based reconstruction (see our perspective on deep imaging (Wang 2016)). Simply speaking, we are interested in developing deep learning methods going from measured features to tomographic images. Currently, deep learning techniques are being actively developed worldwide for tomographic image reconstruction, delivering excellent results (figure 0.6, and also see (Wang *et al* 2019)).

While many of us share optimism about this new wave of tomographic imaging research, there are doubts and concerns regarding deep reconstruction. This conflict of opinions is natural and healthy. In retrospect, at the beginning of the development of analytic reconstruction, there was a major critique that given a finite number of projections, the tomographic reconstruction is not uniquely determined (introducing ghosts in a reconstructed image). Later, this was successfully addressed by regularization methods. When iterative reconstruction algorithms were first developed, it was observed that a reconstructed image was strongly influenced by the penalty term. In other words, it seemed that what one saw was what one wanted to see! Nevertheless, by optimizing the reconstruction parameters, iterative algorithms have been made into commercial scanners. As far as compressed sensing is concerned, it was proved that there is a chance that a sparse solution is not the truth. For example, a tumor-like structure could be introduced, or pathological vessels might be smoothed out if total variation is overly minimized. Similarly, deep learning appears to present issues in practice, such as the interpretability problem.

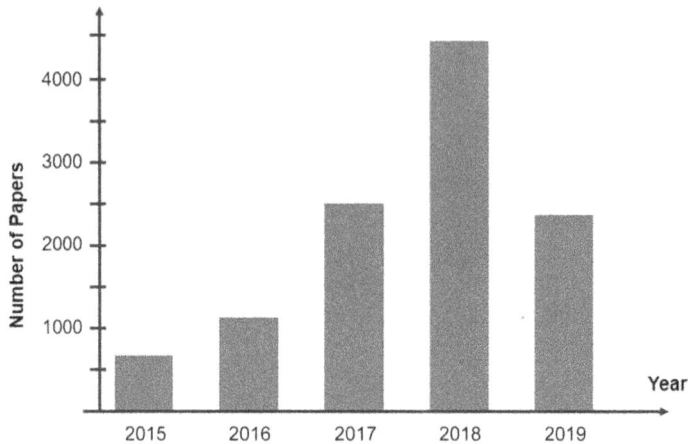

Figure 0.5. A Web of Knowledge search, with 'deep learning', 'medical', and 'imaging' as the topic terms (data collected on 11 July 2019).

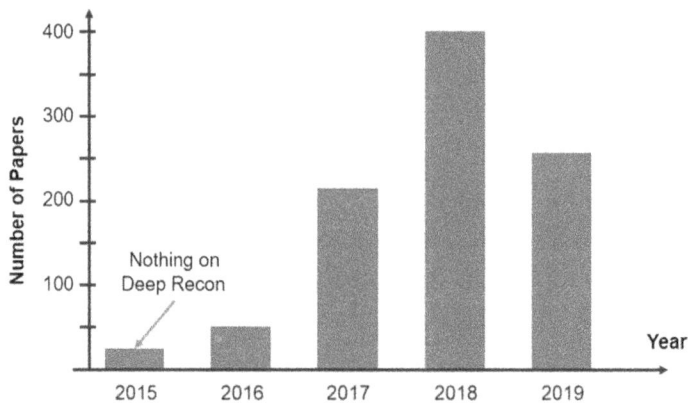

Figure 0.6. Web of Knowledge search, with 'deep learning' in the article title (data collected on 11 July 2019).

No Maxwell equations for deep learning yet exist, and a deep network as a black box is trained to work with big data in terms of parameter adjustment. The interpretability of neural networks is currently a hot topic. Given the rapid progress being made in theoretical and practical aspects, we believe that deep learning algorithms will become the mainstream for medical imaging.

With the encouraging results and insights, some of which are in this book, we are highly confident that, in principle, AI/ML methods for deep reconstruction ought to outperform iterative reconstruction (IR) and compressed sensing (CS) for medical imaging. To convince the reader that AI/ML will dominate tomographic imaging, let us highlight three key arguments: (i) IR/CS can be used as a component in a neural network (such as in our 'LEARN' network in chapter 5); (ii) the result from IR/CS can be used as the baseline (such as for the denoising, despeckling, or de-blurring networks mentioned in several chapters in this book); and (iii) IR/CS reconstruction algorithms can be enhanced or even replaced by powerful neural networks with advanced architectures trained with big data with unprecedented domain priors.

There are a good number of deep learning books of high quality. However, they are either on general deep learning methods or other specific deep learning applications. This book is dedicated to the emerging area of deep reconstruction, representing a new frontier of machine learning, and offers a unified treatment of this theme. In particular, this book is focused on medical imaging, which is a primary example of tomographic imaging that affects all people worldwide, spans a huge business, and remains a major driver for technical innovations.

0.5 The organization of this book

This book reflects the state-of-the-art, since all of the co-authors are active researchers in the deep imaging field. Also, the materials are presented in a reader-friendly way, covering classic reconstruction ideas and human vision inspired insights, naturally leading to deep artificial neural networks and deep tomographic reconstruction. There are four parts in this book, with two to three chapters per part.

The first part consists of chapters 1–3, laying out the foundation for the remaining parts. The first chapter describes general principles for imaging, with an emphasis on the importance of prior information when data are imperfect, inconsistent, or incomplete, either in the Bayesian framework or in the context of the human vision system (HVS). From these perspectives, the concepts of regularization and sparsity naturally arise. The second chapter focuses on regularized image reconstruction in the Bayesian and compressed sensing perspectives, with an emphasis on dictionary learning, whose computational structure can be viewed as a single-layer neural network. As a good example, a statistical reconstruction algorithm is empowered with either a global or adaptive dictionary for low-dose computed tomography (CT). Based on the materials covered in chapters 1 and 2, chapter 3 offers a basic but quite complete presentation of neural network architectures, including the concepts and components of deep neural networks, representative networks such as auto-encoder, VGG, U-Net, ResNet, generative adversarial network (GAN), and graph convolutional network (GCN), as well as training, validation, and testing strategies.

The second part includes chapters 4 and 5, exclusively dedicated to CT. Chapter 4 reviews the CT data acquisition process and the development of CT scanners. Also, both analytic and iterative reconstruction algorithms are exemplified. In addition to analytic and iterative algorithms, chapter 5 covers the latest developments of the new type of reconstruction algorithm that employs deep neural networks. A number of recently published deep learning based methods are presented to show the feasibility, merits, and potential of deep learning techniques in the CT field.

The third part has chapters 6 and 7 on magnetic resonance imaging (MRI), in parallel to chapters 4 and 5. Chapter 6 reviews the MRI data acquisition process and the MRI scanner instrumentation. Fourier transform and compressed sensing algorithms are first presented. Then, classic post-processing algorithms are discussed. Chapter 7 covers various deep-learning-based MRI techniques, including a

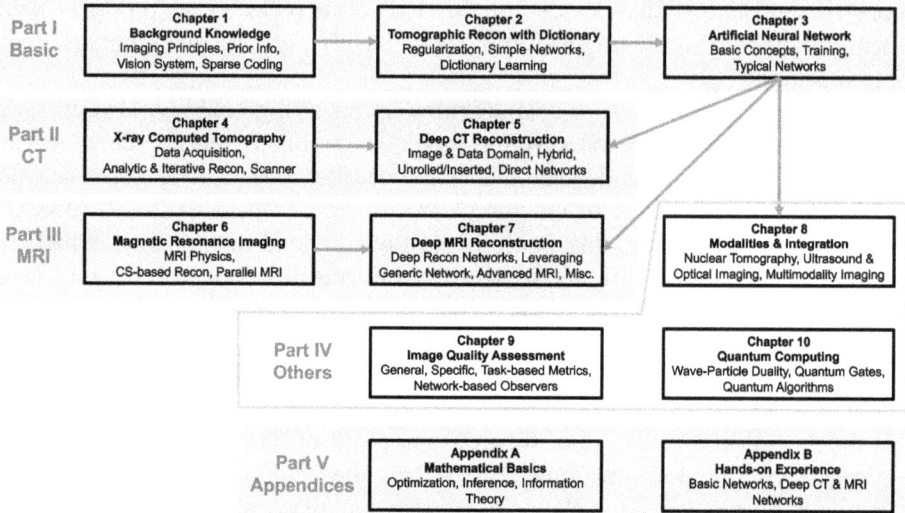

Figure 0.7. Diagram suggesting the order in which the reader reads the components of this book.

variety of deep reconstruction networks with applications to regular MRI, parallel MRI, dynamic MRI, and magnetic resonance fingerprinting (MRF). Miscellaneous topics are also covered, such as optimal k-space sampling and activation functions for complex-valued inputs. Finally, we discuss the integration of MRI data acquisition and image reconstruction with a synergized pulsing and imaging network (SPIN).

In the fourth part, we offer chapters 8–10. Chapter 8 briefly presents other imaging modalities including nuclear imaging, ultrasound imaging, and optical imaging in terms of working principles, and then describes representative neural networks developed for these imaging modalities individually. After that, we mention multi-modality imaging. Chapter 9 discusses image quality for general and task-specific assessment. In this chapter, network-based model observers are presented as a new approach for cost-effective reader studies. Chapter 10 is on quantum computing. We start with wave–particle duality and quantum puzzles, define quantum bits and gates, and touch upon quantum algorithms and quantum machine learning.

For your convenience, the relationships among the four parts and the associated chapters are summarized in figure 0.7, supplemented by appendices A and B. It is underlined that appendix B and associated web resources are under development, and should be invaluable to enhance the learning experience and AI/ML skills. As shown by this book, AI/ML techniques are applicable and instrumental to all tomographic modalities, and promise to unify individual modalities computationally.

0.6 More to learn and what to expect next

As implied by figures 0.5 and 0.6, there are too many relevant papers to read, and the number of such papers is growing rapidly. After reading this book systematically or selectively, you will need more time to master more materials, dive in more deeply, and practice for better skills. According to PwC (https://www.pwc.com/gx/en/issues/data-and-analytics/publications/artificial-intelligence-study.html in June 2017), AI will yield a global GDP increase of 14% in 2030, and a top area AI affects is healthcare. In such an expanding phase of AI/ML R&D, we have no choice but to pursue continuous learning—papers, books, online materials, and hands-on projects.

Despite all the above positive comments on AI/ML, this field has previously experienced two winters, and it is natural to wonder if we will sooner or later enter another winter of AI/ML. All this depends on how much and how quickly we can continue advancing the field and meeting the majority's expectations. Although the future is often unpredictable but sometimes indeed inventable (Virginia Tech's old logo 'Invent the Future'), we tend to be very optimistic about the future of AI/ML in the long run, and are particularly hopeful for several directions of development.

There are two scientific approaches for reasoning—deduction and induction. Accordingly, we have two associated schools of AI/ML. Deduction goes from general to specific, and is a top-down approach. Decades ago, research on rule-based expert systems was popular, and the fifth generation computer was a hot topic. In this context, it was hoped to reason from general rules to specific claims. On the other hand, induction works from data toward knowledge or information. This is bottom-up or data-driven. The recent champions of the ImageNet contest developed their deep learning programs in a data-driven fashion. There are great opportunities to merge these two approaches in the future. Knowledge graphs and self-supervised learning are two ideas along this direction.

Also, AI/ML and neuroscience/psychiatry are closely intertwined and mutually promoting. For example, studies on the human vision system are an integral part of neuroscience research, which played an instrumental role in the development of AI/

Machine Learning for Tomographic Imaging
Ge Wang, Yi Zhang, Xiaojing Ye, & Xuanqin Mou

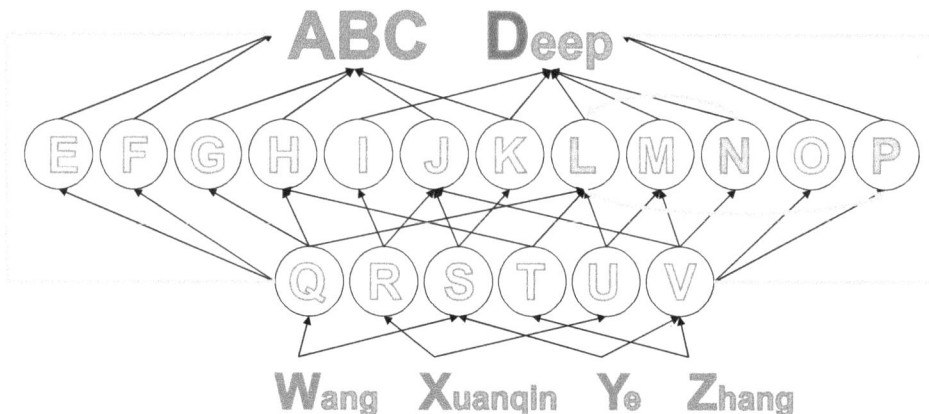

Figure 0.8. NLP is believed to be an important direction for development of AI/ML.

ML and suggested a mechanism for cellular-level processing and interconnection for visual perception and image analysis. New findings and insights in AI/ML and neuroscience will continue promoting each other. Arguably, human intelligence is closely related to natural language understanding and expression. It is believed by many that an important direction of AI/ML development is natural language processing (NLP) (figure 0.8).

Yet another outside-the-box approach is quantum computing, but no-one is sure when its prime time will come. However, if it becomes practical, AI/ML will be revolutionized. For example, our proposed 'SPIN' network may be implemented via quantum computing. A few days ago, Google announced a quantum supremacy, and heated on-going discussions on this topic (https://www.sciencenews.org/article/google-quantum-computer-supremacy-claim). We cannot exclude the possibility that intelligence is essentially a quantum phenomenon and must be implemented through quantum computing. Let us continue making and enjoying our AI/ML related efforts.

References

Wang G 2016 A perspective on deep imaging *IEEE Access* **4** 8914–24

Wang G, Ye J C, Mueller K and Fessler J A 2018 Image reconstruction is a new frontier of machine learning *IEEE Trans. Med. Imag.* **37** 1289–96

Part I

Background

IOP Publishing

Machine Learning for Tomographic Imaging

Ge Wang, Yi Zhang, Xiaojing Ye and Xuanqin Mou

Chapter 1

Background knowledge

1.1 Imaging principles and *a priori* information

1.1.1 Overview

Tomography is an imaging technology that studies an object with externally measured data generated by some physical means such as x-ray radiation, where data are projections in the form of line integrals of an object function from different angles of view. This kind of imaging technique can be used to produce images of hidden 3D structures in an opaque object non-destructively, even though the object is not transparent to the human eye. Seeing through a patient's body is highly valuable for medicine. Hence, modern medicine is, in a sense, enabled by tomography.

With major discoveries in physics, such as x-ray radiation and magnetic resonance, an object can now be imaged using various mechanisms. X-ray photons easily penetrate most every-day materials, including human tissues, and produce line integral information on the linear attenuation coefficient that characterizes the interactions between the materials and x-ray radiation. Various types of materials have different linear attenuation coefficients. From sufficiently many x-ray projections, a cross-sectional image can be reconstructed, which is called computed tomography (CT). In such a reconstructed image of a patient, bone, soft tissue, and fat can be clearly discriminated, and anatomical features can be well defined. Magnetic resonance imaging (MRI) is another important imaging modality. It takes advantage of nuclear spins (in particular spins of hydrogen nuclei) to generate signals when the spins are aligned in an external magnetic field, excited by radio frequency pulses, and then relaxed to their steady states. During the relaxation process, various tissues have different proton densities and take different times to relax, thereby exhibiting distinct characteristics. The relaxation time has two main aspects: T1 (longitudinal relaxation time) and T2 (transverse relaxation time). While CT images enjoy excellent bone–tissue and air–tissue contrasts, MRI images give rich soft tissue contrasts and functional information.

In addition to x-ray CT and MRI, there are multiple other tomographic imaging modalities. Importantly, nuclear emission imaging concentrates on physiological functions such as blood flow and metabolism. In this category, positron-emission tomography (PET) and single-photon emission computed tomography (SPECT) are the primary modes of nuclear tomography (Leigh *et al* 2002, Zeng 2009).

A PET system detects pairs of gamma rays generated by a positron-emitting radionuclide, such as fluorine-18, which is introduced into a patient inside a biologically active molecule called a radioactive tracer. A pair of gamma ray photons is emitted in opposite directions from the radioactive tracer. Hence, the PET scanner utilizes their linearity and simultaneity to detect the emission event along a corresponding line of response defined by two detectors facing each other. Specifically, if two gamma ray photons are simultaneously (within a time window of 10–20 ns) detected, then a signal is recorded for the line of response. From these data, a distribution of the radioactive tracer can be reconstructed. Similarly, SPECT reconstructs distributions of radioactive tracers from gamma ray induced signals, but these tracers only emit gamma ray photons individually, and a metallic collimator is needed to determine the line path along which a gamma ray photon travels. In both the PET and SPECT cases, the tracer nuclides require a patient to contain a radiation source in his/her body, and the detector measures radiation-induced data externally for image reconstruction.

Furthermore, we can also utilize ultrasound and light waves for imaging purposes, which are called ultrasound imaging and optical imaging, respectively. Generally speaking, for any physical measurement on an object of interest, as long as the data do not directly reflect structural features, a so-called 'inverse' process will be needed to estimate these features from the indirect data. Tomography is a very important class of inverse problems, targeting cross-sectional/volumetric image formation. In the next section, we will heuristically explain the CT problem as a specific example.

1.1.2 Radon transform and non-ideality in data acquisition

The Radon transform describes the relationship between an underlying function and a simple indirect measurement process. First studied in 1917, Johann Radon sought to invert this transform and reconstruct the underlying function from these measurements. It is mathematically natural, as it applies an integral operator over a sub-space (for example, a line or a plane) in a high-dimensional space (for example, a 3D Euclidean space). Also, it is physically relevant, as an x-ray signal attenuated by an object can be easily converted into a line or planar integrals after practical approximations.

Without loss of generality, let us consider a function defined on a plane and perform the Radon transform along lines in the plane. Then, the value of the Radon transform along an arbitrary line is equal to the following line integral:

$$Rf(L) = \int_L f(x)\mathrm{d}x, \tag{1.1}$$

where *Rf* denotes projection data that can be acquired by a CT scan and *f* represents an underlying function depicting linear attenuation coefficients inside the object. Technically, x-ray photons go through the object, are attenuated, and are then recorded by detector elements along a 1D array from various viewing angles. X-ray data can be put in a 2D array with respect to the viewing angle and the detector location, forming a so-called sinogram, as shown in figure 1.1.

In this 2D case, the most commonly used analytical formula to recover *f* from its Radon transform is the filtered backprojection (FBP) formula. In the 3D case, the Radon transform typically gives planar integrals, instead of line integrals, and can be inverted using more complicated formulas. In a nutshell, the Radon transform is closely related to the Fourier transform. Thus, if all Radon data is available, then the inverse Radon transform is essentially the inverse Fourier transform. Alternatively, we can view each line or planar integral as a linear equation. With all available x-ray data, we have a system of linear equations. In principle, we can solve this linear system in some way to uncover all unknown pixel/voxel values, i.e. to reconstruct the underlying function/image. We will discuss specific technical details in chapter 4, which is dedicated to CT basics.

In the ideal case, data indirectly measured by many tomographic imaging systems can often be approximated as linear combinations of unknown variables assuming neither noise nor bias. However, in reality there are many practical factors that prevent us from obtaining idealized data. During a measurement process, interactions between a physical probing method and an object to be reconstructed are often stochastic processes, and both the object and the imaging system can be time-variant, introducing uncertainties and inconsistencies in the data. For example, the measurement of x-ray or gamma ray photons contains inherent Poisson noise. Also, current-integrating x-ray detectors have electronic noise.

Given the radiation risk of x-ray radiation, which might carry genetic, cancerous, and other risks, CT scanning with a reduced radiation dose has been a hot topic over the past decade. Currently, the medical community is striving for high-quality CT images at a lowest possible dose level, for example, in the Image Gently campaign on pediatric patients and the Image Wisely campaign on adult patients the 'as low as reasonably achievable' (ALARA) principle has been widely accepted. These efforts

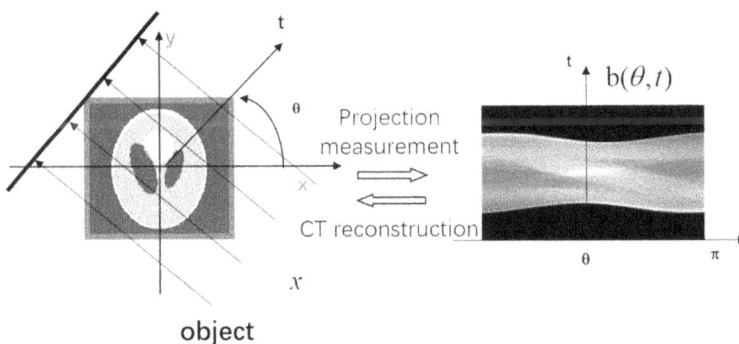

Figure 1.1. Radon transform as a simple example of indirectly measured data, from which an underlying function needs to be estimated or reconstructed.

bring up the well-known low-dose CT challenge. The low-dose condition will severely degrade the image quality, since x-ray imaging is a quantum accumulation process.

As another example, an MRI scan takes much longer than a CT scan does, and needs to be accelerated with fast MRI techniques. As a result, measured MRI data, known as Fourier space (or k-space) samples, cannot fully cover the Fourier space. Simply applying the inverse Fourier transform to the partially measured Fourier spectrum produces an image with strong artifacts, in particular when these data are further compromised with patient and/or physiological motion.

In the above CT and MRI issues, which are among many imaging problems, analytic image reconstruction methods are not suitable, as they demand ideal data and are almost exclusively based on the Fourier formulation. To alleviate these types of problems, iterative image reconstruction methods are advantageous. Iterative reconstruction algorithms use optimization techniques, easily incorporate prior knowledge on the imaging process and the image information content, and solve the imaging problem iteratively at an increased computational cost. Compared to analytic reconstruction algorithms, iterative algorithms reduce image noise and artifacts effectively.

1.1.3 Bayesian reconstruction

As mentioned before, tomography is nothing other than image reconstruction from measurement data indirectly related to hidden structures. In a simple case, such as the Radon transform, data $\mathbf{b}$ is related to an underlying image $\mathbf{x}$ through a linear mapping $\mathbf{A}$. This mapping is called the forward process:

$$\mathbf{b} = \mathbf{Ax}. \tag{1.2}$$

In the case of a CT scan, $\mathbf{b}$ denotes the data acquired by the scan, $\mathbf{x}$ represents the underlying CT image, and $\mathbf{A}$ is the system matrix specific to a CT scanner and the imaging protocol used for the CT scan. Actually, equation (1.2) is a discrete form of the Radon transform. Now, our inverse problem is to calculate the image $\mathbf{x}$ from data $\mathbf{b}$, given the imaging system matrix $\mathbf{A}$.

The analytical solution to the inverse problem in the Fourier domain has a close-form expression but cannot handle image noise and artifacts well. Another straightforward way to resolve this problem is to compute the solution as $\mathbf{x} = \mathbf{A}^{-1}\mathbf{b}$ when $\mathbf{A}$ is nonsingular. However, the matrix $\mathbf{A}$ in tomographic imaging cases always has high condition numbers, hence the inverse problem is ill-posed, making the simplistic matrix inversion impracticable. A better strategy is to use iterative algorithms, which are of non-closed-form and improve an intermediate image gradually with deterministic and/or statistic knowledge.

Based on equation (1.2), the iterative algorithm always targets a criterion, such as minimizing the difference between the measured data and calculated data. One approach is obtained by minimizing the functional:

$$f(\mathbf{x}) = \|\mathbf{Ax} - \mathbf{b}\|_2^2. \tag{1.3}$$

With this functional, the solution is updated iteratively until it converges. Landweber iteration is among the first deterministic iterative methods proposed to solve the inverse problem. It is the fastest descent method to minimize the residual error. The Newton method based on the first and second derivatives can accelerate the search process. The conjugate gradient method is somehow between the gradient descent method and the Newton method. It addresses the slow convergence of the gradient descent method and avoids the need to calculate and store the second derivative required by the Newton method (Bakushinsky and Kokurin 2004, Landweber 1951). For more details, read further on the Bayesian approach to inverse problems.

Here, we illustrate the iterative algorithm that solves the inverse problem in the simplest way: the gradient descent method. The main idea behind the gradient descent search is to move a certain distance at a time in the direction opposite to the current gradient which is equivalent to updating an intermediate image as follows:

$$\mathbf{x}_{k+1} = \mathbf{x}_k - \alpha \nabla f(\mathbf{x}_k), \tag{1.4}$$

where α is the step size controlling the step size in each iteration. The update is repeatedly performed until a stopping condition is met. Then, the optimal result is obtained.

It is inevitable that the measurement data always contain noise or error in practice, which means that the inverse problem can be denoted as

$$\tilde{\mathbf{b}} = \mathbf{A}\mathbf{x} + \mathbf{n}, \tag{1.5}$$

where $\mathbf{n}$ denotes the data noise or error. As presented above, the CT measurement always contains noise and error introduced by non-idealities of the physical imaging model, which may cause inaccuracy and instability of CT image reconstruction. In this situation, the inverse problem may not be uniquely solvable, and depends sensitively on the measurement error and the initialized image as the starting point of the iterative process, given its ill-posed property. Hence, how to compute a practically acceptable solution is the critical issue of any ill-posed inverse problem.

If we still use the same least squares optimization method to find a solution, the minimum of $f(\mathbf{x})$ will only minimize the average energy of noise or error. Thus, there would be a set of solutions all of which are consistent with the measured data. How can we choose the good ones among all these solutions? Clearly, we need additional information to narrow the space of solutions. In other words, the more such prior information/knowledge we have, the smaller the space of solutions, and the greater the chance we can recover the ground truth better (Bertero and Boccacci 1998, Dashti and Stuart 2017, Stuart 2011).

Fortunately, profound yet elegant solutions have been found to settle this problem. Bayesian inference is one of the most important approaches for solving inverse problems. Instead of directly minimizing the error between the calculated and real tomographic data, the Bayesian approach takes the prior knowledge into consideration. Bayesian inference derives the posterior probability from a prior probability and a 'likelihood function' derived from a statistical model for the measured data.

Bayesian inference computes the posterior probability according to Bayes' theorem:

$$P(\mathbf{x}|\mathbf{b}) = \frac{P(\mathbf{b}|\mathbf{x})P(\mathbf{x})}{P(\mathbf{b})}, \tag{1.6}$$

where $P(\mathbf{x})$ is the prior probability, an estimate of the probability of the hypothesis $\mathbf{x}$ before the data, the current evidence $\mathbf{b}$, is measured. $P(\mathbf{x}|\mathbf{b})$ is the probability of a hypothesis $\mathbf{x}$ given the observed $\mathbf{b}$. $P(\mathbf{b}|\mathbf{x})$ is the probability of $\mathbf{b}$ given $\mathbf{x}$, which is usually called the likelihood function. $P(\mathbf{b})$ is sometimes termed the marginal likelihood or 'model evidence'. It is a constant after the data are measured. In Bayesian inference, the recovery of hidden varibles can be simply achieved by maximizing the posterior probability (MAP):

$$\hat{\mathbf{x}}_{\mathrm{MAP}} = \arg\max{}_{\mathbf{x}}P(\mathbf{x}|\mathbf{b}). \tag{1.7}$$

According to Bayes' theorem, we can rewrite this formula as

$$\hat{\mathbf{x}}_{\mathrm{MAP}} = \arg\max{}_{\mathbf{x}}P(\mathbf{b}|\mathbf{x})P(\mathbf{x}). \tag{1.8}$$

In optimization theory, applying a monotone function to the objective function does not change the result of optimization. Hence, we can apply a logarithmic operation to separate the first and second term:

$$\hat{\mathbf{x}}_{\mathrm{MAP}} = \arg\max{}_{\mathbf{x}}(\log(P(\mathbf{b}|\mathbf{x})) + \log(P(\mathbf{x}))). \tag{1.9}$$

In the application of Bayesian inference to solve the inverse problem, Lagrangian optimization is extensively used, and the above objective function can be presented as

$$F(\mathbf{x}) = \phi(\mathbf{x}, \tilde{\mathbf{b}}) + \lambda\psi(\mathbf{x}). \tag{1.10}$$

The first term is the penalty term to measure the data fitting error corresponding to the likelihood function. According to different data models, the first term can be specialized into $\frac{1}{2}\|\mathbf{A}\mathbf{x} - \tilde{\mathbf{b}}\|_2^2$, $\|\mathbf{A}\mathbf{x} - \tilde{\mathbf{b}}\|_1$, $\int(\mathbf{A}\mathbf{x} - \tilde{\mathbf{b}}\ln\mathbf{A}\mathbf{x})\mathrm{d}x$, which are statistically well suited to additive Gaussian noise, impulsive noise, and Poisson noise, respectively. The second term can be interpreted as the regularization functional, which is the the prior probability $P(\mathbf{x})$ in Bayesian inference. Just like the first term, the second term can be also specialized according to the statistical model for $\mathbf{x}$. λ is the regularization parameter which balances these two terms For tomographic imaging in particular, the first term is the logarithmic likelihood term, which mainly describes the statistical distribution of the original measured data, and the second item is the prior information item, which is on some prior distribution of images to be reconstructed.

Perceiving the prior information of an image means that we should extract as much information from the image as possible. In an extreme case, if we have perfect knowledge of the image, the reconstruction process is no longer needed since we have already known everything about the image. In common cases, an image cannot be perfectly known before it is reconstructed. Practically, general information of images can be extracted and then assumed, and this prior information in turn can constrain the candidate images allowed as a reconstruction outcome. In fact, an image as a high-dimensional variable is a point in a high-dimensional space, and natural images just occupy a very small portion of this high-dimensional space, although they vary greatly with dramatically different content. This phenomenon implies that we can represent natural images by exploring their intrinsic distribution properties and utilizing them for image reconstruction. Generally speaking, the intrinsic properties exhibit themselves as correlations or redundancies among image regions, obeying some structured distributions in gray scales or colors.

In Bayesian inference, solving the inverse problem uses intrinsic distribution properties to narrow the search space of unknown variables. These properties also have the ability to suppress image noise or measurement error in the inversion process because the error will disturb the intrinsic properties. Hence, how to extract the intrinsic distribution properties and how to use them for an inverse problem solution are two key aspects of Bayesian inference. With natural image statistics, these key questions can be answered. Natural image statistics is a discipline to figure out the natural image statistics based on a number of statistical models whose parameters are estimated from image samples. This is widely used in image processing fields. In a simple form, natural images can be regarded as a linear combination of features or intrinsic properties. Rather than directly model the statistics in natural images pixel/voxel-wise, an image can be transformed to a feature space and obtain feature statistics to build a prior model. Also, the features, unlike pixel values having corresponding dependence, are independent or nearly independent of each other, which makes the statistical model informative. This concept is the key to all natural image statistics. When it comes to natural image statistics, it is necessary to introduce the human vision system (HVS) because many natural image statistics and analyses are derived based on observations of the HVS. There are two sides to sensing prior information. In the next subsection, we will first briefly introduce the HVS mechanism, and then describe some basic techniques in natural image statistics.

1.1.4 The human vision system

The human vision system (HVS) is an important part of the central nervous system, which enables us to observe and perceive our surroundings (Hyvärinen *et al* 2009). After its long-term adaptation to natural scenes, the HVS is highly efficient in working with natural scenes through multi-layer perceptive operations. Here, natural scenes refer to daily-life inputs to the HVS. Visual perception begins with the pupils which catch light, then the information carried by light photons is processed step by step, and finally analyzed for perception in the brain, as depicted in figure 1.2. This pathway consists of neurons. Typically, a neuron consists of a cell body (soma), dendrites to weight and integrate inputs, and an axon to output a

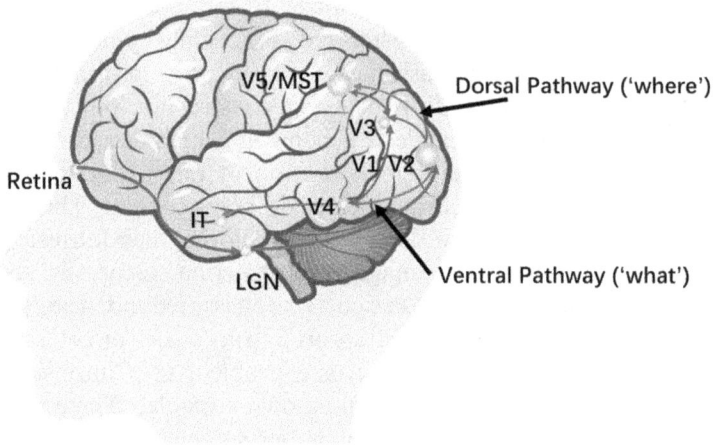

Figure 1.2. A schematic of the HVS pathway.

signal, also referred to as an action potential. In the following, we briefly introduce the multiple layer structure of HVS.

The first stage involves light photons reaching the retina. The retina is the innermost light-sensitive layer of tissue of the eye. It is covered by more than a hundred million photoreceptors, which translate the light into electrical neural impulses. Depending on their function, the photoreceptors can be divided into two types—cone cells and rod cells. Rod cells are mainly distributed in the peripheral area of the fovea, which is sensitive to light and can respond even to a single photon. These cells are mainly responsible for vision in a low-light environment, with neither high acuity nor color sensing. Contrary to rod cells, cone cells are distributed in the fovea region, and are responsible for perception of details and colors in a bright environment, but are light-insensitive.

In the second stage, the electrical signals are transmitted and processed through neural layers. One of the most important cell-types, called Ganglion cells, gather all the information from other cells and send the signal from the eye along their long axons. The visual signals are initially processed in this stage. Neurobiologists have found that the receptive field of ganglion cells is usually centralized or circularly symmetric, with the center either excited or inhibited by light. Such light responses can be simulated by the Laplace of Gaussian (LOG) or zero-phase component (ZCA) operator. We describe two kinds of LOG operator in figure 1.3 from three perspectives: 3D visualization, 2D plane figure, and center profile.

In the HVS, the receptive field of a visual neuron is defined as the specific light pattern over the photoreceptors of the retina which yields the maximum response of the neuron. We illustrate this operation with a vivid example decipted in figure 1.4 with two different operators.

Next, the signal is transmitted to the lateral geniculate nucleus (LGN) of the thalamus, which is the main sensory processing area in the brain. The receptive field of the LGN is also centralized or circularly symmetric. After processing by the

Center off   Center on

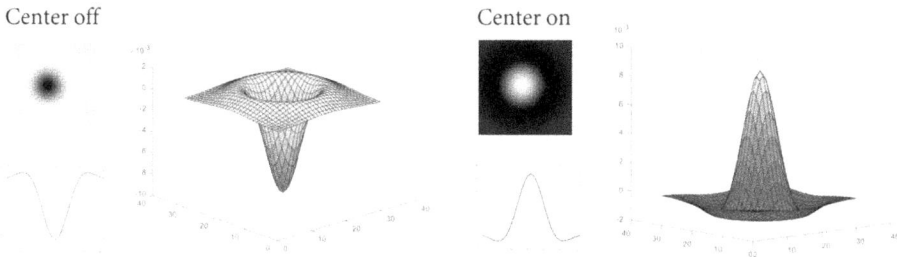

Figure 1.3. Visualization of the LOG operator.

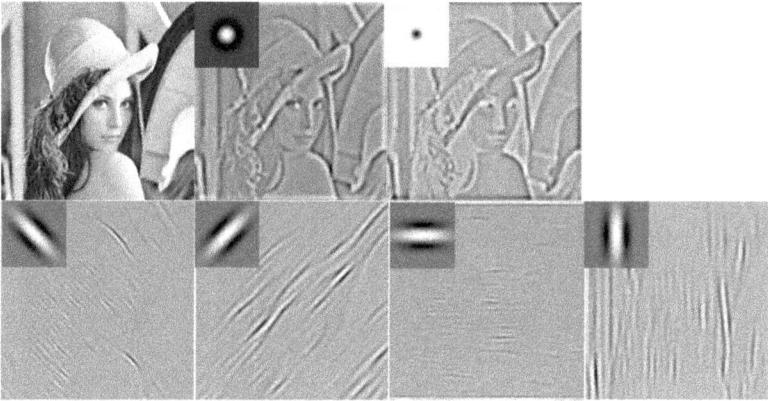

Figure 1.4. Responses of the LOG and Gabor filters, which can be modeled as convolutions with an underlying image. Lena image © Playboy Enterprises, Inc.

LGN, the signal is transmitted to the visual cortex at the back of the brain for subsequent processing steps. It is worth mentioning that, different to the retina, the number of ganglion or LGN cells is not great, only just over a million. That is to say, they work with the compressed features from the retina after reducing the redundancy in original data.

The first place in the cortex where most of the signals go is the primary visual cortex, or V1 for short. One type of cell in V1, which we understand the best, is the simple cells, whose receptive fields are well understood (Ringach 2002). Simple cells have responses that depend on the direction and spatial frequency of the stimulus signal. These responses can be modeled as a Gabor function or Gaussian derivative. Hence, the receptive fields of simple cells are interpreted as Gabor-like or directional band-pass filters. The Gabor function can be regarded as a combination of Gaussian and sine functions. There are several parameters to control the shape of a Gabor function. Similarly to LOG visualization, we also describe the Gabor function in figure 1.5 with different parameter settings. Observe how the parameters affect the Gabor function.

With selective characteristics, hundreds of millions of simple cells work together in V1. Neurobiologists have found that only a few cells are activated when a signal is inputted, which means that simple cells implement a sparse coding scheme. After

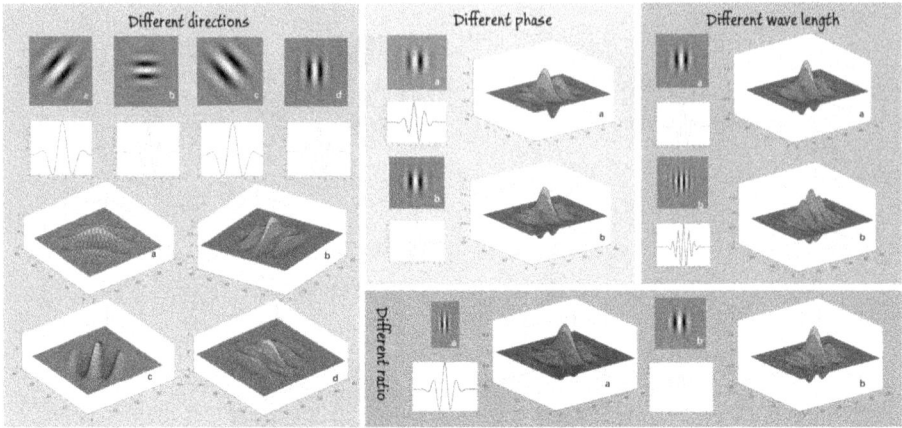

Figure 1.5. Visualization of the Gabor function.

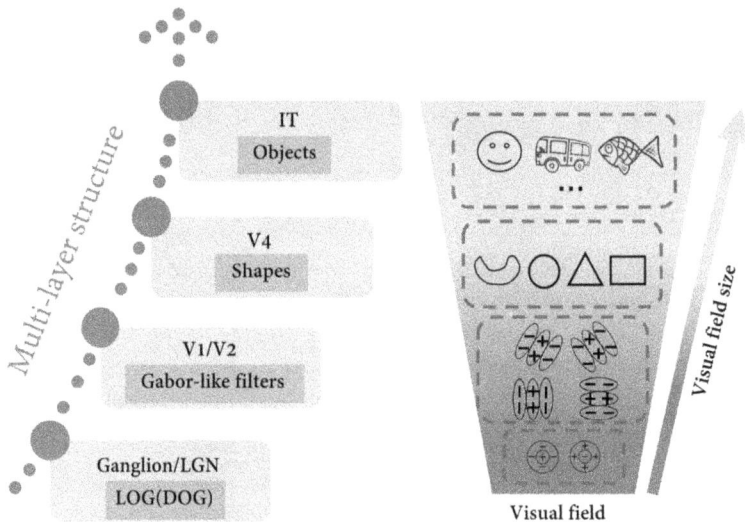

Figure 1.6. Multi-layer structures of HVS, perceiving the world in multiple stages from primitive to semantic.

being processed in V1, the signal is transformed to multiple destinations for further processing in the cortex. The destinations can be categorized into 'where' and 'what' pathways. The 'where' pathway is also known as the dorsal pathway going from V1/V2 through V3 to V5. It distinguishes moving objects and helps the brain to recognize where objects are in space. The 'what' pathway, namely the ventral pathway, begins from V1/V2 to V4 and inferior temporal cortex, IT, where the HVS performs content discrimination and pattern recognition (Cadieu *et al* 2007). Given the emphasis of this book on medical imaging, we emphasize the 'what' pathway that is modeled as multi-layer perceptive operations from simple to complex when the visual field becomes increasingly larger, as illustrated in figure 1.6.

In addition to the simple cells, there are also other kinds of visual neurons in the HVS. Another kind of visual cell we have studied extensively is the complex cells, which are mainly distributed in V1, V2, and V3. Complex cells integrate the outputs of nearby simple cells. They respond to specific stimuli located within the receptive field. In addition, there are also hypercomplex cells, called end-stopped cells, which are located in V1, V2, and V3, and respond maximally to a given size of stimuli in the receptive field. This kind of cell is recognized to perceive corners and curves, and moving structures.

To date, the investigation of our brains has been far from sufficient. We only have some partial knowledge of these areas, in particular of deeper layers such as V4 and the posterior regions. Generally speaking, visual cells in V1 and V2 detect primary visual features with selectivity for directions, frequencies, and phases. Some specific cells in V2 also provide stereopsis based on the difference in binocular cues, which helps recover the surface information of an object. In V4, the visual cells perceive the simple geometric shapes of objects in receptive fields larger than that of V2. This shape-oriented analysis capability is due to the selectivity of V4 cells for complex stimuli and is invariant with respect to spatial translation. In posterior regions of the visual pathway, such as the IT, image semantic structures are recognized, which depend on much larger receptive fields than that of V4. In general, billions of various visual neurons construct the hierarchically sophisticated visual system that analyzes and synthesizes visual features for observing and perceiving the outside world. Figure 1.6 illustrates the hierarchy of the HVS.

Fred Attneave and Horace Barlow realized that the HVS perceives surroundings in an 'economical description' or 'economical thought' that compresses the information redundancy in the visual stimuli. Actually, this point of view suggests an opportunity for us to consider extracting prior information in the HVS perspective. Specifically, in neurophysiological studies Barlow proposed the efficient coding hypothesis in 1961, as a theoretical model of sensory coding in the human brain. In the brain, neurons communicate with one another by sending electrical impulses or spikes (action potentials), which represent and process information on the outside world. Since among the hundreds of millions of neurons in the visual cortex only a few neurons are activated in response to a specific input, Barlow hypothesized that a neural code formed by the spikes represents visual information efficiently; that is, the HVS has the sparse representation ability. HVS tends to minimize the number of spikes needed to transmit a given signal, which can be modeled as an optimization problem. In his hypothesis, the brain uses an efficient coding system suitable for expressing the visual information of different scenes. Barlow's model treats the sensory pathway as a communication channel, in which neuronal spikes are sensory signals, with the goal to maximize the channel capacity by reducing the redundancy in a representation. They thought that the goal of the HVS is to use a collection of independent events to explain natural images. To form an efficient representation of natural images, the HVS uses pre-processing operations to get off first- and second-order redundancy. In natural image statistics, first-order statistics gives the direct current (DC), which is average luminance, and the second order describes variance and covariance, i.e. the contrast of the image. The heuristics is that image recognition should not be changed by the average luminance and contrast scale. In mathematics, this pre-processing can be modeled as zero-phase component analysis (ZCA). Interestingly, it was found that the responses of ganglion and LGN cells are similar to features obtained with natural image statistics techniques such as ZCA.

Inspired by the mechanism of the HVS, researchers have worked to mimic the HVS by reducing the redundancy of images so as to represent them efficiently. In this context, machine learning techniques were used to obtain similar features as observed in the HVS. In figure 1.7, we explain the relationship between an artificial neural network (to be explained in chapter 3) and the HVS. Furthermore, in the HVS feature extraction and representation, high-order redundancy is also reduced. Specifically, the receptive field properties are accounted for with a strategy to sparsify the output activity in response to natural images. The 'sparse coding' concept was introduced to describe this phenomenon. Olshausen and Field, based on neurobiological observations, used a network to code image patches in an over-complete basis to capture image structures under sparse constraints. They found that the features have local, oriented, receptive fields, essentially the same as V1 receptive fields. That is to say, the HVS and natural image statistics are closely related, both of which are very relevant to prior information extraction.

In the following sub-sections, we will introduce several HVS models, and describe how to learn features from natural images in the light of visual neurophysiological findings.

1.1.5 Data decorrelation and whitening

How can one represent natural images with their intrinsinc properties? One of the widely used methods in natural image statistics is principal component analysis (PCA). PCA considers the second-order statistics of natural images, i.e. the variances of and covariances among pixel values. Although PCA is not a sufficient

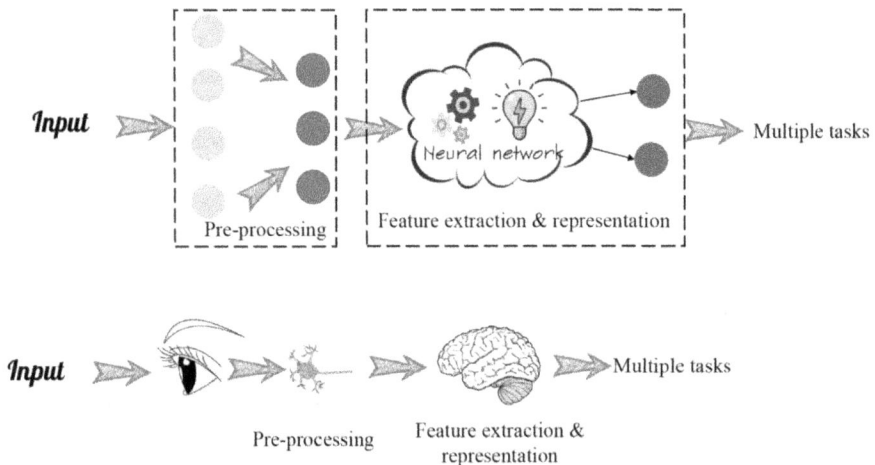

Figure 1.7. The relationship between an artificial neural network and the HVS.

model for the HVS, it is the foundation for the other models, and is usually applied as a pre-processing step for further analysis (Hyvärinen *et al* 2009). It can map original data into a set of linearly decorrelated representations of each dimension through linear transformation of the data, identifying the main linear components of the data.

During the linear transformation, we would like to make the transformed vectors as dispersed as possible. Mathematically, the degree of dispersion can be expressed in terms of variance. The variance of data provides information about the data. Therefore, by maximizing the variance, we can obtain the most information, and we define it as the first principle component of the data. After obtaining the first principal component, the next linear feature must be orthogonal to the first one and, more generally, a new linear feature should be made orthogonal to the existing ones. In this process, the covariance of vectors is used to represent their linear correlation. When the covariance equals zero, there is no correlation between the two vectors. The goal of PCA is to diagonalize the covariance matrix: that is, minimizing the amplitudes of the elements other than the diagonal ones, because diagonal values are the variances of the vector elements. Arranging the elements on the diagonal from top to bottom according to their amplitude, we can achieve PCA. In the following, we briefly introduce a realization of the PCA method.

Usually, before calculating PCA we remove the DC component in images (the first-order statistical information, often containing little structural information for natural images). Let $\mathbf{X} \subseteq \mathbb{R}^{n \times m}$ denote a sample matrix with DC removed, n be the data dimension, and m be the number of samples. Then, the covariance matrix can be computed as follows:

$$\Sigma = \frac{1}{m}\mathbf{X}\mathbf{X}^\mathsf{T}. \tag{1.11}$$

By singular value decomposition (SVD), the covariance matrix can be expressed as

$$\Sigma = \mathbf{U}\mathbf{S}\mathbf{V}, \tag{1.12}$$

where $\mathbf{U}$ is an $n \times n$ unitary matrix, $\mathbf{S}$ is an $n \times n$ eigenvalue matrix, and $\mathbf{V} = \mathbf{U}^\mathsf{T}$ is also an $n \times n$ unitary matrix. The magnitude of the eigenvalues reflects the importance of the principal components. Arranging the eigenvalues from top to bottom in descending order, PCA can be realized with the following formula:

$$\mathbf{X}_{\mathrm{PCA}} = \mathbf{U}^\mathsf{T}\mathbf{X}. \tag{1.13}$$

Figure 1.8 depicts 64 weight matrices for Lena image patches of 8×8 pixels. The descending order of variance is from left to right along each row, and from top to bottom row-wise. PCA has been widely applied as a handy tool to compress data. In figure 1.9, we show a simple experiment of PCA compression. It can be seen that a natural image can be represented by a small number of components, relative to its original dimensionality. This means that some data redundancy in natural images can be removed by PCA.

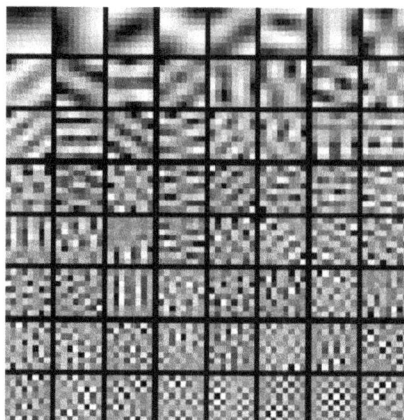

Figure 1.8. 64 weighting matrices for Lena image patches of 8×8 pixels.

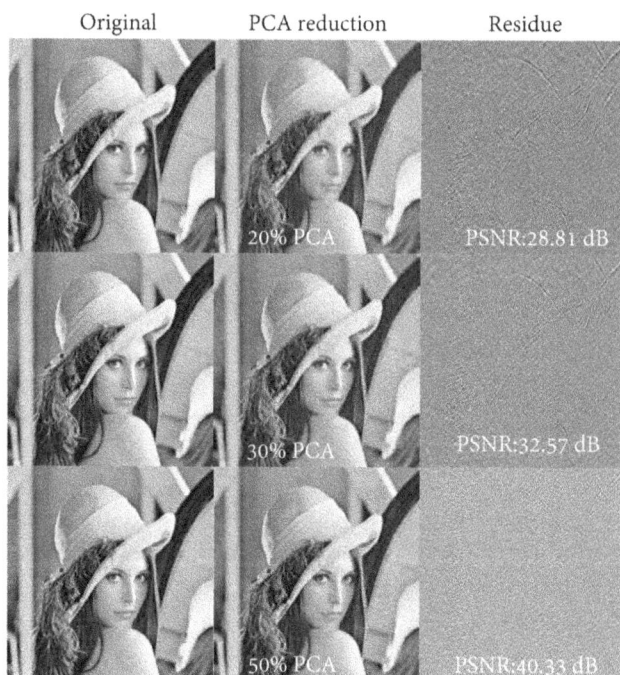

Figure 1.9. Image compressed with PCA. Lena image © Playboy Enterprises, Inc.

There is an important pre-processing step related to PCA, which is called whitening. It removes the first- and second-order information which, respectively, represent the average luminance and contrast information, and allows us to focus on higher-order statistical properties of the original data. Whitening is also a basic processing function of the retina and LGN cells (Atick and Redlich 1992). The data exhibit the following properties after the whitening operations: (i) the features are uncorrelated and (ii) all features have the same variance. In the patch-based

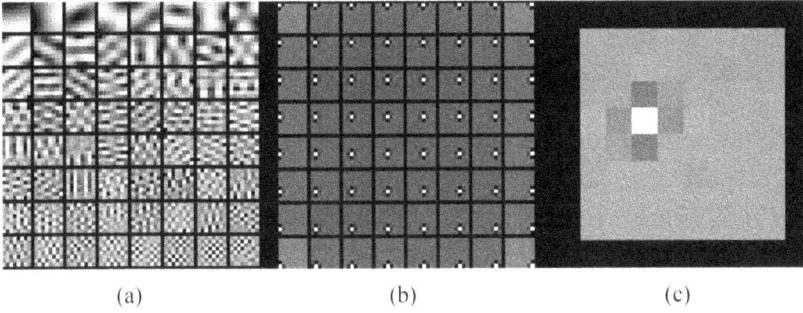

Figure 1.10. Basis functions obtained with PCA and ZCA, respectively. (a) PCA whitening basis functions, (b) ZCA whitening basis functions (with size 8 × 8), and (c) an enlarged view of a typical ZCA component in which significant variations happen around a specific spatial location.

whitening process, it is worth mentioning that the whitening process works well with PCA or other redundancy reduction methods. After PCA, the only thing we need to do for whitening data is to normalize the variances of the principal components. Thus, PCA with whitening can be expressed as follows:

$$\mathbf{X}_{\text{PCA}_{\text{white}}} = \mathbf{S}^{-\frac{1}{2}}\mathbf{U}^{\mathsf{T}}\mathbf{X}, \tag{1.14}$$

where $\mathbf{S}^{-\frac{1}{2}} = \text{diag}(\frac{1}{\sqrt{\lambda_1}}, \ldots, \frac{1}{\sqrt{\lambda_n}})$, λ_i is the eigenvalues.

After the whitening process, we have nullified the second-order information. That is, PCA with whitening remove the first- and second-order redundancy of data. Whitening, unlike PCA that is solely based on image patches, can also be performed by applying a filter.

Based on PCA, we can apply another component analysis algorithm called zero-phase component analysis, abbreviated as ZCA. ZCA is accomplished by transforming PCA data into the original data space:

$$\mathbf{X}_{\text{ZCA}_{\text{white}}} = \mathbf{U}\mathbf{X}_{\text{PCA}_{\text{white}}}, \tag{1.15}$$

where $\mathbf{U}$ is the unitary matrix with the same definition as the SVD, $\mathbf{U}\mathbf{U}^{\mathsf{T}} = \mathbf{I}$ (also referred to as the 'Mahalanobis transformation'). It can be shown that ZCA attempts to keep the transformed data as close to the original data as feasible. Hence, compared to PCA, data whitened by ZCA are more related to the original in terms of preserving structural information, except for luminance and contrast data. Figure 1.10 illustrates the global and local behaviors of PCA and ZCA, respectively. Since natural image features are mostly local, decorrelation or whitening filters can also be local. For natural images, high frequency features are commonly associated with small eigenvalues. The luminance and contrast components take up most of the energy of the image. In this context, ZCA is a simple yet effective way to highlight these structural features by removing the luminance and contrast components that account for little structural information in the image.

In the HVS, the receptive field is tuned to a particular light pattern for a maximum response, which is achieved via local precessing. The receptive field of ganglion cells in the retina is a good example of a local filtering operation, so is the field of view of ganglion and LGN cells.

If the HVS had to transmit each pixel value to the brain separately, it would not be cost-effective. Fortunately, local neural processing yields a less redundant representation of an input image and then transmits the compressed code to the brain. According to the experimental results with natural images, the whitening filters for centralized receptive fields are circularly symmetric and similar to the LOG function, as shown in figure 1.3. Neurobiologists have verified that compared to the millions of photoreceptors in the retina, the numbers of ganglion and LGN cells are quite small, indicating a compression operation is performed on the original data.

1.1.6 Sparse coding

In the previous subsection, we have introduced several models on natural image statistics, which produce results similar to the responses of the HVS retina and LGN. These models only get rid of the first- and second-redundancy in images. Now, we will introduce two models that are the first successful attempts to give similar results to those found in simple cells in the visual cortex. These models suppress higher-order redundancy in images. These models interpret visual perception, the data are pre-processed by whitening and DC is removed, being consistent with the HVS pre-processing step for LGN and ganglion cells.

Although these two models are milestones in mimicking the responses of simple cells to natural images, their computational methods are quite time-consuming. Here, we only focus on the main idea behind these models, and in chapter 4 we will introduce some efficient methods to obtain the same results.

The first model was proposed by Olshausen and Field (Olshausen and Field 1996). They used a one-layer network and trained it with natural image patches to extract distinguished features for natural image coding. According to this study, a simple cell of V1 contains about 200 million cells, while the number of ganglion and LGN cells responsible for visual perception is only just over 1 million. This indicates that sparse coding is an effective strategy for data redundancy reduction and efficient image representation.

Sparse coding means that a given image may be typically described in terms of a small number of suitable basis functions chosen out of a large training dataset. A heavy-tailed distribution of representation coefficients is often observed, as illustrated in figure 1.10. For instance, if we consider a natural image patch as a vector x, as shown in figure 1.11, this vector can be represented just by two components, i.e. numbers 3 and 6 out of the 12 features in total. To generalize this problem, a typical sparse encoding strategy is to approximate an image as a linear combination of basis functions:

$$x = \sum_{i=1}^{K} \alpha_i \phi_i, \qquad (1.16)$$

where α_i is a representation coefficient or an active coefficient for the ith basis function, and ϕ_i is the ith basis function.

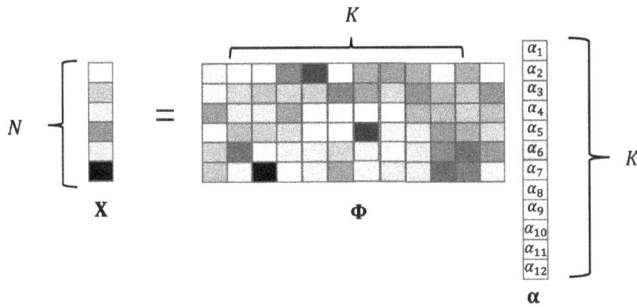

Figure 1.11. An image modeled as a linear superposition of basis functions. The sparse encoding is to learn basis functions which capture the structures efficiently in a specific domain, such as natural images. Adapted from figure 1 in Baraniuk (2007) with permission. Copyright 2007 IEEE.

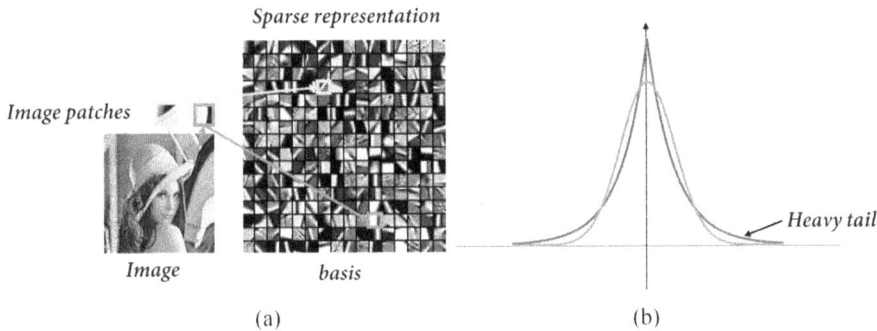

Figure 1.12. Sparse representation characterized by a generalized Gaussian distribution of representation coefficients, which generates sparse coefficients in terms of an over-complete dictionary. (a) An image is represented by a small number of 'active' code elements and (b) the probability distribution of its 'activities'. Lena image © Playboy Enterprises, Inc.

In their work published in 1996 (Olshausen and Field 1996), they trained a feed-forward artificial neural network on natural image patches in terms of the sparse representation with an over-complete basis. In sparse coding, the search process should achieve a match as closely as possible between the distribution of images described by the linear image model under sparse constraints and the corresponding training targets (figure 1.12). For this purpose, Lagrangian optimization was used to describe this problem. Then, the final optimization function can be formularized as follows:

$$\min_{\alpha,\phi} \sum_{j=1}^{m} \left\| x_j - \sum_{i=1}^{K} \alpha_{j,i}\phi_i \right\|^2 + \lambda \sum_{i} S(\alpha_{j,i}), \tag{1.17}$$

where x_j is an image patch extracted from a natural image, $\alpha_{j,i}$ is the representation coefficient for basis function ϕ_i in image patch x_j, S is a sparsity measure, and λ is a weighting parameter. This formula contains two components: the first term computes the reconstruction error while the second term imposes the sparsity penalty.

Although this formula is quite simple and easy to comprehend, it has an open question: how does one measure sparseness mathematically? As a reference point, the distribution of a zero-mean random variable can be compared to the Gaussian distribution with the same variance and mean. The rationale for selection of the Gaussian distribution as the reference is that the Gaussian distribution has the largest entrophy relative to all probability distributions for the same variance. Thus, if the distribution of interest is more concentrated than the Gaussian distribution, it can be regarded as being sparse. Based on this consideration, the measurement of sparseness can be heuristically calculated. The criteria for a sparsity function to work as intended are to emphasize values that are close to zero or values that are much larger than a positive constant, such as 1 for a normalized/whitened random variable. A sparse function satisfying these two requirements is often heavy tailed, i.e. many coefficients are insignificant, and significant coefficients are sparse so that a resultant image representation is sparse. Interestingly, if we use $S(x) = |x|$, the coding process is to solve a Lasso problem, which means that the regularization term is in the L1 norm. This explains why we often use the L1 norm for a sparse solution.

By training their network with image patches of 12 by 12 pixels, they obtained 144 basis functions, as shown in figure 1.13. Recall that the patches were whitened before feeding into the network. The basis functions obtained by sparse coding of natural images are Gabor-like, similar to the responses of the receptive fields of simple cells in V1. Hence, these basis functions model the receptive fields of simple cells in V1 very well.

The second model (Bell and Sejnowski 1997) was proposed based on the independent component analysis (ICA) principle. ICA is an approach to solving

Figure 1.13. Basis functions learned by the sparse coding algorithm. All were normalized, with zero always represented by the same gray level. Reproduced with permission from Olshausen and Field (1997). Copyright 1997 Elsevier.

the blind source separation (BSS) problem (figure 1.14). In natural image statistics, let $\mathbf{X} = \{\mathbf{x}_i | i = 1,\ldots, N\}$ represent N independent source signals forming a column vector, $\mathbf{Y} = \{\mathbf{y}_1, \mathbf{y}_i,\ldots | i = 1,\ldots, \mathbf{M}\}$ representing M image patches also forming a column vector, and $\mathbf{W}$ is the mixing matrix of $N \times M$ dimensions. The BSS problem is to invert the measurement $\mathbf{Y} = \{\mathbf{y}_j | j = 1,\ldots, \mathbf{M}\}$,

$$\mathbf{Y} = \mathbf{WX}, \quad M \geqslant N, \tag{1.18}$$

for both $\mathbf{W}$ and $\mathbf{X}$ subject to uncertainties in amplitudes and permutations of independent source signals. ICA helps us find the basis components $\mathbf{X} = \{\mathbf{x}_i | i = 1,\ldots, N\}$ which have representative features of image patches.

The premise of ICA is based on statistical independence among hidden data sources. In information theory (see appendix A for more details), we use the mutual information to measure the relationship between two signals. Let $H(X)$ and $H(Y)$ represent the self-information, which solely depend on the probability density functions of X and Y, respectively. $H(X, Y)$ is the joint information, which represents the amount of information generated when X and Y occur together. $I(X, Y)$ is the mutual information that is the information we have when a certain X or Y is known (figure 1.15). For example, if we know the information about Y, we only need to have the amount of $H(X) - I(X, Y)$ information to determine X completely. When two signals are independent, the mutual information is zero (Chechile 2005).

We consider the ICA operation as a system, $\mathbf{Y}$ as the input and $\mathbf{X}$ as the output. When the output information of the system reaches its maximum, it indicates the minimum mutual information between output components. That is to say, the output components are as independent of each other as possible, since any non-trivial linear combination will compromise data independency. This is a simplified description of the infomax principle.

In 1997, Bell and Sejnowski applied ICA using the information theoretic approach in the case of natural images, and found that ICA is a special sparse coding method. They explained the results obtained with the network proposed by Olshausen and Field in the ICA framework. ICA on natural images produces decorrelating filters that are sensitive to both phase and frequency, similar to the cases with transforms involving oriented Gabor functions or wavelets. Representative ICA filters generated from natural images are shown in figure 1.16. It can be seen that ICA can also model the Gabor-like receptive fields of simple cells in V1.

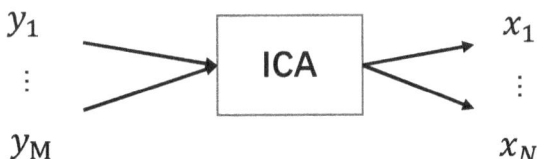

Figure 1.14. ICA to find both embedded independent components and the mixing matrix that blends the independent components.

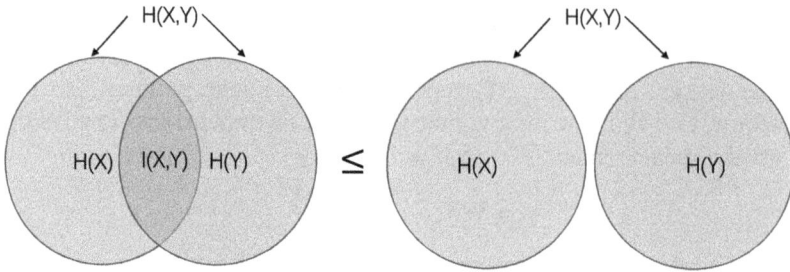

Figure 1.15. Joint information determined by the signal information $H(X)$ and $H(Y)$ as well as mutual information $I(X, Y)$.

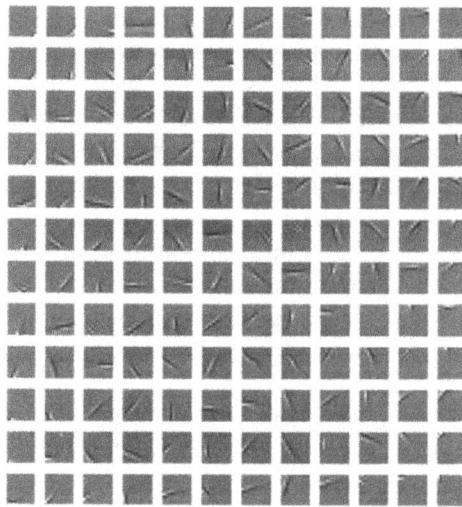

Figure 1.16. A matrix of 144 filters obtained using ICA on ZCA-whitened natural images. Reproduced with permission from Bell and Sejnowski (1997). Copyright 1997 Elsevier.

In this chapter, we have provided a general explanation on how to reduce data redundancy and form a sparse representation in the HVS. Multiple types of cells, such as ganglion and LGN cells, are involved to normalize first- and second-order statistics and remove the associated redundancy. In the HVS, higher-order redundancy is eliminated with simple cells. From the viewpoint of biomimicry, the mechanism of simple cells is the basis for sparse representation. In addition, from the natural image perspective, we can use a sparsifying transform or model to obtain similar results as are observed in the HVS. It is noted that deep neural networks (to be formally explained in chapter 3) exhibit workflows similar to that of the HVS, such as multi-resolution analysis. As a second example, the whitening process is used to pre-process data for both the HVS and machine learning. Yet another example is that higher-order redundancy operations share Gabor-like characteristics observed in the HVS and machine learning. It will become increasingly more clear that machine learning imitates the HVS in major ways. Now, we have the tools to extract features constrained by or in reference to natural image statistics. How could we use these features to help solve practical problems? This question naturally leads us to our following chapters.

References

Atick J J and Redlich A N 1992 What does the retina know about natural scenes? *Neural Comput.* **4** 196–210

Bakushinsky A B and Kokurin M Y 2004 *Iterative Methods for Approximate Solution of Inverse Problems* (Berlin: Springer)

Baraniuk R G 2007 Compressive sensing *IEEE Signal Process. Mag.* **24** 118–21

Bell A J and Sejnowski T J 1997 The 'independent components' of natural scenes are edge filters *Vis. Res.* **37** 3327–38

Bertero M and Boccacci P 1998 *Introduction to Inverse Problems in Imaging* (Boca Raton, FL: CRC Press)

Cadieu C, Kouh M, Pasupathy A, Connor C E, Riesenhuber M and Poggio T 2007 A model of V4 shape selectivity and invariance *J. Neurophysiol.* **98** 1733–50

Poggio T 2007 A model of V4 shape selectivity and invariance *J. Neurophysiol.* **98** 1733–50

Chechile R A 2005 Independent component analysis: a tutorial introduction *J. Math. Psychol.* **49** 426

Dashti M and Stuart A M 2017 *The Bayesian Approach to Inverse Problems* (Berlin: Springer)

Hyvärinen A, Hurri J and Hoyer P O 2009 *Natural Image Statistics* (London: Springer)

Landweber L 1951 An iteration formula for Fredholm integral equations of the first kind *Am. J. Math* **73** 615–24

Leigh P N, Simmons A, Williams S, Williams V, Turner M and Brooks D 2002 Imaging: MRS/MRI/PET/SPECT: summary *Amyotro. Later. Sclero. Other Motor Neur. Disord.* **3** S75–80

Olshausen B A and Field D J 1996 Emergence of simple-cell receptive field properties by learning a sparse code for natural images *Nature* **381** 607–9

Olshausen B A and Field D J 1997 Sparse coding with an overcomplete basis set: a strategy employed by V1? *Vis. Res.* **37** 3311–25

Ringach D L 2002 Spatial structure and symmetry of simple-cell receptive fields in macaque primary visual cortex *J. Neurophysiol.* **88** 455

Stuart A M 2011 Bayesian approach to inverse problems *LMS-EPSRC Short Course (University of Oxford, 3–8 April)*

Zeng G L 2009 *Medical Image Reconstruction* (Berlin: Springer)

IOP Publishing

Machine Learning for Tomographic Imaging

Ge Wang, Yi Zhang, Xiaojing Ye and Xuanqin Mou

Chapter 2

Tomographic reconstruction based on a learned dictionary

2.1 Prior information guided reconstruction

In the first chapter, we have discussed the acquisition and representation of prior knowledge on images from two perspectives, which are natural image statistics and neurophysiological HVS functions. In this chapter, we will discuss the computational methods to solve the inverse problem iteratively, aided by prior information.

Previously, we have presented the approach for solving inverse problems in the Bayesian inference framework, which is equivalent to minimizing the objective function via Lagrangian optimization:

$$\hat{\mathbf{x}} = \arg\min_{\hat{\mathbf{x}} \in \mathbf{x}} \{\phi(\mathbf{x}, \mathbf{y}) + \lambda\psi(\mathbf{x})\}, \tag{2.1}$$

where $\phi(\mathbf{x}, \mathbf{y})$ is the data fidelity term to encourage that an estimated CT image $\mathbf{x}$ be consistent with observed projection data $\mathbf{y}$; $\phi(\mathbf{x}, \mathbf{y})$ is also a logarithmic function that reflects the noise statistics. As a result, the fidelity term can be characterized by some norm to model the type of statistical noise. For instance, if the noise obeys a Gaussian distribution, we use the L2-norm such as in the form of $\frac{1}{2}\|\mathbf{Ax} - \mathbf{y}\|_2^2$. In the case of Poisson noise, an informational measure is proper in the form of $\int(\mathbf{Ax} - \mathbf{y}\ln\mathbf{Ay})$. If the inverse problem is subject to impulsive noise, such as salt-and-pepper noise, the L1-norm can be used, which is expressed as $\|\mathbf{Ax} - \mathbf{y}\|_1$. On the other hand, the regularization term $\psi(\mathbf{x})$ promotes solutions with some desirable properties. As explained in the first chapter, this term reflects the characteristic of natural images, which can be obtained by removing the redundancy of images. As mentioned before, by using either principal component analysis (PCA) whitening basis functions or zero-phase component analysis (ZCA) whitening basis functions prior to independent component analysis (ICA), the first- and second-order redundancies of images can be removed and the whitened features are uncorrelated.

doi:10.1088/978-0-7503-2216-4ch2

Moreover, a number of excellent basis functions can be learned using a sparse coding technique to capture statistically independent features. To go a step further along this direction, with the use of a multi-layer neural network, we can extract and represent structural information or semantics, which could promote the effectivity and efficiency of the solution to inverse problems. This neural network perspective will be focused on in this book.

It is underlined that Bayesian inference is a classic approach to solve inverse problems. In the Bayesian framework, an image prior is introduced to constrain the solution space for suppression of measurement noise and image artifacts. This strategy generally requires an iterative algorithm. Indeed, there are many optimization methods to minimize a regularized objective function, the optimal result is almost always iteratively obtained for a balance between the data fidelity and the regularization term. In other words, the fidelity term is not necessarily equal to zero. It is so because the observation $\mathbf{y}$ contains both the ideal signal we want to obtain and noise/error that cannot be avoided in practice. Then, a regularizer or prior knowledge can be used to guide the search path for an optimal solution $\hat{\mathbf{x}}$ as a trade-off between imperfect measurement and desirable image properties.

In the following, let us intuitively explain regularized image reconstruction. Without loss of generality, let us consider the objective function with the fidelity term in the L2-norm:

$$\hat{\mathbf{x}} = \arg\min_{\hat{\mathbf{x}} \in \mathbf{x}} \left\{ \|\mathbf{A}\mathbf{x} - \mathbf{y}\|_2^2 + \lambda \psi(\mathbf{x}) \right\}. \tag{2.2}$$

It is assumed that the observation $\mathbf{y}$ is related to the unknowns of interest $\mathbf{x}$ through a degradation operator $\mathbf{A}$ and, at the same time, $\mathbf{y}$ is corrupted by non-ideal factors in the data acquisition process, such as Gaussian noise, and is modeled as

$$\mathbf{y} = \mathbf{A}\mathbf{x} + \mathbf{n} \qquad \mathbf{y}, \mathbf{x}, \mathbf{n} \in \mathbb{R}^N. \tag{2.3}$$

Indeed, this imaging model is totally different from conventional image processing tasks which are from images to images or from images to features. The quintessence of image processing is how to discriminate the signal from its noisy or some error contaminated background. It is often not an easy task in the spatial domain because an image appearing as a collection of pixel values does not provide the signal and the noise/error separately. Fortunately, we can transform the image to a feature space in which the image signal and noise/measurement error can be much more easily discriminated. The workflow involved consists of the following three steps. First, an original image is transformed from the spatial domain into a feature space, where a specific aspect of physical properties of the image can be well presented. It should be noted that the transform is invertible and conservative, which means that transformational loss is zero. The Fourier transform and wavelet transform are good examples. With such a lossless transform, structures, errors, and noise in the image are all preserved but exist in a different form. In the second step, according to statistical rules in the transformed feature space, noise and errors can be suppressed by modifying features so that they satisfy the underlying statistical laws; for example, by means of either soft thresholding in the wavelet domain or frequency filtering in the Fourier domain. The former makes refinements in reference to the prior information in the wavelet domain that the structural components should have a sparse distribution of significant coefficients, while the image noise has a broad and weak spectrum of amplitudes in the wavelet domain. The latter is based on the fact that the frequency components of the image concentrate to a low frequency band while the noise spreads over the whole Fourier space. Finally, in the third step the output is obtained by the corresponding inverse transform of the modified features.

Different to the transform method, the key ingredient of the regularization method is to estimate an underlying image by leveraging prior knowledge on desirable features/structures of images, while eliminating the noise and error. The use of a regularizer constrains the image model. In this way, it is convenient and effective to promote favorable properties of images such that the learned model will represent the image characteristics we want, such as sparseness, low rank, smoothness, and so on. Mathematically, the regularizer can be expressed as a norm that measures the image x in a way that is optimal for the inverse problem of interest.

It is worth mentioning that both the above strategies utilize natural image statistics but in different ways. To be specific, the Bayesian framework uses prior knowledge as the constraint in the objective function, while the transform approach uses the statistical distributions of signals and noise/errors in the feature space, which can be also regarded as prior knowledge.

We have discussed the regularization term in the previous chapter. In the following, we will give an intuitive example to show the impact of the regularizer on the efficiency of the solution to the inverse problem. More specifically, the L1-norm-based regularized solution will be compared with that based on the L2-norm.

We will focus on the regularization issue in the context of the machine learning-based inverse problem solution in this book. Mathematically, the L2-norm-based term is called Tikhonov regularization, or ridge regression, which is a commonly used method. The L1-norm-based term is called Lasso regression, which is essentially a sparsity constraint. In the previous chapter, we have shown that the sparse constraint favors an efficient information representation. Next, let us elaborate the effects of the two regularization terms, respectively.

Based on equations (2.2) and (2.3), we want to estimate the original signal, i.e. the image x from the observation y. When n is the Gaussian white noise with zero mean and a variance of σ^2, the noise has a constant power spectral density in the frequency domain. Equation (2.3) can be written as $\|Ax - y\|^2 = \sigma^2$. The expected solution space of x is on the surface of the hyper-ball, as shown in figure 2.1.

How to find the optimal solution? It is well known that the L1-norm is defined as the summation of absolute component values of a vector, while the L2-norm is the square root of the sum of the squared element values. Therefore, the geometric meanings of the L1- and L2-norm-based regularization are quite different, with the L1-based regularization clearly favoring a sparse solution, as shown in figure 2.1.

Let us consider the optimization problem as expanding the manifold defined by the regularizer until it intersects the manifold defined by $\|Ax - y\|^2 = \sigma^2$ and then

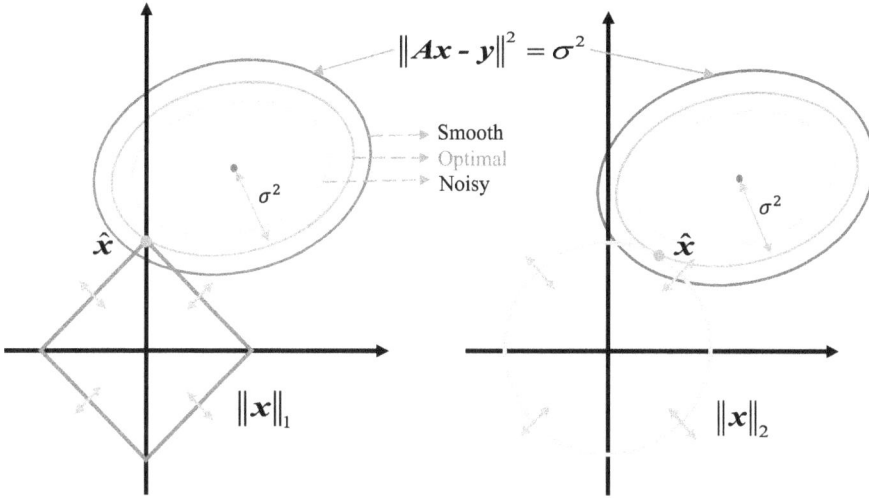

Figure 2.1. Under the data fidelity constraint, the minimized L2 distance corresponds to a non-sparse solution, and the minimized L1 distance most likely gives a sparse solution. Adapted with permission from Hastie *et al* (2009). Copyright 2009 Springer.

returning the touching point or singling out any point in the intersection. When the touching point does not find the optimal solution, the result can be a smooth version or noisy version, as is depicted in figure 2.1. Since our data fidelity term is quadratic, it is very likely that the touching point in the L1 case is at a corner of the square, cube or hyper-square defined by $\|\mathbf{x}\|_1$. On the other hand, in the L2 case, the circle or ball or hyper-ball defined by $\|\mathbf{x}\|_2$ does not have such a property. Since a hyper-ball has no corner at all, the probability for the touching point to arrive at a sparse position becomes very slim. Intuitively, if the square/cubic boundary expands in the L1 case, the only way to reach a non-sparse solution is for the L1 boundary to meet the solution subspace on the 'side' of the titled square/cube. But this only happens if the solution subspace is in parallel to the square/cubic boundary, which could happen, such as in an under-determined case, but with a very low probability. Thus, the L1 solution will almost certainly be sparse, which explains why the L1 solution tends to give a sparse solution. Accordingly, the L1-norm minimization is widely performed to satisfy the sparse constraint.

The L1-norm-based regularization shows a high utility in the inverse problem solution. Theoretically, the convergence of this solution is also important. An iterative solution to equation (2.2) with total variation minimization that is measured with an L1-norm has been proved to be convergent (Chambolle 2004). Nevertheless, the best operator for the sparsest solution should be established based on the L0-norm minimization that counts the number of nonzero components of the solution vector, leading to the highest representation efficiency. Unfortunately, the L0-norm solution is an NP problem and computationally impracticable. In 2006, it was demonstrated that the L1-norm prior is equivalent to the L0-semi-norm prior in signal recovery for some important linear systems (Dandes 2006). Thus, the

L1-norm-based sparse regularization is popular in solving the optimization problem. Moreover, being capable of extracting structural information efficiently in images, various machine learning-based techniques have been applied in the optimization framework, which exhibit excellent performance, superior to conventional sparse operators such as total variation minimization (Chambolle and Lions 1997).

2.2 Single-layer neural network

In this section, we will introduce two typical dictionary learning algorithms both of which are single-layer neural networks. Although these algorithms are theoretically equivalent to the sparsity constraint method (Olshausen and Field 1996) and ICA method (Bell and Sejnowski 1997) as we introduced in the previous chapter, the algorithms introduced in this chapter are more data-driven and have been widely used in practice.

Previously, we introduced Bayesian reasoning, in which the prior information of images is of great importance and must be analytically expressed. Thus, it is an essential task to find an efficient measure to express different information contents. This points to the field of image representation and sensory coding. The HVS tends to ignore the first-order and second-order features when sensing structures in natural scenes. Since the HVS has an efficient strategy for conveying as much information as possible, features used by the HVS should be statistically independent of each other. Researchers suggested a number of de-correlating strategies to remove statistical redundancy from natural images and represent the original information by a dictionary consisting of over-complete atoms.

Sparse dictionary learning is a representation method aiming at finding a sparse representation of an input signal or image as a linear combination of basic building units. One of the most important applications of sparse dictionary learning is in the field of compressed sensing or signal recovery. In the context of compressed sensing, a high-dimensional input signal can be recovered with only a few dictionary components provided that the signal is sparse. A sparse representation of that input signal is an important assumption since not all signals satisfy the sparsity condition. One of the key principles behind dictionary learning is that the dictionary must be generated from data. In other words, the dictionary is not constructed according to a principle or principles in advance. The dictionary learning methods are developed to represent an input signal with a minimized number of dictionary components and at maximized representation accuracy.

Let a dictionary be denoted as $D \in \mathbb{R}^{n \times K}$, each column of D denoted as d_k is a component of the dictionary and also referred to as an atom of the dictionary. Without loss of generality, the dictionary D has K atoms and each atom consists of n elements. A signal or image (after vectorization) $y \in \mathbb{R}^{n \times 1}$ can be represented sparsely as a linear combination of atoms in the dictionary D, which is formulated as $y = D\alpha$, or approximated as $y \approx D\alpha$, and $\|y - D\alpha\|_p \leqslant \varepsilon$ for some small value ε and a certain Lp-norm. In order to seek a sparse solution, a variety of regularization schemes were proposed, which reflect prior information of coding coefficients $\alpha \in \mathbb{R}^{K \times 1}$, into the optimization problem, such as the L0-norm constraint

$$\min_{\mathbf{x}} \|\boldsymbol{\alpha}\|_0 \quad \text{s.t.} \quad \mathbf{y} = \mathbf{D}\boldsymbol{\alpha} \tag{2.4}$$

or

$$\min_{\mathbf{x}} \|\boldsymbol{\alpha}\|_0 \quad \text{s.t.} \quad \|\mathbf{y} - \mathbf{D}\boldsymbol{\alpha}\|_2 \leqslant \varepsilon, \tag{2.5}$$

where $\|a\|_0$ denotes the total number of nonzero entries in the vector $\boldsymbol{\alpha}$. The use of the L0-norm as a measure of sparsity makes the problem nonconvex and, as mentioned earlier, solving the equation directly is an NP-hard problem (Elad 2010). The choice of these atoms is crucial for efficient sparse representation, but it is not trivial. Therefore, it is informative to study what kind of dictionary expression or visual coding is accurate and efficient.

Different from the dictionary learning and sparse representation methods presented in chapter 1, in the following sections we will introduce more efficient methods to complete the tasks. In general, the methodology of dictionary learning and sparse representation contains two tasks. The first is to learn an over-complete dictionary $\mathbf{D}$ in equations (2.4) and (2.5) from a set of data which captures plenty of essential structures, i.e. the atoms of the dictionary such that any single datum can be sparsely or independently represented by a few numbers of the structures of the dictionary. The second is sparse coding which means finding, for a specific signal, the representation vector $\boldsymbol{\alpha}$ regarding the dictionary $\mathbf{D}$ to meet the sparse constraint. Let us first learn a classical sparse coding method, i.e. the matching pursuit algorithm (Mallat and Zhang 1993), given that the dictionary $\mathbf{D}$ is provided.

2.2.1 Matching pursuit algorithm

For the purpose of finding a least set of atoms and determining the corresponding coefficients to represent a signal, the essence of the MP algorithm aims to find the atom with the highest correlation with the signal and removes the component of this atom from the signal. This process is repeated iteratively until the stop condition is met. More specifically, let us take a closer look at the steps to use MP for signal representation. k represents the index of iteration number, $\mathbf{y}_k = \mathbf{D}\boldsymbol{\alpha}_k$ is the current approximation, the signal is denoted as $\mathbf{y}$, and the current residual is $\mathbf{R}_k\mathbf{y}$:

Initialization: $\mathbf{R}_0\mathbf{y} = \mathbf{y} \quad \mathbf{y}_0 = \mathbf{0} \quad k = 1$

Step 1. Calculate the inner product of the signal $\mathbf{y}$ with each column (atoms) in the dictionary $\mathbf{D}$, denoted as $\{\langle \mathbf{R}_0\mathbf{y}, \mathbf{d}_k \rangle\}_k$.

Step 2. Find an atom in the dictionary which yields the maximum inner product satisfying $|\langle \mathbf{R}_0\mathbf{y}, \mathbf{d}_{r_0} \rangle| \geqslant \sup_{j \in (1, \ldots, K)} |\langle \mathbf{R}_0\mathbf{y}, \mathbf{d}_j \rangle|$, where r_0 is the column index of the corresponding atom in $\mathbf{D}$.

Hence, in the first iteration, an intermediate representation of the single $\mathbf{y}_0$ consists of two parts, as follows:

$$\mathbf{y}_0 = \langle \mathbf{R}_0\mathbf{y}, \mathbf{d}_{r_0} \rangle \mathbf{d}_{r_0} + \mathbf{R}_1\mathbf{y}. \tag{2.6}$$

In this equation, the first part is the orthogonal projection of the most matching atom $\mathbf{d}_{r_0}$ on the span of $\langle \mathbf{R}_0\mathbf{y}, \mathbf{d}_{r_0} \rangle$, and the second part is the residue.

Step 3. Increment k, repeat steps 1 and 2 for the kth residue $R_k y$.

So, in the kth iteration, the signal is represented as follows:

$$R_k y = \langle R_k y, d_{r_{k+1}} \rangle d_{r_{k+1}} + R_{k+1} y. \tag{2.7}$$

Step 4. Stop the iterations until the residue satisfies the given convergence criterion. The signal is finally approximated as $y \approx \sum_{n=0}^{k} \langle R_n y, d_{r_n} \rangle d_{r_n}$.

Let us take a very simple example to explain the searching process for the MP algorithm. Suppose that the observed data are $y = (-1, -2)^T$ and a dictionary $D = \begin{pmatrix} 1.3 & -1 & 0 \\ 0.3 & 0.8 & -1 \end{pmatrix}$, each column of which is one atom. We also set the coding errors Er = 0.8. Figure 2.2 shows the intuitive process of the MP algorithm to find optimal α to represent y, which contains four steps. Under such a coding error vector, the observed data y are finally represented as $y \approx -1.3 \cdot d_2 + 2 \cdot d_3$ by using two of three atoms in the dictionary.

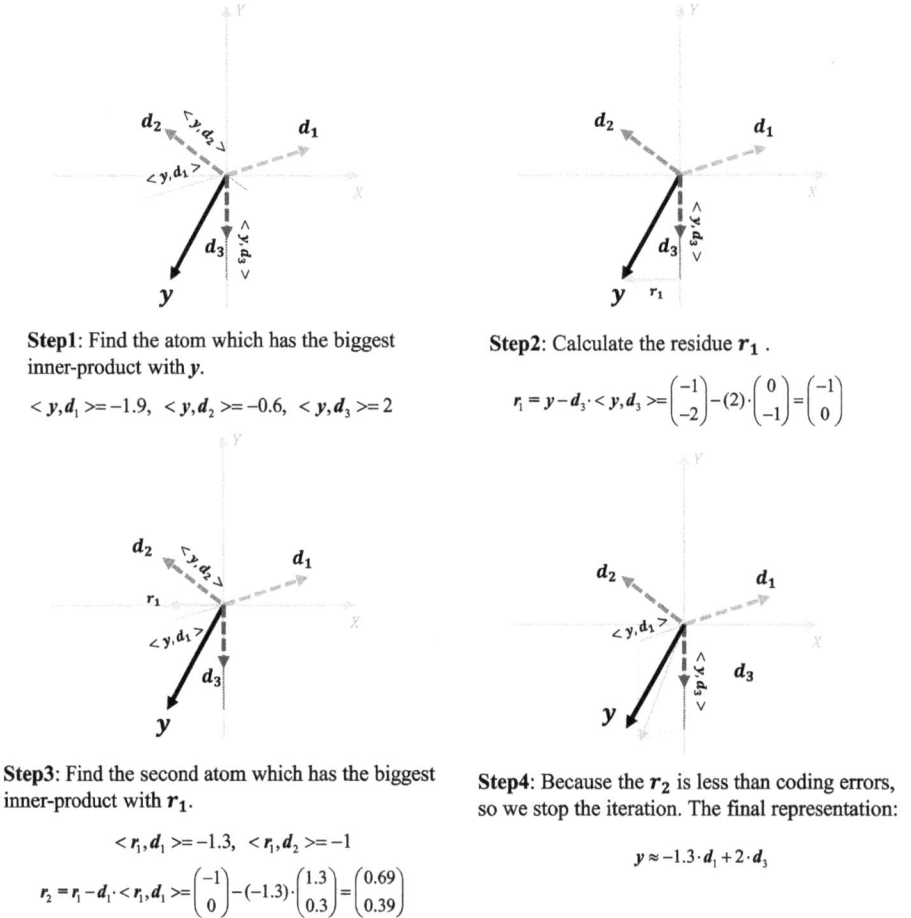

Step1: Find the atom which has the biggest inner-product with y.

$< y, d_1 > = -1.9, \ < y, d_2 > = -0.6, \ < y, d_3 > = 2$

Step2: Calculate the residue r_1.

$r_1 = y - d_3 \cdot < y, d_3 > = \begin{pmatrix} -1 \\ -2 \end{pmatrix} - (2) \cdot \begin{pmatrix} 0 \\ -1 \end{pmatrix} = \begin{pmatrix} -1 \\ 0 \end{pmatrix}$

Step3: Find the second atom which has the biggest inner-product with r_1.

$< r_1, d_1 > = -1.3, \ < r_1, d_2 > = -1$

$r_2 = r_1 - d_1 \cdot < r_1, d_1 > = \begin{pmatrix} -1 \\ 0 \end{pmatrix} - (-1.3) \cdot \begin{pmatrix} 1.3 \\ 0.3 \end{pmatrix} = \begin{pmatrix} 0.69 \\ 0.39 \end{pmatrix}$

Step4: Because the r_2 is less than coding errors, so we stop the iteration. The final representation:

$y \approx -1.3 \cdot d_1 + 2 \cdot d_3$

Figure 2.2. An example to illustrate the searching process of the MP algorithm when observed signal y and a dictionary D are given, r_k denotes the residue of the kth iteration.

One of the problems with MP is that it may pick up the same atom multiple times. Orthogonal matching pursuit (OMP) (Pati *et al* 1993) corrects this issue by updating all the nonzero coefficients according to a least-squares criterion in each step, which makes the residual orthogonal to the chosen atoms. It means that each atom would not be picked more than once and hence this strategy results in a faster convergence.

So, at the kth iteration, the OMP gives the signal representation as follows:

$$y = \sum_{n=1}^{k} \alpha_n^k d_n + R_k y, \text{ with } \langle R_k y, d_n \rangle = 0, n = 1, .., k. \tag{2.8}$$

Still, OMP is a greedy approach that iteratively finds the locally optimal choice in hope of approximating the global minimum. This is a reasonable compromise for the sparse approximation problem, given that the global solution for the NP-hard problem is practically unreachable. Instead, sequentially refining the entries in α to reduce the approximation error is very computationally efficient. Various extensions were proposed to accelerate the convergence of OMP (such as compressive sampling OMP (CoSaMP) (Needell and Tropp 2009), regularized OMP (ROMP) (Needell and Vershynin 2010), and stagewise OMP (StOMP) (Donoho *et al* 2012)).

Currently, sparse approximation algorithms are widely used in image processing, audio processing, machine learning, and many other fields. An unknown signal is usually modeled as a sparse combination of a few atoms in a given dictionary. The use of sparsity-inspired models often yields state-of-the-art results in a variety of applications.

2.2.2 The K-SVD algorithm

As discussed before, OMP works well for a fixed dictionary. On the other hand, it should work better if we optimize the dictionary to fit the data, which brings us to the second problem, i.e. how to train a dictionary from data. K-SVD is a classical dictionary learning algorithm for designing an over-complete dictionary for a sparse representation via a singular value decomposition. It was first introduced by Aharon *et al* (2006). The kernel function of K-SVD can be regarded as a direct generalization of the K-means, in which each input signal is allowed to be represented by a linear combination of dictionary atoms.

Theoretically an over-complete dictionary $D \in \mathbb{R}^{M \times K}$ can be directly chosen as a pre-specified transform matrix, such as over-complete wavelets, curvelets, steerable wavelet filters, short-time Fourier transforms, and others. However, dictionary learning driven by training data is a novel route, and has attracted major attention recently due to its amazing performance in terms of both the effectiveness and efficiency of the resultant sparse representation. Specifically, given the training signals $Y \in \mathbb{R}^{M \times N}$ that contain N samples, the goal is to find the dictionary $D \in \mathbb{R}^{M \times K}$ that can be used for sparse representation of Y with a coefficient matrix $\varphi \in \mathbb{R}^{K \times N}$. It is noted that different from equation (2.4) in which the y and α are the column vectors, Y and φ are both matrices, representing a number of vectorized samples together to train the dictionary, which is expressed as

$$\min_{D,\varphi} \| Y - D\varphi \|_F^2 \qquad \text{s.t.} \| \varphi_i \|_0 \leqslant \varepsilon \forall i. \tag{2.9}$$

It is also a nonconvex problem over the dictionary D and sparse coefficients φ. Correspondingly, the measure metric on the representation residue $\|\cdot\|_F^2$ is called the Frobenius-norm, which calculates the sum of squares of all the elements of the matrix. Moreover, finding a sparse code and constituting a dictionary for a sparse representation have to be performed in a coordinated fashion. Next, let us introduce K-SVD as a typical example. The objective function of K-SVD is formulated as

$$\min_{D,\varphi} \{ \| Y - D\varphi \|_F^2 \} \qquad \text{s.t.} \ \forall i, \| \varphi_i \|_0 \leqslant T_0. \tag{2.10}$$

The K-SVD algorithm consists of two steps, namely, sparse coding and dictionary updating.

Step 1: The sparse coding stage.
Before the sparse coding procedure, an initial dictionary is generated by a traditional data sparsifier, such as an over-complete discrete cosine transform (DCT) dictionary. Then, we can seek the sparse coding vector φ for the current dictionary. Figure 2.3 depicts the sparse coding stage, in which red rectangles at the $\varphi \in \mathbb{R}^{K \times N}$ matrix present the coding of selected atoms. Usually, the OMP algorithm is used to compute the coefficients, as long as it can supply a solution with a predetermined number of nonzero entries T_0. The OMP method has been discussed in the previous section; here we will focus on the next stage.

Step 2: The dictionary updating stage.
In this stage, the total number of K atoms in the dictionary is updated independently with φ fixed. In practice, the K atoms are updated one by one, which means that in one step, only one atom is updated while the others are fixed. As is seen in figure 2.4(1), the kth column of D, denoted as $\mathbf{d}_k$, $k \in (1, 2,.., K)$, and the coefficients that correspond to it, i.e. the kth row in φ, denoted as φ_T^k (note that this is not the vector φ^k that is the kth column in φ) are taken, for example, to be updated while fixing the rest of the atoms, the objective function is rewritten as

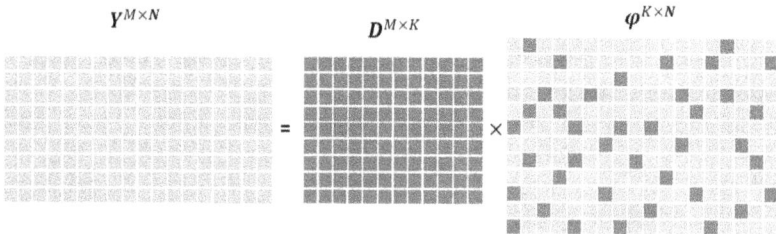

Figure 2.3. The sparse coding stage of the dictionary learning method.

$$\|Y - D\varphi\|_F^2 = \left\| Y - \sum_{j=1}^{K} d_j \varphi_T^j \right\|_F^2$$

$$= \left\| \left(Y - \sum_{j \neq k} d_j \varphi_T^j \right) - d_k \varphi_T^k \right\|_F^2 \qquad (2.11)$$

$$= \left\| E_k - d_k \varphi_T^k \right\|_F^2 .$$

In this equation, $D = [d_1, d_2, \ldots, d_K] \in R^{M \times K}$ contains K atoms, and E_k stands for the approximation error when atom d_k is removed from the dictionary. With this variable separation, $D\varphi$ is decomposed into the K matrices with a rank of 1 (denoted as rank-1), in which only the kth column remains in question for updating.

Based on the refined objective function in equation (2.11), whose goal is to search for the closest d_k with its coefficient φ_T^k to maximally approximate E_k, the SVD algorithm could be a straightforward method to solve the problem of updating only one column of D and φ_T^k. The SVD decomposition is done with a factorization of the form $E_k = U \Delta V^\top$, where both U and V are an orthogonal matrix, and Δ is a diagonal matrix. The key point of the SVD method in this study is that only a few principal components in the diagonal matrix can closely approximate the matrix. Based on this, d_k is taken as the first column of U while φ_T^k is taken as the first column of V multiplied by the first element of the diagonal matrix Δ denoted as $\Delta(1, 1)$.

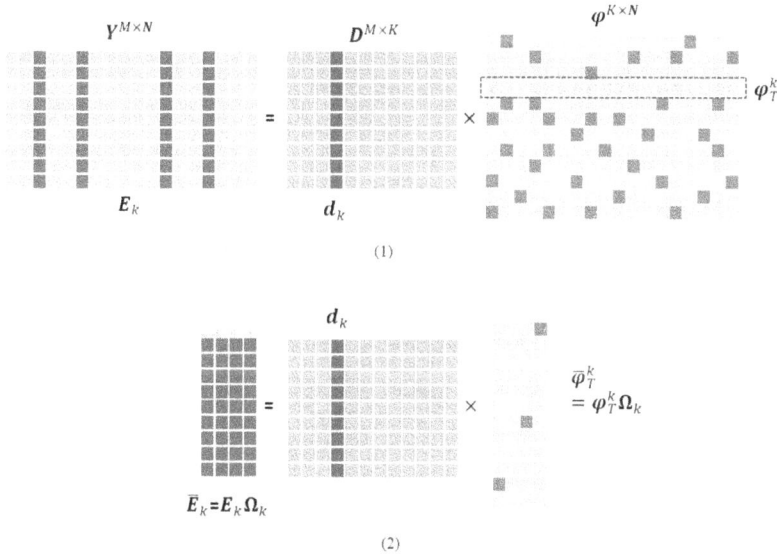

Figure 2.4. Dictionary update stage. (1) Independent update of each column of the dictionary and (2) shrink operation.

However, directly decomposing E_k by SVD for the update of d_k and the corresponding φ_T^k cannot ensure the sparsity of the updated φ_T^k, because there is no such sparsity control regularizer in SVD (Sahoo and Makur 2013); in other words, the position and value of the zero elements in the update φ_T^k may change. To address the loss of sparsity issue, K-SVD considers those columns of E_k by extracting the corresponding φ_T^k that is nonzero, and obtain a new E_k, denoted as $\bar{E}_k$. A simple solution is to restrict E_k to obtain by a shrink operation $\bar{E}_k = E_k \Omega_k$, whose process is illustrated in figure 2.4 by a shrink operation Ω_k. In particular, Ω_k is defined as a matrix to restrict the E_k by discarding the zero entries. Similarly, we manage to obtain the $\bar{\varphi}_T^k$ by $\bar{\varphi}_T^k = \varphi_T^k \Omega_k$, as well. Hence, by rewriting the objective function in equation (2.11), we have an expected $\hat{d}_k$ and the corresponding $\hat{\varphi}_T^k$:

$$\left(\hat{d}_k, \hat{\varphi}_T^k\right) = \arg\min_{d_k, \varphi_T^k} \left\| E_k \Omega_k - d_k \varphi_T^k \Omega_k \right\|_F^2 = \arg\min_{d_k, \bar{\varphi}_T^k} \left\| \bar{E}_k - d_k \bar{\varphi}_T^k \right\|_F^2. \quad (2.12)$$

In order to solve it by approximating the $\bar{E}_k$ term with a rank-1 matrix, the SVD method is suggested to find alternative d_k and $\bar{\varphi}_T^k$. In detail, $\bar{E}_k$ is SVD decomposed into $U \Delta V^\top$, setting d_k to the first column of the matrix U and $\bar{\varphi}_T^k$ to the first column of V multiplied by $\Delta(1, 1)$, and updating the whole dictionary, the process solves φ and D iteratively. Once all the K atoms of D are updated column by column in the same fashion to obtain an expected dictionary, we fix this dictionary and go to the sparse coding stage until it reaches the stopping condition. It is noted that the K-SVD algorithm is flexible to use other methods in the sparse coding stage. The workflow of the K-SVD algorithm can be summarized in algorithm 2.1.

Algorithm 2.1 The K-SVD algorithm for dictionary learning. Reprinted with permission from Aharon *et al* (2006). Copyright 2019 IEEE.

Input: A set of training data $Y = \{y_i\}_{i=1}^N \in \mathbb{R}^{M \times N}$
Output: An over-complete dictionary $D \in \mathbb{R}^{M \times K}$ and a sparse coding vector $\varphi \in \mathbb{R}^{K \times N}$
1: Initialize the dictionary D with K randomly selected training signals
2: **while** not converge **do**
3:     **Sparse coding:**
4:     **for** each training signal $y_i \in Y$, use OMP to compute the representation vectors φ_i, $i = 1, 2,\ldots, N$: **do**
5:       $\min_\alpha \left\{ \left\| y_i - D\varphi_i \right\|_F^2 \right\}$ s.t. $\forall i$, $\left\| \varphi_i \right\|_0 \leqslant T_0$
6:     **end for**
7:     **Dictionary updating:**
8:     **for** $k = 1, 2,\ldots,K$ update the kth atom d_k of D and the kth row of the coding coefficients φ_T^k: **do**
9:       Compute the representation error matrix: $E_k = Y - \sum_{j \neq k} d_j \varphi_T^j$
10:       Obtain $\bar{E}_k$ and $\bar{\varphi}_T^k$ by discarding zero entries with a shrink operation $\bar{E}_k = E_k \Omega_k$ and $\bar{\varphi}_T^k = \varphi_T^k \Omega_k$

(Continued)

11:	Apply SVD decomposition $\bar{E}_k = U\Delta V^T$. Update the atom d_k with the first column of U. The corresponding coefficient vector $\bar{\varphi}_T^k$ is together updated with the first column of V multiplied by $\Delta(1, 1)$
12:	**end for**
13:	**end while**

Despite the fact that the K-SVD algorithm converges quickly, it is still computationally expensive as the SVD decomposition must be performed K times, and all N training signals are used for sparse coding at each iteration. This task demands a heavy memory use, in particular when the set of training data is large. Some modifications to the original algorithm were presented that alleviate the computational challenge; for example, the use of batch-OMP for K-SVD presented by Rubinstein *et al* (2008). In the next section, we will introduce an application of the dictionary learning approach for CT reconstruction.

2.3 CT reconstruction via dictionary learning

It is suggested that dictionary learning for a sparse representation effectively mimics the HVS perceiving structural/semantic information in natural images. In this section, we will use low-dose CT reconstruction as an example to illustrate how the dictionary learning method is applied in the CT field. This case is described in two parts. In the first part, we will introduce a statistic iterative reconstruction framework (SIR) extensively used in the CT field, which exemplifies the application of the Bayesian approach for CT imaging. In the second part, based on SIR we will illustrate the application of dictionary learning for CT reconstruction to preserve subtle features in reconstructed images from low x-ray dose data. This example will show the power of dictionary learning as a machine learning technique.

2.3.1 Statistic iterative reconstruction framework (SIR)

X-ray photons recorded by the detectors can be regarded as evidence. A reconstructed image can be seen as a hypothesis. In the Bayesian framework, given our statistical knowledge and evidence, we should make a hypothesis that is most reasonable/plausible.

X-ray photons are emitted from an x-ray source. During their propagation, they interact with tissues inside a patient. As a result, some x-ray photons are absorbed or re-directed, while the rest of them penetrate the patient, reaching an x-ray detector. Assuming a monochromatic x-ray beam, Beer's law is expressed as

$$I_i = I_0 e^{-\int_{L_i} x(l)\mathrm{d}l}, \tag{2.13}$$

where I_i represents the recorded x-ray intensity at the ith element of the detector, I_0 is the initial flux, and x is the linear attenuation coefficient along the ith x-ray path L_i

from which the intensity I_i is produced. This equation can be converted into the line integral and is discretized as

$$b_i = \sum_{j=1}^{N} a_{ij}x_j, \tag{2.14}$$

where $b_i = \ln I_0/I_i = \int_{L_i} x(l)dl$ is the line integral value along the ith x-ray path and a_{ij} denotes the contribution of the jth pixel to the ith line integral b_i. Then, we have a discretization version of Beer's law

$$I_i = I_0 \exp(-b_i) = I_0 \exp(-[\mathbf{A}\mathbf{x}]_i), \tag{2.15}$$

where $\mathbf{A} = \{a_{ij} | i = 1, 2,..., I, j = 1, 2,..., J\}$ and $[\mathbf{A}\mathbf{x}]_i$ denotes the ith element of $\mathbf{A}\mathbf{x}$.

With a CT scanner, an original signal contains an inherent quantum noise. Approximately, the measured data follow the Poisson distribution

$$I_i \sim \text{Poisson}(\bar{I}_i), \tag{2.16}$$

where I_i denotes a specific measurement of the Poisson noise compromised line integral.

Since the measured data along different paths are statistically independent from each other, according to the Poisson model, the joint probability distribution of the measured data can be written as follows,

$$P(\mathbf{I}|\mathbf{x}) = \prod_{i=1}^{I} \frac{\bar{I}_i^{I_i} \exp(-\bar{I}_i)}{I_i!}, \tag{2.17}$$

and the corresponding likelihood function is expressed as the following:

$$L(\mathbf{I}|\mathbf{x}) = \log(P(\mathbf{I}|\mathbf{x})) = \sum_{i=1}^{I}(I_i \log(\bar{I}_i) - \bar{I}_i - \log(I_i!)) \tag{2.18}$$

or

$$L(\mathbf{I}|\mathbf{x}) = \sum_{i=1}^{I}(I_i \log(I_0 e^{-[\mathbf{A}\mathbf{x}]_i}) - I_0 e^{-[\mathbf{A}\mathbf{x}]_i} - \log(I_i!)). \tag{2.19}$$

Ignoring the constants, we rewrite the above formula as

$$L(\mathbf{I}|\mathbf{x}) = -\sum_{i=1}^{I}(I_i[\mathbf{A}\mathbf{x}]_i + I_0 e^{-[\mathbf{A}\mathbf{x}]_i}). \tag{2.20}$$

Statistically, it makes a perfect sense to maximize the posterior probability:

$$P(\mathbf{x}|\mathbf{I}) = P(\mathbf{I}|\mathbf{x})P(\mathbf{x})/P(\mathbf{I}). \tag{2.21}$$

Considering the monotone increasing property of the natural log operation, the image reconstruction process is equivalent to maximizing the following objective function:

$$\tilde{\Phi}(\mathbf{x}) = L(\mathbf{I}|\mathbf{x}) + \log(P(\mathbf{x})). \tag{2.22}$$

In equation (2.22), $\log(P(\mathbf{x}))$ denotes the prior information and can be conveniently expressed by changing $-\log(P(\mathbf{x}))$ to $\beta\psi(\mathbf{x})$. Then, the image reconstruction problem can be expressed as the following optimization problem:

$$\Phi(\mathbf{x}) = -L(\hat{\mathbf{I}}|\mathbf{x}) - \log(P(\mathbf{x})) = \sum_{i=1}^{I}(I_i[\mathbf{Ax}]_i + I_0 e^{-[\mathbf{Ax}]}) + \beta\psi(\mathbf{x}). \tag{2.23}$$

It is noted that $\psi(\mathbf{x})$ is the regularizer and β is used as a regularization parameter. Because the objective function is not easy for numerical optimization, we transform the first term into its second-order Taylor expansion around the measurement-based estimate of the line integral $b_i = \ln(I_0/I_i)$ and obtain a simplified objective function

$$\Phi(\mathbf{x}) = \sum_{i=1}^{I} \frac{w_i}{2}([\mathbf{Ax}]_i - b_i)^2 + \beta\psi(\mathbf{x}), \tag{2.24}$$

where $w_i = I_i$ is also known as the statistical weight. For the deduction, see Elbakri and Fessler (2002).

This objective function has two terms, for data fidelity and image regularization, respectively. In the fidelity term, the statistical weight modifies the discrepancy between the measured and estimated line integrals along each x-ray path. Heuristically, there would be fewer photons along a more attenuating path, resulting in a greater uncertainty in estimating the line integral along the path, thus having a smaller weight assigned to the line integral along that path. In contrast, a less attenuating path will be given a larger weight. With the regularization term, a statistical model can be enforced to guide the image reconstruction process so that a reconstructed image will look more like what we expect based on a statistical model of images.

2.3.2 Dictionary-based low-dose CT reconstruction

As we discussed in the first chapter, a reconstructed CT image suffers from degraded image quality in the case of low-dose scanning. How to maintain or improve the diagnostic performance is the key issue associated with low-dose CT. Inspired by compressive sensing theory, the sparse constraint in terms of dictionary learning is developed as an effective way for a sparse representation. Recently, a dictionary learning-based approach for low-dose x-ray CT was proposed by Qiong Xu *et al* (2012), in which a redundant dictionary is incorporated into the statistical reconstruction framework. The dictionary can be either predetermined before image reconstruction or adaptively defined during an image reconstruction process. We will describe both the modes for dictionary learning-based low-dose CT in the following.

2.3.2.1 Methodology

Recall the objective function for statistical reconstruction equation (2.24), the regularization term $\psi(\mathbf{x})$ represents prior information on reconstructed images. By utilizing the sparsity constraint in terms of a learned dictionary as the regularizer, the objective function can be rewritten as

$$\min_{\mathbf{x},\alpha,(\mathbf{D})} \sum_{i=1}^{I} \frac{w_i}{2}([\mathbf{Ax}]_i - b_i)^2 + \lambda \left(\sum_s \|\mathbf{E}_s\,\mathbf{x} - \mathbf{D}\alpha_{s,}\|_2^2 + \sum_s v_s \|\alpha_{s,}\|_0 \right), \qquad (2.25)$$

where $\mathbf{E}_s = \left\{ e_{nj}^s \right\} \in \mathbb{R}^{N \times J}$ represents an operator to extract an image patch from an image, and $b_i = \ln \frac{I_0}{I_i}$ denotes a line integral. It is worth mentioning that they proposed two strategies for dictionary learning: a global dictionary learned before a statistical iterative reconstruction (GDSIR) and an adaptive dictionary learned in the statistical iterative reconstruction (ADSIR). The former uses a predetermined dictionary to sparsely represent an image while the later learns a dictionary based on intermediate images obtained during the iterative reconstruction process and only uses the dictionary for sparse coding each intermediate image. In this subsection, let us look at the two update processes for GDSIR and ADSIR, respectively.

For GDSIR, a redundant dictionary was trained for a chest region in the baseline image, as shown in figure 2.5. Then, the image reconstruction process is equivalent to solving the following optimization problem, which contains two variables $\mathbf{x}$ and α:

$$\min_{\mathbf{x},a} \sum_{i=1}^{I} \frac{w_i}{2}([\mathbf{Ax}]_i - b_i)^2 + \lambda \left(\sum_s \|\mathbf{E}_s\mathbf{x} - \mathbf{D}\alpha_s\|_2^2 + \sum_s v_s \|\alpha_s\|_0 \right). \qquad (2.26)$$

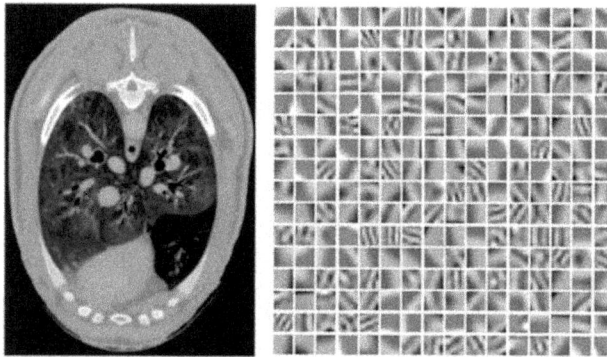

Figure 2.5. Global dictionary learned using the online dictionary learning method (Mairal *et al* (2009). Reprinted with permission from Xu *et al* 2012). Copyright 2012 IEEE.

An alternate minimization is performed with respect to $\mathbf{x}$ and α alternately. First, the sparse expression $\tilde{\alpha}$ is fixed, and the current image is updated. At this point, the objective function becomes

$$\min_{\mathbf{x}} \sum_{i=1}^{I} \frac{w_i}{2}([\mathbf{A}\mathbf{x}]_i - b_i)^2 + \lambda \sum_{s} \|\mathbf{E}_s\,\mathbf{x} - \mathbf{D}\tilde{\alpha}_s\|_2^2. \qquad (2.27)$$

Using the separable paraboloid alternative method (Elbakri and Fessler 2002), the objective function is optimized iteratively:

$$\mathbf{x}_j^t = \mathbf{x}_j^{t-1} - \frac{\sum_{i=1}^{I}\left(a_{ij}w_i([\mathbf{A}\mathbf{x}^{t-1}]_i - b_i)\right) + 2\lambda \sum_s \sum_{n=1}^{N} e_{nj}^s([\mathbf{E}_s\mathbf{x}^{t-1}]_n - [\mathbf{D}\tilde{\alpha}_s]_n)}{\sum_{i=1}^{I}\left(a_{ij}w_i \sum_{k=1}^{J} a_{ik}\right) + 2\lambda \sum_s \sum_{n=1}^{N} e_{nj}^s \sum_{k=1}^{J} e_{nk}^s}, \qquad (2.28)$$

$$j = 1, 2, \ldots, J$$

where $t = 1, 2, \ldots, T$ indexes the iterations.

After obtaining an intermediate image $\mathbf{x}^t$, it is re-coded for a sparse representation. The objective function is changed to

$$\min_{a} \sum_{s} \|\mathbf{E}_s\mathbf{x}^t - \mathbf{D}\alpha_s\|_2^2 + \sum_{s} v_s\|\alpha_s\|_0. \qquad (2.29)$$

This equation represents a sparse coding problem, which can be solved using the OMP method described in subsection 2.2.1.

For ADSIR, the image reconstruction process is equivalent to solving the following optimization problem:

$$\min_{\mathbf{x},\mathbf{D},\alpha} \sum_{i=1}^{I} \frac{w_i}{2}([\mathbf{A}\mathbf{x}]_i - b_i)^2 + \lambda\left(\sum_{s} \|\mathbf{E}_s\mathbf{x} - \mathbf{D}\alpha_s\|_2^2 + \sum_{s} v_s\|\alpha_s\|_0\right). \qquad (2.30)$$

In ADSIR, the dictionary $\mathbf{D}$ is regarded as an additional variable along with α and $\mathbf{x}$. The objective function is solved in an alternating fashion, by keeping all variables $\mathbf{D}$, α, and $\mathbf{x}$ fixed except for one at a time, gradually converging to a satisfactory sparse representation α, a final image $\mathbf{x}$, and the associated dictionary $\mathbf{D}$. As mentioned above, dictionary learning-based reconstruction consists of two steps: sparse coding and image updating. It is noted that the updating step for an intermediate reconstructed image $\mathbf{x}$ with a fixed sparse representation $\tilde{\alpha}$ and the current dictionary $\tilde{\mathbf{D}}$ is exactly the same as that for GDSIR expressed in equation (2.28). For the sparse coding step, the $\mathbf{D}$ and α are estimated with a fixed reconstruction image $\mathbf{x}^t$, whose objective function is thus as follows:

$$\min_{\mathbf{D},\alpha} \sum_{s} \|\mathbf{E}_s\mathbf{x}^t - \mathbf{D}\alpha_s\|_2^2 + \sum_{s} v_s\|\alpha_s\|_0. \qquad (2.31)$$

This equation is the generic dictionary learning and sparse representation problem, which can be solved with respect to α and $\mathbf{D}$ alternatingly using the K-SVD algorithm described in subsection 2.2.2.

In summary, the GDSIR and ADSIR algorithms are summarized in algorithms 2.2 and 2.3, respectively. In order to improve the convergence speed, a fast online algorithm is used to train a dictionary from patches extracted from an intermediate image.

Algorithm 2.2. Workflow for the GDSIR algorithm. Reprinted with permission from Xu *et al* (2012). Copyright 2012 IEEE.

Global dictionary learning
  1: Choose parameters for dictionary learning.
  2: Extract patches to form a training set.
  3: Construct a global dictionary.
Image reconstruction
  4: Initialize $\mathbf{x}^0$, $\boldsymbol{\alpha}^0$, and $t = 0$.
  5: Set parameters λ, ε, and L_0^S.
  6: **while** a stopping criterion is not satisfied **do**
  7:    Update $\mathbf{x}^{t-1}$ to $\mathbf{x}^t$ using equation (2.29).
  8:    Represent $\mathbf{x}^t$ with a sparse code $\boldsymbol{\alpha}^t$ using OMP.
  9: **end while**
 10: Output the final reconstruction.

Algorithm 2.3. Workflow for the ADSIR algorithm. Reprinted with permission from Xu *et al* (2012). Copyright 2012 IEEE.

1: Choose λ, ε, L_0^S and other parameters.
2: Initialize $\mathbf{x}^0$, $\mathbf{D}^0$, $\boldsymbol{\alpha}^0$, and $t = 0$.
3: **while** a stopping criterion is not satisfied **do**
4:    Update $\mathbf{x}^{t-1}$ to $\mathbf{x}^t$ using equation (2.29).
5:    Extract patches from $\mathbf{x}^t$ from the training set.
6:    Construct a dictionary $\mathbf{D}^0$ from the training set.
7:    Represent $\mathbf{x}^t$ with a sparse code $\boldsymbol{\alpha}^t$ using OMP.
8: **end while**
9: Output the final reconstruction.

2.3.2.2 Experimental results

The dictionary learning algorithms produced quite exciting image reconstructions from low-dose data. Compared to the traditional regularizers widely used for CT reconstruction, the dictionary learning results are superior in terms of edge preservation and noise suppression. Under few views and/or low-dose conditions, the dictionary learning framework performed very well, with two examples shown in figures 2.6 and 2.7.

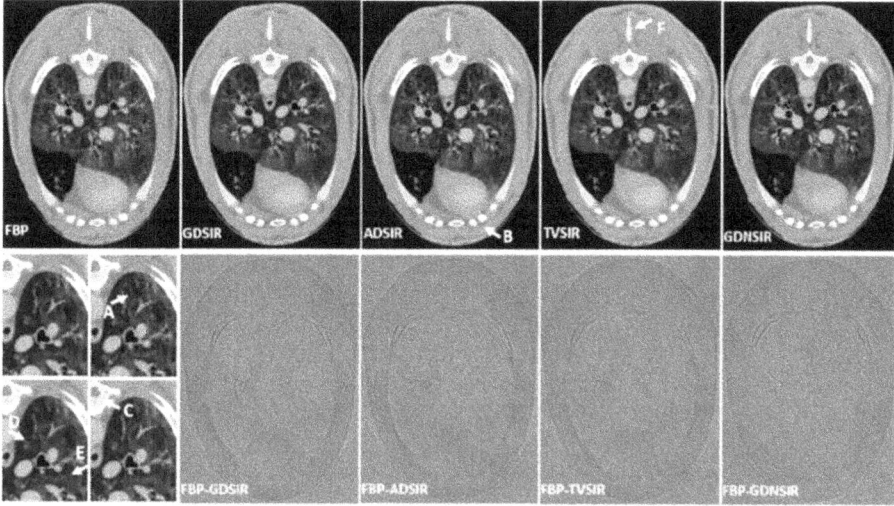

Figure 2.6. Reconstructed images from a low-dose sinogram collected in a sheep lung CT perfusion study. The upper row shows the images reconstructed using the FBP, GDSIR, ADSIR, TVSIR, and GDNSIR methods (from left to right), respectively. The magnified local regions are shown below the FBP result (upper left, upper right, lower left, and lower right correspond to GDSIR, ADSIR, TVSIR, and GDNSIR, respectively). The display window is [−700, 800] HU. The second to fifth images in the bottom row are the difference images between the FBP image and the results obtained using the GDSIR, ADSIR, TVSIR, and GDNSIR methods, respectively, in the display window [−556, 556] HU. Reproduced with permission from Xu *et al* (2012). Copyright 2012 IEEE.

In addition to using the sheep dictionary to reconstruct images of the same sheep, Xu *et al* also used this sheep dictionary to reconstruct human chest CT images. Excitingly, the global dictionary trained from images of the sheep performed well in the case of human chest scans, which implies that CT images have similar low-level structural information just as natural images. In fact, the dictionary learning approach can transfer natural image statistics for medical image reconstruction, which has a profound impact on tomographic imaging in general (Xu *et al* 2012).

The proposed L0-norm-based dictionary learning method is highly nonconvex and computationally nontrivial, although pilot results were reported through L0-norm-based dictionary learning. Motivated by this challenge, an algorithm for low-dose CT via L1-norm-based dictionary learning should be popular to avoid non-convexity. It has been verified that the L1-norm-based sparse constraint has a similar performance as that of GDSIR and ADSIR (Mou *et al* 2014).

2.4 Final remarks

Tomographic image reconstruction is an inverse problem. Due to the non-ideal factors in the data acquisition, such as noise and errors introduced in practice, as well as possibly the data incompleteness, image reconstruction can be a highly ill-posed problem. Thanks to the Bayesian inference, prior knowledge can be utilized to solve the inverse problem. Compared with the simplest iterative algorithms that only address the data fidelity term, the regularized/penalized reconstruction techniques take advantage of the posterior probability and naturally incorporate prior

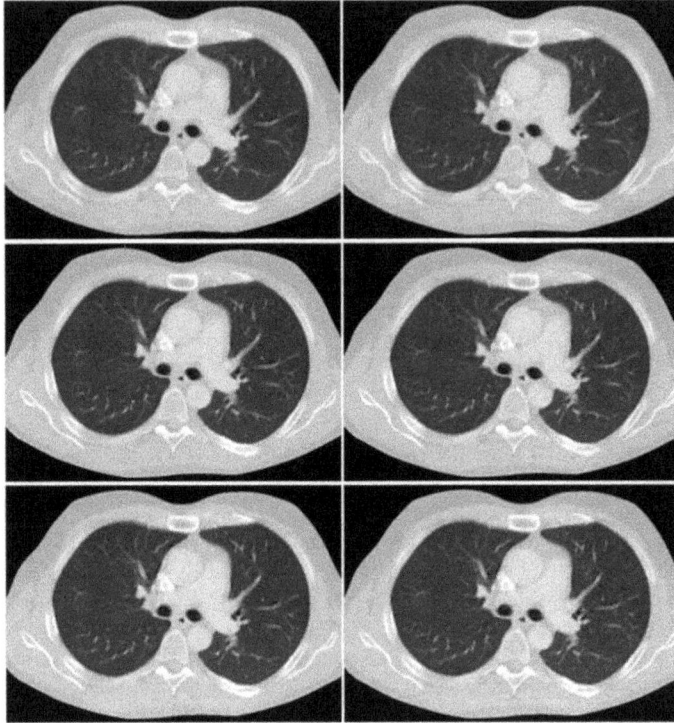

Figure 2.7. Reconstruction results using the FBP, GDSIR, and ADSIR methods, respectively (from top to bottom), with the left and right columns representing the results from 1100 and 550 projection views, respectively, in the display window [0, 2]. Reprinted with permission from Xu *et al* (2012). Copyright 2012 IEEE.

knowledge into the reconstruction process. Consequently, a regularized solution must be in a much narrower searching space of the unknown variables. In determining how to extract the prior knowledge, or in other words, the intrinsic distribution properties of images, the theory of natural image statistics and the HVS give us great inspiration. Both anatomical and neurophysiological studies indicate a multi-layer HVS architecture which supports a sparse representation to eliminate image redundancy as much as possible and use statistical independent features when we perceive natural scenes. Inspired by this discovery, machine learning, such as a typical single-layer neural network or the dictionary learning method, is capable of extracting the prior information and representing it sparsely. The building block of a neural network is an artificial neuron which takes multiple variables as the input, computes an inner product between the input vector and a weighting vector, and thresholds the inner product in some way as the output of the neuron.

Furthermore, the multi-layer neural mechanism exists in HVS, which inspires the development of multi-layer neural networks, which is commonly known as deep learning. Along this direction, a set of non-linear machine learning algorithms were developed for modeling complex data representations. It is the multi-layer artificial neural architecture that allows the learning of high-level representations of data through multi-scale analysis from low-level primitives to semantic features. It is comforting that this type of multi-layer neural networks resembles the multi-layer mechanism in the HVS. In the next chapter, we will cover the basics of artificial neural networks.

References

Aharon M *et al* 2006 K-SVD: an algorithm for designing overcomplete dictionaries for sparse representation *IEEE Trans. Signal Process.* **54** 4311

Bell A J and Sejnowski T J 1997 The 'independent components' of natural scenes are edge filters *Vis. Res.* **37** 3327–38

Chambolle A 2004 An algorithm for total variation minimization and applications *J. Math. Imaging Vis.* **20** 89–97

Chambolle A and Lions P-L 1997 Image recovery via total variation minimization and related problems *Numer. Math.* **76** 167–88

Dandes E 2006 Near-optimal signal recovery from random projections: universal encoding strategies *IEEE Trans. Inform. Theory* **52** 5406–25

Donoho D L, Tsaig Y, Drori I and Starck J 2012 Sparse solution of underdetermined systems of linear equations by stagewise orthogonal matching pursuit *IEEE Trans. Inform. Theory* **58** 1094–121

Elad M 2010 *Sparse and Redundant Representations: From Theory to Applications in Signal and Image Processing* (Berlin: Springer)

Elbakri I A and Fessler J A 2002 Statistical image reconstruction for polyenergetic x-ray computed tomography *IEEE Trans. Med. Imaging* **21** 89–99

Hastie T, Tibshirani R and Friedman J 2009 *The Elements of Statistical Learning: Data Mining Inference and Prediction* vol 1 (New York: Springer)

Mairal J, Bach F, Ponce J and Sapiro G 2009 Online dictionary learning for sparse coding *Proc. of the 26th Annual Int. Conf. on Machine Learning* (New York: ACM), 689–96

Mallat S G and Zhang Z 1993 Matching pursuits with time-frequency dictionaries *IEEE Trans. Signal Process.* **41** 3397–415

Mou X, Wu J, Bai T, Xu Q, Yu H and Wang G 2014 Dictionary learning based low-dose x-ray CT reconstruction using a balancing principle *Proc. SPIE* **9212** 921207

Needell D and Tropp J A 2009 COSAMP: Iterative signal recovery from incomplete and inaccurate samples *Appl. Comput. Harmon. Anal.* **26** 301–21

Needell D and Vershynin R 2010 Signal recovery from incomplete and inaccurate measurements via regularized orthogonal matching pursuit *IEEE J. Sel. Topics Signal Process.* **4** 310–316

Olshausen B A and Field D J 1996 Emergence of simple-cell receptive field properties by learning a sparse code for natural images *Nature* **381** 607–9

Pati Y C, Rezaiifar R and Krishnaprasad P S 1993 Orthogonal matching pursuit: recursive function approximation with applications to wavelet decomposition *Proc. of 27th Asilomar Conf. on Signals Systems and Computers* (Piscataway, NJ: IEEE), 40–4

Sahoo S K and Makur A 2013 Dictionary training for sparse representation as generalization of *K*-means clustering *IEEE Sign. Process Lett.* **20** 587–90

Rubinstein R, Zibulevsky M and Elad M 2008 Efficient implementation of the K-SVD algorithm using batch orthogonal matching pursuit *Technical report* CS Technion http://www.cs.technion.ac.il/~ronrubin/Publications/KSVD-OMP-v2.pdf

Xu Q, Yu H, Mou X, Zhang L, Hsieh J and Wang G 2012 Low-dose x-ray CT reconstruction via dictionary learning *IEEE Trans. Med. Imaging* **31** 1682–97

IOP Publishing

Machine Learning for Tomographic Imaging

Ge Wang, Yi Zhang, Xiaojing Ye and Xuanqin Mou

Chapter 3

Artificial neural networks

3.1 Basic concepts

In chapter 2, we have introduced tomographic reconstruction based on a learned structural dictionary in which the prior information of low-level image features is expressed as atoms, which are over-complete basis functions. This prior information is actually a result of image information extraction. Indeed, it is an essential task to find an efficient measure to express the information for various images contents. In the development of deep learning techniques, it has become a common belief now that multi-layer neural networks extract image information from different semantic levels, thereby representing image features effectively and efficiently, which is consistent with the principle of the human vision system (HVS) perceiving natural images. Therefore, in this chapter we will focus on the basic knowledge of artificial neural networks, providing the foundation for feature representation and reconstruction of medical images using deep neural networks.

3.1.1 Biological neural network

Artificial neural networks originated from mimicking biological neural systems. It is necessary to understand the connection between the artificial neural network and the biological neural network before one is introduced to deep learning.

The hierarchical structure of the HVS is shown in figure 1.2. In the HVS, features are extracted layer by layer. As described in chapter 1, in the 'what' pathway, the V1/V2 area is mainly sensitive to edges and lines, the V4 area senses object shapes, and finally the IT region completes the object recognition. In this process, the receptive field is constantly increasing in size, and the extracted features are increasingly more complex.

To a large degree, the artificial neural network attempts to duplicate the biological neural network from the perspective of information processing. As a result, an artificial neural network serves as a simple mathematical model, and different networks are defined by different interconnections among various numbers of artificial neurons. A neural network is a computational framework, including a large number of neurons as basic computing units connected to each other with varying connection strengths known as weights. This feed-forward, layered architecture is mainly intended to reflect the process of extracting visual features with biological neurons layer by layer. Not being totally the same as the HVS, the artificial neural network is composed of the following elements: neurons with weights and activation functions, and network topology or connections, which ought to be trained with training data according to some learning rules. The neuron is the node of the network; each connection transfers the output of a neuron to the next neuron as its input, weights are used to modify input data before summation, and the learning rules guide the parametric modification for the neural network to perform optimally. The components of the artificial neural network will be described in detail below.

Figure 3.1 correlates the convolutional neural network (CNN) with the HVS, in which we map the components of the CNN to the corresponding components of the HVS (Yamins *et al* 2014). This correspondence helps us understand the neural network. First, the input to a CNN is usually either 3D (RGB) or 1D (gray scale) pixel values after a preprocess of normalization. This roughly corresponds to part computations performed by the retina and lateral geniculate nucleus. The convolutional operations create feature maps that have a spatial layout, such as the retinotopic maps for visual information processing, and each of the artificial neurons can only process data within a receptive field of a limited extent. The convolutional filters define feature maps, which can be grouped into a number of different types.

In a good sense, artificial neural networks are engineered copies of the HVS. At the same time, an artificial neural network is not an exact replica of the human vision system or the human brain. In the biological neural networks such as the HVS or more generally the human brain, the learning process is much more complicated

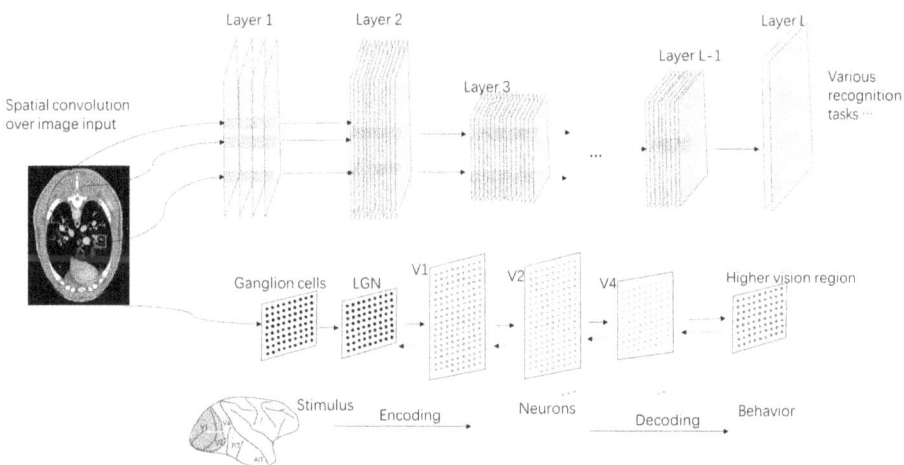

Figure 3.1. Relating the CNN to the HVS consisting of the brain areas responsible for a sequence of visual information processing tasks.

and is achieved by the combination of many factors, such as the surrounding environment, interest/attention/drive, mode, internal representations, and so on. While weights in an artificial network are initialized randomly, connections between biological neurons are genetically derived, and then reinforced or weakened in the learning process. Unlike the biological neural network, the artificial neural network is commonly trained by relevant data from the application domain. At present, the network topology is pre-specified based on a designer's experience, does not change in the training process, and weights are randomly initialized and adjusted using an optimization algorithm to map input stimuli to desired output values. The single-layer neural network introduced in chapters 1 and 2 is a good example.

When we design an artificial neural network, we should select a network topology for feature extraction, a training procedure to update the network parameters, and so on. In order to output a desirable result, an artificial neural network should have certain properties. By construction, we must ensure that the artificial neural network is sufficiently expressive, and can be trained with relevant data toward a converged configuration. Convergence means that the process of training the neural network converges to a limit. When the neural network training process converges, the model has the tendency of stabilizing over time, predicting consistent, meaningful outputs from new data. As explained late in this chapter, several related problems must be addressed to have a convergent and favorable outcome.

3.1.2 Neuron models

Neurons are a special type of biological cell, serving as the organic computing units in a neuronal system. In previous studies, researchers built a computational model mirroring neurons in the biological system. In order to build such a model, it was essential to emulate the biological mechanism of the neuron.

According to biological research, a neuron consists of three key parts: the cell body, dendrites, and axons, as shown in figure 3.2. The cell body offers the energy supply for neuronal activities, where metabolism and other biochemical processes are carried out, dendrites are the ports to receive information from other neurons, and axons are the gateways that transmit the excitatory information to other

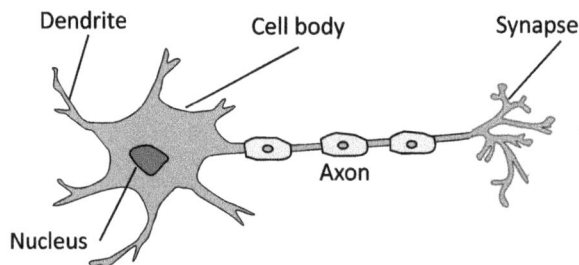

Figure 3.2. Key elements of a biological neuron.

neurons. In addition, the synapse is the structure in which one neuron interacts with another neuron for communication.

According to neurobiological research results, the information processing and transmission mechanism of the biological neuron has the following characteristics:

1. *Information integration*: A neuron can integrate different neurotransmitters transmitted by other neurons into an overall response.
2. *Potential difference*: Defined as the difference between the electrical potentials inside and outside the cell membrane, and the differences in neuronal state.
3. *Transmission threshold*: The membrane potential changes constantly when a neuron receives information. If the fixed potential value is exceeded, an action potential is transmitted along an axon as an electrical pulse. Since the threshold is used, the neuron is a nonlinear system.

In reference to the characteristics of a biological neuron and its functions, an artificial neuron is an approximation to a biological neuron, which is a simple mathematical model illustrated in figure 3.3.

Mimicking a biological neuron, the mathematical model of an artificial neuron can be represented as follows: the input vector x represents the signals (from the dendrites), the weight vector w corresponds to the strength of the pathway (dendrite and synapse), and the summation node $\sum$ and the activation function $\varphi(\cdot)$ represent the integration and activation of the input signals (the cell body) and the thresholder output y (along the axon), respectively. Such an artificial neuron model can be formulated as follows:

$$\begin{cases} v = \sum_{i=1}^{m} x_i w_i + b \\ y = \varphi(v) \end{cases}, \tag{3.1}$$

where w_i represents the weight for the input signal component x_i, b is a bias, v is the inner product of the input vector and the weight vector, and y represents the output of the neuron after a nonlinear activation. Please note that the neuron based on the inner product is not the only type of neuron in artificial neural systems. For example,

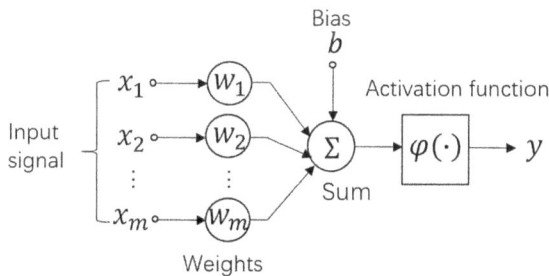

Figure 3.3. Mathematical model of an artificial neuron.

there are also quadratic neurons dedicated to extracting nonlinear features directly (Fan *et al* 2017a, 2017b). Discussion of these is outside the scope of this chapter.

Given the mathematical model of the neuron, the aforementioned single-hidden layer neural network can be formulated into equation (3.2). The structure of the single-hidden layer neural network is shown in figure 3.4. The corresponding mathematical formulas are as follows:

$$\left[\begin{array}{l} h^{(1)} = \varphi^{(1)}\left(\sum_{i=1}^{m} x_i w_i^{(1)} + b^{(1)}\right) \\ y = \varphi^{(2)}\left(\sum_{j=1}^{n} h_j^{(1)} w_j^{(2)} + b^{(2)}\right) \end{array}\right], \tag{3.2}$$

where $x \in \boldsymbol{R}^m$ is an input vector, $h^{(1)} \in \boldsymbol{R}^n$ is an output vector of the hidden layer, $w^{(1)} \in \boldsymbol{R}^{m \times n}$ and $b^{(1)} \in \boldsymbol{R}^n$ are the weight matrix and bias for the input to the hidden layer, and $\varphi^{(1)}$ and $\varphi^{(2)}$ are the corresponding activation functions, respectively. A multi-hidden-layer neural network can be obtained when the number of hidden layers exceeds 1, which is an extension of the single-hidden-layer neural network.

3.1.3 Activation function

The activation function is an essential part of an artificial neuron, which determines the output behavior of the neuron. The activation function empowers the network with the nonlinear mechanism. This nonlinearity enables the artificial neural network to learn a complex nonlinear mapping from input to output signals. Without a nonlinear activation function, the network will be a linear system whose information processing capability will be very limited. Mathematically, even with a single-hidden-layer neural network, we can approximate all continuous functions when the activation function is nonlinear.

For an activation function to perform satisfactorily, it should satisfy the following conditions: (i) differentiability, which is necessary for the gradient descent method to work for optimization of a network and (ii) monotonicity, which is biologically motivated for the neuron to be in either a prohibitory or an excitatory state. Only

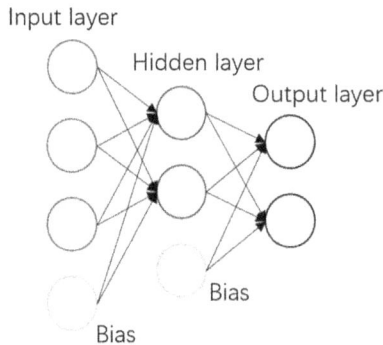

Figure 3.4. Architecture of the neural network with a single-hidden layer.

when the activation function is monotonic can a single-hidden-layer network be optimized as a convex problem.

Generally speaking, the activation function is of great importance since it delivers a single number via a 'soft' thresholding operation as the final result of the information processing processed by the neuron. Several commonly used activation functions are described in the following subsections.

Sigmoid

Sigmoid is a commonly used activation function, as shown in figure 3.5 (Han and Moraga 1995). The Sigmoid function sets the output value in a range between 0 and 1, where 0 represents not activated at all, and 1 represents fully activated. This binary nature simulates the two states of a biological neuron, where an action potential is transmitted only if the accumulated stimulation strength is above a threshold. A larger output value means a stronger response of the neuron. The mathematical expression is as follows:

$$\sigma(x) = \frac{1}{1 + e^{-x}}.$$ (3.3)

The sigmoid function was very popular in the past since it has a clear interpretation as the firing rate of a neuron. However, the sigmoid nonlinearity is rarely used now, because it has the following major drawback: when the activation saturates at either tail of 0 or 1, the gradient in these regions is almost zero, which is undesirable for optimization of the network parameters. During the backpropagation-based training process (see section 3.1.7 for details on backpropagation), a nearly zero gradient for a neuron will be multiplied with other factors according to the chain rule to compute an overall gradient for parametric updating, but such a diminishing gradient will effectively 'kill' the overall gradient, and almost no information will flow through the neuron to its weights and recursively to its data. In addition, we need to initialize the weights of sigmoid neurons carefully to avoid saturation. For example, when some neurons' weights are set as too large initially, they will become saturated and not learn significantly. Also, sigmoid outputs are not zero-centered.

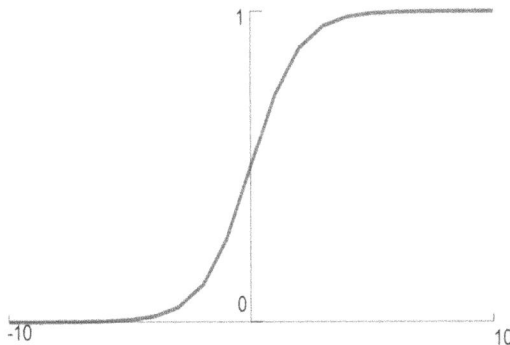

Figure 3.5. The sigmoid function is differentiable and monotonic with the range [0, 1] and the number axis as the domain.

This means that the output is always greater than 0, which will make the input values to the next layers all positive. Then, in the gradient derivation for backpropagation, elements in the weighted matrix change in a biased direction, compromising the training efficacy. In addition, the sigmoid function involves the exponential operation, which is computationally demanding.

Tanh

Tanh is also a commonly used nonlinear activation function, as shown in figure 3.6 (Fan 2000). Although the sigmoid function has a direct biological interpretation, it turns out that sigmoid leads to a diminishing gradient, undesirable for training a neural network. Like sigmoid, the tanh function is also 's'-shaped, but its output range is $(-1, 1)$. Thus, negative inputs to tanh are mapped to negative outputs, and only a zero input is mapped to zero. These properties make it better than sigmoid. The tanh function is defined as follows:

$$\tanh(x) = \frac{e^x - e^{-x}}{e^x + e^{-x}}. \tag{3.4}$$

Tanh is nonlinear and squashes a real-valued number to the range $[-1, 1]$. Unlike sigmoid, the output of tanh is zero-centered symmetrically. Therefore, tanh is often preferred over sigmoid in practice.

ReLU

Currently, the rectified linear unit (ReLU) function has become very popular, and is shown in figure 3.7. Instead of sigmoid/tanh, ReLU outputs 0 if its input is less than 0; otherwise, it just reproduces the input. The mechanism of ReLU is more like the biological neurons in the visual cortex. ReLU allows some neurons to output zero while the rest of the neurons respond positively, often giving a sparse response to alleviate overfitting and simplify computation. In the brain, only when there is a suitable stimulus signal, do some specialized neurons respond at a high frequency.

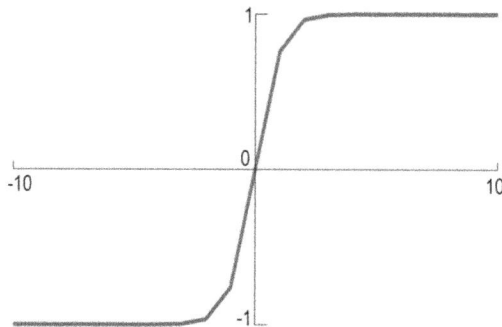

Figure 3.6. The tanh function is similar to sigmoid in shape but has the symmetric range $[-1, 1]$.

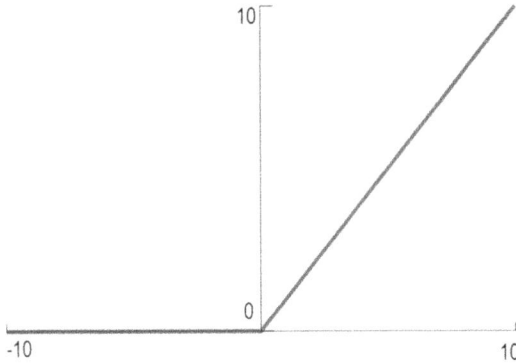

Figure 3.7. The ReLU function, which is equal to zero for a negative input, and otherwise reproduces the input.

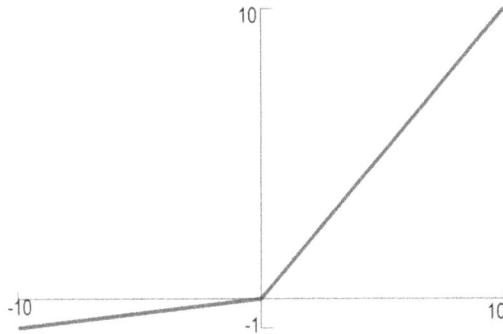

Figure 3.8. The leaky ReLU function, which greatly attenuates a negative input but still records the negative information (the slope for the negative input is set to $\alpha = 0.1$).

Otherwise, the response frequency of the neuron is no more than 1 Hz, which is just like being processed by a half-wave rectifier. The formula of ReLU is as follows:

$$\text{ReLU}(x) = \max(0, x). \tag{3.5}$$

As shown in figure 3.7, the ReLU activation is easy to calculate, which simply thresholds the input value at zero. There are several merits of the ReLU function: (i) there is no saturation zone for a positive stimulation, without any gradient diminishing issue; (ii) there is no exponential operation so that the calculation is most efficient; and (iii) in the network training process, the convergence speed of ReLU is much faster than that of sigmoid/tanh. On the other hand, the ReLU function is not perfect. The output of ReLU is not always informative, which affects the efficiency of the network training process. Specifically, the ReLU output is always zero when $x < 0$. As a result, the related network parameters cannot be updated with a zero gradient, leading to the phenomenon of 'dead neurons'.

Leaky ReLU
The leaky ReLU function is an improved ReLU, as shown in figure 3.8 (Maas and Hannun 2013). In order to save 'dead neurons' when the input is less than 0, leaky

ReLU responds to a negative input in such a way that the negative input is greatly attenuated but the information on the negative input is still kept. Compared to ReLU, leaky ReLU is written as follows:

$$\text{Leaky ReLU}(x) = \begin{cases} \alpha x, & x < 0 \\ x, & x \geqslant 0 \end{cases}, \tag{3.6}$$

where α is a small positive constant, which is usually set to 0.1. Leaky ReLU gives all negative values a small positive slope to prevent the information loss, effectively solving the gradient diminishing problem.

ELU

The exponential linear unit (ELU) is also a variant of ReLU, as shown in figure 3.9 in Clevert *et al* (2015). When the input is less than 0, the ELU is expressed in an exponential form, and the output saturation is controlled by the parameter α to ensure a smooth transition from the deactivated to activated state. Compared to RELU, ELU has negative values that push the mean output closer to zero. Mean shifting toward zero helps speed up learning because of a reduced bias. The ELU function is defined as follows:

$$\text{ELU}(x) = \begin{cases} x, & x > 0 \\ \alpha(e^x - 1), & x \leqslant 0 \end{cases}. \tag{3.7}$$

ELU inherits major advantages of leaky ReLU, and is small at the system origin, which means a smoother/more robust performance with respect to noise than leaky ReLU. However, the computational overhead for ELU is greater than that for leaky ReLU due to the exponential factor in ELU.

3.1.4 Discrete convolution and weights

It is well known that a convolution is a linear operation, which is of great importance in mathematics. A discrete convolution is a weighted summation of components of a vector/matrix/tensor. In the signal processing field, the convolution

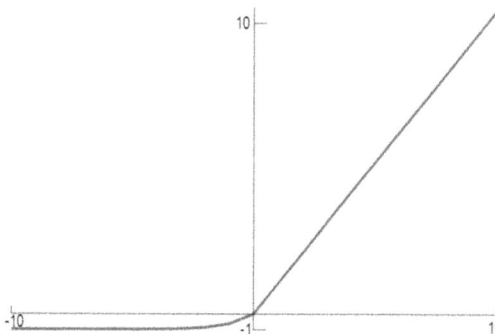

Figure 3.9. ELU function, which gives a more negative area controlled by a parameter (usually $\alpha = 0.5$) to balance the positive area over positive input values.

is used to recognize a local pattern in an image by extracting local features and integrating them properly. There are often local correlations in images, and the convolution is to find a local linear correlation. It will become clear below that the multi-layer convolution network is a powerful multi-resolution analysis, being consistent with the inner-working of the HVS. The three most common types of convolution operations for signal processing are full convolution, same convolution, and valid convolution. Without loss of generality, in the 1D case let us assume that an input signal $x \in \mathbb{R}^n$ is a one-dimensional vector, and the filter $w \in \mathbb{R}^m$ is another one-dimensional vector, the convolution algorithm can be categorized into:

1. Full convolution

$$
\begin{cases}
y = \text{conv}(x, w, \text{"full"}) = (y(1), \ldots, y(t), \ldots, y(n + m - 1)) \in \boldsymbol{R}^{n+m-1} \\
y(t) = \sum_{i=1}^{m} x(t - i + 1) \cdot w(i), \; t = 1, 2, \ldots, n + m - 1
\end{cases}, \quad (3.8)
$$

where zero padding is applied as needed.

2. Same convolution

$$
y = \text{conv}(x, w, \text{"same"}) = \text{center}(\text{conv}(x, w, \text{"full"}), n) \in \boldsymbol{R}^n. \quad (3.9)
$$

The result of the same convolution is the central part of the full convolution, which is of the same size as the input vector x.

3. Valid convolution

$$
\begin{cases}
y = \text{conv}(x, w, \text{"valid"}) = (y(1), \ldots, y(t), \ldots, y(n - m - 1)) \in \boldsymbol{R}^{n-m+1} \\
y(t) = \sum_{i=1}^{m} x(t + i - 1) \cdot w(i), \; t = 1, 2, \ldots, n - m + 1
\end{cases}, \quad (3.10)
$$

where $n > m$. In contrast to the full and same convolutions, no zero padding is involved in the valid convolution.

The ideas behind the one-dimensional convolutions can be extended to the 2D case. Assuming that a two-dimensional input image is $X \in \mathbb{R}^{n \times m}$ and the two-dimensional filter is $W \in \mathbb{R}^{s \times k}$. Then, the discrete two-dimensional convolution operation can be represented as follows:

$$
Y(p, t) = (X * W)(p, t) = \sum_{i} \sum_{j} X(i, j) \cdot W(p - i, t - j), \quad (3.11)
$$

where $*$ represents convolution and $\cdot$ represents multiplication. Likewise, the convolution operations (full, same, and valid) can be defined in higher dimensional cases.

In contrast to the convolution formulas given above, cross-correlation functions can be defined in nearly the same way as the convolution functions:

$$
Y(p, t) = (X * W)(p, t) = \sum_{i} \sum_{j} X(p + i, t + j) \cdot W(i, j). \quad (3.12)
$$

The difference between cross-correlation and convolution is whether the filter W is flipped or not. It is not common in the machine learning field to use convolution exactly, but instead we often process an image with a cross-correlation operation; that is, we do not flip the filter W. Without flipping W, we also call the operation convolution (rigorously, cross-correlation).

Figure 3.10 illustrates an example of a convolution operation (without flipping) on a 2D image.

In the neural network, a convolution operation is specified with two accessory parameters, namely, stride and zero padding. Stride refers the step increment with which the filter window jumps from its current position to the next position. For example, in figure 3.10 the initial position of the window is at the first pixel, and then the second position is at the second pixel, thus stride $= 2 - 1 = 1$. Zero padding refers to the number of zeros appended to the original data along a dimensional direction. Generally speaking, when a valid convolution operation is combined with stride and zero padding, the output size is calculated as follows (without loss of generality, in the 2D case):

$$\begin{cases} Y = X * W \in R^{u \times v} \\ u = \left\lfloor \dfrac{n - s + 2 \cdot \text{zero padding}}{\text{stride}} \right\rfloor + 1 \\ v = \left\lfloor \dfrac{m - k + 2 \cdot \text{zero padding}}{\text{stride}} \right\rfloor + 1 \end{cases}, \qquad (3.13)$$

where '$\lfloor\rfloor$' represents a downward rounding.

In the early neural networks, the connection between layers is in a fully connected form; that is, each neuron is connected to all neurons in the previous layer, needing a large number of parameters. Improving upon the fully connected network, convolutional neural networks rely on convolutions, greatly reducing the number of parameters. The core of the convolution operation is that it reduces unnecessary weighting links, only keeps local connections, and shares weights across the field of

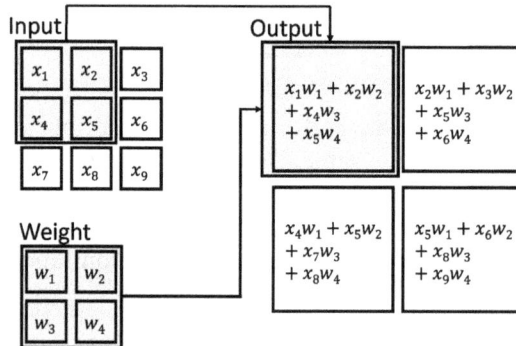

Figure 3.10. Example of a 2D convolution operation (weight without flipping) on a 2D input image.

view. Since the convolution operation is shift-invariant, the learned features tend to be robust without overfitting.

Actually, the convolution is an operation of feature extraction in the premise of specific weights, such as the redundancy-removed ZCA and ICA features presented in the previous chapters. Not limited to the low-level feature space, higher level features can also be obtained in this way for representing the image information semantically.

A special convolution: 1 × 1 convolution

Now, let us introduce a special convolution kernel, which is of 1×1. As mentioned above, convolution is a local weighted summation. In the 1×1 convolution, the local receptive field is 1×1. Therefore, 1×1 convolution is equivalent to the linear combination of feature maps. In the case of multi-channel and multiple convolution kernels, a 1×1 convolution mainly has two effects:

1. 1×1 convolution can lead to dimension reduction. If a 1×1 convolution is applied after the pooling layer, its effect is also dimension reduction. Moreover, it can reduce the redundancy in feature maps, which are obtained after the processing in each layer of the network. In reference to Olshausen and Field's work (Olshausen and Field 1996), the learnt sparse features can be considered as a linear combination of ZCA features, which is an example of feature scarcity.

2. Under the premise of keeping the feature scale unchanged (i.e. without loss of resolution), the activation function applied after 1×1 convolution can greatly increase the nonlinearity of the network, which helps to deepen the network.

Figure 3.11 helps illustrate the effects of the 1×1 convolution.

In figure 3.11, the number of input feature maps is 2, and their sizes are all 5×5. After three 1×1 convolution operations, the number of the output feature maps is 3, and their sizes are 5×5. It is seen that the 1×1 convolutions realize the linear combinations of multiple feature maps while keeping the feature map size intact, realizing cross-channel interaction and information integration.

Furthermore, in figure 3.12, we combine the 3×3 convolution with the 1×1 convolution. Assuming that the number of the input feature maps of $w \times s$ is 128,

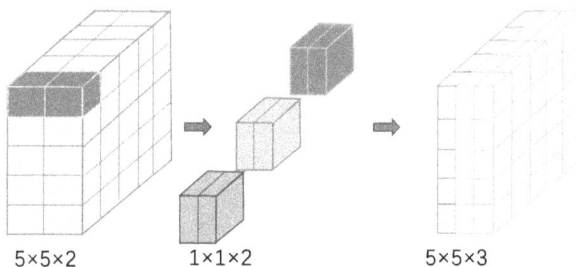

5×5×2   1×1×2   5×5×3

Figure 3.11. Example of a 1 × 1 convolution operation.

the computational complexity on the left is $w \times h \times 128 \times 3 \times 3 \times 128 = 147\,456$ $\times w \times h$, and that on the right is $w \times h \times 128 \times 1 \times 1 \times 32 + w \times h \times 32 \times 3 \times 3 \times 32 + w \times h \times 32 \times 1 \times 1 \times 128 = 17\,408 \times w \times h$. The number of parameters on the left is approximately 8.5 times that on the right. Therefore, the use of 1×1 convolution causes dimension reduction and reduces the number of parameters.

In addition, after 1×1 convolution a new activation function for nonlinear transformation takes effect. Therefore, 1×1 convolution makes the neural network deeper and its nonlinear fitting ability stronger. Such a network can extract and express more complex and higher dimensional features.

Transposed convolution

The transposed convolution performs a transformation from the opposite direction of a normal convolution, i.e. transforms the output of a convolution into something similar to its input. The transposed convolution constructs the same connection pattern as a normal convolution, except that this is connected from the reverse direction. With the transposed convolution, one can expand the size of an input for up-sampling of feature maps (from a low to high resolution feature map).

To explain the transposed convolution, we take an example shown in figure 3.13. It is already known that a convolution can be expressed as a matrix multiplication. If an input X and an output Y are unrolled into column vectors, and the convolution kernels are represented as a sparse matrix C (normally, the convolution kernel is local), then the convolution operation can be written as

$$Y = CX, \tag{3.14}$$

where the matrix C can be written as

$$C = \begin{pmatrix} w1 & w2 & 0 & w3 & w4 & 0 & 0 & 0 & 0 \\ 0 & w1 & w2 & 0 & w3 & w4 & 0 & 0 & 0 \\ 0 & 0 & 0 & w1 & w2 & 0 & w3 & w4 & 0 \\ 0 & 0 & 0 & 0 & w1 & w2 & 0 & w3 & w4 \end{pmatrix}. \tag{3.15}$$

Using this representation, the transposed matrix $C^\top$ is easily obtained for transposed convolution. Then, we have the output X' of the transposed convolution expressed as

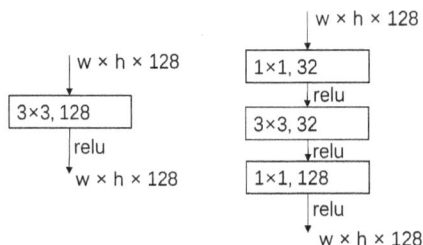

Figure 3.12. Original 3×3 convolution, and improved 3×3 convolution combined with two 1×1 convolutions.

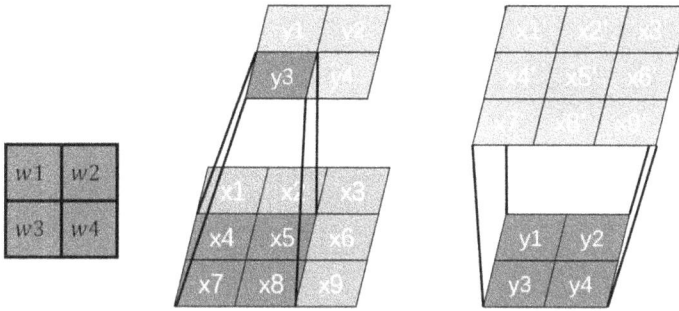

Figure 3.13. Convolution kernel (left), normal convolution (middle), and transposed convolution (right). The input is in blue and the output is in green.

$$X' = C^\mathsf{T} Y. \qquad (3.16)$$

It is worth mentioning that the output X' of the transposed convolution does not need to be equal to the input X, but they maintain the same connectivity. In addition, the actual weight values in the transposed convolution do not necessarily copy those for the corresponding convolution. When training a convolution neural network, the weight parameters of the transposed convolution can be iteratively updated.

3.1.5 Pooling strategy

Essentially, a pooling operation executes the aggregation of feature types, reducing the dimensionality of the feature space. In neurological terms, neurons aggregate and process bioelectrical signals of various bioelectricity rates from other neurons which is equivalent to pooling. The max pooling rate is to process the signal which has the highest bioelectricity rate, while the mean pooling gives the average of involved signals. Similarly, the pooling strategy in artificial neural networks is to compress features, accelerate the computation, allow translational invariance, and reduce the risk of overfitting. Pooling operations can be in many forms, such as max pooling, mean pooling, stochastic pooling, etc.

1. *Max pooling*: Select the maximum value within an image window as the value of the corresponding pixel/voxel.
2. *Mean pooling*: Calculate the average value of an image window as the value of the corresponding pixel/voxel.
3. *Stochastic pooling*: Stochastic pooling first computes the probabilities for each region (Zeiler and Fergus 2013). In a simple way, the probability for each pixel can be calculated by dividing the pixel value by the sum of the values in the pooling window. Then, it randomly selects one value within each pooling region according to the probability distribution. Among these values, the one with the largest probability will be selected, but it is not to say that the largest value must be selected.

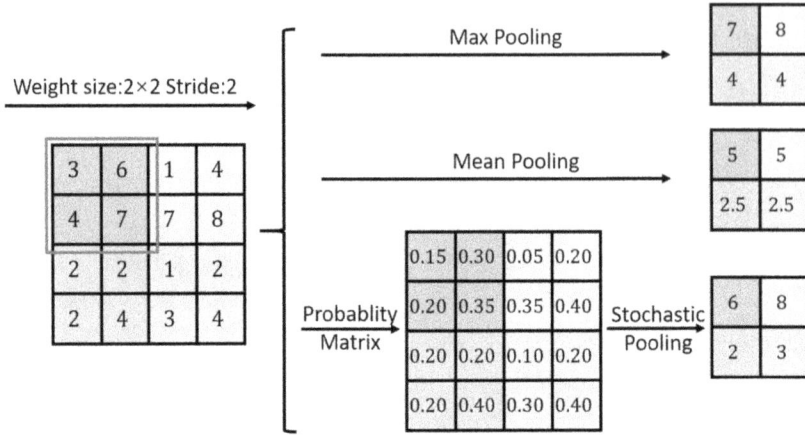

Figure 3.14. Illustration of the three types of pooling strategies.

Generally speaking, mean pooling often retains the overall characteristics of the data and protrudes the background information, max pooling can reveal the textural information, and stochastic pooling has the advantages of max pooling and partially avoids the excessive distortion caused by max pooling. Figure 3.14 illustrates these three pooling strategies.

3.1.6 Loss function

The loss function is critical to guide the process of training an artificial neural network. The loss is used to measure the discrepancy between a predicted value $\hat{y}$ and the corresponding label y. It is a non-negative function, whose minimization drives the performance of the network reaching convergence in the training stage. Training a neural network is to update the network parameters so that $\hat{y}$ approaches y as closely as possible by some certain measure. The local slope or more general gradient by which the loss value changes at a current parametric setting will tell us how to update the parameters for a reduced loss. That is, we use the loss function to compute a clue by which we refine our parameters. The loss function is defined in terms of labels as follows:

$$L(\theta) = \frac{1}{n} \sum_{i=1}^{n} L(y^{(i)}, f(x^{(i)}, \theta)), \tag{3.17}$$

where $[x^{(i)} = x_1^i, x_2^i, \ldots, x_m^i] \in \mathbb{R}^m$ denotes a training sample, $y^{(i)}$ denotes the corresponding label or a gold standard, θ is a set of parameters to be learned, and $f(\cdot)$ is the model function. The loss function can take a variety of forms as the definition of discrepancy is not unique. Next, we introduce several commonly used loss functions.

Mean squared error/L2

The mean squared error (MSE) is widely used in linear regression as the performance measure. The method for minimizing MSE is called ordinary least squares (OSL), which minimizes the sum of squared distances from data points to the regression line. The MSE function takes all errors into consideration with the same or different weights. The standard form of the MSE function is defined as

$$L = \frac{1}{n}\sum_{i=1}^{n}(y^{(i)} - \hat{y}^{(i)})^2, \qquad (3.18)$$

where $(y^{(i)} - \hat{y}^{(i)})$ is also referred to residual and is used to minimize the sum of squared residuals. Note that more rigorously, the normalizing factor $1/n$ should be $1/(n-1)$ to eliminate any bias of the estimator L.

Mean absolute error/L1

The mean absolute error (MAE) (Chai and Draxler 2014) is computed as

$$L = \frac{1}{n}\sum_{i=1}^{n}|y^{(i)} - \hat{y}^{(i)}|, \qquad (3.19)$$

where $|\cdot|$ denotes the absolute value. Although both MSE and MAE are used in predictive modeling, there are several differences between them. First, MAE is more complicated for computing the gradient than MSE. Also, MSE focuses more on large errors whose consequences are much larger than smaller ones, due to the squaring operation. In practice, some larger errors could be outliers that should be ignored. Instead, MAE treats all errors linearly so that it is more robust to outliers.

Similar to MAE, the L1 loss function is the sum of absolute differences between actual and predicted values. L1 does not have the normalizing factor n (or $n-1$). That is, the L1 loss is defined as

$$L = \sum_{i=1}^{n}|y^{(i)} - \hat{y}^{(i)}|. \qquad (3.20)$$

It is underlined that the L1 loss is extremely important in the field of compressed sensing. The minimization of the L1 loss leads to a sparse solution, which is considered a major breakthrough in the signal processing field.

Mean absolute percentage error

The mean absolute percentage error (MAPE) (De Myttenaere *et al* 2016) is a variant of MAE, which is computed as

$$L = \frac{1}{n}\sum_{i=1}^{n}\left|\frac{y^{(i)} - \hat{y}^{(i)}}{y^{(i)}}\right| \cdot 100. \qquad (3.21)$$

MAPE is used to measure the percentage error between predicted and real values. The idea of MAPE is quite simple and convincing, but it is not commonly used in practical applications due to some major drawbacks. For instance, it cannot be used if there are zero values, which would mean a division by zero. Moreover, it is not upper bounded.

Cross entropy

The cross entropy (De Boer *et al* 2002) is often used in the case of binary classification, in which labels are assumed to take values 0 or 1. As a loss function, the cross entropy is computed by the following equation when y and $\hat{y}$ are probability functions:

$$L = -\frac{1}{n}\sum_{i=1}^{n}[y^{(i)}\log(\hat{y}^{(i)}) + (1 - y^{(i)})\log(1 - \hat{y}^{(i)})]. \tag{3.22}$$

The cross entropy measures the divergence between the probability distributions of predicted and real results. A large cross entropy value means that the two distributions are clearly distinct, while small cross entropy values suggest that the two distributions are close to each other. Compared to the quadratic cost function, the cross entropy often enjoys fast convergence and is more likely associated with the global optimization.

The idea of cross entropy can be extended to the cases of multi-classification tasks, leading to the concept of the multi-class cross entropy (see appendix A for more details).

Poisson

The Poisson loss function offers a measure on how the predicted distribution diverges from the expected distribution. The Poisson distribution is widely used for modeling photon-count data, which is highly relevant to medical imaging signals. The Poisson loss function is computed as follows:

$$L = \frac{1}{n}\sum_{i=1}^{n}(\hat{y}^{(i)} - y^{(i)} \cdot \log(\hat{y}^{(i)})). \tag{3.23}$$

Cosine proximity

The cosine proximity function calculates the angular distance between the predicted and real vectors in terms of their inner product. It is well known that the inner product of the two unit vectors gives the cosine value of the angle between them. The formula is as follows:

$$L = -\frac{y \cdot \hat{y}}{\|y\|_2 \cdot \|\hat{y}\|_2} = -\frac{\sum_{i=1}^{n}y^{(i)} \cdot \hat{y}^{(i)}}{\sqrt{\sum_{i=1}^{n}(y^{(i)})^2} \cdot \sqrt{\sum_{i=1}^{n}(\hat{y}^{(i)})^2}}. \tag{3.24}$$

It is also called cosine similarity. In this context, the vectors are maximally 'similar' if they are parallel and maximally 'dissimilar' if they are orthogonal.

3.1.7 Backpropagation algorithm

For a multi-layer neural network, the initial weights are randomly initialized. As such, the network does not immediately perform well on data. We need to iteratively adjust these weights based on the error between training labels and predicted values by the network so as to obtain a converged/trained neural network with an appropriate internal representation, which can then map from an input to a desirable output. Such a training process is almost exclusively done using the backpropagation algorithm (Rumelhart *et al* 1986).

Backpropagation is a shorter term for 'error backward propagation', and essentially the gradient descent optimization. In backpropagation, the error or loss is calculated at the output port of the network, and then distributed backwards throughout the network layers. How does the backpropagation work exactly? Let us explain it now, illustrating the backpropagation process with a simple neural network.

Backpropagation is achieved by iteratively calculating the gradient of the loss function layer by layer using the chain rule. In order to illustrate the principle of backpropagation, we rewrite the loss function equation (3.14), as follows:

$$J(W; b) = \arg \min_{\theta} \frac{1}{n} \sum_{i=1}^{n} L(W, b; x^{(i)}, y^{(i)}). \tag{3.25}$$

Compared to equation (3.14), the loss function equation (3.25) is a more specific formula that contains all parameters to be trained in the neural network, as well as training data and labels. Each gradient descent iteration updates the parameters W, b in the matrix/vector form as follows:

$$w_{ij}^{(l+1)} = w_{ij}^{(l)} - \alpha \frac{\partial}{w_{ij}^{(l)}} J(W, b) \tag{3.26}$$

$$b_i^{(l+1)} = b_i^{(l)} - \alpha \frac{\partial}{b_i^{(l)}} J(W, b), \tag{3.27}$$

where α is the learning rate, $w_{ij}^{(l)}$ denotes the weight between a neuron i in a layer l and the neuron j in the layer $(l - 1)$, and $b_i^{(l)}$ denotes the bias for the neuron i in the layer l. The key step is to compute the above partial derivatives. The following paragraphs describe the backpropagation algorithm, which is an efficient way to compute these partial derivatives.

As an example, consider a neural network which consists of two input units, one output unit, and no hidden unit, and each neuron only outputs the weighted sum of its input data, as shown in figure 3.15.

For such a simple neural network, the loss function is as follows:

$$L = \frac{1}{2}(y - \hat{y})^2, \tag{3.28}$$

where L is the squared error, y is the target output for a training sample, and $\hat{y}$ is the actual output.

With the sigmoid function as the activation function, for each neuron j, its output o_j is equal to

$$o_j = \sigma(s_j) \qquad (3.29)$$

$$s_j = \sum_{k=1}^{n} w_{kj} o_k, \qquad (3.30)$$

where the inner product s_j inside a neuron is generally the weighted sum of outputs o_k of the previous neurons, and when the neuron in the first layer after the input layer s_j is the weighted sum of input data x_k, $k = 1, 2$. The number of input variables to the neuron is often denoted by n, and σ is the sigmoid function defined by equation (3.3).

The activation function σ is nonlinear and differentiable, whose derivative is as follows:

$$\frac{d\sigma}{dx}(x) = \sigma(x)(1 - \sigma(x)). \qquad (3.31)$$

Then, using the chain rule to calculate the partial derivative of the error to the weight w_{kj} is done using the chain rule:

$$\frac{\partial L}{\partial w_{ij}} = \frac{\partial L}{\partial o_j} \frac{\partial o_j}{\partial s_j} \frac{\partial s_j}{\partial w_{ij}}. \qquad (3.32)$$

Since in the last factor on the right-hand side only one term in the sum s_j depends on w_{ij}, we have

$$\frac{\partial s_j}{\partial w_{ij}} = \frac{\partial}{\partial w_{ij}} \left(\sum_{k=1}^{n} w_{kj} o_k \right) = \frac{\partial}{\partial w_{ij}} (w_{ij} o_i) = o_i. \qquad (3.33)$$

As just mentioned, if the neuron is in the first layer after the input layer, o_i is the same as x_i. The derivative of the output of neuron j to its input is simply the partial derivative of the activation function:

$$\frac{\partial o_j}{\partial s_j} = \frac{\partial}{\partial s_j} \sigma(s_j) = \sigma(s_j)(1 - \sigma(s_j)). \qquad (3.34)$$

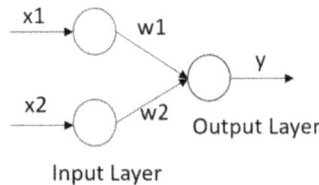

Figure 3.15. Toy neural network with two input units and one output unit.

In equation (3.32), the first term can be easy to evaluate when the neuron is in the output layer, because in that case $o_j = \hat{y}$ and

$$\frac{\partial L}{\partial o_j} = \frac{\partial L}{\partial \hat{y}} = \frac{\partial}{\partial \hat{y}}\left(\frac{1}{2}(y - \hat{y})^2\right) = \hat{y} - y. \tag{3.35}$$

However, if j is in a hidden layer of the network, finding the derivative L with respect to o_j is less obvious. Let $M = u, v, \ldots, w$ be the set of indices for all neurons that receive inputs from neuron j, we have

$$\frac{\partial L(o_j)}{\partial o_j} = \frac{\partial L(s_u, s_v, \ldots, s_w)}{\partial o_j}. \tag{3.36}$$

Then, let us take the total derivative with respect to o_j; a recursive expression for the derivative is obtained as follows:

$$\frac{\partial L(o_j)}{\partial o_j} = \sum_{m \in M}\left(\frac{\partial L}{\partial s_m}\frac{\partial s_m}{\partial o_j}\right) = \sum_{m \in M}\left(\frac{\partial L}{\partial o_m}\frac{\partial o_m}{\partial s_m}W_{jm}\right). \tag{3.37}$$

Therefore, the derivative with respect to o_j can be calculated if all the derivatives with respect to the outputs o_m of the next layer, and those for the neurons in the output layer are known. The overall computational process can be expressed as follows:

$$\frac{\partial L}{\partial w_{ij}} = \delta_j o_i$$

$$\delta_j = \frac{\partial L}{\partial o_j}\frac{\partial o_j}{\partial s_j} = \begin{cases} (o_j - \hat{y}_j)o_j(1 - o_j) & \text{if } j \text{ is an output neuron} \\ \left(\sum_{m \in M} w_{jm}\delta_m\right)o_j(1 - o_j) & \text{if } j \text{ is a hidden neuron} \end{cases} \tag{3.38}$$

In the gradient descent method, the adjustment of w_{ij} can be calculated according to the partial derivatives expressed in equation (3.38). In the $(n + 1)$th iteration, the update of w_{ij} can be computed as follows:

$$w_{i,j}^{n+1} = w_{i,j}^n + \Delta w_{i,j}^n, \tag{3.39}$$

The change in weights reflects how much L will be changed by adjusting w_{ij}. If $\frac{\partial L}{\partial w_{ij}} > 0$, an increase in w_{ij} leads to an increased L value. On the other hand, if $\frac{\partial L}{\partial w_{ij}} < 0$, an increment in w_{ij} means a decrement in L. Clearly, to specify the updating speed or learning rate, a coefficient η is needed:

$$\Delta w_{ij} = -\eta\frac{\partial L}{\partial w_{ij}} = -\eta\delta_j o_i. \tag{3.40}$$

One should choose a learning rate $\eta > 0$. The product of a proper learning rate and the gradient, multiplied by -1 will guarantee that w_{ij} is refined in a way that will decrease the loss function. In other words, the change of w_{ij} by $-\eta\frac{\partial L}{\partial w_{ij}}$ will reduce the L value. For more details on the backpropagation algorithm, see Rumelhart *et al* (1986).

3.1.8 Convolutional neural network

The convolution neural network (CNN) is a popular kind of artificial neural network, which recently became a research focus in the fields of speech analysis and image recognition. In the HVS, there exist several information processing stages, from the V1 area, where simple cells have selective responses for directional structures, to the V4 area, where complex curvatures are identified. In the layer-by-layer process, the receptive field is gradually enlarged, and the image characteristics to which the neurons respond become more and more complicated. Inspired by the HVS, when an image is processed, the activities in the artificial neural network are made similar to those in the HVS. The CNN provides an excellent model for this mechanism of the HVS. That is to say, the convolution neural network extracts features layer by layer. The deeper the layer in the network, the more complex and higher dimensional the feature maps are.

History
The CNN dates back to the papers by Hubel and Wiesel in the late 1960s (Hubel and Wiesel 1968), in which they claimed that the visual cortex of cats and monkeys contains neurons that react individually to directional structures. Visual stimulus can affect a neighborhood of a single neuron, known as the receptive field. Adjacent cells have similar and overlapped receptive fields, the size and position of which vary, forming a complete visual spatial map. This justifies the use of local receptive fields in CNNs.

In 1980, neocognition was proposed, marking the birth of the CNN, which introduced the concept of the receptive field in the artificial neural network (Fukushima 1980).

In 1988, the shift-invariant neural network was proposed to improve the performance of the CNN, which can successfully complete the object recognition in the existence of displacements or slight deformations of objects (Waibel *et al* 1989). The feed-forward architecture of CNN was then extended in the neural abstraction pyramid by lateral and feedback connections. The resultant recurrent convolutional network allows for incorporation of contextual information to resolve local ambiguities iteratively. In contrast to the previous models, image-like outputs at high resolution were generated.

Finally, in 2005, there was a GPU implementation of CNN, making CNN much more effective and efficient (Steinkraus *et al* 2005). As a result, CNN entered its prime.

Architecture
CNN consists of input and output layers and a number of hidden layers. The hidden layers can be categorized by convolution, pooling, activation, and full connection.

The input layer is generally a vector, matrix, or tensor. A convolutional layer is used to convolve an input layer and extract features at a higher level, while a pooling layer is for a sample to reduce the amount of data while maintaining critical information. An activation layer introduces nonlinear features. A fully connected layer integrates features obtained by convolution and pooling. Finally, the output layer produces the final output.

Distinguishing features

CNN has three distinguishing features: local connectivity, shared weight, and multiple feature maps. According to the concept of receptive fields, CNN exploits the spatial locality by enforcing a local connectivity between neurons of adjacent layers. This architecture ensures that the learned 'filters' produce the strongest responses to spatially local input patterns of relevance. Stacking many such layers together forms nonlinear filters that become increasingly global as the depth goes deeper (i.e. responsive to an increasingly larger region) so that the network first creates representations for local and primitive features of the input, and then assembles them for semantic and global features.

In CNN, each filter is replicated across the entire visual field. These replicated units share the same parameters; that is, the same weight vector and bias is repeatedly used to produce a feature map. In a given convolutional layer, the features of interest for all neurons can be analyzed by a shift-invariant correlation. Replicating units in this way allows for the same feature to be detected regardless of their position in the visual field.

Each CNN layer has neurons arranged in three dimensions: width, height, and depth. The width and height represent the size of a feature map. The depth represents the number of feature maps over the same receptive field, which offers different structural features in the same visual scope to respond to the visual stimuli of various types, respectively. Finally, different types of layers, both locally and completely connected, are stacked to form the CNN architecture.

Together, these properties allow CNNs to achieve impressive trainability and generalizability on visual information processing problems. Local connectivity enforces the fact that correlations are often local. Weight sharing reflects the prior knowledge on shift invariance, dramatically reducing the number of free parameters, lowering the memory requirement for training the network, and enabling larger and deeper networks.

A CNN example: LeNet-5

Let us use the famous LeNet-5 network as an example to showcase the convolution neural network. Yan LeCun *et al* proposed the LeNet-5 model in 1998 as shown in figure 3.16 (Lecun *et al* 1998). This network is the first convolutional neural network with a classic result in the field. It is deep and very successful for handwritten character recognition. It is widely used by banks in the United States to identify handwritten digits on checks.

LeNet-5 has seven layers in total, each of which contains trainable parameters. Each layer produces multiple feature maps, and each feature map extracts features through convolution. The input data are a handwritten dataset, which is divided into a training set of 60 000 images in ten classes, and a testing set of 10 000 images in the same ten classes. The network outputs probabilities corresponding to the ten classes respectively, in the final layer, allowing it to predict a digit images class using the softmax function. More specifics on LeNet-5 are as follows.

0. Input layer: The input image is uniformly normalized to be 32 × 32 in size.
1. C1 layer: The first convolution layer operates upon the input image with six convolution filters 5 × 5 in size, producing six feature maps 28 × 28 in size.
2. S2 layer: Pooling with six 2 × 2 filters for down-sampling. The pooling layer is to sum the pixel values in each 2 × 2 moving window over the C1 layer. The S2 layer produces six feature maps of 14 × 14.
3. C3 layer: The C3 layer performs convolutions after the S2 pooling layer. The filter size is 5 × 5. In total, 16 feature maps of 10 × 10 are obtained by the C3 layer. Each feature map is a different combination of the feature maps from the S2 layer.
4. S4 layer: Similar to the S2 layer, S4 is a pooling layer, with a 2 × 2 pooling window to obtain 16 feature maps of 5 × 5.
5. C5 layer: The C5 layer is another convolutional layer. The filter size is 5 × 5. In total, 120 feature maps of 1 × 1 are produced by this layer.
6. F6 layer: The F6 layer is a fully connected layer consisting of 84 nodes. It represents a stylized image of the corresponding character class in a 7 × 12 bitmap.
7. Output layer: The output layer is also a fully connected layer, with ten nodes representing digits 0 to 9, respectively. The minimum output of a node indicates the positive identification result for that node, i.e. if the value of node i is the minimum among all the values of the output neurons, the recognition result for the digit of interest would be i. In any case, only a one-digit class will be assigned to the current image.

Specifically, all the aforementioned three features of a CNN can be found in the LeNet-5 network, which are the local connectivity, shared weight, and multiple feature maps. Since a convolution neural network is close to the real biological

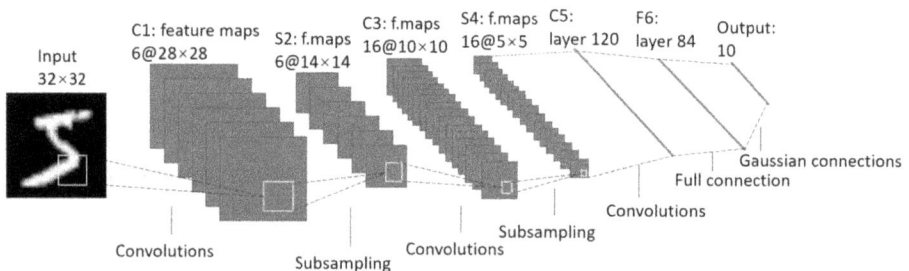

Figure 3.16. LeNet-5 network for digit recognition. Adapted from Lecun *et al* (1998). Reproduced with permission. Copyright IEEE 1998.

neural system in terms of information processing workflow, a CNN analyzes the structural information of digit images well.

3.2 Training, validation, and testing of an artificial neural network

In this section, we will describe a number of key concepts and critical skills related to practical applications of an artificial neural network, selecting a dataset to train the network, cross-validating the model, and testing the trained model. We will also present closely related topics, including overfitting, bias, dropout, pruning, data augmentation, and so on. To better understand these basic concepts, let us take a look at the overall construction, training, and testing process of a convolutional neural network. The whole process is shown in figure 3.17, and can be divided into the following stages:

1. *Designing*: Given a specific task, we can design or select a convolutional neural network architecture based on our experience, including the topology and details on convolution, activation, pooling, loss function, and so on. Then, the weights of the neural network will be initialized.
2. *Forward propagation*: Training samples are fed into the network to produce the corresponding outputs. The input samples are those samples that are in the training dataset used to train the network, and the rest of the samples in the training dataset are used for validation, i.e. to verify the convergence of the network.
3. *Backpropagation*: The weights of the neural network are updated using the above-described backpropagation method.
4. *Iteration*: Repeat steps 2 and 3 until the network converges.
5. Measure the performance of the trained network on the testing dataset.

3.2.1 Training, validation, and testing datasets

In the context of neural network (also referred to as a model) based machine learning, the training data (in a general sense) used to build the final model are usually divided into three types for three inter-related purposes: training (in a specific sense), validation, and testing. The training dataset is used to estimate the model, the validation dataset is used to determine the network structure or the parameters that control the complexity of the model, and the test dataset tests the performance of the final selected optimal model. Brian D Ripley gave the definition of these three words

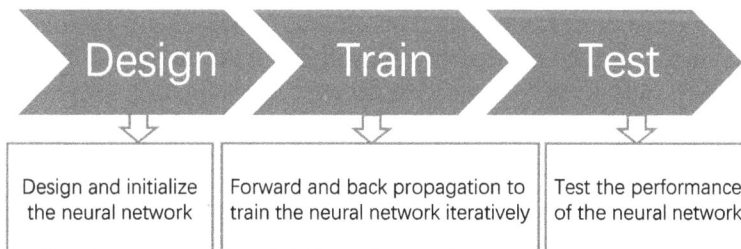

Design	Train	Test
Design and initialize the neural network	Forward and back propagation to train the neural network iteratively	Test the performance of the neural network

Figure 3.17. Workflow of designing, training, and testing an artificial neural network.

in his classic monograph *Pattern Recognition and Neural Networks* (Bell and Sejnowski 1997).

1. *Training set*: A set of samples used for minimizing the loss function. The training set is also used to compute the gradient of the loss function and then adjust the parameters (i.e. weights) of the network.

2. *Validation set*: A set of samples used to evaluate the trained network and avoid overfitting. In the case of overfitting, the network will perform poorly on the validation set.

   If the network performs poorly on either training or validation data, we can train and validate further or modify the network architecture. Currently, there is no governing theory on training, validation, and network architectural design. Therefore, practical experience is quite important in the field of machine learning. Also, training and validation data can switch their roles as needed so that the network can be trained up to an optimal performance. Multiple training–validation cycles can be used in a network training process.

3. *Testing set*: A set of samples never used for training and validation. These testing samples can be processed by the network that has been well trained and validated through steps 1 and 2. The performance of the final network is characterized with the testing set.

In summary, the training set is used to train the network or determine the parameters of the model; the validation set is used for model validation, modification, or selection; and the testing set is purely used to characterize the final model.

3.2.2 Training, validation, and testing processes

The aforementioned three types of datasets are used at different stages of the network development for a specific application. This model was initially applied to the training set to fit model parameters. A model is trained on training data using a supervised learning method such as gradient descent or random gradient descent search. In practice, a training set typically consists of input vectors (or scalars) and their corresponding output vectors (or scalars), which are also called targets, labels, or markers. Normally, we run the current model on the training set, generate a result for each input in the set, and then compare it to the target. The parameters of the model are adjusted based on the comparison using a particular learning algorithm. Model fitting can be done from scratch or via transfer learning (to be detailed later).

The fitted model is used to process a second dataset called the validation dataset and produce the outputs accordingly. In general, the validation set is coupled with the training set to arrive at an accurate and robust neural network. In other words, in the training process the training set is used to update the parameters of the model while the validation set is used to sense the convergence of the model. The error rate of the final model on the validation set is usually smaller than the true error rate since the validation set is used to confirm, modify, or select the final model. One needs to stop training the network when the validation error increases, as this is a sign of overfitting to the training set. This simple process, called early stopping, is complicated in practice because errors in the validation data can fluctuate during the training process, depending on details of the training protocol in use, typically yielding multiple local minima. This complexity has led to the development of several special rules to detect signs of overfitting.

Finally, the test set provides an unbiased estimate of the final model after training and validation. Since the data in the test set have never been used before, they provide the true error rate as the main performance index of the final model.

It is common to leave a small portion of the data to perform the test, and the K-fold cross-validation method is then applied to the remaining data (Kohavi 2001). That is, the remaining N samples are evenly divided into K parts, the $(K-1)$ parts are designated as the training set, and the remaining part is used as the validation set. This training and validation process will be iterated K times so that each part is used for validation once. The squared errors from the K training–validation cycles are averaged to provide the basis for model verification, modification, or selection. In particular, when K is equal to N, the training method is called 'leave one out'. If available data are limited, the K-fold cross-validation method can be repeatedly used to utilize the data fully and update the network progressively. This duplicated use of data is described as an epoch, which means one cycle of presenting all the available data to a neural network of interest. It is not uncommon in practice that many epochs are involved in the training process.

3.2.3 Related concepts

In order to train the neural network well, speeding up the training process and preventing the model overfitting, we have several established strategies. These include batch normalization, dropout, early stopping, pruning, and data argumentation. Let us describe them one by one.

Gradient management

When training an artificial neural network via backpropagation, there are gradient exploding and vanishing issues. The cause of gradient exploding or vanishing is that the backpropagation method computes gradients by the chain rule. This has the possibility of continuously multiplying small or large gradients of the 'front' layers together, which means that the gradient will decrease to zero or increase to infinity exponentially with respect to the number of layers. Specifically, the product will explode if the gradients of the 'front' layers are larger than 1, or vanish if they are smaller than 1 in amplitude.

There are a number of methods to address the gradient exploding or vanishing problem. Currently, the most popular method is the use of ReLU. Since the derivative of the ReLU activation function is 1 for a positive input, the ReLU function is more advantageous than the sigmoid function for gradient-based searching. Also, it is helpful to initialize the network weights carefully. Furthermore, the gradient amplitude can be in a threshold to prevent it from becoming too large. The weights of the network can be regularized with a penalty term in the loss function to discourage large weights and avoid gradients exploding.

Overfitting

In statistics, overfitting is 'generating an analysis that is too close or completely corresponding to a particular dataset, such that it may not be possible to fit other data or reliably predict future observations' (Hawkins 2004). The mechanism of overfitting is to unconsciously extract some residual changes (i.e. noise) as if the change represented an underlying structure of the data. In contrast, under-fitting occurs when the statistical model does not adequately capture the underlying structure of the data.

One possibility of overfitting is when the criteria used to select the model are different from the criteria used to judge its suitability. For example, models can be selected by maximizing the performance of certain training datasets, but their applicability may depend on their ability to perform well on other data. Then, when the model begins to 'remember' training data rather than 'learn', overfitting occurs when new data are processed.

The direct consequence of overfitting is the poor performance of the network on the validation set. To reduce the likelihood of overfitting, several techniques can be used, such as cross validation, regularization, dropping out, early stopping, pruning, data argumentation, and so on.

Batch normalization

When a neural network with many layers is used, which is referred to as deep learning, mini-batch training means taking samples of a batch size to train the neural network during each iteration, instead of all available samples at the same time. Batch normalization is used to normalize the distribution of each layer's data to speed up the training speed. In batch normalization, the input training data are processed in a batch, instead of for all the data. Batch normalization seeks to reduce complication in deep neural network training: normally, training is complicated because the distribution of each layer's parameters could change as the parameters of previous layers. The constant changes in the distribution of each layer's inputs make the network training unstable. This slows down the training speed, and thus requires lower learning rates and suitable careful parameter initialization, making it difficult to train a model with saturated nonlinearities. This phenomenon is referred to as internal covariate shift, and can be addressed by batch normalization.

The operation flow of batch normalization is shown in figure 3.18. The input data x for each layer is before activation over a mini-batch. The output data y is each layer's normalized value, which is the value to be activated.

Dropout

Dropout can be used as a trick to train a deep neural network (Srivastava *et al* 2014). In each iteration, overfitting can be significantly reduced by ignoring part of the hidden neurons (by setting their node values to 0). Simply speaking, for forward propagation, let the activation values of some neurons stop working with a certain

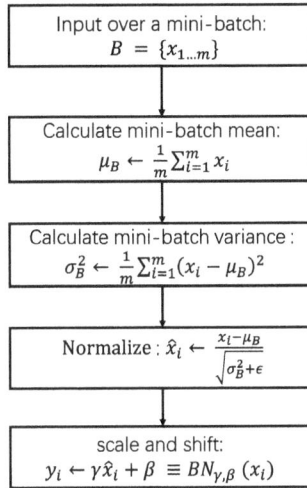

$$\boxed{\begin{array}{c} \text{Input over a mini-batch:} \\ B = \{x_{1...m}\} \end{array}}$$

$$\boxed{\begin{array}{c} \text{Calculate mini-batch mean:} \\ \mu_B \leftarrow \frac{1}{m} \Sigma_{i=1}^{m} x_i \end{array}}$$

$$\boxed{\begin{array}{c} \text{Calculate mini-batch variance:} \\ \sigma_B^2 \leftarrow \frac{1}{m} \Sigma_{i=1}^{m} (x_i - \mu_B)^2 \end{array}}$$

$$\boxed{\begin{array}{c} \text{Normalize}: \hat{x}_i \leftarrow \frac{x_i - \mu_B}{\sqrt{\sigma_B^2 + \epsilon}} \end{array}}$$

$$\boxed{\begin{array}{c} \text{scale and shift:} \\ y_i \leftarrow \gamma \hat{x}_i + \beta \equiv BN_{\gamma, \beta}(x_i) \end{array}}$$

Figure 3.18. Batch normalization applied to activation of x over a mini-batch.

probability, which generally makes the model not depend too much on some local features. Therefore, when applying dropout in the iterative process of updating model parameters, the final neural network does not rely too much on specific neurons, effectively preventing the final network from overfitting.

There are two reasons why dropout can solve overfitting. First, in every iteration, dropout suppresses different hidden neurons, which is similar to training different structures of networks. The whole process of dropout is equivalent to 'averaging' many different neural networks. Different networks produce different overfittings, and 'opposite' fittings cancel each other out to reduce the overfitting as a whole. Second, by dropout the complex co-adaptation relationship is reduced between neurons, because the dropout causes two neurons not necessarily to appear in the network with dropout each time. Such updates of weights no longer rely on the interaction of nodes with fixed relationships, facilitating the network to learn more robust features.

Early stopping
Early stopping is a widely used method to prevent overfitting (Prechelt 2000). The basic idea is to calculate the performance of the model on the validation set in the training process. When the performance of the model on the validation set begins to decline, training is stopped to avoid the problem of overfitting, which will become significant with continued training. The main steps can be as follows:

1. Divide the original training dataset into a training set and a validation set according to some rule (many divisions are possible, each of which are associated with two specific datasets to be used for one training cycle).
2. Train only on the training set and calculate the error of the model on the verification set for each cycle.

3. Stop training when the error for the model on the validation set is worse than the error with the previous training result.
4. Use the parameters from the last iteration as the final parameters of the model.

However, in practice the error of the model on the validation set does not decrease monotonically. We need a stopping criterion to implement the early stopping method. Specifically, the stopping criteria can be divided into three classes:

1. *Threshold-based*: Stop as soon as the validation error exceeds a certain threshold.
2. *Combination-based*: Assuming that overfitting is not a significant problem until the training error decreases rather slowly. Stop training when both the training error cannot be significantly reduced and the validation loss exceeds a certain threshold.
3. *Sequence-based*: Stop when the validation error increased in N successive trainings where N is a pre-specified number.

With a proper stopping criterion, we hope that the training process will produce the lowest validation error and also have the best performance. That is, we seek an optimal balance between training time and validation error. Then, the trained model will be characterized with the test set to make sure.

Pruning

The rationale for pruning a neural network is that there are many parameters in the neural network, but some of them are redundant and make little contribution to the final output (Molchanov *et al* 2016). The pruning is to eliminate redundant parameters. For that purpose, the neurons of the model are first sorted out according to their contributions to the final output. Then, those neurons with low contributions are discarded to make the model less complex, more robust, and computationally more efficient.

The index for ranking the contributions can be based on the average of the L1/L2 regularization term over the parameters, the average output of the activation function, or other indicators. Pruning weaker neurons, the accuracy of the model should have an insignificant loss. Generally, the pruned model usually requires more training data to ensure a satisfactory performance.

Data argumentation

For deep learning, a large number of training samples is critical to achieve satisfactory results with robustness. However, if the training set is not sufficiently large, or a certain type of data is missing, data augmentation is a good choice to prevent overfitting and make the trained model more robust. Data augmentation should generate as large variability as feasible from a relatively small number of available training samples.

For example, several common methods for data argumentation are as follows:

1. *Color jittering*: Data argumentation is performed by coloring, including image brightness, saturation, and contrast.
2. *PCA jittering*: PCA is performed on a set of RGB pixel values or gray values throughout the training set. Then, apply the principal component to each training image, the magnitude of which is proportional to the corresponding eigenvalues, multiplied by a random variable drawn from a Gaussian function with mean zero and a small standard deviation, e.g. 0.1.
3. *Random scale and crop*: Scale transformation with randomized scale factors, and then crop the image up to a certain size.
4. *Horizontal/vertical flip*: More generally, flipping can be performed with respect to any axis.
5. *Shift*: Translation transformation, or more generally deformable transformation.
6. *Rotation/reflection*: Rotation/affine transformation.
7. *Noising/blurring*: Adding various types of noise and blurring.

3.3 Typical artificial neural networks

In this part, we will introduce several classic neural network architectures used in medical image reconstruction and analysis. These networks have distinct characteristics, which are worth highlighting to show how to design a good neural network. Through studies on these networks, we will not only understand the artificial neural network and its intrinsic connection with the HVS, but also appreciate the deep neural network based medical applications to be explained in the following chapters.

Auto-encoder

First, we introduce a classic deep neural network, which is the auto-encoder. As a special case of artificial neural networks, an auto-encoder performs feature representation and then signal reconstruction. In a traditional network for wavelet de-noising, the information processing consists of two steps. First, the wavelet transform is applied to reveal the spatial-frequency components of an input signal. In the transform domain, the original signal and the noise have different characteristics. Hence, the features of the original signal can be preserved while the noise components are discarded. Second, the wavelet inverse transform is performed to estimate the noise-free version of the original signal. The auto-encoder network contains encoding and decoding networks, corresponding to the above-mentioned two steps. The encoding process is equivalent to feature extraction, while the decoding process is equivalent to signal reconstruction. In the auto-encoder network, the encoder automatically learns a feature representation through multiple layers without any need to manually extract the features, thereby eliminating errors and noise in the signal. How the auto-encoder encodes efficiently depends on the strong nonlinear fitting ability of the deep network and the special architecture of the auto-encoder.

'Bottleneck' architecture

The encoding part of an auto-encoder aims to learn a representation for a set of data, typically for dimensionality reduction, and compresses data in an unsupervised manner. The decoding part of the auto-encoder tries to reconstruct the original signal as faithfully as possible from the encoded latent variables.

Architecturally, the auto-encoder consists of the encoder and decoder sub-networks in series, which can be formulated as transitions θ and φ such that:

$$\begin{aligned} &\theta: X \to F \\ &\varphi: F \to X' \\ &\theta, \varphi = \arg \min_{\theta, \varphi} \|X - (\theta \circ \varphi)X\|^2. \end{aligned} \tag{3.41}$$

Now, let us take an example to illustrate how the auto-encoder works. In the simplest auto-encoder, there is only one hidden layer, and the encoder stage of the auto-encoder receives the input $x \in \mathbb{R}^d$ and maps it to $z \in \mathbb{R}^p$:

$$z = \sigma(Wx + b). \tag{3.42}$$

This output z is often referred to as the code, latent variable, or potential representation, where σ is an activation function, such as a sigmoid function, W is a weight matrix, and b is a bias vector. Then, the decoder stage maps z to the reconstructed x' of the same shape as x,

$$x' = \sigma'(W'z + b'), \tag{3.43}$$

where σ', W', and b' for the decoder may differ in general from the counterparts for the encoder, depending on the design of the auto-encoder.

The auto-encoder must be trained to minimize the reconstruction error (for example, the mean squared error), which is also referred to as the loss function,

$$L(x, x') = \|x - x'\|^2 = \|x - \sigma'(W'(\sigma(Wx + b)) + b')\|^2, \tag{3.44}$$

where x is usually averaged over a training dataset.

In practice, the auto-encoder is generally not such a single-layer network, but a deep convolutional neural network. Figure 3.19 shows the architecture of an auto-encoder, which contains encoding and decoding blocks, each of which contains many layers. There are two main applications of the auto-encoder demonstrated in the literature. The first one is data de-noising, while the other is dimension reduction. With appropriate dimension information and sparse regularization, the auto-encoder can learn a data representation more effectively than other technologies such as PCA.

Some details of the auto-encoder

Building an auto-encoder requires three steps: building an encoder, building a decoder, and setting a loss function to measure the information lost due to compression. The encoder and decoder are generally parametric equations. The optimization of these parameters is guided by the minimization of the loss function.

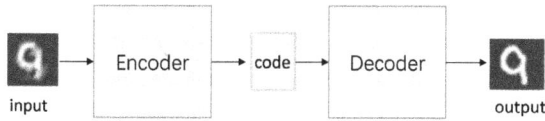

Figure 3.19. The network architecture of an auto-encoder consisting of the encoder and decoder. The input data to the auto-encoder is the same as the output label for the auto-encoder for unsupervised learning.

An auto-encoder is a data compression algorithm in which the compression and decompression of data are automatically learned from these data themselves. In the context of deep learning, the functions of compression and decompression are implemented via neural networks. There are three main characteristics of the auto-encoder:

1. The auto-encoder is data-dependent, which means that it can only compress data that are of the same type as the training data. For example, an auto-encoder trained by face images performs poorly when compressing other images, such as trees, because the features it learned are just related to faces.

2. The auto-encoder is lossy, meaning that the decompressed output is degraded compared to the original input, as are the compression algorithms such as MP3 and JPEG. This is different to lossless compression algorithms. However, by learning enough samples, the auto-encoder focusses on removing irrelevant components from the input signal, such as noise and artifacts.

3. The auto-encoder performs unsupervised learning; that is, it is easy to train the encoder and decoder together with un-labeled data.

Typical auto-encoders

In this section, several commonly used auto-encoders are explained as working examples. Each of the examples has some unique merits. Based on them, new auto-encoders can be designed.

1. The de-noising auto-encoder (DAE) takes a corrupted input and is trained to recover the original undistorted input by predicting the original uncorrupted signal as the output (Vincent *et al* 2010a). The DAE is based on the regular auto-encoder. In order to prevent overfitting, noise can be added to the input data (that is, the input layer of the network) such that the learned encoder is robust. The DAE is similar to the human visual perception process. For example, when one observes an object, we can still recognize it even if a small part of the object is hidden.

2. The sparse auto-encoder produces sparse feature maps in the hidden layers, as shown in figure 3.20 (Makhzani and Frey 2013). In a sparse auto-encoder, there are often more hidden units than the number of input variables, but only a small part of the hidden units is activated at the same time. During training, sparsity could be implemented by additional terms in the loss function or by manually zeroing all but the few strongest hidden unit activations.

3. The variational auto-encoder (VAE) (Doersch 2016) inherits the generic auto-encoder architecture and makes a specific assumption on the

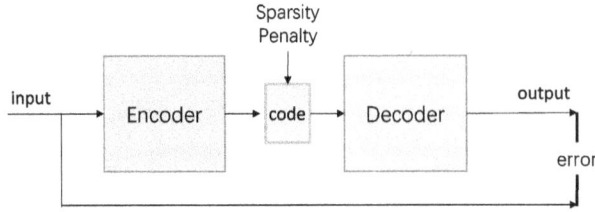

Figure 3.20. Architecture of a sparse auto-encoder whose loss function incorporates both a reconstruction error and a sparsity penalty.

distribution of latent variables, using an additional loss component to minimize the distribution of latent variables and the target distribution such as the Gaussian distribution. While the reconstruction error is often measured as MSE, the latent loss can be quantified using the KL divergence. For more details, see Frans (2016). Based on the target distribution, a random code can be generated to produce a new output, which means that the VAE is a generative model.

4. The contractive auto-encoder (CAE) (Rifai *et al* 2011) adds an explicit regularizer in the loss function that forces the model to learn a function, being robust to slight variations of training data. This regularizer is the Frobenius norm of the Jacobian matrix in the encoder activations to the input, where the Frobenius norm is defined as the sum of squared elements of the Jacobian matrix.

5. The stacked auto-encoder (SAE) (Vincent *et al* 2010b) is a cascade of multiple auto-encoders to complete the task of encoder-by-encoder feature extraction. The training process of the SAE is greedy in the sense that the auto-encoders are trained sequentially, i.e. after the first auto-encoder training is completed, the extracted features of the encoder are used as the input to the second auto-encoder, and so on and so forth. The features of the last auto-encoder are the final output. After the encoder-by-encoder training, a fine-tuning process is initiated. The general idea is as follows. All the decoding layers of the SAE are discarded, and the last hidden layer is linked to a classifier such as the softmax function. The classification error is minimized with respect to all the parameters of the encoding layers. The structure and training procedure of the SAE is illustrated in figure 3.21.

3.3.1 VGG network

The VGG network is a deep convolutional neural network designed by the Visual Geometry Group with Oxford University in 2014 (Simonyan and Zisserman 2014). They proposed a series of models and configurations of deep CNNs, named VGG-N, where N represents layers of the network. Among them, VGG16 and VGG19 won first and second place, respectively, in the localization and classification tasks in the ImageNet Challenge 2014. The main implication of the work on the

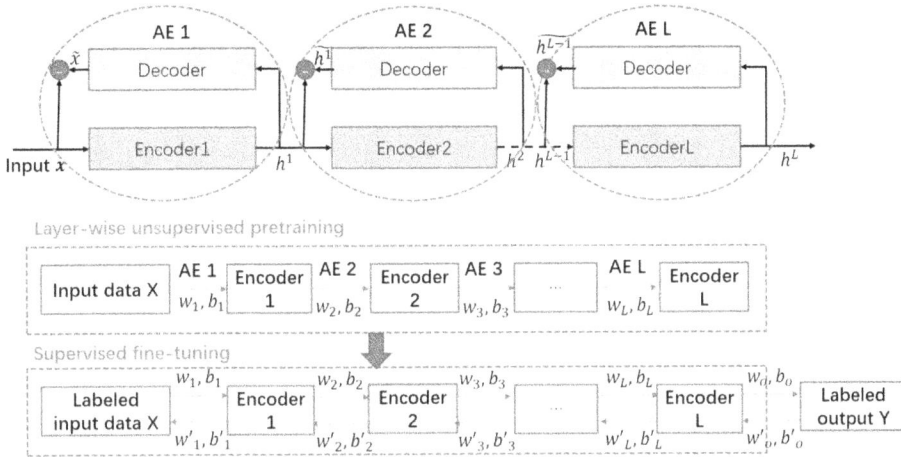

Figure 3.21. Structure and training procedure of the stacked auto-encoder. Adapted with permission from Yuan *et al* (2018). Copyright 2018 IEEE.

VGG networks is that the depth of the convolutional neural network and the size of the convolution kernel have a great effect on the final performance of the network.

The VGG network is very representative. It expresses natural image features effectively and efficiently. It was dedicated to a specific image processing task classification, by training over the ImageNet dataset consisting of tens of millions of natural images. The trained network can be used to perceive multi-layer semantic features of images in various applications. The multi-layer convolution operation is consistent with the human vision system, which extracts increasingly higher order features layer by layer to perform specific perception tasks. In this process, with the layer-by-layer feature extraction, the size of the receptive field is gradually enlarged, and the image characteristics are more and more complicated. In the VGG network, the convolution neural network also extracts features layer by layer. The deeper a layer goes, the more complex and high-dimensional a feature map the layer will produce. Typically, the features learned by a convolutional neural network are discriminative for specific recognition tasks. In the first several layers of the network, low-level features are extracted, such as colors, lines, edges, and corners. Gradually, the features learned by the network become increasingly more complex, such as texture features, and more distinctive patterns. Finally, the key features are obtained to complete a specific task.

VGG architecture
First, there are a series of frameworks for the VGG network, which are called A, A-LRN, B, C, D, and E, respectively, as shown in figure 3.22. The internal configuration and implementation details of the VGG network are also included in the figure.

For training purposes, the input to the VGG network is a fixed-size 224 × 224 RGB image, which is pre-processed by subtracting the mean RGB value. The size of

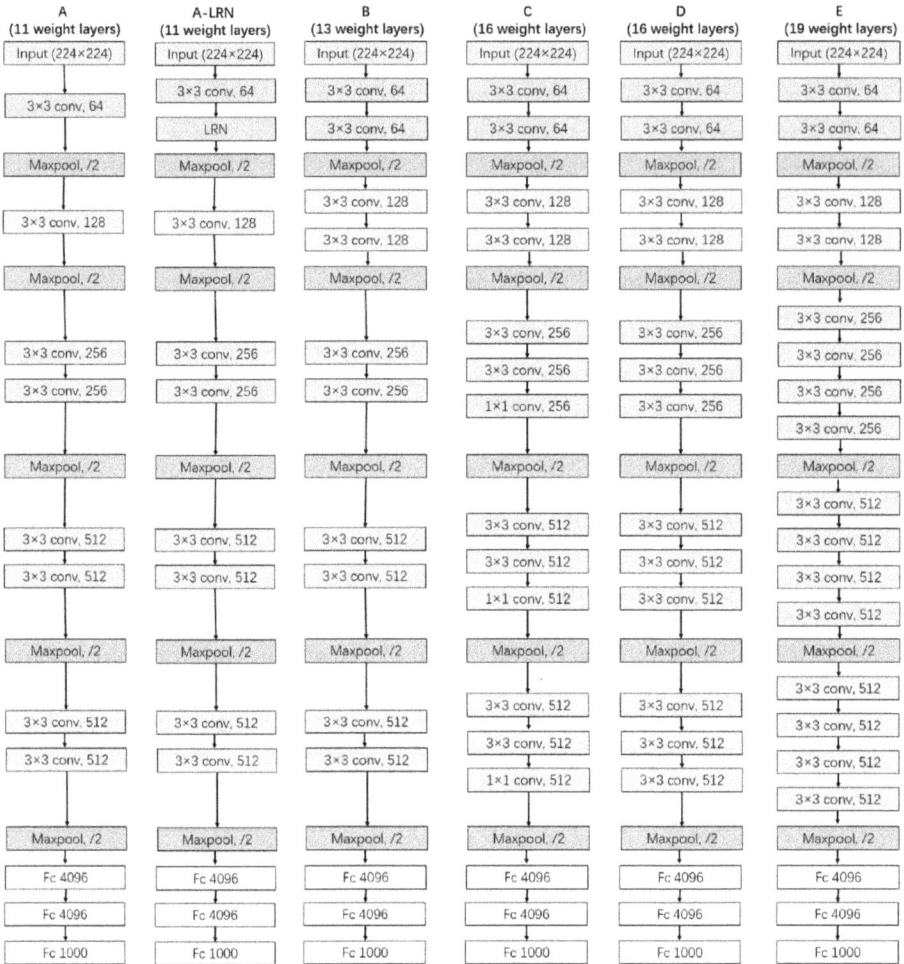

Figure 3.22. Typical configurations of the VGG network, denoted as A, A-LRN, B, C, D, and E, respectively, where LRN means a local response normalization layer which normalizes over local input regions.

convolution filters is 3 × 3, covering a rather small receptive field. In the configuration C, 1 × 1 convolution filters are also used. The convolution stride is fixed to 1 pixel. The max pooling with stride 2 is carried out over a 2 × 2 window. A stack of convolutional layers is followed by three fully connected layers: the first two contain 4096 channels each, the third contains 1000 channels. The final layer is the softmax layer. The configuration of the fully connected layers is the same for all the network configurations. All the hidden layers are equipped with the nonlinear ReLU activation function. The network weights of the configuration A are randomly initialized. Then, when training other deeper architectures, their weights are initialized with the parameters of a well-trained network A.

Key points of the VGG

1. *3 × 3 convolution kernels.* Convolutions with larger spatial kernels tend to be computationally expensive and subject to overfitting. As a model network, the VGG network uses successive 3 × 3 convolution kernels instead of larger convolution kernels such as 11 × 11, 7 × 7, and 5 × 5. For a given receptive field, the use of stacked small convolution kernels is demonstrated to be superior to the configuration with large convolution kernels, and increasing the network depth ensures more complex patterns can be learned with small kernels, which means a decreased number of parameters. Specifically, in the VGG network, three 3 × 3 convolution kernels are used instead of the 7 × 7 convolution kernel, and two 3 × 3 convolution kernels are used instead of the 5 × 5 convolution kernel. This practice shows the idea of shrinking parametric dimensionality by deepening the network architecture. Why could the two 3 × 3 convolution kernels replace a 5 × 5 convolution kernel? As shown in figure 3.23, two 3 × 3 convolution kernels and a 5 × 5 convolution kernel show receptive fields of identical size. On the left side of this figure, the output is produced by a fully connected network with a 5 × 5 window sliding over the input image. Since we are dealing with a visual information processing task, it seems natural to assume translation invariance and replace the fully connected layer by a two-layer convolutional architecture: the first layer is a 3 × 3 convolution, the second is also a 3 × 3 layer on top of the 3 × 3 output grid of the first layer. Sliding this small network over the input grid boils down to replacing the 5 × 5 convolution with two 3 × 3 convolutional layers, as shown on the right-hand side of the figure.

2. *1 × 1 convolution kernels.* In the VGG16 network C of 16 layers, there are a number of 1 × 1 convolution kernels. As described previously, 1 × 1 convolution realizes cross-channel interaction for information integration. With a linear combination of multiple feature maps, cross-channel pooling can be performed. The

Figure 3.23. Mini-network replacing the 5 × 5 convolutional layer with two 3 × 3 convolutional layers. The combination of two 3 × 3 convolutional layers has an effective receptive field of 5 × 5, which in turn simulates a larger filter while keeping the benefits of smaller filter sizes. In this case, we have 2 × 3 × 3 parameters in the former case and 5 × 5 parameters in the latter case. Generally speaking, a major benefit with the use of smaller filters is a decreased number of parameters, facilitating the network training and the performance regularization. Also, with two convolutional layers we use two ReLU layers instead of one, which increases the nonlinear fitting ability. This mechanism is similar to that of the biological neural network, implementing a large receptive field by increasing the network depth and extracting high-order features in a bottom-up manner, with small filters in each step of the visual information processing workflow.

activation function after 1×1 convolution greatly increases the nonlinear character-istics and computational efficiency.

3. *Network depth.* In early network designs, the depth of the network was only a few layers; for example, the AlexNet network has a total of eight layers. A major feature of the VGG network is its great depth (i.e. a large number of layers). In the VGG network, deep convolutional networks (up to 19 weight layers) were evaluated for image classification. It was demonstrated that a decent depth is beneficial for state-of-the-art performance on the ImageNet dataset. Impressively, such network models generalize well to a wide range of tasks and datasets, matching or outperforming more complex recognition pipelines built for less deep image representations.

In summary, VGG is one of the most influential networks, pioneering the use of successive 3×3 convolution kernels and 1×1 convolution kernels in a deep network configuration to increase the nonlinear fitting ability. It indicates that convolutional neural networks can be made deep to have a hierarchical representation of visual data and perform image recognition and other similar tasks well. More importantly, the VGG network trained over a very large database ImageNet can be used to perceive a multi-layer semantic knowledge of images in various applications, such as image classification, image de-noising, image quality assessment, transfer learning, and even the optimization of the GAN network.

3.3.2 U-Net

The U-Net is a convolutional neural network which was initially developed for biomedical image segmentation at the University of Freiburg, Germany (Ronneberger *et al* 2015). The U-Net is a generic deep learning solution for frequently occurring quantification tasks such as cell detection and shape measurements in biomedical images. There are many great examples of U-Net for biomedical image segmentation, such as brain image segmentation and liver image segmentation.

The U-Net shares similarities with the auto-encoder, both of which are processes combining feature representation and signal reconstruction. In the feature repre-sentation stage, the U-Net utilizes a 3×3 convolution and a 2×2 max pooling to extract increasingly more complicated features layer by layer. The difference is that during the reconstruction stage, the U-Net not only takes advantage of complex features in higher convolutional layers, but also exploits simple features in lower convolutional layers. While deeper convolutional layers extract high-order global features, the U-Net contains links to lower convolutional layers, which preserve spatial resolution and information on small details. After both of the lower and deeper features are combined, the signal/image reconstruction is completed in high spatial resolution.

U-Net architecture
Initially, the U-Net architecture was built as a full convolution network without any fully connected layer, and then modified in a way that it yields better medical image segmentation results. Compared to basic convolution networks, a key feature of U-Net is that it is symmetric, with the corresponding layers linked by the skip

connections across the down-sampling path and the up-sampling path. In this way, both low- and high-level features are simultaneously used. Each skip connection is implemented with a concatenation operator, instead of an addition operator to combine input feature maps.

The U-Net is named to reflect its symmetric architecture, which is shown in figure 3.24 and consists of a contracting part (left side) and an expansive part (right side).

The contracting part follows the typical architecture of a convolutional network. It repeatedly applies two 3×3 convolutions without padding, each followed by the ReLU activation and a 2×2 max pooling operation with stride 2 for down-sampling. At each down-sampling step, U-Net doubles the number of feature channels.

In the expansive part, each step is for 2×2 up-convolution, which is known as transposed convolution, as described in subsection 3.1.4. By the 2×2 up-convolution, it finishes up-sampling feature maps and halves the number of feature channels. Then, U-Net concatenates these up-sampling feature maps with the correspondingly cropped feature maps from the contracting part of the U-Net. The operation of cropping is necessary because of the loss of border pixels during convolution. Next, two 3×3 convolutions are followed by the activation ReLU. At the final layer, a 1×1 convolution is employed to map each 64-component feature vector to the desired number of classes. As a result, the network has 23 convolutional layers in total.

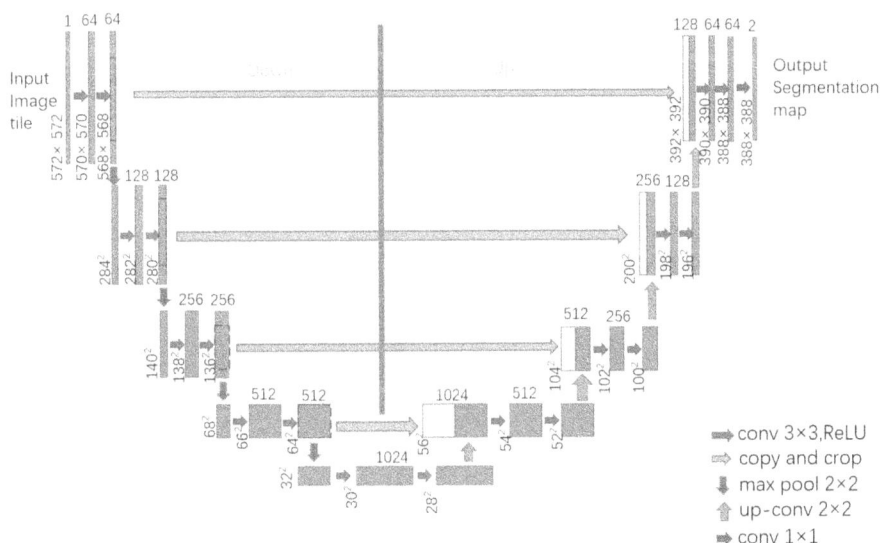

Figure 3.24. Initial architecture of U-Net. Adapted with permission from Ronneberger *et al* (2015). Copyright 2015 Springer Nature.

In addition, in the initial study on the U-Net there was very little training data available so that excessive data augmentation was done by applying elastic deformations to the available training images, which allows the network to learn invariance in the data with such realistic deformations.

Some details of U-Net

The U-Net architecture consists of three parts: the contracting/down-sampling part (also referred to as the down-sampling path), the bottleneck, and the expanding/ up-sampling part (or up-sample path):

1. *Contracting/down-sampling path*

    The contracting path consists of four blocks, each of which is composed of two 3 × 3 convolutional layers coupled with nonlinear activation and batch normalization operations, followed by a 2 × 2 max pooling operation.

    Note that the number of feature maps during every pooling operation is doubled. That is to say, there are 64 feature maps for the first block, 128 for the second, and so on. The purpose of this contracting path is to capture the multi-layer semantic features of an input image, which can be transferred to the up-sampling path through skip connections.

2. *Bottleneck*

    The bottleneck is located between the contracting and expanding paths. The bottleneck includes two convolutional layers with nonlinear activation and batch normalization. Also, dropout layers are used in the original U-Net.

3. *Expanding/up-sampling path*

    The expanding path includes four blocks. Each block is combined with an up-convolutional layer with stride 2, which is concatenated with the corresponding cropped feature map from the contracting path, two 3 × 3 convolutional layers with nonlinear activation and batch normalization. Importantly, the up-convolution means the deconvolution, which is the transposed convolution to finish up-sampling. The concatenation operation allows that multiple input feature maps are concatenated to combine multi-layer semantic features. For example, suppose that as the input to a specific layer we have two groups of feature maps with the size of $w \times h \times d1$ and $w \times h \times d2$, respectively. After concatenation, the size of the output of the layer feature maps is $w \times h \times (d1 + d2)$. Note that the concatenation operation requires input feature maps of the same width and height. Hence, the U-Net crops the feature maps from the contracting path to concatenate with the feature maps from the expanding path which are of a smaller size.

Key points of U-Net

The three key points established in the U-Net network studies are summarized as follows:

1. *Multi-scale synergy*: The expanding path combines high-order semantic features with low-level structural details from the contracting path. It is this combination that greatly facilitates the network training, the performance

robustness and generalizability. This combination allows necessary to excellent segmentation and classification results.

2. *Practical adaptability*: In the training of the original U-Net, the input image size was 572×572. Once the training is completed, images of different sizes can be used as the input, since the only parameters to learn on convolutional layers are the kernel filters, whose size and structure are independent of the size of input images (supposing that they are not too small).

3. *Data augmentation*: The use of massive data augmentation is important in applications such as biomedical segmentation, since it is usually limited to getting a number of annotated samples.

Variants of U-Net

Since U-Net produced excellent segmentation results in the area of medical image segmentation, several improvements were made on the original U-Net, leading to variants of U-Net:

1. V-Net (Milletari *et al* 2016) is a full convolutional neural network for volumetric medical image segmentation. It was developed for 3D image segmentation based on the U-Net network, and trained on magnetic resonance image volumes with the prostate as the target organ of interest. V-Net was trained in an end-to-end fashion to perform volumetric segmentation.

2. M-Net (Mehta and Sivaswamy 2017) is a convolutional neural network for MRI-based brain segmentation. M-Net is similar to U-Net except that there are two down-sampling and up-sampling branches. M-Net utilizes only 2D convolution operations although it operates on 3D images. Thanks to the use of 2D filters, M-Net is memory-efficient.

3. W-Net (Xia and Kulis 2017) is a deep model for unsupervised image segmentation, which was proposed to revisit the unsupervised image segmentation problem. It borrows the ideas behind supervised semantic segmentation methods, in particular by concatenating two U-Net networks together into an auto-encoder, one for encoding and one for decoding. The encoder is used to segment the image and the decoder is used to reconstruct the segmented image to restore the original image. W-Net combines the encoding and decoding parts to realize an unsupervised image segmentation.

4. U-Net++ (Zhou *et al* 2018b) is an advanced architecture for medical image segmentation. It is essentially a deeply supervised encoder–decoder network, where the encoder and decoder sub-networks are connected through a series of nested convolutional blocks to offer the semantic continuity between the feature maps of the encoder and those of the decoder.

3.3.3 ResNet

ResNet was introduced in 2015 and won first place in the ImageNet classification contest (He *et al* 2016). It is arguably one of the most important ground-breaking technologies in the computer vision/deep learning community that has been developed in the past few years. Due to ResNet, a network with hundreds or even thousands of layers can be trained and consistently achieve competitive or state-of-the-art performance. Compared to other convolutional neural networks, the unique feature of ResNet is that it uses residual connections to extend the depth of the neural network; for example, up to 152 layers when it was first put forward. Prior to ResNet, the depth of the neural network was at most dozens of layers. As the network deepens, the problem of gradient exploding or vanishing will become troublesome or unmanageable in the training process. In ResNet a residual connection is used to solve the gradient degradation problem. In a deeper network, residual connections allow gradient descending directly to where the optimization of parameters is needed. Furthermore, by deepening the network, its performance will generally become better in a sufficiently challenging task, since such a deep network is capable of extracting rather complex features.

Degradation of deep networks

For conventional deep learning networks such as the VGG network, convolutional layers and fully connected layers are usually applied in various image processing applications. Such conventional neural networks without any skip or shortcut connection pass data forward layer by layer and back-propagate errors layer by layer as well. When such a plain network is really deep, i.e. with many layers, the problem of vanishing/exploding gradients occurs.

It is well known that the depth of the network is critical to the performance of the model. The network extracts complex feature patterns when the number of network layers is large. If the model is deeper, theoretically better results can be achieved. However, are deeper networks really better? The experimental results show that the gradient degradation is a major problem in training a deep network: when the depth of a conventional network is increased to be large enough, the network accuracy will be saturated or even decreased. This phenomenon is illustrated in figure 3.25: the 56-layer conventional network performs worse than the 20-layer counterpart in the training and test processes. This is not an overfitting problem because the training error of the 56-layer network is high. It is known that deep networks suffer from the problem of gradients disappearing or exploding, which makes deep learning models difficult to train. Although there are already some ways, such as batch normalization, to alleviate this problem, the degradation of deep networks is still severe when a plain network goes too deep.

Residual learning

It has been proved that conventional deep networks are not easy to train, which is indicated by the gradient degradation problem. With the growth of the network depth, the accuracy will become saturated and then degraded. Considering a shallower architecture and its deeper counterpart that has more layers, these added layers should in principle not degrade the network performance, because we could simply stack identity mappings (layers that do nothing) upon the current network,

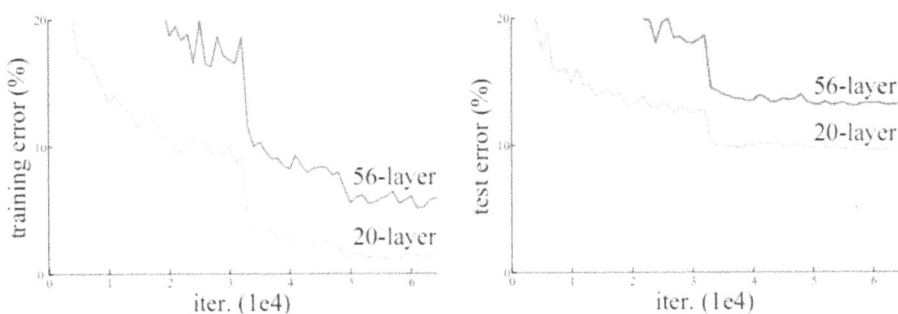

Figure 3.25. Performance of conventional networks of 20 and 50 layers, respectively, on CIFAR10. Adapted with permission from He *et al* (2016). Copyright 2016 IEEE.

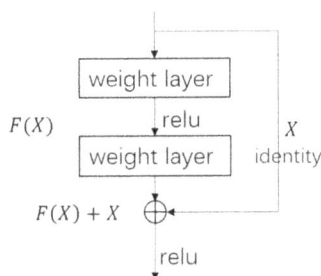

Figure 3.26. Residual learning featured by a residual block enabled by a shortcut connection. Adapted with permission from He *et al* (2016). Copyright 2016 IEEE.

and the architecture would produce the same result. That is, the deeper model should not produce a training error higher than its shallower counterparts. However, the experiments show that it is difficult for current solvers to find good parameters for deeper conventional networks.

To solve the gradient degradation problem in training conventional deep networks, the residual learning idea was proposed, which is shown in figure 3.26. For a structure with several stacked layers, the input is denoted as x, and the desired output mapping is denoted as $H(x)$. Instead of letting stacked layers directly fit $H(x)$, we explicitly let these stacked nonlinear layers fit a residual mapping $F(x) := H(x) - x$. The original output mapping is then obtained as $F(x) + x$. Extensive studies suggest that it is easier to optimize the residual mapping than to optimize the original mapping. For example, when an identity mapping was optimal, it would be easier to drive the residual to zero than to fit an identity mapping using a stack of nonlinear layers. The flowchart of residual learning is somewhat similar to the 'short circuit', which is called a shortcut connection. Identity shortcut connections do not add extra parameters or computational complexity. The entire ResNet network can be still trained by backpropagation. Clearly, by the chain rule for backpropagation, identity shortcut connections alleviate the gradient degradation problem.

ResNet architecture

The ResNet network was first developed based on the VGG19 network, where the residual blocks were added through the shortcut connections as shown in figure 3.27. Modified from the VGG19 network, ResNet directly uses the convolution of stride = 2 for down-sampling, and replaces the fully connected layer with the global average pooling layer. In ResNet, the convolutional layers mostly have 3×3 filters and follow two simple design rules: (i) for the same output feature map size, the layers have the same number of filters, and (ii) when the feature map size is reduced by half, the number of feature maps is doubled. As shown in figure 3.27, ResNet adds a shortcut connection across every two layers of the plain network for residual learning. In that figure, each dotted curve indicates that the number of feature maps is not matched. In the unmatched cases, we can use one of the two solutions: (i) the shortcut is still used for the identity mapping, with extra zero entries padded for the same dimensionality, and (ii) the 1×1 convolution operation is performed to keep the dimensions consistent.

Since the network is now very deep, the computational complexity is high. ResNet takes advantage of a bottleneck design to reduce the complexity, as shown in figure 3.28. The block diagram on the left corresponds to a residual block for ResNet34 with a depth of 34 layers, while the diagram on the right corresponds to a 'bottleneck' residual block for ResNet-50/101/152, where the depth of layers is 50/101/152, respectively. In the bottleneck design, a stack of three layers instead of two is used. The three layers perform 1×1, 3×3, and 1×1 convolutions, respectively, where the two 1×1 layers are responsible for changing dimensions as needed, leaving the 3×3 layer the bottleneck with smaller input/output dimensions. The use of 1×1 convolution and identity shortcut for the bottleneck designs leads to an efficient model with significantly increased depth and decreased complexity compared to the plain network counterpart.

Variants and interpretation of ResNet

As ResNet becomes more popular in the deep learning community, the architecture of ResNet is being improved into interesting variants. In the following, a few exemplary variants are introduced:

1. *ResNeXt*

   The ResNeXt (Xie *et al* 2017) architecture is shown in figure 3.29. This block diagram reflects a 'split-transform-merge' paradigm. First, splitting functions are learned with multi-branch convolutional networks. Then, a set of transformations are performed in the same way. Finally, the outputs of different paths are merged by adding them together.

   In the ResNeXt, a hyper-parameter is introduced, which is called cardinality, i.e. the number of convolution paths, to provide a new way of adjusting the model capacity. Experimental results show that the accuracy can be improved more efficiently by increasing the cardinality than by increasing the depth or the width of the original ResNet. This novel

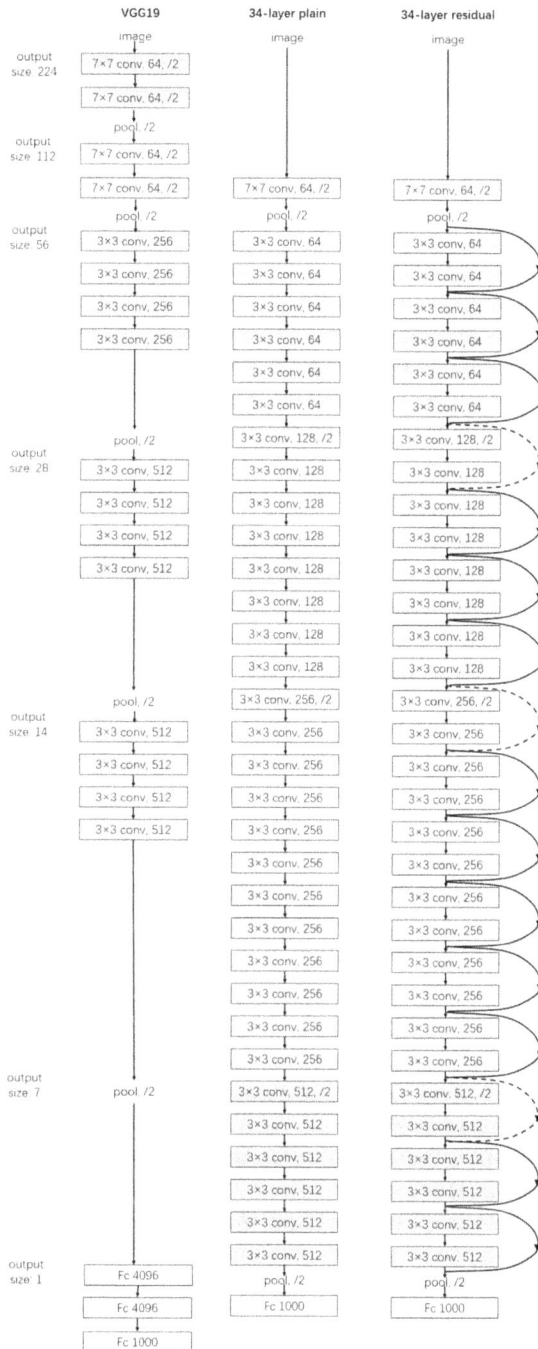

Figure 3.27. Example network architectures tested on the ImageNet dataset. The left column is the VGG19 model as the reference, the middle column is a plain network with 34 layers, and the right column is the residual network corresponding to the middle network, where each dotted curve indicates that the number of feature maps is not matched, and /2 means a stride of 2. Adapted with permission from He *et al* (2016). Copyright 2016 IEEE.

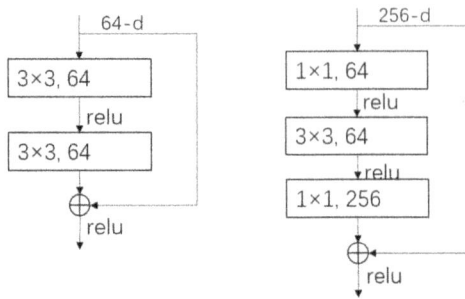

Figure 3.28. Deepening the residual function from a residual block for ResNet34 (left) to a 'bottleneck' residual block for ResNet-50/101/152 (right). Adapted with permission from He *et al* (2016). Copyright 2016 IEEE.

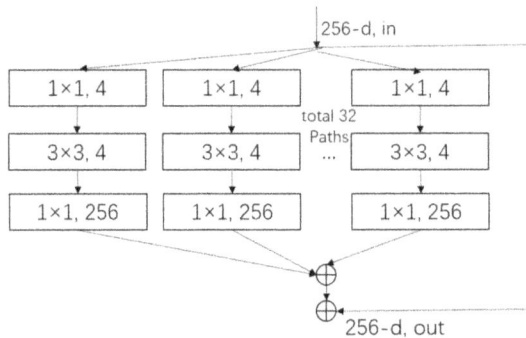

Figure 3.29. ResNeXt architecture following a 'split-transform-merge' paradigm.

architecture can easily be adapted to a new program, because it has a simple structure and only one hyper-parameter that needs adjustment.

2. *Densely connected convolutional neural network*

DenseNet (Huang *et al* 2016a) is a novel architecture that further exploits the effects of shortcut connections so that all layers are directly connected with each other, as shown in figure 3.30. In DenseNet, the input of each layer consists of the feature maps of all earlier layers, and the output is passed to each subsequent layer. The feature maps are aggregated through depth-concatenation. Other than tackling the vanishing/exploding gradients problem, DenseNet encourages feature reuse, which makes the network highly parameter-efficient. DenseNet introduces a hyper-parameter called the growth rate k to prevent the network from growing too large. The growth rate is equal to the number of output channels of a layer.

3. *Deep network with stochastic depth*

Although ResNet has been proved to be powerful in many applications, a deeper network usually requires a long time for training, which makes it practically undesirable or even infeasible. In order to tackle this issue, a

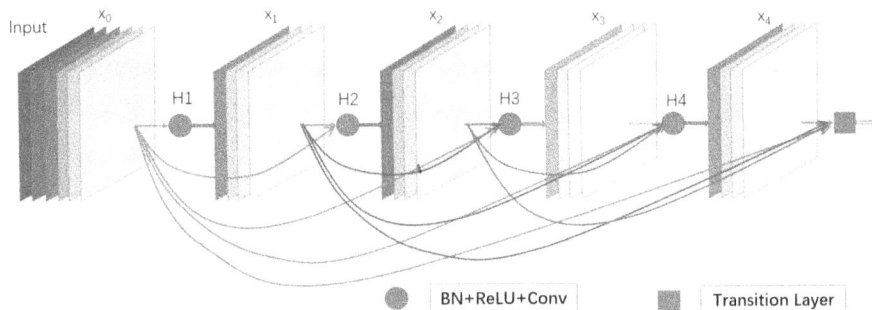

Figure 3.30. DenseNet block of five layers with a growth rate of $k = 4$. Adapted with permission from Huang *et al* (2016a). Copyright 2016 IEEE.

method is introduced that randomly drops layers during training (Huang *et al* 2016b). The deep network with stochastic depth can be coupled with residual blocks. Then, the input flows through both the identity shortcut and the weight layers. During the training process, each layer has a probability to be randomly dropped. During the validation and testing processes, the full network is employed, with all blocks being kept active.

3.3.4 GANs

The generative adversarial network (GAN) became widely known due to the paper published by Goodfellow *et al* (2014). It is a powerful learning framework inspired by game theory. In a GAN network, two models are simultaneously trained to outperform each other, which eventually yields a generative model to capture the data distribution of interest, and the discriminative model that estimates the probability that a sample is from the training data.

Unlike the previously described neural networks, GAN is a unique network that consists of two competing models, instead of a single model. The generative model imitates and learns the distribution of real data. The discriminative model determines whether an input sample is from the real data distribution or not. Through the constant competition between these two models, the ability to generate a realistic sample and discriminate between real and fake samples is significantly improved in the training stage. After the training, a balance is achieved between the generative model and the discriminative model.

GANs provide an effective way to learn deep representations without a large amount of annotated training data. The training task is performed through deriving backpropagation signals through a competitive process involving the two networks in the contest. The representations learned by GANs can be used in a variety of other similar applications, such as image synthesis, image segmentation, classification, and image super-resolution.

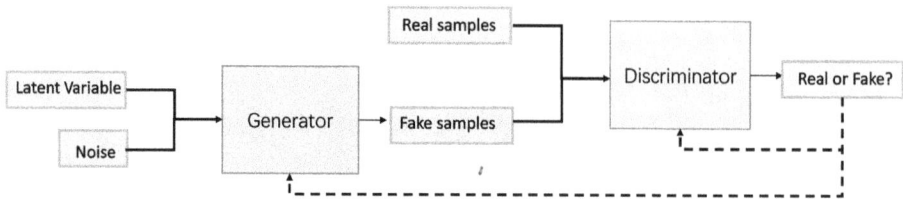

Figure 3.31. Overall GAN architecture consisting of a generator and a discriminator in completion with each other to learn the distribution of real samples.

Discriminative versus generative models

It is of great importance to further comprehend the difference between the generative and discriminative models before GANs are adapted for real-world applications.

A discriminative model learns a function that maps input features (x) to some desired output class label (y). In probabilistic terms, the conditional distribution $P(y|x)$ is learned to make a judgment on which class the input belongs to. On the other hand, a generative model cares about how to produce features x as realistically as possible. It learns the probability distribution of real data from input S. Such a combination of discriminative and generative models is very desirable when working on data modeling problems.

GAN principle

The generative model and the discriminative model are briefly referred to as the generator and discriminator, respectively. The discriminative model decides whether an instance of data belongs to the actual training dataset or not, while the generative model learns the distribution of real data so as to generate new data instances.

For example, we can use the GAN to generate animal pictures. In this context, a generative model generates a new animal picture which is not in the training dataset, and a discriminative model will tell whether the generated picture is real or faked. A high likelihood of success of the discriminative model against the generative model suggests the need for improvement on the generative model. On the other hand, a high performance of the generative model demands an enhancement of the discriminative model. The generative and discriminant models are integrated to form the overall GAN architecture, as shown in figure 3.31.

Now, let us explain the basic principle of GAN further. In the case of generating animal pictures through adversarial learning, the generator (G) and discriminator (D) must be used. G is a network that generates animal pictures, it receives a random noise z, generates a picture from this noise as a random code, and records it as $G(z)$. Meanwhile, D is a discriminating network that discriminates whether a picture is 'real' or not. Its input parameter is x. x represents a picture and the output $D(x)$ represents the probability that x is a real picture. If the output is 1, it means that the confidence level is 100% for the input to be a real picture, while the output 0 means that it is impossible to be a real picture.

In the training process, the purpose of the generative network G is to generate a picture or sample as realistic as possible to deceive the discriminant network D, while the purpose of the discriminant network D is to identify fake and real images or samples as reliably as possible. Thus, the interaction of G and D is a dynamic adversarial process. The overall goal of generating realistic images and the conflicting interests of the two networks G and D lead to the name of the generative adversarial network: each of the two networks G and D tries to beat the other, and by doing so they both improve performance. The competition between them makes these two networks evolve toward better selves.

In the most ideal state, G can generate pictures $G(z)$ indistinguishable from real pictures using D. Then, given such an input to D, the output probability of D will be no different from a random guess, i.e. $D(G(z)) = 0.5$. This means the generator produces data from the exact targeted distribution, and the discriminator predicts 'true' or 'generated' with equal probabilities.

When the GAN model converges, the generated data have the same distribution as that of real data. The loss function of the GAN model is expressed as follows:

$$\min_{G} \max_{D} V(D, G) = E_{x \sim p_{\mathrm{data}}(x)}[\log D(x)] + E_{z \sim p_z(z)}[\log(1 - D(G(z)))], \quad (3.45)$$

where x represents a real picture, z represents a noise input to the G network, $G(z)$ represents the picture generated by the G network, and $D(\cdot)$ represents the probability that the D network determines that the picture is real. This loss function combines two opposite goals, which are (i) for the discriminator D is to maximize the possibility of assigning the accurate label to training and generated samples, and (ii) for the generator G to minimize $\log(1 - D(G(z)))$ so as to maximize $D(G(z))$ to fool the discriminator.

Variants of GAN

1. The Wasserstein generative adversarial network (WGAN)

   WGAN (Arjovsky *et al* 2017) improves the original GAN in terms of an improved loss function. Instead of using cross entropy in the loss function, WGAN introduces the Wasserstein distance to measure how close the generated distribution is to the real one, which helps avoid unstable training results. In particular, the training of WGAN does not require a careful balance between steps for the discriminator and the generator, respectively.

2. The super-resolution generative adversarial network (SRGAN)

   SRGAN (Ledig *et al* 2016) is an influential variant of the original GAN. It was designed to de-blur images into high quality counterparts. The discriminator of SRGAN incorporates the VGG19 network, while the generator contains a series of residual blocks. The loss function of SRGAN includes two parts, which are the adversarial loss and the content loss. The adversarial loss maps the image to a manifold, and uses the discriminant network to discriminate between the de-blurred image and the original one. The content loss is based on the perceptual similarity instead of the pixel-based similarity. By comparing the features of a generated image and the features of the target through a convolutional neural network, the generated image and the target one are made similar in terms of semantics and style.

3. The deep convolutional generative adversarial network (DCGAN)

DCGAN (Yang *et al* 2017) modifies the architecture of the original GAN to improve the training quality and convergence speed. Specifically, to extract features more efficiently, in the discriminant and generative networks, convolution with stride is applied, discarding the pooling operation. In order to improve the convergence and stability of the training process, batch normalization is implemented in the generative and discriminant networks. Then, the fully connected layer is removed after the last convolutional layer to reduce the number of parameters. In addition, all layers use the ReLU activation function in the generative network, except that the last layer uses the tanh function. Finally, in the discriminant network all layers use the leaky ReLU activation function.

4. The least square generative adversarial network (LSGAN)

LSGAN (Mao *et al* 2017) replaces cross entropy in the loss function with the least squares measure. Experimental results demonstrate that LSGAN performs more stably and generates higher quality images than the original GAN.

5. The boundary equilibrium generative adversarial network (BEGAN)

BEGAN (Berthelot *et al* 2017) is a new powerful GAN model, which converges quickly and stably. In BEGAN, the discriminator is an auto-encoder, which encodes and decodes both generated data and real data. In the loss function of the discriminator, the Wasserstein distance is used to measure the distribution of errors before and after the auto-encoder and compute the similarity of error distributions for real and generated data, respectively. Also, the generative network is trained to maximize the similarity of error distributions between real and generated data, respectively.

3.3.5 RNNs

Recurrent neural networks (RNNs) are a family of neural networks for processing sequential data, where outputs from the previous step are fed as the input to the current step. RNNs are used for the development of models for deep learning and simulation of neuronal activities in the human brain. RNNs are particularly powerful in situations where dynamic, correlative, and causal information is important to predict outcomes. They are different from other types of artificial neural networks because they process a sequence of data by feedback loops, which allow previous information to persist, and the effect of which is often described in terms of memory. The most important feature of RNN is the hidden states, which remember some information about a sequence.

What is RNNs?

Figure 3.32 shows a simple recurrent neural network, which consists of an input layer, a hidden layer, and an output layer.

On the left of figure 3.32, it is the whole architecture of a simple recurrent neural network, where x is a vector representing the input layer with a specified array of data; s is a vector representing the hidden layer, which represents the 'memory' of the network; U is the weight matrix from the input layer to the hidden layer; o is a vector representing the output layer; and V is the weight matrix from the hidden layer to the output layer. The states of the hidden layer of the recurrent neural network depend not only on the current input x but also on the values of the previous hidden layer, the weight matrix W is the weight matrix mapping from the previous hidden layer to the current hidden layer.

The right-hand side of figure 3.32 shows the unfolded view of the left counterpart with respect to time steps, which illustrates the idea behind RNNs in a straightforward way. When the network receives the input x_t at time step t, the value of the hidden state is s_t, and the output value is o_t. We can use the following formulas to represent the forward process of the recurrent neural network:

$$o_t = f(V \cdot s_t) \tag{3.46}$$

$$s_t = f(U \cdot x_t + W \cdot s_{t-1}). \tag{3.47}$$

The result from the output layer can be calculated by equation (3.46), where $f(\cdot)$ represents the nonlinear activation function such as tanh or ReLU. Furthermore, the result from the hidden layer can be calculated by equation (3.47). If we repeatedly take (3.47) into (3.46), we will have

$$
\begin{aligned}
o_t \cdot &= f(V \cdot s_t) + \\
\cdot &= f(V \cdot f(U \cdot x_t + W \cdot s_{t-1})) \\
\cdot &= f(V \cdot f(U \cdot x_t + W \cdot f(U \cdot x_{t-1} \\
&\quad + W \cdot f(U \cdot x_{t-2} + W \cdot s_{t-3})))) \\
\cdot &= f(V \cdot f(U \cdot x_t + W \cdot f(U \cdot x_{t-1} \\
&\quad + W \cdot f(U \cdot x_{t-2} + W \cdot f(U \cdot x_{t-3} + \ldots))))) \\
\cdot &= f(V \cdot f(U \cdot x_t + W \cdot f(U \cdot x_{t-1} \\
&\quad + W \cdot f(U \cdot x_{t-2} + W \cdot f(U \cdot x_{t-3} + \ldots))))).
\end{aligned}
\tag{3.48}
$$

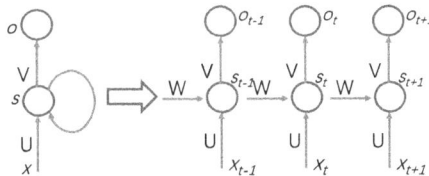

Figure 3.32. A simple recurrent neural network and its unfolding view with respect to computational steps.

As seen from the above, the output of the recurrent neural network is affected by the previous states, which is why the recurrent neural network can process sequential data.

The basic principles of the RNN are not difficult, but there are two points to we need to notice:

1. Compared to traditional deep neural networks that use different parameters in each layer, the RNN shares the same parameters (U, V, W above) in all steps. Such a design of RNN means that we perform the same task at each step, just using different inputs, which greatly reduces the total number of parameters to be learned.

2. At each step, there are outputs in the above diagram. However, it may not be necessary to have an output after each step, depending on the task. For example, when predicting the sentiment of a sentence, we probably only focus on the final output. Similarly, we may not need an input for each step. Some variants of RNNs are shown in figure 3.33. The blocks in green are hidden states of an RNN, which capture some information about a sequence. The blocks in red are inputs at each step, which can be a sequence. The blocks in blue are outputs at each step.

How to train RNNs?

Training a recurrent neural network is similar to training a traditional neural network. A typical training process is exemplified as follows:

1. Provide the input at a time step to the network.
2. Calculate its current hidden state from the current input and the previous hidden state vector.
3. The current hidden state vector s_t becomes s_{t-1} for the next step t.
4. Complete the calculation for all the steps, and use the final hidden state vector to calculate the output.
5. Calculate the error between the target output and the actual output.
6. Back-propagate the error in the network to update the weights and hence the network (RNN) is trained.

The backpropagation of RNN is called backpropagation through time (BPTT). Since the parameters (U, V, W) are shared by all time steps in the RNN, the gradient calculation depends on parametric values at both the current and previous steps. For

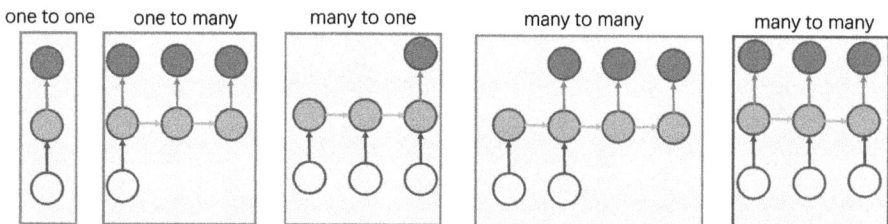

Figure 3.33. Different types of RNNs.

example, when calculating the gradient at $t = 5$, we need to back-propagate four steps to compute the gradients.

RNN extensions

1. *Bidirectional RNNs*

Bidirectional RNNs (Schuster and Paliwal 1997) are proposed based on the idea that the output at a given time step may depend on not only the previous information in the sequence but also on future information. For example, when predicting a missing word in a data sequence, we need to look at both the left and the right context. The structure of a bidirectional RNN is quite simple, which is shown in figure 3.34. It contains two RNNs, which are stacked symmetrically. Finally, the output can be calculated according to the hidden state of both RNNs.

2. *Deep (bidirectional) RNNs*

Deep (bidirectional) RNNs (Graves *et al* 2013) are similar to bidirectional RNNs, except that deep RNNs have multiple hidden layers per time step, which is shown in figure 3.35. The deeper the structure, the higher the learning capacity of the network, and the more training data are needed.

3. *Long short-term memory (LSTM) networks*

One drawback of a standard RNN is the gradient vanishing and exploding problem, and the performance of the neural network is compromised if it cannot be trained properly. Such a case happens when a neural network has deep layers to process complicated data. The LSTM units (Hochreiter and Schmidhuber 1997) can solve this problem. With LSTM units, RNNs can process data with short-term and long-term memory cells, which enables RNNs to figure out what kind of data is important and should be remembered and applied in the network, and other data can be forgotten.

The main module of the LSTM network is shown in figure 3.36. In the diagram, each blue line carries an entire vector, from the output of one node to the inputs of others. The yellow boxes are learned neural network layers, while the purple circles denote pointwise operations, such as vector addition.

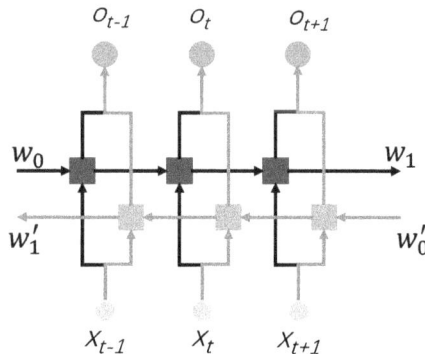

Figure 3.34. Structure of a bidirectional recurrent neural network.

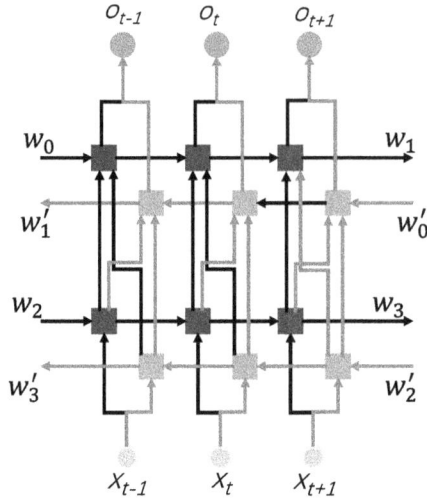

Figure 3.35. Structure of a deep bidirectional recurrent neural network.

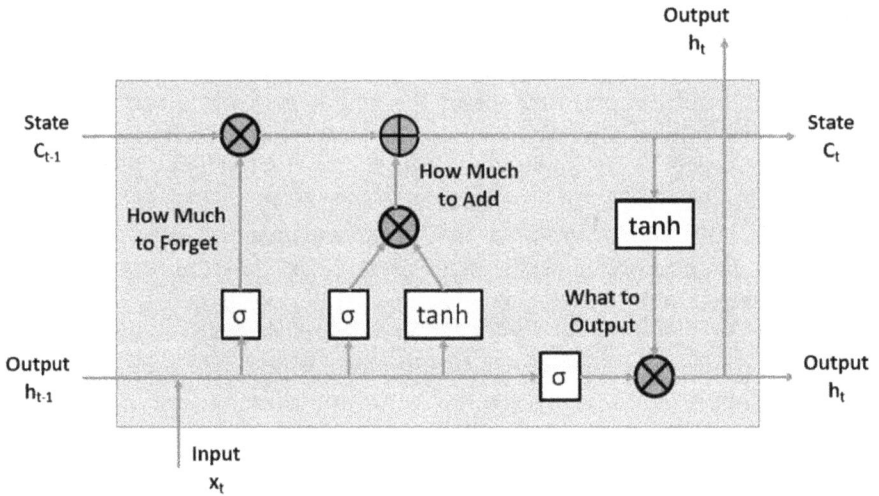

Figure 3.36. Representative configuration of an LSTM network, where the left yellow box is the forget gate, the second yellow box is the input gate that works with the tanh gate as the third yellow box to update states, and the last yellow box takes the states into account to output a result.

The merging of lines denotes concatenation, while a line forking denotes its content being copied and directed to different locations.

The key idea of LSTM is the control of memory cell states. In fact, LSTM controls what to keep in and what to erase from its memory by what are called gates, which are composed with a sigmoid activation layer and a pointwise multiplication operation. Gates are a way to optionally let information through. An LSTM has three of these gates to coordinate the information processing flow optimally.

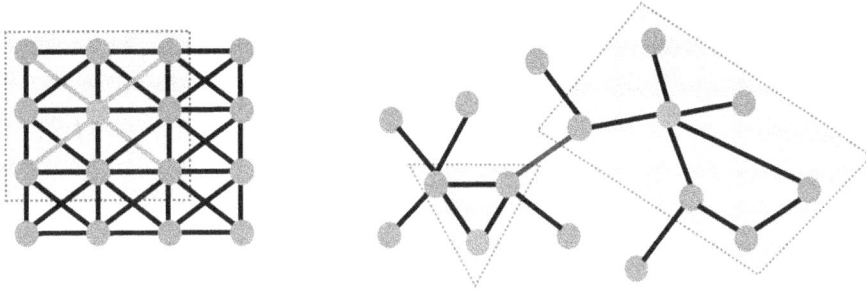

Figure 3.37. Comparison between traditional convolution (left) and graph convolution (right).

3.3.6 GCNs*

Currently, most convolutional neural networks can handle regular Euclidean data such as vectors and images. In reality, there is a large amount of data with arbitrary structures; for example, the World Wide Web, protein-interaction networks, physical interactions such as the dynamics of multiple objects, and knowledge in the form of graphs (Zhou *et al* 2018a). Compared with the images that are Euclidean structures and admit traditional convolution operations, graphs are non-Euclidean structures with complicated topological features.

As is shown in figure 3.37, while a conventional image is shown on the left, with each pixel as a node, a simple undirected graph is given on the right. The shaded portion in each case represents a convolution kernel. There are two differences between traditional and graph convolution operations:

1. In the Euclidean space represented by an image, the number of immediate neighbors of any node is fixed. For example, there are eight neighbors of the green node (the nodes on the edge can be filled with padding), and the number of neighbors of another node is the same. However, in the non-Euclidean space represented by a graph, the number of neighbors is not fixed. One green node has two neighbors while the other green node has five neighbors.
2. The convolution operation in the Euclidean space actually extracts a feature of an image with a fixed-size learnable convolution kernel. However, because the number of the neighboring nodes in a graph is generally not fixed, the traditional convolution kernel cannot be directly used to extract a feature from the graph.

Clearly, the graph is a more general and more advanced abstraction of data structures involving the degree of a node, direction of a link, and neighborhood, and many other features. It is underlined that the graph reflects a topological map of a system and provides more information than conventional images. The characteristics and complexity of graph-based data represents challenges and opportunities in the deep learning field.

In recent years, graph convolutional networks (GCNs) as powerful neural network architectures have attracted increasing attention, which utilize a generalized notion of convolution so that it deals with graph data meaningfully. The first spectral convolution neural network (spectral CNN) was proposed by Bruna *et al* (2014) which performs graph convolutions based on spectral graph theory. Since then, a number of studies were performed to develop GCNs in multiple applications, such as graph classification (Atwood and Towsley 2016, Kipf and Welling 2017) and graph representations (Niepert *et al* 2016).

The fundamentals of GCNs

Let us first formulate the convolutions on graphs, which are often categorized as spectral-based models and spatial-based models. First of all, to get an overall heuristic idea behind generalized convolutions for graph networks, we highly recommend this blog (10 minute read): https://towardsdatascience.com/how-to-do-deep-learning-on-graphs-with-graph-convolutional-networks-7d22250723780, which helps us understand both spectrally-based and spatially-based models, although it is more related to spatially-based models than spectrally-based models. Having had an insight from the above blog, let us now describe some fundamentals of spectrally-based and spatially-based models.

We define a graph as $G = (V, E, A)$, where $V = \{v_1, .., v_n\}$ represents a set of nodes/vertexes, E is the set of edges, and $\epsilon_{ij} = (v_i, v_j) \in E$ represents an edge. An adjacency matrix A describes the graph structure in a matrix form with a size of $N \times N$, in which $A_{i,j} = w_{i,j} > 0$ means there is an edge from node v_j to node v_i with a weight $w_{i,j}$ which is often set to 1, and otherwise $A_{i,j} = 0$. A matrix D of the same size is a degree matrix in which each diagonal element represents the number of neighbors of each node in the graph. Base on these notations, the Laplace matrix could be defined as $L = D - A$ and the normalized Laplacian is $\bar{L} = D^{-\frac{1}{2}}LD^{-\frac{1}{2}} = I - D^{-\frac{1}{2}}AD^{-\frac{1}{2}}$. Let us take a simple graph as an example to illustrate the calculation of the Laplacian matrix in figure 3.38.

In a spectrally based model, the eigen-factorization of a Laplacian matrix in the graph theory is used to generalize the convolution operation to a graph. As seen in the Laplacian matrix example, it is real and symmetric such that it can be represented by an eigenvalue decomposition of $L = U\Lambda U^T$. In this equation, $U = [u_0, u_1, .., u_{n-1}]$ is the matrix of eigenvectors, and $\Lambda = \text{diag}(\lambda_1, \lambda_2, .., \lambda_n)$ represents the diagonal matrix of eigenvalues. Then, we define the graph Fourier transform of a graph signal x by $F(x) = U^T x$, and the inverse transform is thus $F^{-1}(x) = U\tilde{x}$, where $\tilde{x} = F(x)$ is the transformed signal in the graph transform domain. Hence, the input graph signal is projected to the orthonormal space with

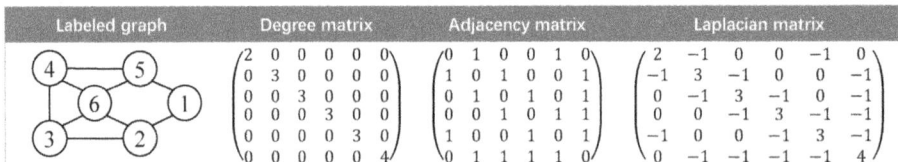

Figure 3.38. Example of a graph, its degree, adjacency, and Laplacian matrices.

the basis formed by eigenvectors of the Laplacian matrix (Wu *et al* 2019). The graph convolution of a signal with a filter g could be defined as

$$
\begin{aligned}
x_G^* g &= F^{-1}(F(x) \odot F(g)) \\
&= U(U^T x \odot U^T g).
\end{aligned}
\tag{3.49}
$$

Alternatively,

$$
x_G^* g = U \, \mathrm{diag}(g(\lambda_1), g(\lambda_2), \ldots, g(\lambda_n)) U^T x,
\tag{3.50}
$$

where $_G^*$ denotes the graph convolution operation, and $\odot$ is the Hadamard product. A simplified graph convolution is formulated as $x_G^* g_\theta = U_\theta U^T x$ when we denote a filter $g_\theta = \mathrm{diag}(U^T g)$. A variety of spectral-based GCNs can be defined based on the graph convolution with different choices of the filter g_θ. For example, in the spectral CNN (Bruna *et al* 2014), the graph convolution kernel layer is designed as $y_{\text{output}} = \sigma(U g_\theta U^T x)$, where $\sigma(\cdot)$ is a nonlinear transformation and the filter $g_\theta = \mathrm{diag}(\theta)$ is a diagonal matrix of eigenvalues. Also, there are a number of papers on spectral-based GCNs, such as Chebyshev's spectral CNN (ChebNet) (Defferrard *et al* 2016), 1stChebNet (Kipf and Welling 2017), and the adaptive graph convolution network (AGCN) (Li *et al* 2018). However, the learned filter in spectral-based models depends on the specific structure and needs to load the whole graph into the memory for the eigenvalue decomposition. As a result, it is a challenging task in the cases of big graphs.

In the spatial-based models, the way to extract the spatial features in the graph is to search for the neighbors of a specific node and aggregate feature information from neighbors, which is more similar to the convolutional kernel in CNNs. Next, let us introduce the main idea of spatial-based GCN with an example in figure 3.39.

The process of extracting features of the GCN is divided into three steps:

1. *Send*: Each node in the graph sends its feature information to the neighboring nodes after a feature transform process.

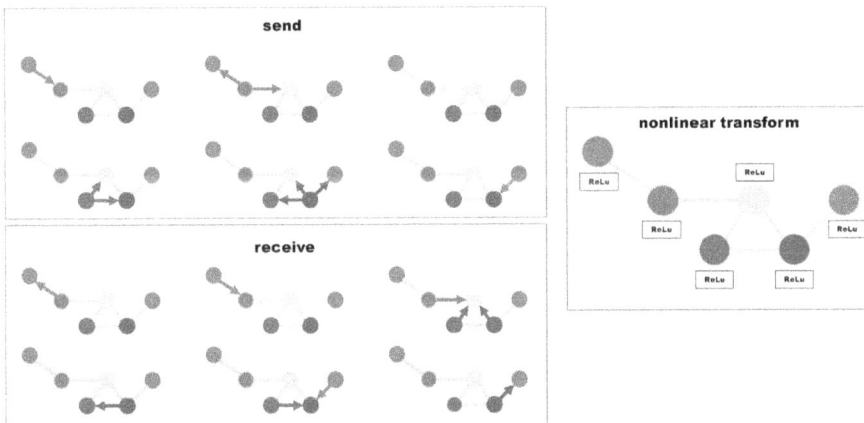

Figure 3.39. Exemplary hidden layer in a spatial-based GCN.

2. *Receive*: Each node aggregates the feature information from neighboring nodes. It is an information fusion step sensing the local structural information.
3. *Nonlinear transform*: Perform a nonlinear transform just like the nonlinear activation used in conventional neural networks.

It is noted that the receptive field on a graph (the same as the size of convolutional kernel in a regular CNN) can be expanded over more neighbors, but in this case we just consider the first-order neighbors for simplicity.

Mathematically, given a graph $G = (\mathbf{V}, \mathbf{E}, \mathbf{A})$, the input of the current l-th hidden layer consists of an $N \times M$ feature matrix $\mathbf{X}$ (N denotes the number of nodes and M denotes the input features for each node) and the adjacency matrix $\mathbf{A}$ of the graph. Thus, the graph convolutional layer-wise propagation rule can be written as follows:

$$H^{(l+1)} = f(H^{(l)}, A),$$ (3.51)

where $H^{(0)} = \mathbf{X}$. Each layer $H^{(l)}$ corresponds to an $N^l \times K^l$ feature matrix in which each row is a feature representation of a node. At each layer, these features are aggregated to form the features for the next layer using the propagation rule $f(\cdot)$. In this way, features become increasingly more abstract layer by layer. In this spatial-based framework, GCNs differ only in the choice of the propagation rule $f(\cdot)$.

For example, a very simple form of a layer-wise propagation rule can be written as

$$f(H^{(l)}, A) = \sigma(A H^{(l)} W^{(l)}),$$ (3.52)

where $W^{(l)}$ is the weight matrix with a size of $K^l \times K^{l+1}$ for the l-th layer, and $\sigma(\cdot)$ is an nonlinear activation function (such as the ReLU function).

In practice, a symmetric normalized adjacency matrix, $D^{-\frac{1}{2}} A D^{-\frac{1}{2}}$, is used to normalize the feature vectors. Then, we have a scaled propagation rule as

$$f(H^{(l)}, A) = \sigma(\hat{D}^{-\frac{1}{2}} \hat{A} \hat{D}^{-\frac{1}{2}} H^{(l)} W^{(l)})$$ (3.53)

with $\hat{A} = A + I$ to take each node itself into account in addition to its neighbors, with $\hat{D}$ being the degree matrix of $\hat{A}$.

The operation on a GCN is similar to a filtering operation on a CNN but with basic elements defined on the graph. GCN is a generalization of CNN. By extending deep learning to graph data, there are multiple methods already published, such as the message passing neural network (MPNN) (Gilmer *et al* 2017), GraphSage (Hamilton *et al* 2017), diffusion convolution neural network (DCNN) (Atwood and Towsley 2016), PATCHY-SAN (Niepert *et al* 2016), and other networks. This is currently a very active area, since a wide array of diverse knowledge types can be expressed as knowledge graphs, and processed using GCNs, or general graph networks. Interested readers should see Zhou *et al* (2018a) and Wu *et al* (2019) for more information.

References

Arjovsky M, Chintala S and Bottou L 2017 *Wasserstein GAN* ArXiv: abs/1701.07875

Atwood J and Towsley D 2016 Diffusion-convolutional neural networks *Advances in Neural Information Processing Systems (NIPS 2016)*

Bell A J and Sejnowski T J 1997 The 'independent components' of natural scenes are edge filters *Vis. Res.* **37** 3327–38

Berthelot D, Schumm T and Metz L 2017 *BEGAN: Boundary Equilibrium Generative Adversarial Networks* ArXiv: abs/1703.10717

Bruna J, Zaremba W, Szlam A and Lecun Y 2014 Spectral networks and locally connected networks on graphs *Int. Conf. on Learning Representations*

Chai T and Draxler R 2014 Root mean square error (RMSE) or mean absolute error (MAE)? *Geosci. Model Dev.* **7** 1247–50

Clevert D-A, Unterthiner T and Hochreiter S 2015 Fast and accurate deep network learning by exponential linear units (ELUS) *Comput. Sci.* arXiv: abs/1511.07289

De Boer T P, Kroese P D, Mannor S and Rubinstein Y R 2002 A tutorial on the cross-entropy method *Ann. Op. Res.* **134** 19–67

De Myttenaere A, Golden B, Le Grand B and Rossi F 2016 Mean absolute percentage error for regression models *Neurocomputing* **192** 38–48

Defferrard M, Bresson X and Vandergheynst P 2016 Convolutional neural networks on graphs with fast localized spectral filtering *Neural Inf. Process. Syst. (NIPS)* arXiv: abs/1606.09375

Doersch C 2016 Tutorial on variational autoencoders arXiv: abs/1606.05908

Fan E 2000 Extended tanh-function method and its applications to nonlinear equations *Phys. Lett.* A **277** 212–18

Fan F, Cong W and Wang G 2017a General backpropagation algorithm for training second-order neural networks *Int. J. Numer. Meth. Biomed. Eng.* **34** e2956

Fan F, Cong W and Wang G 2017b A new type of neurons for machine learning *Int. J. Numer. Meth. Biomed. Eng.* **34** e2920

Frans K 2016 Variational autoencoders explained *Kevin Frans Website v3* http://kvfrans.com/variational-autoencoders-explained/

Fukushima K 1980 Neocognitron: a self-organizing neural network model for a mechanism of pattern recognition unaffected by shift in position *Biol. Cybern.* **36** 193–202

Gilmer J, Schoenholz S S, Riley P, Vinyals O and Dahl G E 2017 Neural message passing for quantum chemistry *Int. Conf. on Machine Learning* pp 1263–72

Goodfellow I, Pouget-Abadie J, Mirza M, Xu B, Warde-Farley D, Ozair S, Courville A and Bengio Y 2014 Generative adversarial networks arXiv: abs/1406.2661

Graves A, Mohamed A-r and Hinton G 2013 Speech recognition with deep recurrent neural networks 2013 *IEEE Int. Conf. on Acoustics, Speech and Signal Processing* https://ieeexplore.ieee.org/document/6638947

Hamilton W L, Ying Z and Leskovec J 2017 Inductive representation learning on large graphs *NIPS'17 Proc. of the 31st Int. Conf. on Neural Information Processing Systems* pp 1025–35

Han J and Moraga C 1995 The influence of the sigmoid function parameters on the speed of backpropagation learning *From Natural to Artificial Neural Computation* (Berlin: Springer), pp 195–201

Hawkins D M 2004 The problem of overfitting *J. Chem. Inf. Comput. Sci.* **44** 1–12

He K M, Zhang X Y, Ren S Q and Sun J 2016 Deep residual learning for image recognition *2016 IEEE Conf. on Computer Vision and Pattern Recognition (CVPR)* pp 770–8

Hochreiter S and Schmidhuber J 1997 Long short-term memory *Neural Comput.* **9** 1735–80

Huang G, Liu Z and Weinberger K 2016a Densely connected convolutional networks *2017 IEEE Conf. on Computer Vision and Pattern Recognition (CVPR)* 2261–69

Huang G, Sun Y, Liu Z, Sedra D and Weinberger K Q 2016b Deep networks with stochastic depth *Computer Vision—ECCV 2016 Pt. 4* vol 9908 pp 646–61

Hubel D H and Wiesel T N 1968 Receptive fields and functional architecture of monkey striate cortex *J. Physiol.* **195** 215–43

Kipf T and Welling M 2017 Semi-supervised classification with graph convolutional networks *Int. Conf. on Learning Representations (ICLR)* arXiv: abs/1609.02907

Kohavi R 2001 A study of cross-validation and bootstrap for accuracy estimation and model selection *Proc. of the 14th Int. Joint Conf. on Artificial intelligence* vol 2 pp 1137–43

Lecun Y, Bottou L, Bengio Y and Haffner P 1998 Gradient-based learning applied to document recognition *Proc. IEEE* **86** 2278–324

Ledig C, Theis L, Huszar F, Caballero J, Aitken A, Tejani A, Totz J, Wang Z and Shi W 2016 Photo-realistic single image super-resolution using a generative adversarial network *IEEE Conf. on Computer Vision and Pattern Recognition (CVPR)* 105–14

Li R, Wang S, Zhu F and Huang J 2018 Adaptive graph convolutional neural networks *National Conf. on Artificial Intelligence* pp 3546–53

Maas A L and Hannun A Y 2013 Rectifier nonlinearities improve neural network acoustic models *Proc. of the 30th Int. Conf. on Machine Learning (ICML) Workshop on Deep Learning for Audio, Speech and Language Processing* vol 30, No 1 p 3

Makhzani A and Frey B 2013 k-sparse autoencoders *Int. Conf. on Learning Representations (ICLR)* arXiv: abs/1312.5663

Mao X D, Li Q, Xie H R, Lau R Y K, Wang Z and Smolley S P 2017 Least squares generative adversarial networks *2017 IEEE Int. Conf. on Computer Vision (ICCV)* pp 2813–21

Mehta R and Sivaswamy J 2017 M-net: A convolutional neural network for deep brain structure segmentation *IEEE 14th Int. Symp. on Biomedical Imaging (ISBI)* pp 437–40

Milletari F, Navab N and Ahmadi S-A 2016 V-Net: Fully convolutional neural networks for volumetric medical image segmentation *Fourth Int. Conf. on 3D Vision (3DV)* pp 565–71

Molchanov P, Tyree S, Karras T, Aila T and Kautz J 2016 Pruning convolutional neural networks for resource efficient transfer learning arXiv: abs/1611.06440

Niepert M, Ahmed M O and Kutzkov K 2016 Learning convolutional neural networks for graphs *Int. Conf. on Machine Learning* pp 2014–23

Olshausen B A and Field D J 1996 Emergence of simple-cell receptive field properties by learning a sparse code for natural images *Nature* **381** 607–9

Prechelt L 2012 *Neural Networks: Tricks of the Trade* 2nd edn ed G Montavon *et al* (Berlin: Springer) pp 53–67

Rifai S, Vincent P, Muller X, Glorot X and Bengio Y 2011 Contractive auto-encoders: explicit invariance during feature extraction *Proc. of the 28th Int. Conf. on Machine Learning* pp 833–40

Ronneberger O, Fischer P and Brox T 2015 U-net convolutional networks for biomedical image segmentation *Proc. of Medical Image Computing and Computer-Assisted Intervention - MICCAI* pp 234–41

Rumelhart D E, Hinton G E and Williams J R 1986 Learning representations by back-propagating errors *Nature* **323** 533–6

Schuster M and Paliwal K 1997 Bidirectional recurrent neural networks *IEEE Trans. Signal Proc.* **45** 2678–81

Simonyan K and Zisserman A 2014 Very deep convolutional networks for large-scale image recognition *Int. Conf. on Learning Representations (ICLR)* arXiv: abs/1409.1556

Srivastava N, Hinton G, Krizhevsky A, Sutskever I and Salakhutdinov R 2014 Dropout: a simple way to prevent neural networks from overfitting *J. Machine Learn. Res.* **15** 1929–58

Steinkraus D, Buck I and Simard P 2005 Using GPUs for machine learning algorithms *Eighth Int. Conf. on Document Analysis and Recognition (ICDAR'05)* **2** 1115–20

Vincent P, Larochelle H, Lajoie I, Bengio Y and Manzagol P A 2010a Stacked denoising autoencoders: learning useful representations in a deep network with a local denoising criterion *J. Machine Learn. Res.* **11** 3371–408

Vincent P, Larochelle H, Lajoie I, Bengio Y and Manzagol P-A 2010b Stacked denoising autoencoders: learning useful representations in a deep network with a local denoising criterion *J. Machine Learn. Res.* **11** 3371–408

Waibel A, Hanazawa T, Hinton E G, Shikano K and Lang K J 1989 Phoneme recognition using time-delay neural networks *IEEE Trans. Acoust. Speech Signal Process.* **37** 328–39

Wu Z, Pan S, Chen F, Long G, Zhang C and Yu P S 2019 A comprehensive survey on graph neural networks, arXiv:1901.00596

Xia X and Kulis B 2017 *W-Net:* A deep model for fully unsupervised image segmentation arXiv: abs/1711.08506

Xie S N, Girshick R, Dollar P, Tu Z W and He K M 2017 Aggregated residual transformations for deep neural networks *30th IEEE Conf. on Computer Vision and Pattern Recognition (CVPR 2017)* pp 5987–95

Yamins D L K, Hong H, Cadieu C F, Solomon E A, Seibert D and DiCarlo J J 2014 Performance-optimized hierarchical models predict neural responses in higher visual cortex *Proc. Natl Acad. Sci USA* **111** 8619–24

Yang Y, Gong Z, Ping Z and Shan J 2017 Unsupervised representation learning with deep convolutional neural network for remote sensing images *Proc. of Int. Conf. on Image Graphics* pp 97–108

Yuan X, Huang B, Wang Y, Yang C and Gui W 2018 Deep learning-based feature representation and its application for soft sensor modeling with variable-wise weighted SAE *IEEE Trans. Ind. Inform.* 1

Zeiler M D and Fergus R 2013 Stochastic pooling for regularization of deep convolutional neural networks *Int. Conf. on Learning Representations (ICLR)* arXiv: abs/1301.3557

Zhou J, Cui G, Zhang Z, Yang C, Liu Z and Sun M 2018a Graph neural networks: a review of methods and applications, arXiv:1812.08434

Zhou Z, Rahman Siddiquee M M, Tajbakhsh N and Liang J 2018b UNet++: A nested U-net architecture for medical image segmentation *Proc. of Deep Learning in Medical Image Analysis and Multimodal Learning (DLMIA)* pp 3–11

Part II

X-ray computed tomography

IOP Publishing

Machine Learning for Tomographic Imaging

Ge Wang, Yi Zhang, Xiaojing Ye and Xuanqin Mou

Chapter 4

X-ray computed tomography

Tomography originates from the Greek word 'tomos', which means a section, a slice, or a cut. Tomography is a technology for imaging a cross-section through an object. After x-rays were discovered in 1895, x-ray imaging was immediately applied in medical imaging, and then computed tomography (CT) was invented, which not only revolutionized medical diagnosis but also generated huge effects on other fields, such as industrial inspection and security screening. Inspired by x-ray CT, gamma ray CT (also known as emission tomography or nuclear tomography) was developed, and then other tomographic imaging modalities emerged, including magnetic resonance imaging (MRI), ultrasound, and optical tomography in various modes. Now, CT has become an indispensable part of medicine. In this chapter, we will introduce the imaging principles and typical methods of CT. Of course, since the focus of this book is machine learning in the tomography imaging context, we will only introduce some basic concepts and algorithms. More detailed works are available for further study (Natterer 2001, Kak *et al* 2002, Barrett and Swindell 1996, Zeng 2010, Bushberg and Boone 2011). To obtain a bigger picture, the history and applications of CT will also be mentioned for interested readers. As a warm-up, we recommend you watch the TED-Ed lesson on x-ray imaging, https://ed.ted.com/lessons/how-x-rays-see-through-your-skin-ge-wang, by Wang *et al*.

4.1 X-ray data acquisition

4.1.1 Projection

When x-rays are in contact with matter, the photoelectric, Compton, and electron pair effects all occur. The intensity of x-rays, which is positively correlated with the photon numbers, will be attenuated by the absorption of matter caused by the three effects. Since different materials have different linear coefficients, we can image an object based on linear attenuation data.

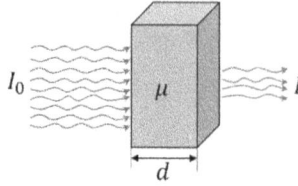

Figure 4.1. Schematic diagram of x-rays passing through a homogeneous substance.

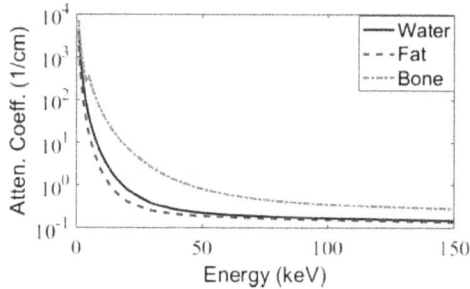

Figure 4.2. Linear attenuation coefficients of different substances.

As shown in figure 4.1, when an x-ray beam of intensity I_0 passes through a uniformly distributed object, the attenuation can be formulated using Lambert Beer's law as

$$I = I_0 e^{-\mu d}, \tag{4.1}$$

where d is the length of the ray passing through the object and μ represents the linear attenuation coefficient of the object. I denotes the intensity of the ray departing the object, which can be measured by an x-ray detector, while I_0 can be obtained by an air scanner (the attenuation in air is nearly 0). μ is determined by the property of the object and the energy of the x-ray. If we use Z and ρ respectively to denote the equivalent atomic number and density of the matter, and E denotes the energy of the x-ray, then the attenuation coefficient can be parameterized to $\mu(E, Z, \rho)$. Figure 4.2 displays attenuation coefficient curves for various materials at different x-ray energies (Hubbell and Seltzer 1995).

In equation (4.1), I, I_0, and d can all be measured, therefore the linear attenuation coefficient can be obtained by $\mu = \ln(I_0/I)/d$. The attenuation coefficient, which reflects the physical properties of different substances, can be correlated with the pixel value of the CT image.

When an x-ray passes through an object consisting of multiple uniformly distributed substances of attenuation coefficients $\mu_1, \mu_2, \cdots, \mu_n$ and of thickness $d_1, d_2, \cdots, d_n$ as shown in figure 4.3, the intensity of the x-ray can be formulated as

$$I = I_0 e^{-(\mu_1 d_1 + \mu_2 d_2 + \cdots + \mu_n d_n)}. \tag{4.2}$$

Then, we can obtain a multivariate equation as follows:

$$\sum_{i=1}^{n} u_i d_i = \ln\left(\frac{I_0}{I}\right). \tag{4.3}$$

With data for multiple x-rays, attenuation coefficients can be calculated.

In practice, the substances in an object are not evenly distributed. Let $d_i \rightarrow 0$, then equation (4.2) is converted to an integral form. The intensity variation along the x-ray path L is as follows:

$$I = I_0 e^{-\int_L \mu \mathrm{d}l}. \tag{4.4}$$

The projection along the x-ray path L is defined as the line integral of the attenuation coefficients of materials along L, defined as

$$p = \int_L \mu \mathrm{d}l = \ln\left(\frac{I_0}{I}\right). \tag{4.5}$$

A simple example is used here to illustrate the projection. As shown in figure 4.4, an object is represented as a 2×2 array, where $x_j (j = 1, 2, 3, 4)$ denotes the linear attenuation coefficient value of the jth pixel. Four parallel rays pass through the object and four projections are measured. $p_i (i = 1, 2, 3, 4)$ denotes the projection of the ith ray path. a_{ij} represents the length of the ith ray passing through the jth pixel. Then, each projection can be calculated as

$$p_i = \sum_{j=1}^{4} a_{ij} x_j, \tag{4.6}$$

$$\begin{cases} p_1 = a_{11}x_1 \\ p_2 = a_{23}x_3 + a_{21}x_1 + a_{22}x_2 \\ p_3 = a_{33}x_3 + a_{34}x_4 + a_{32}x_2 \\ p_4 = a_{44}x_4 \end{cases} \tag{4.7}$$

Another example of continuous integration is shown in figure 4.5, where there is a disc with uniform linear density ρ on the x–y plane. The center of the disc is at the origin. The projection of this object can be calculated as the chord length t times the linear density ρ. That is,

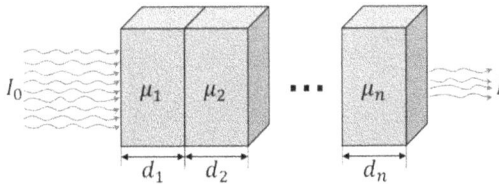

Figure 4.3. Schematic diagram of x-rays passing through multiple homogeneous substances.

$$p(s) = \begin{cases} \rho t = 2\rho\sqrt{R^2 - s^2}, & |s| < R \\ 0, & |s| \geqslant R \end{cases}. \tag{4.8}$$

The projections in the above examples are all shown in the same view angle. In practice, the projections are at different view angles. For simplicity, we use a point source to describe this. As shown in figure 4.6, there is a point source on the x-axis with a distance r from the origin. We set the projection value to a pulse. At the angle

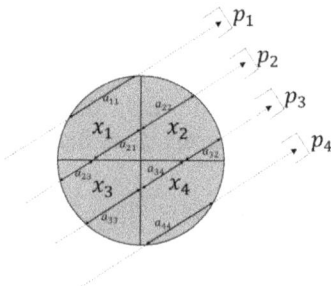

Figure 4.4. A discrete projection on a 2×2 array.

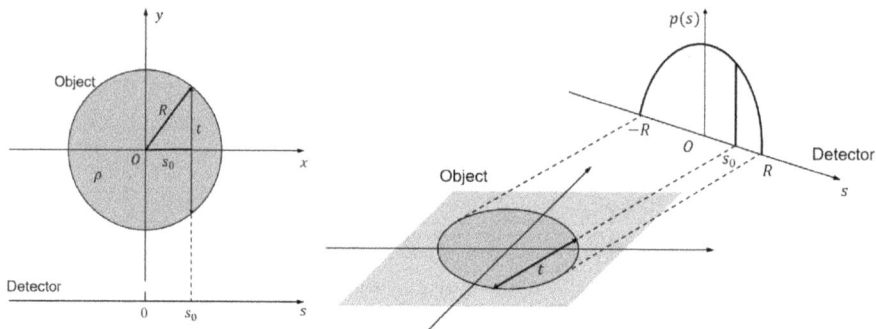

Figure 4.5. A continuous projection of a disc.

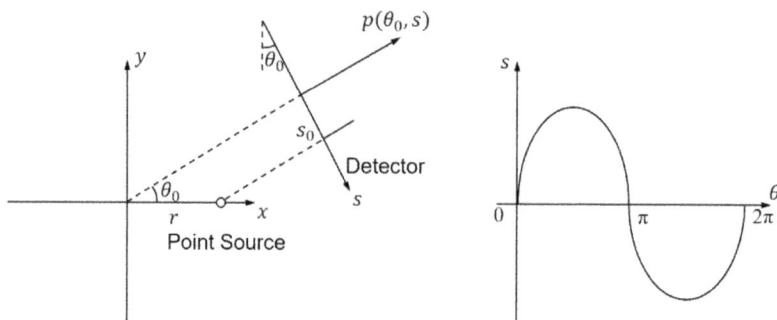

Figure 4.6. Projection of a point source at different view angles.

θ_0, the pulse projection appears at a position s_0 away from the origin. We can observe that there is a relationship between θ_0 and s_0 as follows:

$$s_0 = r \sin \theta_0. \tag{4.9}$$

With the variation of angle θ, the variation curve of the pulse projection position s becomes a sine wave (see figure 4.6, right). Based on this observation, the projection data are also called a sinogram. Figure 4.7 shows sinograms with 360 view angles using a point source and image, respectively. For different points, the obtained sinograms have differences in amplitude and phase. The distance between the point and origin determines the amplitude. And if a point is located on the negative x-axis, there will be a half cycle phase shift.

Next we have the Radon transform, which plays an important role in the progress of CT imaging. Before that, we need to introduce the Dirac δ-function (Thaller 2013).

The Dirac δ-function is a generalized function or a distribution function. It has some useful properties:

1. $\delta(x) = 0, \quad$ when $x \neq 0$,

2. $\int_{-\infty}^{\infty} \delta(x) \mathrm{d}x = 1$,

3. $\int_{-\infty}^{\infty} f(x)\delta(x)\mathrm{d}x = f(0)$,

4. $\delta(x - x_0) = 0, \quad$ when $x \neq x_0$,

5. $\int_{-\infty}^{\infty} f(x)\delta(x - x_0)\mathrm{d}x = f(x_0)$.

For 2D space, $\delta(x, y)$ has the property

$$\int_{-\infty}^{\infty}\int_{-\infty}^{\infty} f(x, y)\delta(x, y)\mathrm{d}x\mathrm{d}y = f(0, 0).$$

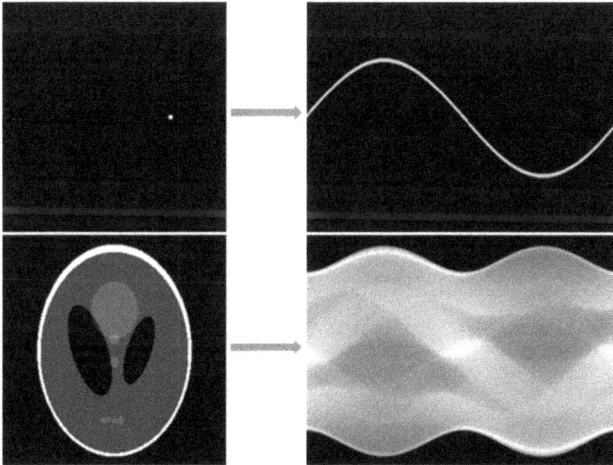

Figure 4.7. A sinogram of a point source and an image.

If this is difficult to understand here, simply treat the δ-function as a sampling function. $\int_{-\infty}^{\infty} \int_{-\infty}^{\infty} f(x, y)\delta(x, y)\mathrm{d}x\mathrm{d}y$ is the display expression of sampling the point $f(0,0)$. If we want to sample the point $f(x_0, y_0)$, the display expression is

$$\int_{-\infty}^{\infty} \int_{-\infty}^{\infty} f(x, y)\delta(x - x_0, y - y_0)\mathrm{d}x\mathrm{d}y = f(x_0, y_0). \tag{4.10}$$

Radon transform is a method of converting an x–y coordinate system into another coordinate system (Radon 1986). Consider an object situated in a 2D space, which has the attenuation coefficient distribution $f(x, y)$. There is an x-ray path, which can be formulated as $s_0 = x \cos \theta_0 + y \sin \theta_0$, θ_0 is the angle between the orientation of the detector and the positive orientation of the x-axis, while s_0 denotes the distance from the origin to the ray. Figure 4.8 shows the specific situation. The projection can be accumulated as the attenuation coefficients along the ray as follows:

$$p(\theta_0, s_0) = \int_{-\infty}^{\infty} \int_{-\infty}^{\infty} f(x, y)\delta(x \cos \theta_0 + y \sin \theta_0 - s_0)\mathrm{d}x\mathrm{d}y. \tag{4.11}$$

Therefore, for all x-rays determined by (θ, s), the projection data can be obtained using

$$p(\theta, s) = \int_{-\infty}^{\infty} \int_{-\infty}^{\infty} f(x, y)\delta(x \cos \theta + y \sin \theta - s)\mathrm{d}x\mathrm{d}y. \tag{4.12}$$

4.1.2 Backprojection

In the previous subsection, we have learned how the projections are obtained. In this subsection, we will introduce the methods to obtain the image. The process of obtaining a tomographic image from projection is called reconstruction.

After we obtain the projection $p(\theta, s)$, our purpose is to recover the distribution $f(x,y)$ of the object, which is imaging. This can be obtained by solving equations with

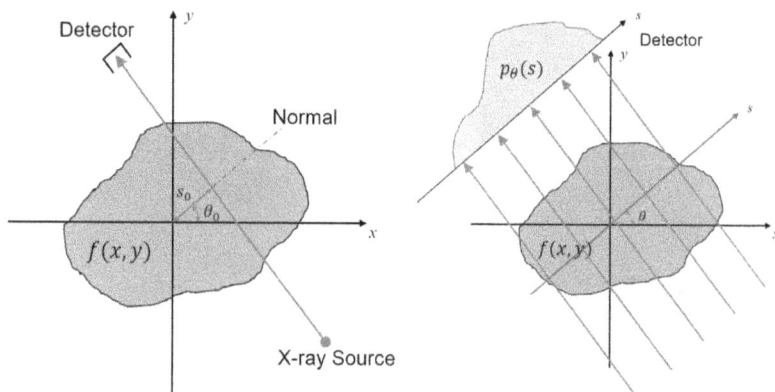

Figure 4.8. Projection under Radon transform.

sufficient projections. However, there are some problems with this approach, which will be discussed in a later section. Here, we introduce the backprojection method.

Let us use a digital example to discuss it. As shown in figure 4.9, an image is represented as a 2×2 array, and $x_i(i = 1, 2, 3, 4)$ denotes the linear attenuation coefficient value of the ith pixel. We obtain four projection data from two views. Then, we recover the object from the projection. We backproject the projections back into the image along the opposite direction of the projection direction as shown in figure 4.10.

Obviously, the result is different from the original image. So, the backprojection is not the inverse operation of projection. However, the result reflects the distribution of the original image. The result will have larger values at the position where the value of the original image is large, and vice versa. If the projection views are sufficient, the result will be more accurate. However, there is a big problem. We will discuss it using an experiment.

As shown in figure 4.11, we generate a standard Shepp–Logan phantom (Shepp and Logan 1974). Data from 10, 30, and 90 views are collected for backprojection. With an increasing number of views, the artifacts in the result gradually decrease. But inevitably, the image becomes blurred and the edges are not clear.

Actually, this is related to the optical system to some extent. To explain this, we use a point source system to describe the backprojection process. As shown in figure 4.12, assume that there is a point source in a 2D x–y coordinate system, and the detector rotates around the origin and collects projections. The measured projection data are a pulse with a height of 1. Then, we backproject these pulses along the opposite directions of the projection directions. We will obtain a high pulse in the original position. As the number of backprojected pulses becomes large, we will obtain a cone-like image like that in figure 4.12(d). This is because the pulses cannot be accurately concentrated at one point during backprojection. Actually, we will obtain a blurry image of the pulse. The closer the spatial location is to the center, the higher the value at that point. The reader can think about whether something similar happened when we took a picture of a light source. We often use a point spread function (PSF) to describe this process. The phenomenon is similar in CT

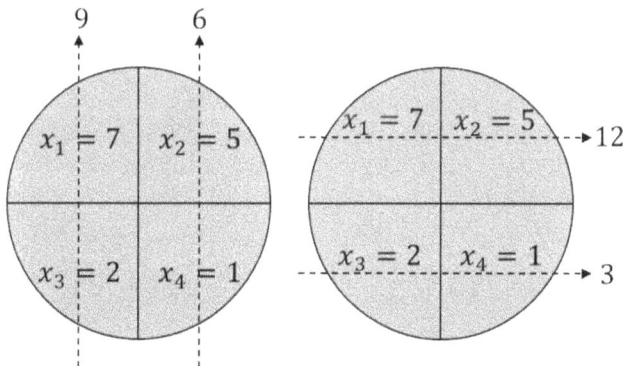

Figure 4.9. Projections of vertical and horizontal directions.

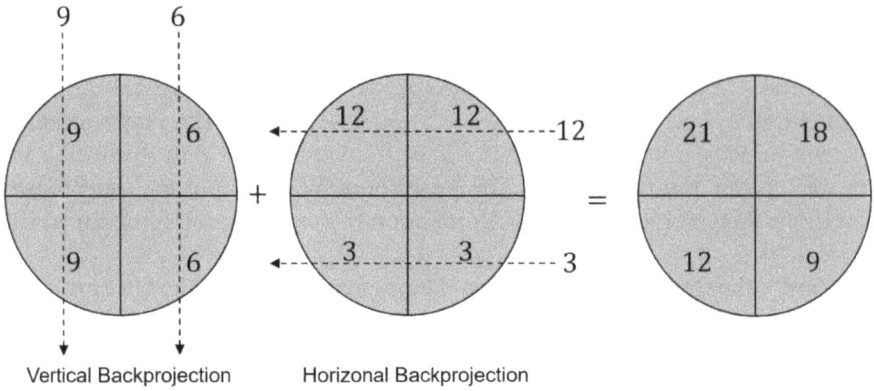

Figure 4.10. Backprojection.

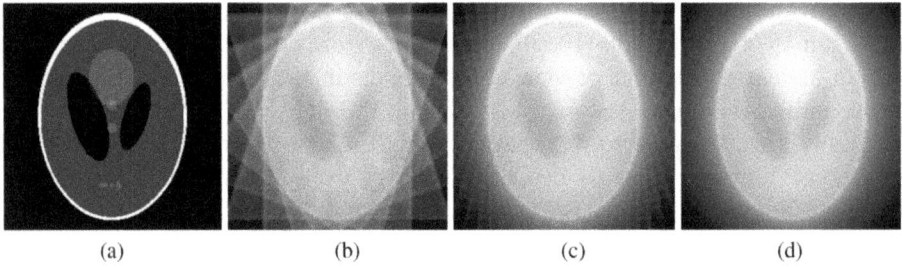

Figure 4.11. The backprojections with different numbers of projection views. (a) Original image. (b) Result with 10 views. (c) Result with 30 views. (d) Result with 90 views.

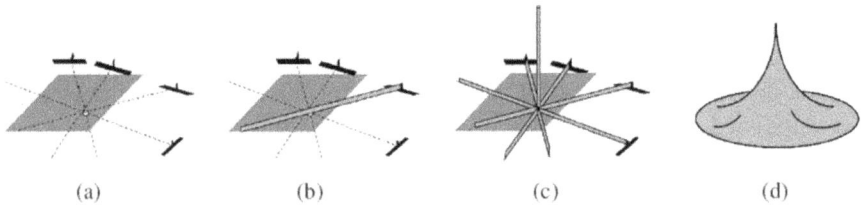

Figure 4.12. Reconstruction of a point source. (a) Projection of a point source. (b) Backprojection from one view. (c) Backprojection from a few views. (d) Backprojection from all views. Reproduced with permission from Zeng (2010). Copyright 2010 Springer.

reconstruction, in particular when only backprojection is performed, and normally we backproject hundreds of projections over a field of view, which leads to decreased contrast and blurred edges.

The backprojection can be expressed as

$$b(x, y) = \int_0^\pi p(\theta, s)|_{s=x \cos\theta + y \sin\theta} d\theta. \tag{4.13}$$

Since we cannot have an accurate reconstruction through backprojection, here we use $b(x,y)$ to represent the backprojection result.

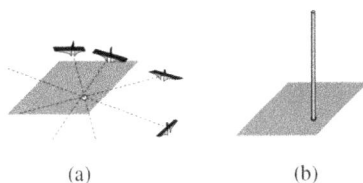

Figure 4.13. Reconstruction by backprojection with filtered data. (a) Add negative wings. (b) Backprojection of modified data. Reproduced with permission from Zeng (2010). Copyright 2010 Springer.

How can we make the backprojected blurry image clear? This is an excellent question. Heuristically, we can add negative wings around the pulse in each projection before backprojection (see figure 4.13(a)). The procedure of adding negative wings into each projection of the pulse is called filtering. The back-projection using filtered projections is called filtered backprojection (FBP) which will be introduced in detail later (see figure 4.13(b)).

4.1.3 (Back)Projector

We already know what is going on with projection and backprojection, these processes are essentially a line integral and its transpose. When we consider the issue from different perspectives, we will have different approaches to obtain the (back) projection. In this subsection, we will introduce typical methods to perform these operations. The model for such a purpose is called a (back)projector. We divide the main (back)projectors into three types, which are pixel-driven, ray-driven and distance-driven (Peters 1981, Siddon 1985, De Man and Basu 2004), respectively. In the following we will take some examples to visually illustrate them.

In practice, we have to discretize objects to be reconstructed and x-rays through them. We regard the human body as a grid or matrix. Similarly, we also divide a continuous x-ray beam into discrete rays. Figure 4.14 shows a simple pixel-driven model. We consider the (back)projection from the perspective of each pixel of the grid. Because of the finite sizes of the pixel and detector element, we define the coordinates of their centers as their positions. Along the direction from the source toward the pixel, we can find the position the ray falls on the detector. It usually lays between two detector elements, and their positions are shown in figure 4.14. We denote the pixel value as μ, the projection values of the k and $k + 1$ th detector elements as p_k and p_{k+1}, respectively. Then, we can calculate the projection components due to this pixel as follows

$$
\begin{aligned}
p_k &= \frac{t - d_k}{d_{k+1} - d_k}\mu \\
p_{k+1} &= \frac{d_{k+1} - t}{d_{k+1} - d_k}\mu.
\end{aligned}
\tag{4.14}
$$

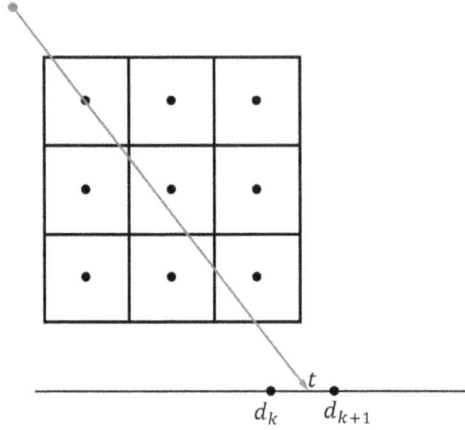

Figure 4.14. Illustration of the pixel-driven (back)projector.

We denote the backprojection value as b, it can be immediately calculated as

$$b = \frac{t - d_k}{d_{k+1} - d_k} p_k + \frac{d_{k+1} - t}{d_{k+1} - d_k} p_{k+1}. \tag{4.15}$$

The ray-driven approach is similar to the line integrals we introduced earlier. In this method, each ray is determined by the position of the source and the detector element. Then, we calculate the length of the ray inside each pixel. We take the example shown in figure 4.4, the projections has been presented in equation (4.7). Here we give the backprojection

$$
\begin{aligned}
b_1 &= a_{11}p_1 + a_{21}p_2 \\
b_2 &= a_{22}p_2 + a_{32}p_3 \\
b_3 &= a_{23}p_2 + a_{33}p_3 \\
b_4 &= a_{34}p_3 + a_{44}p_4.
\end{aligned}
\tag{4.16}
$$

The distance-driven approach is very different from the previous two methods. In the pixel or ray driven method, we take the centers of the pixel and detector element to build the rays. However, in distance-driven model, we take the boundaries of the pixel and detector element into account. As shown in the left part in figure 4.15, we map the boundaries of pixels and detector elements onto x-axis. Notice here the points on the detector and grid are all boundaries. Then, we get a set of points on the x-axis as shown on the right part in figure 4.15 where t is the position mapped from the pixel boundary, and d is the coordinate mapped from the detector element boundary, μ and p are the pixel value and detector reading, respectively. We can perform the projection as follows:

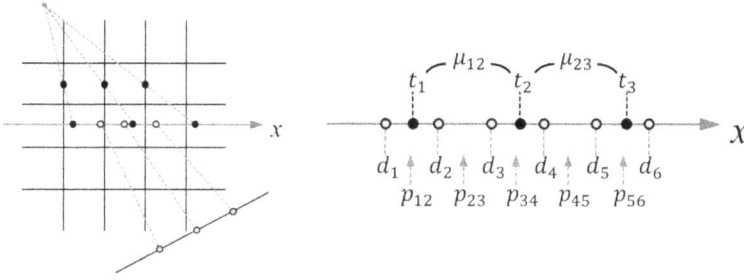

Figure 4.15. Illustration of the distance-driven (back)projector.

$$p_{12} = \frac{d_2 - t_1}{d_2 - d_1}\mu_{12}$$

$$p_{23} = \mu_{12}$$

$$p_{34} = \frac{(t_2 - d_3)\mu_{12} + (d_4 - t_2)\mu_{23}}{d_4 - d_3}$$

$$p_{45} = \mu_{23}$$

$$p_{56} = \frac{t_3 - d_5}{d_6 - d_5}\mu_{23},$$

(4.17)

and the backprojection values can be computed

$$b_{12} = \frac{(d_2 - t_1)p_{12} + (d_3 - d_2)p_{23} + (t_2 - d_3)p_{34}}{t_2 - t_1}$$

$$b_{23} = \frac{(d_4 - t_2)p_{34} + (d_5 - d_4)p_{45} + (t_3 - d_5)p_{56}}{t_3 - t_2}.$$

(4.18)

The pixel-driven method is easy to understand and simple to implement. But it is easy to introduce the artifacts during reconstruction. In contrast, the image quality reconstructed with the ray-driven model is better. However, the ray-driven method has a low usage of the cache in the parallel system, which affects the computational efficiency. The distance driven method can alleviate artifacts and is simple to implement. It is well suited for hardware implementation.

4.2 Analytical reconstruction

4.2.1 Fourier transform

We know that white light is made up of the complete spectrum. We can use a prism to decompose white light into colored light with different frequencies, as shown in figure 4.16. We can also use a prism to fuse colored light into white light. In mathematics, we also have a tool to decompose and fuse the signals, which is Fourier transform (Bracewell and Bracewell 1986).

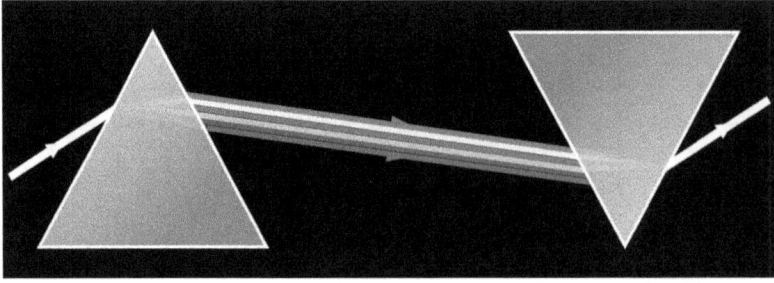

Figure 4.16. White light can be decomposed into colored light and this colored light can be fused into white light.

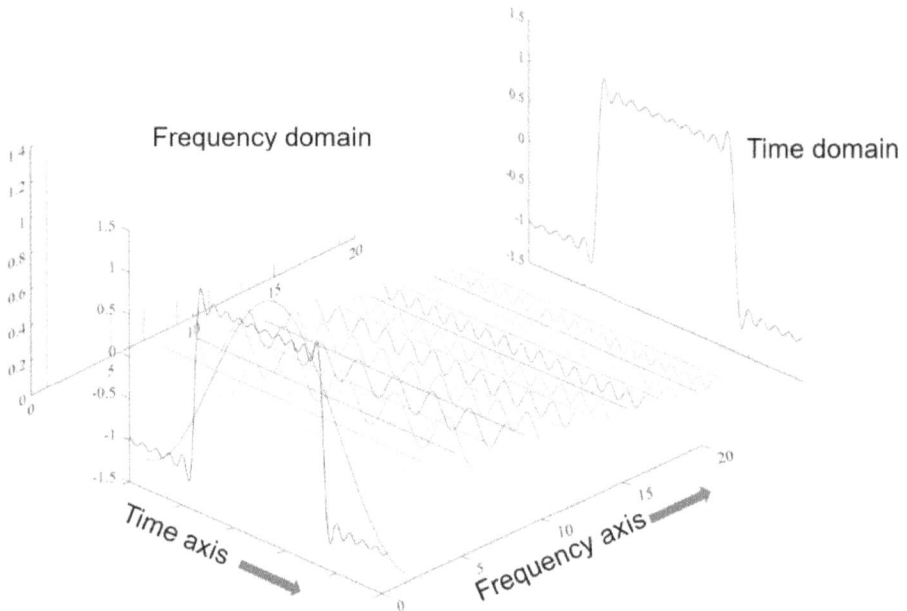

Figure 4.17. Graphic of the Fourier transform.

Fourier analysis is not only a powerful mathematical tool but also brings a new perspective on transform domain analysis. It is also the basis of CT analytical reconstruction. We will introduce the Fourier transform before discussing the analytical reconstruction in detail.

In Fourier analysis, all the functions can be decomposed into a series of sinusoidal functions. As shown in figure 4.17, we can use a series of sinusoidal functions to approximate a rectangle function. These functions have diverging frequencies and can be seen as the frequency components of the original function. The coefficient of each sinusoidal function represents the amplitude and also measures the energy of frequency components. As the number of sinusoidal functions increases, the rectangle will be more accurate. Then, with a Fourier transform, we can analyze the original function in the frequency domain. Now we provide the formulation.

For a signal in the time domain $f(t)$, we can obtain its amplitude in the frequency domain using the Fourier transform

$$F(\omega) = \int_{-\infty}^{\infty} f(t)e^{-j2\pi\omega t}dt, \tag{4.19}$$

where $F(\omega)$ is the amplitude of the sinusoidal function with frequency ω, which is usually a complex number. Its real part determines the cosine function component, while the imaginary part affects the sine function component. We can also restore the original function perfectly from the frequency domain by the inverse transform

$$f(t) = \int_{-\infty}^{\infty} F(\omega)e^{j2\pi\omega t}d\omega. \tag{4.20}$$

For a 2D signal, we can also use the Fourier transform

$$F(\mu, v) = \int_{-\infty}^{\infty}\int_{-\infty}^{\infty} f(x, y)e^{-j2\pi(\mu x + vy)}dxdy, \tag{4.21}$$

and its inverse transform

$$f(x, y) = \int_{-\infty}^{\infty}\int_{-\infty}^{\infty} F(\mu, v)e^{j2\pi(\mu x + vy)}d\mu dv. \tag{4.22}$$

For an image signal, the Fourier transform can extract its frequency information. For images, the frequency corresponds to the gradient of the pixel values. In the edge of large color blocks, the pixel values change rapidly, reflecting high-frequency information. In the color block area, the pixel values are smoothly transitioned, reflecting low-frequency information. However, the signal is discrete in the image, so we need the discrete Fourier transform. For an image of size $M \times N$, the discrete Fourier transform is defined as

$$F(\mu, v) = \sum_{x=0}^{M-1}\sum_{y=0}^{N-1} f(x, y)e^{-j2\pi(\mu x/M + vy/N)}, \tag{4.23}$$

and its inverse transform is defined as

$$f(x, y) = \frac{1}{MN}\sum_{\mu=0}^{M-1}\sum_{v=0}^{N-1} F(\mu, v)e^{j2\pi(\mu x/M + vy/N)}. \tag{4.24}$$

Frequency analysis is a powerful tool in image processing. We use the image 'Lena' as an example. Figure 4.18(a) displays the original image, and we can use the discrete Fourier transform to obtain the frequency spectrum. The center is the zero-frequency, and with the expansion, the frequency is getting higher and higher. It can be seen that most of the energy is concentrated in the low-frequency region. We can use the inverse transform to obtain a lossless image. If we set the spectrum value of the middle part to 0, the inverse transformed image will lose the color block information, leaving only the edge structure, as shown in figure 4.18(b). This operation is called high-pass filtering. If we perform the opposite operation, the

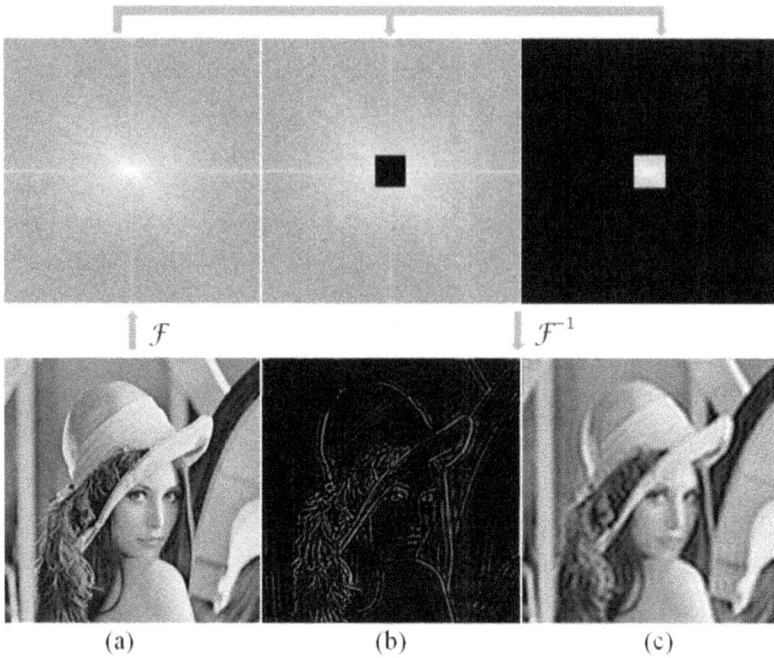

Figure 4.18. Comparison of the recovered results for low-frequency and high-frequency data loss. Lena image © Playboy Enterprises, Inc.

outlines of the inverse transformed image will become blurred, as shown in figure 4.18(c). This operation is called low-pass filtering.

4.2.2 Central slice theorem

Central slice theorem is the foundation of analytic reconstruction. Its mathematical form is cumbersome and difficult to understand, so let us illustrate it first.

As shown in figure 4.19, the projection data $p_\theta(s)$ can be regarded as a 1D signal and its spectrum $P_\theta(\omega)$ can be obtained by Fourier transform. Now suppose we have imaged the object and obtain $f(x,y)$. We can use the 2D Fourier transform to obtain its spectrum $F(\mu, v)$. Here is the most amazing moment: if we take the central slice on the 2D spectrum along the angle θ, we will find that the profile is the same as $P_\theta(\omega)$.

If we rotate the detector around the object by at least 180°, all the central slices in the 2D Fourier transform $F(\mu, v)$ will cover the whole 2D Fourier space. Then, the original function can be obtained by inverse Fourier transform. This is the basic idea of the theory. Let us take a look at the mathematical expressions.

First, the projection at angle θ can be obtained

$$p_\theta(s) \int_{-\infty}^{\infty} \int_{-\infty}^{\infty} f(x, y)\delta(x\cos\theta + y\sin\theta - s)\mathrm{d}x\mathrm{d}y, \tag{4.25}$$

and its Fourier transform is

$$P_\theta(\omega) = \int_{-\infty}^{\infty} p_\theta(s)\mathrm{e}^{-j2\pi\omega s}\mathrm{d}s. \tag{4.26}$$

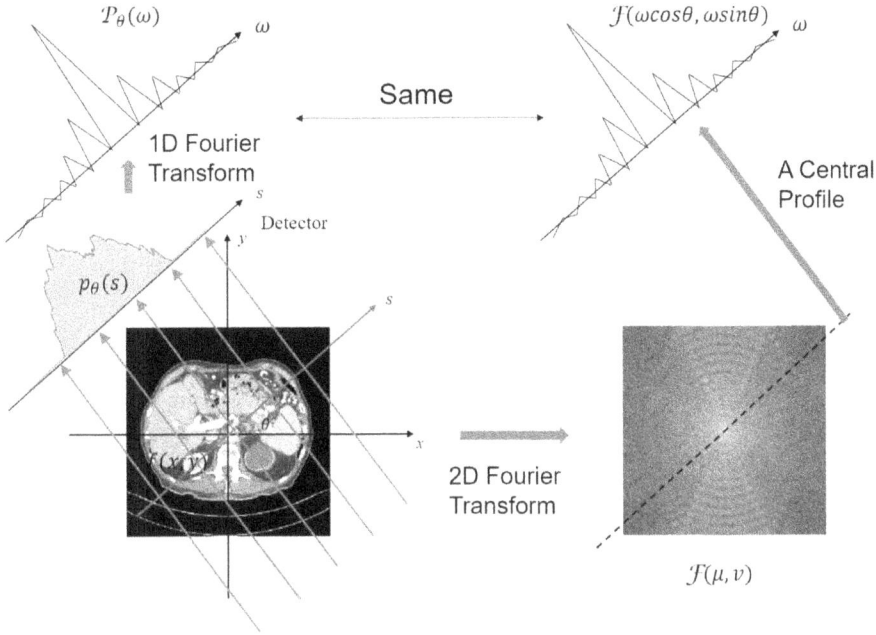

Figure 4.19. Illustration of the central slice theorem.

The 2D Fourier transform of $f(x,y)$ is

$$F(\mu, v) = \int_{-\infty}^{\infty} \int_{-\infty}^{\infty} f(x, y)e^{-j2\pi(\mu x + vy)}dxdy. \qquad (4.27)$$

With the 2D Fourier transform,

$$P_\theta(\omega) = F(\mu, v)|_{\mu=\omega\cos\theta, v=\omega\sin\theta}. \qquad (4.28)$$

The proof of the above formulations is given as follows (if you do not want to spend more time on the formula, you can skip the following and it will not affect your reading):

$$P_\theta(\omega) = \int_{-\infty}^{\infty} p_\theta(s)e^{-j2\pi\omega s}ds$$

$$= \int_{-\infty}^{\infty}\left[\int_{-\infty}^{\infty}\int_{-\infty}^{\infty} f(x, y)\delta(x\cos\theta + y\sin\theta - s)dxdy\right]e^{-j2\pi\omega s}ds$$

$$= \int_{-\infty}^{\infty}\int_{-\infty}^{\infty}\left[\int_{-\infty}^{\infty} f(x, y)\delta(x\cos\theta + y\sin\theta - s)e^{-j2\pi\omega s}ds\right]dxdy$$

$$= \int_{-\infty}^{\infty}\int_{-\infty}^{\infty} f(x, y)e^{-j2\pi\omega(x\cos\theta + y\sin\theta)}dxdy$$

$$= \int_{-\infty}^{\infty}\int_{-\infty}^{\infty} f(x, y)e^{-j2\pi(x\omega\cos\theta + y\omega\sin\theta)}dxdy$$

$$= F(\omega\cos\theta, \omega\sin\theta)$$

$$= F(\mu, v)|_{\mu=\omega\cos\theta, v=\omega\sin\theta}.$$

4.2.3 Parallel-beam image reconstruction

From the previous subsection, we already know the central slice theorem, which indicates that the 1D Fourier transform of each angular projection is equal to the corresponding slice of the 2D Fourier transform of the image. If we rotate around the object and obtain all the slices, we can obtain the complete 2D Fourier transform of the image. An intuitive way of reconstruction is to use this principle. We place the 1D Fourier transform results on the 2D plane. We put these slices into the pixel grid by interpolation and obtain the complete 2D Fourier transform data. Then, the inverse transform can help us to obtain the perfect image. However, interpolation will have problems. For example, a point far from the center will have sparser neighboring data than in the center area, resulting in a larger interpolation error, which will be reflected in the high-frequency part. The result of the inverse transform will lose some details.

The filtered backprojection (FBP) algorithm is now commonly used to reduce interpolation errors. Imagine that when we put the slices back into their original position without filtration the center will be superimposed multiple times. We know the center corresponds to the low-frequency component, and when the low frequency is over-weighted, the image will become blurred. It is consistent with the result of backprojection.

In order to deblur, we need to rectify the spectrum. This operation is called filtering. We need a filter to decrease the amplitude of the low frequency. We multiply the 1D Fourier transform of projection by the 1D filter $H(\omega)$ and obtain the rectified result. The inverse transform is used to obtain the corrected projection, and then the backprojection is performed to obtain the image. Let us use formulas to describe this process. First, use a 1D filter to correct the spectrum:

$$Q_\theta(\omega) = P_\theta(\omega) \cdot H(\omega). \tag{4.29}$$

In the correction, the theoretically derived filter is a ramp filter:

$$H(\omega) = |\omega|. \tag{4.30}$$

Then, the inverse transform is imposed to obtain the corrected projection:

$$q_\theta(s) = \int_{-\infty}^{\infty} Q_\theta(\omega) e^{j2\pi\omega s} d\omega. \tag{4.31}$$

Finally, we use the backprojection to obtain the image:

$$f(x, y) = \int_0^\pi q_\theta(s)|_{s=x\cos\theta+y\sin\theta} d\theta. \tag{4.32}$$

The whole reconstruction process is shown in figure 4.20.

In fact, the filtering operation can be imposed not only in the frequency domain but also in the image domain. According to the Fourier transform theorem, multiplication in the frequency domain corresponds to convolution in the image domain, which can be simply expressed as

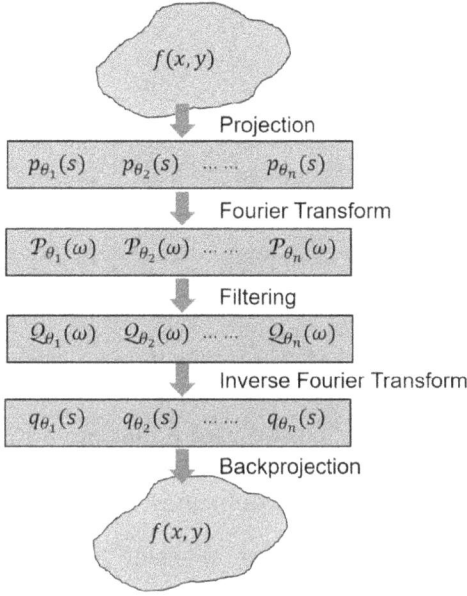

$f(x,y)$

↓ Projection

$p_{\theta_1}(s) \quad p_{\theta_2}(s) \quad \cdots\cdots \quad p_{\theta_n}(s)$

↓ Fourier Transform

$P_{\theta_1}(\omega) \quad P_{\theta_2}(\omega) \quad \cdots\cdots \quad P_{\theta_n}(\omega)$

↓ Filtering

$Q_{\theta_1}(\omega) \quad Q_{\theta_2}(\omega) \quad \cdots\cdots \quad Q_{\theta_n}(\omega)$

↓ Inverse Fourier Transform

$q_{\theta_1}(s) \quad q_{\theta_2}(s) \quad \cdots\cdots \quad q_{\theta_n}(s)$

↓ Backprojection

$f(x,y)$

Figure 4.20. Flow chart of the FBP algorithm.

$$P_\theta(\omega) \cdot H(\omega) \overset{\mathcal{F} \text{ and } \mathcal{F}^{-1}}{\Longleftrightarrow} p_\theta(s) * h(s). \tag{4.33}$$

So, we can use the inverse Fourier transform to calculate the image domain form of the ramp filter, and then use the convolution operation to perform the filtering. Now let us introduce the filter, which is an important part of FBP. The selection of filter directly affects the quality of the reconstruction results.

The R–L filter. In theory, FBP requires the filter $H(\omega) = |\omega|$, which is called the ramp filter. However, it is just an ideal filter with an infinite band, and since it will diverge when $\omega \to \infty$, it is impossible to create the ramp filter according to the Perry–Wiener criterion. In practice, we will use a rectangular window function to truncate this ramp filter and set the value outside the window to 0. It was proposed by Indian researchers Ramachandran and Lakshminarayanan, hence it is also called the R–L filter (Ramachandran and Lakshminarayanan 1971). It can be formulated as

$$H_{\text{R–L}}(\omega) = |\omega| \, \text{rect}\left(\frac{\omega}{2f_m}\right) = \begin{cases} |\omega|, & |\omega| \leqslant f_m, \\ 0, & \text{others} \end{cases}, \tag{4.34}$$

where

$$\text{rect}(x) = \begin{cases} 1, & |x| \leqslant \dfrac{1}{2} \\ 0, & |x| > \dfrac{1}{2} \end{cases}. \tag{4.35}$$

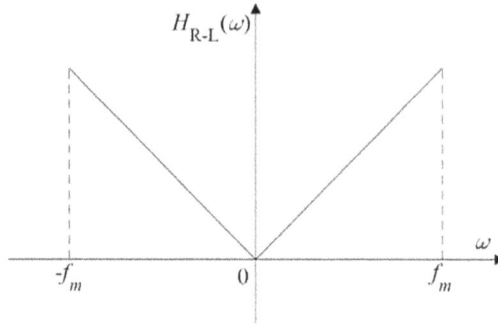

Figure 4.21. The R–L filter in the frequency domain.

As can be seen from the formula, we used a maximum frequency to cut off the ramp filter. The curve of the R–L filter is shown in figure 4.21.

Its convolutional filter in the image domain is

$$h_{R-L}(s) = f_m^2 \left[2\text{sinc}(2\pi f_m s) - \text{sinc}^2(\pi f_m s) \right], \tag{4.36}$$

where

$$\text{sinc}(x) = \frac{\sin(x)}{x}. \tag{4.37}$$

In practical problems, the data are discretely sampled. According to the Nyquist theorem, when sampling interval $\tau \leqslant \frac{1}{2f_m}$, the discrete form of $h_{R-L}(s)$ is

$$h(n\tau) = \begin{cases} \dfrac{1}{4\tau^2}, & n = 0 \\ 0, & n \text{ is even}. \\ -\dfrac{1}{(n\pi\tau)^2}, & n \text{ is odd} \end{cases} \tag{4.38}$$

Figure 4.22 shows the continuous and discrete forms of the R–L filter in the image domain, respectively.

The R–L filter has a simple form and the outline of the result which is reconstructed by the R–L filter is clear. However, there will be an obvious oscillation response, such as the Gibbs phenomenon, because the filter is truncated by the ideal rectangular function. We can go back to figure 4.18, in which we use a rectangular filter, and we can see that the resultant image has oscillating waves. This is because of the Gibbs effect.

The S–L filter. The Gibbs phenomenon of the R–L filter is caused by the ideal rectangular function. Therefore, in order to mitigate the oscillation effect, we can use a smoother window function. Shepp–Logan proposed using the sinc function as the window function (Shepp and Logan 1974):

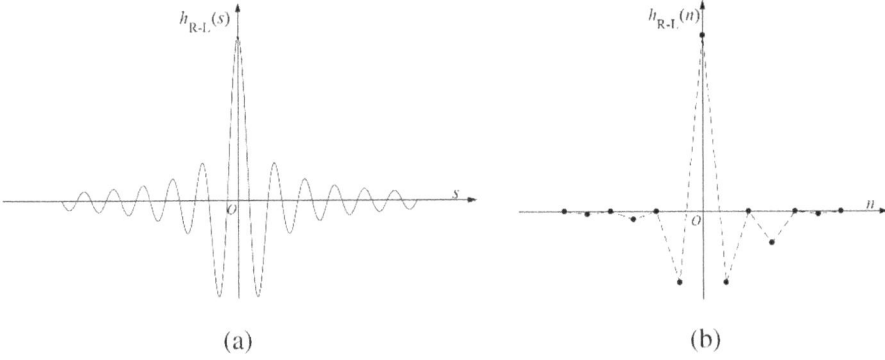

(a)   (b)

Figure 4.22. (a) Continuous R–L filter in the image domain. (b) Discrete R–L filter in the image domain.

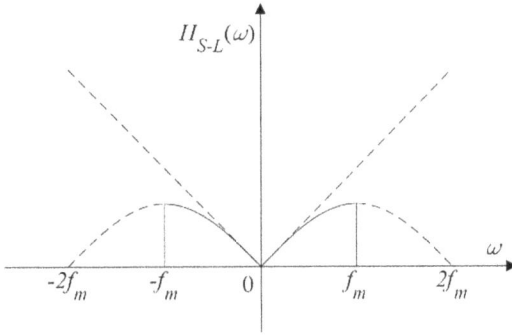

Figure 4.23. S–L filter in the frequency domain.

$$H_{S-L}(\omega) = |\omega| \, \text{sinc}\!\left(\frac{\omega\pi}{2f_m}\right)\text{rect}\!\left(\frac{\omega}{2f_m}\right) = \left| \frac{2f_m}{\pi} \sin\!\left(\frac{\omega\pi}{2f_m}\right) \right|, \quad -f_m \leqslant \omega \leqslant f_m. \quad (4.39)$$

Figure 4.23 shows the S–L filter curve.

The convolutional and discrete forms are as follows:

$$h_{S-L}(s) = \frac{1}{2}\left(\frac{4f_m}{\pi}\right)^2 \frac{1 - 4f_m s \, \sin(2\pi f_m s)}{1 - (4f_m s)^2}, \quad (4.40)$$

$$h(n\tau) = -\frac{2}{\pi^2 \tau^2 (4n^2 - 1)}, \quad n = 0, \pm 1, \pm 2, \cdots \quad (4.41)$$

The curves of these are shown in figure 4.24.

The image reconstructed by the S–L filter has smaller oscillation than the R–L filter. In particular, the S–L filter can achieve better quality with noisy projection. However, because the filter deviates from the ideal ramp filter in the high-frequency

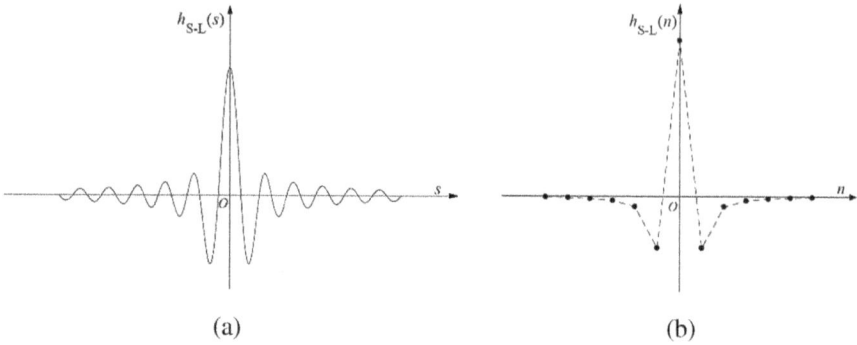

$h_{\text{S-L}}(s)$

s

O

(a)

$h_{\text{S-L}}(n)$

n

O

(b)

Figure 4.24. (a) Continuous S–L filter in the image domain. (b) Discrete S–L filter in the image domain.

band, the S–L filter has a worse response in the area corresponding to the high-frequency band.

In addition to the S–L filter, we can also use other window functions, such as Hanning and Hamming filters, according to different requirements, which are convolved with the R–L filter by the Hanning and Hamming window functions respectively (Nuttall 1981).

4.2.4 Fan-beam image reconstruction

With the development of CT technology, the fan-beam technique gradually replaced the parallel beam. The fan-beam scanner has many advantages over parallel beam. The fan-beam scanner with its single-source and multi-detector geometry has faster data acquisition efficiency. Only the tubes and detectors need to be rotated during CT scanning, which reduces the complexity of the machine. Schematic diagrams for the comparison of parallel and fan-beam geometries are shown in figure 4.25.

Fan-beam scanners can be divided into two types depending on the detector: flat detector and curved detector fan beams. Figure 4.26 shows the two types of fan-beam detectors. For the flat detector, the data are sampled with equal distance Δs intervals. For the curved detector, the data are sampled with equal angle $\Delta \gamma$ intervals.

However, for fan-beam tomography, we cannot directly use the FBP algorithm. One idea is to convert the projection of the fan-beam into a parallel beam and then use the rectified FBP algorithm. An easy way to achieve this is to reorganize the projection. We group parallel rays into a group, which represents the projection at an angle. Here, we discuss the relationship between the two types of fan-beam detectors and the parallel beams.

Figure 4.27 displays the relationship between the curved detector fan-beam system and the parallel-beam system. The red line represents the x-ray path, and the projection of the parallel-beam system is $p(\theta, s)$. θ determines the angle of the ray and s determines each ray in a view. In the fan-beam system, we use β and γ to describe the view angle and detector elements. The projection in the fan-beam

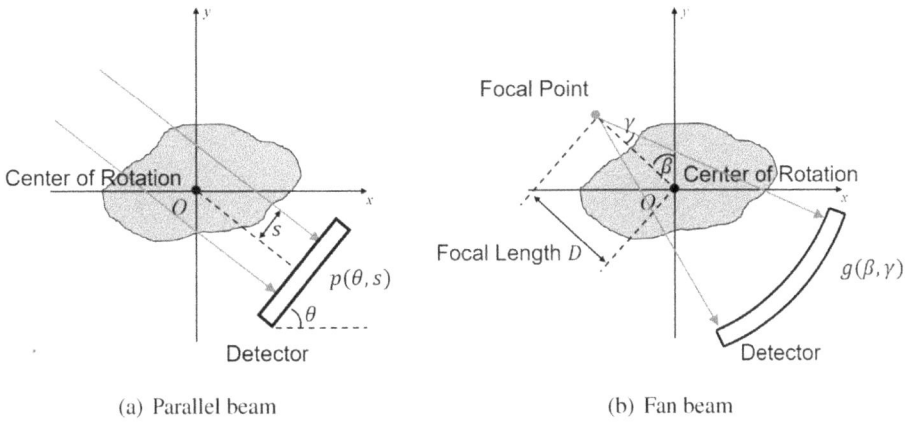

(a) Parallel beam   (b) Fan beam

Figure 4.25. Comparison of the parallel-beam and fan-beam geometries.

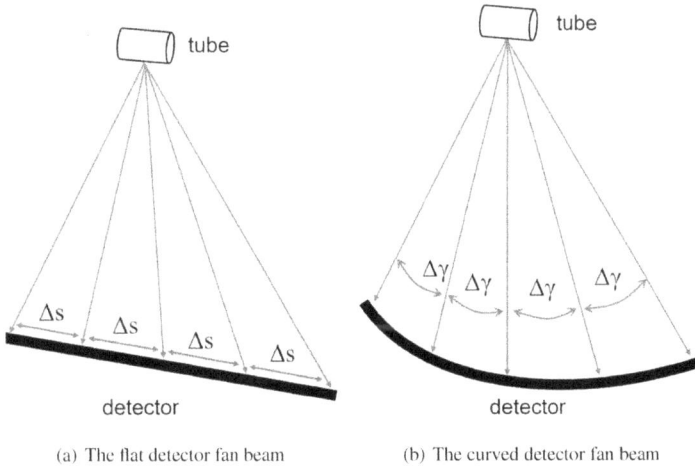

(a) The flat detector fan beam   (b) The curved detector fan beam

Figure 4.26. Comparison of the parallel-beam and fan-beam geometries.

system is $g(\beta, \gamma)$. Obviously, both representations can represent the same ray, they just need to satisfy the following formula:

$$\begin{cases} \theta = \gamma + \beta \\ s = D \sin \gamma \end{cases},$$ (4.42)

where D is the distance between the x-ray source and the center of rotation.

Figure 4.28 displays the relationship between the flat detector fan beam and the parallel beam. $g(\beta, t)$ is the projection expression of the flat detector fan-beam system, where β determines the view angle and t determines each ray in a view scan. For convenience, we can construct a virtual detector. Obviously, $t' = Dt/L$,

$$p(\theta, s) = g(\beta, \gamma)$$

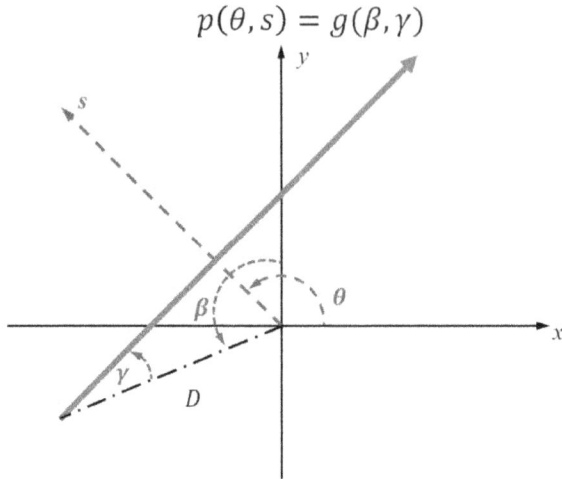

Figure 4.27. The relationship between the curved detector fan beam and parallel beam data.

$$p(\theta, s) = g(\beta, \gamma) = g(\beta, t') = g(\beta, t)$$

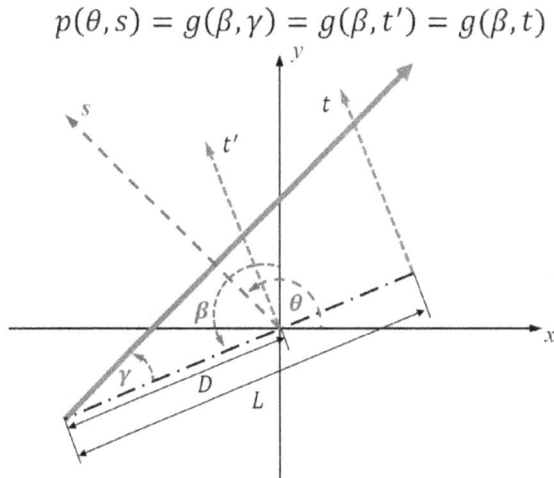

Figure 4.28. Relationship between the flat detector fan beam and parallel beam.

$g(\beta, t') = g(\beta, t)$. $g(\beta, t')$ and $p(\theta, s)$ represent the same projection when the coordination system meets the following relationship:

$$\begin{cases} \theta = \beta + \gamma = \beta - \arctan \dfrac{t'}{D} \\ \dfrac{s}{t'} = \dfrac{D}{\sqrt{t'^2 + D^2}} \end{cases} . \tag{4.43}$$

After reorganizing the fan-beam data into the form of parallel-beam data, the parallel-beam image reconstruction algorithm can be used to reconstruct the image.

However, there are some problems with the practical use of this method. The coordinate transformation, which requires interpolation, is performed when data are rearranged. Interpolation may yield inaccurate results. However, the coordinate transformation is an inspirational method that can be used to replace variables rather than rearrange them. Now, we will introduce the reconstruction method for the curved detector fan beam data.

First, let us review the expression of the parallel-beam reconstruction algorithm:

$$f(x, y) = \frac{1}{2} \int_0^{2\pi} \int_{-\infty}^{\infty} p(\theta, s)h(x \cos \theta + y \sin \theta - s)\mathrm{d}s\mathrm{d}\theta, \tag{4.44}$$

where $\int_{-\infty}^{\infty} p(\theta, s)h(x \cos \theta + y \sin \theta - s)\mathrm{d}s = p(\theta, s) * h(s)$ is the integral expression of convolution filtering. Then, use backprojection to obtain the image.

Here, we need to use polar coordinates, i.e. $x = r \cos \varphi$, $y = r \sin \varphi$, then $x \cos \theta + y \sin \theta = r \cos(\theta - \varphi)$. The polar coordinate form of equation (4.44) is

$$f(r, \varphi) = \frac{1}{2} \int_0^{2\pi} \int_{-\infty}^{\infty} p(\theta, s)h(r \cos(\theta - \varphi) - s)\mathrm{d}s\mathrm{d}\theta, \tag{4.45}$$

where s and θ are replaced with new coordinates (β, γ). We know that $\theta = \gamma + \beta$, $s = D \sin \gamma$, and then perform variable substitution:

$$\frac{\partial(s, \theta)}{\partial(\gamma, \beta)} = J = \begin{vmatrix} \dfrac{\partial s}{\partial \gamma} & \dfrac{\partial s}{\partial \beta} \\ \dfrac{\partial \theta}{\partial \gamma} & \dfrac{\partial \theta}{\partial \beta} \end{vmatrix} = \begin{vmatrix} D \cos \gamma & 0 \\ 1 & 1 \end{vmatrix} = D \cos \gamma, \tag{4.46}$$

$$\mathrm{d}s\mathrm{d}\theta = |J| \, \mathrm{d}\gamma\mathrm{d}\beta = D \cos \gamma\mathrm{d}\gamma\mathrm{d}\beta, \tag{4.47}$$

$$\begin{aligned} f(r, \varphi) &= \frac{1}{2} \int_{-\gamma}^{2\pi-\gamma} \int_{-\pi/2}^{\pi/2} p(\theta, s)h(r \cos(\gamma + \beta - \varphi) - D \sin \gamma)D \cos \gamma\mathrm{d}\gamma\mathrm{d}\beta \\ &= \frac{1}{2} \int_0^{2\pi} \int_{-\pi/2}^{\pi/2} g(\beta, \gamma)h(r \cos(\beta + \gamma - \varphi) - D \sin \gamma)D \cos \gamma\mathrm{d}\gamma\mathrm{d}\beta. \end{aligned} \tag{4.48}$$

In this way, we basically complete the FBP reconstruction algorithm of the fan beam by simple coordinate transformation. However, there is a problem. The above equation is not in convolutional form. If we can convert it into a convolutional form, then this algorithm will be implemented faster. Let us turn it into a convolution integral. As shown in figure 4.29, we define two variables D' and γ'. It can be easily verified that $r \cos(\gamma + \beta - \varphi) - D \sin \gamma = D' \sin(\gamma' - \gamma)$. Then, equation (4.48) can be rewritten as

$$f(r, \varphi) = \frac{1}{2} \int_0^{2\pi} \int_{-\pi/2}^{\pi/2} g(\beta, \gamma)h(D' \sin(\gamma' - \gamma))D \cos \gamma\mathrm{d}\gamma\mathrm{d}\beta. \tag{4.49}$$

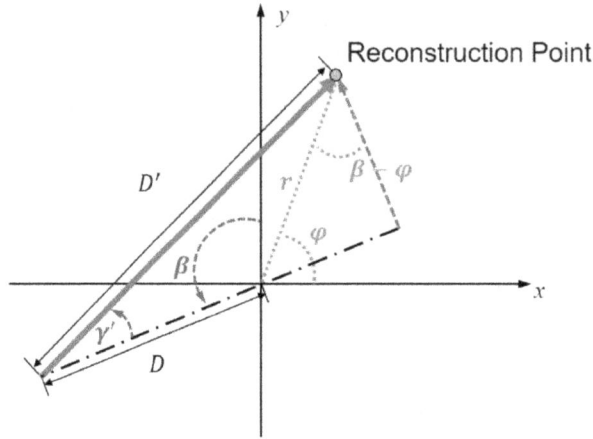

Figure 4.29. The reconstruction point (r, φ) defines the angle γ' and distance D'.

The ramp filter has a property (proof is omitted)

$$h(D' \sin \gamma) = \left(\frac{\gamma}{D' \sin \gamma} \right)^2 h(\gamma). \tag{4.50}$$

We denote $h'(\gamma) = \frac{D}{2} (\frac{\gamma}{\sin \gamma})^2 h(\gamma)$, then we have:

$$h(D' \sin \gamma)D = \frac{1}{(D')^2} D \left(\frac{\gamma}{\sin \gamma} \right)^2 h(\gamma) = \frac{2}{(D')^2} h'(\gamma). \tag{4.51}$$

The final expression of the FBP reconstruction is

$$f(r, \varphi) = \int_0^{2\pi} \frac{1}{(D')^2} \int_{-\pi/2}^{\pi/2} \cos \cdot \gamma g(\beta, \gamma) h'(\gamma' - \gamma) \mathrm{d}\gamma \mathrm{d}\beta. \tag{4.52}$$

The above formula is the entire reconstruction algorithm for the curved detector fan beam data.

For the flat detector fan-beam system, we use the same strategy. For convenience, we use the coordinate on the virtual detector to denote the ray. As shown in figure 4.30, the virtual detector is situated at the center of rotation.

We know that $\theta = \beta + \arctan(t/D)$ and $s = tD/\sqrt{D^2 + t^2}$, then perform the variable substitution. We need to calculate the Jacobi factor first:

$$\frac{\partial(s, \theta)}{\partial(t, \beta)} = J = \begin{vmatrix} \dfrac{\partial s}{\partial t} & \dfrac{\partial s}{\partial \beta} \\ \dfrac{\partial \theta}{\partial t} & \dfrac{\partial \theta}{\partial \beta} \end{vmatrix} = \frac{D^3}{(D^2 + t^2)^{\frac{3}{2}}}. \tag{4.53}$$

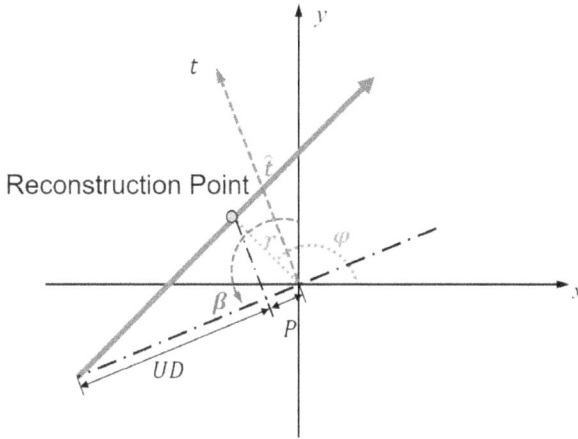

Figure 4.30. The reconstruction point (r, φ) in the flat detector fan-beam system.

We also have the relationship $r\cos(\theta - \varphi) - s = (\hat{t} - t)UD/\sqrt{D^2 + t^2}$. The definitions of UD and $\hat{t}$ are shown in figure 4.30. Then, equation (4.46) can be rewritten as

$$f(r, \varphi) = \frac{1}{2}\int_0^{2\pi}\int_{-\infty}^{\infty} g(\beta, t)h\left(\frac{UD}{\sqrt{D^2 + t^2}}(\hat{t} - t)\right)\frac{D^3}{(D^2 + t^2)^{\frac{3}{2}}}\mathrm{d}t\mathrm{d}\beta. \qquad (4.54)$$

The ramp filter has the property $h(at) = \frac{1}{a^2}h(t)$, therefore:

$$f(r, \varphi) = \frac{1}{2}\int_0^{2\pi}\frac{1}{U^2}\int_{-\infty}^{\infty}\frac{D}{\sqrt{D^2 + t^2}}g(\beta, t)h(\hat{t} - t)\mathrm{d}t\mathrm{d}\beta. \qquad (4.55)$$

In fact, $D/\sqrt{D^2 + t^2}$ is the cosine of the ray angle γ. Therefore, the process of the FPB algorithm of these two fan-beam systems can be divided into the following steps:

1. The projection data is preprocessed and multiplied by the cosine function $\cos\gamma$.
2. The processed data is filtered using a 1D ramp filter.
3. Backproject the weighted projection. The weight depends on the distance from the reconstruction point to the focus.

4.2.5 Cone-beam image reconstruction*

With the development of CT technology, two-dimensional detector arrays are becoming increasingly more popular, such as flat panel detectors and multi-row CT detectors. As shown in figure 4.31, a two-dimensional detector can simultaneously receive data through a volumetric segment. A cone-beam projection consists of many fan-beam projections, all of which are acquired in parallel.

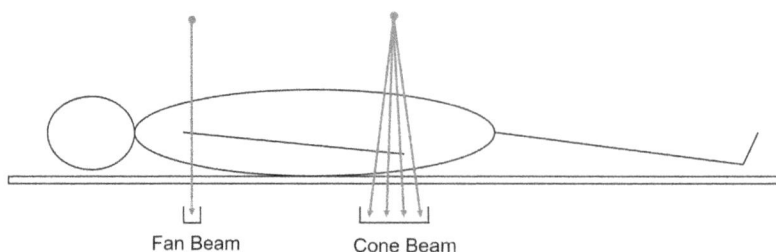

Figure 4.31. Comparison of fan-beam and cone-beam imaging geometries.

Hence, cone-beam geometry is an extended fan-beam geometry. The most important cone-beam projection data acquisition modes are (i) circular cone-beam (Feldkamp *et al* 1984) and (ii) spiral/helical cone-beam scans (Wang *et al* 1992, 1993), respectively. Actually, a spiral/helical cone-beam scan generalizes a circular cone-beam scan, and a circular cone-beam scan is a special case of a spiral/helical cone-beam scan. Also, there are other cone-beam scans, even along quite an arbitrary trajectory. For more details, see two reviews on this topic (Kudo *et al* 2004, Wang *et al* 2007).

First, approximate cone-beam image reconstruction can be performed using the FDK algorithm in the case of circular scanning (Feldkamp *et al* 1984) and using a generalized FDK algorithm in the case of spiral/helical scanning (Wang *et al* 1992, 1993). The key idea can be explained as follows. In a nutshell, the generic FDK or its generalized variants are essentially the fan-beam FBP reconstruction. As shown in figure 4.32, we add the z-axis so that a fan-beam in the x–y plane can be tilted out of the plane. To reconstruct any voxel in the 3D space, we still want to perform a fan-beam reconstruction of the slice that contains the voxel of interest. The key step is explained as follows. Each tilted fan-beam projection that intersects the voxel needs to be first corrected with a multiplicative factor which is equal to the cosine of the tilting angle relative to a horizontal plane. Then, the corrected projection is filtered as if it is an original fan-beam projection.

Recalling the fan-beam FBP reconstruction formula, we preprocess the data by multiplying the cosine of the angle between the ray and the line connecting the x-ray focal spot to the center of rotation. Similarly, in the cone-beam reconstruction we also perform such a correction. In a 2D system, the cosine is $D/\sqrt{D^2 + t^2}$, and in a 3D system, the cosine factor becomes $D/\sqrt{D^2 + t^2 + \hat{z}^2}$, and note that we interpret the coordinate system in a projection-specific fashion, so that the x-ray focal spot is on the x-ray plane to accommodate a 3D scanning locus. Then, we still use the ramp filter to filter the projection profile by profile. Actually, the filtering direction is not very sensitive, for example, we can just convolve data along the direction of the t-axis or along the tangential direction of the scanning locus. Finally, weighted backprojection is performed, with the same weight as used in the fan-beam reconstruction, which is equal to $[D/(D - P)]^2$, as shown in figure 4.32 and formulated as follows:

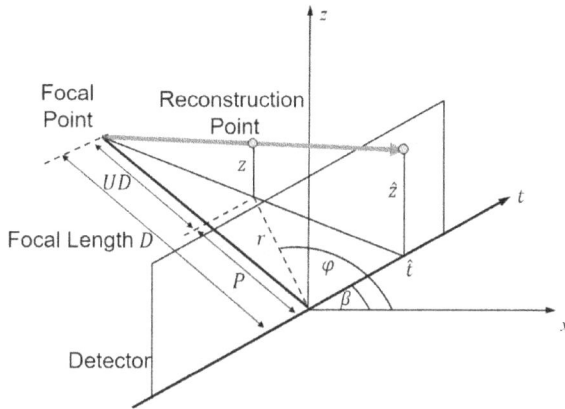

Figure 4.32. Approximate image reconstruction of point form cone-beam projections collected along a 3D scanning locus.

$$f(r, \varphi, \hat{z}) = \frac{1}{2} \int_0^{2\pi} \left(\frac{D}{D - P} \right)^2 \int_{-\infty}^{\infty} \frac{D}{\sqrt{D^2 + t^2 + \hat{z}^2}} g(\beta, t, \hat{z}) h(\hat{t} - t) dt d\beta. \quad (4.56)$$

The above generalized FDK algorithm is simple, robust, and efficient. For more general settings (with varying focal lengths) and discussions on the exactness properties of the approximate reconstruction, see the original spiral/helical cone-beam CT paper (Wang *et al* 1992, 1993).

When the cone angle is relatively small or moderate (for example, less than 20°), the approximate reconstruction is usually satisfactory. Hence, FDK-type algorithms are widely used in practice. However, with more challenging 3D image reconstruction applications, approximate reconstruction can no longer meet the needs. Since the initial cone-beam reconstruction algorithms were formulated in the circular and spiral/helical scanning cases, the researchers conducted a series of studies for accurate cone-beam image reconstruction.

The exact cone-beam reconstruction problem can mainly be defined in two scenarios: (i) a cone-beam projection covers the object completely and (ii) a cone-beam projection is longitudinally truncated. The Grangeat algorithm (Grangeat 1991) is a classic result, revealing a solution to the first case. However, it is rarely used for practical CT scans because not only it is complicated, but it also assumes no data truncation. In 2002, Katsevich proposed an accurate FBP algorithm for spiral/helical cone-beam image reconstruction (Katsevich 2002a, b, 2004), which inspired a number of follow-up studies; see (Wang *et al* 2007) for an overview.

The analytical reconstruction algorithms are advantageous in terms of computational efficiency, but they are subject to certain conditions. For instance, they assume no noisy data and require a quite regular ray distribution (for example, no holes in a projection). Otherwise, the image quality may be seriously damaged. Therefore, under many special conditions such as low-dose CT scanning, the analytical reconstruction algorithms are not suitable. Then, iterative reconstruction algorithms were developed to address these issues at a high computational cost. We will introduce typical iterative reconstruction algorithms in the next subsection.

4.3 Iterative reconstruction

4.3.1 Linear equations

This section will touch upon many scalars and vectors. Therefore, let us unify the expression. In this section, we will use lowercase italic letters to indicate scalars, lowercase bold italic letters to represent vectors, and uppercase italic letters to represent matrices. First, let us treat the projection as linear equations. As shown in figure 4.9, we discretize the object into a 2×2 array. Four x-rays pass through the object and generate four projection data points. We use equations to represent the process, that is

$$\begin{cases} x_1 + x_2 = p_1 \\ x_3 + x_4 = p_2 \\ x_1 + x_3 = p_3 \\ x_2 + x_4 = p_4 \end{cases} . \tag{4.57}$$

We fill the pixels and write them in a matrix:

$$\begin{cases} x_1 + x_2 + 0x_3 + 0x_4 = p_1 \\ 0x_1 + 0x_2 + x_3 + x_4 = p_2 \\ x_1 + 0x_2 + x_3 + 0x_4 = p_3 \\ 0x_1 + x_2 + 0x_3 + x_4 = p_4 \end{cases} . \tag{4.58}$$

Then, we have

$$A\boldsymbol{x} = \boldsymbol{p}, \tag{4.59}$$

where

$$\boldsymbol{x} = [x_1, x_2, x_3, x_4]^T, \ \boldsymbol{p} = [p_1, p_2, p_3, p_4]^T, \ A = \begin{bmatrix} 1 & 1 & 0 & 0 \\ 0 & 0 & 1 & 1 \\ 1 & 0 & 1 & 0 \\ 0 & 1 & 0 & 1 \end{bmatrix}.$$

Each row in A represents a ray. Assuming that the size of A is $M \times N$, then the total number of object pixels is N and the number of rays is $M = M_\theta \times M_d$, where M_θ is the number of angle views and M_d is the number of detector elements. The element a_{ij} in A represents the contribution of the jth pixel to the ith projection data.

In this way, we can obtain CT images by solving linear equations. However, there are still some problems. In most cases, matrix A is not a square matrix. The linear system $A\boldsymbol{x} = \boldsymbol{p}$ has solution(s) if and only if $\boldsymbol{p}$ is in the column space of A, and the solution is unique if A has full column rank (otherwise there can be infinitely many solutions).

For this non-square matrix case, we can solve it using the least square method. First, establish a least squares objective function:

$$J(x) = \|Ax - p\|^2, \tag{4.60}$$

This objective function is very intuitive. When we obtain the solution x, the error between the projection obtained by x and the actual projection p is the smallest. We need to find the partial derivative of x and set the partial derivative to 0 to solve this problem, that is,

$$\frac{\partial J(x)}{\partial x} = 2A^T(Ax - p) = 2A^T Ax - 2A^T p = 0. \tag{4.61}$$

The closed-form solution for this equation is

$$\hat{x} = (A^T A)^{-1} A^T p. \tag{4.62}$$

However, the matrix A is commonly very large, so its inverse matrix is difficult to calculate and store. Therefore, a more practical approach is to use an iterative method to find an approximate solution of a system of equations. There are two commonly used methods, the algebraic iterative reconstruction algorithm and statistical iterative reconstruction algorithm. We will introduce both methods in the following sections.

4.3.2 Algebraic iterative reconstruction

The ART algorithm. The algebraic reconstruction technique (ART) (Gordon *et al* 1970), which was proposed by Kaczmarz in 1937 (Karczmarz 1937), was originally used to solve compatible linear equations. It was introduced into the field of image reconstruction by Gordon *et al* and was adopted by the Hounsfield CT scanner. After many years of theory and practice, ART has been proven to have the advantages of analytic reconstruction, and is suitable for situations where projection data are insufficient, the angle views are missing, the projection interval is uneven, etc.

The basic principle of ART is to correct the solution on a ray-by-ray basis. It is difficult to understand directly through the formulas, so we will try to explain this from a two-dimensional space. As shown in figure 4.33, we obtain the solution of the kth iteration x^k. Then, we intend to use the ith ray to rectify it. We use A_i to denote the ith row of A, i.e. the coefficient of the ith ray. As shown in figure 4.33, we want to make the result close to the ideal value $\hat{x}$. How do we correct x^k when we do not know the ideal vector $\hat{x}$? In fact, we can only use the projection to correct it. We first obtain the projection of x^k and denote it as p_i^k. We can obtain the error between it and the actual projection p_i. Let us denote the error as p_{error}. Note here that since we only use one ray, p_i^k, p_i, and p_{error} are all scalars. We divide them by the module of A_i, then obtain $(p_i^k)'$, p_i' and p'_{error} respectively, whose definitions are shown in figure 4.34. This projection deviation can be eliminated if we heuristically correct the

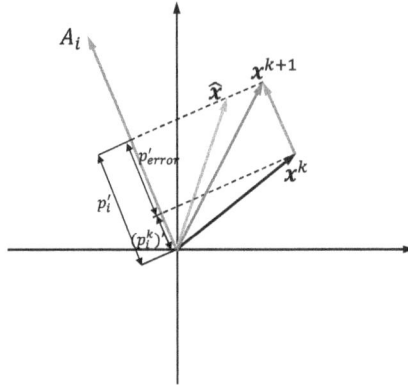

Figure 4.33. Illustration of one iteration of ART.

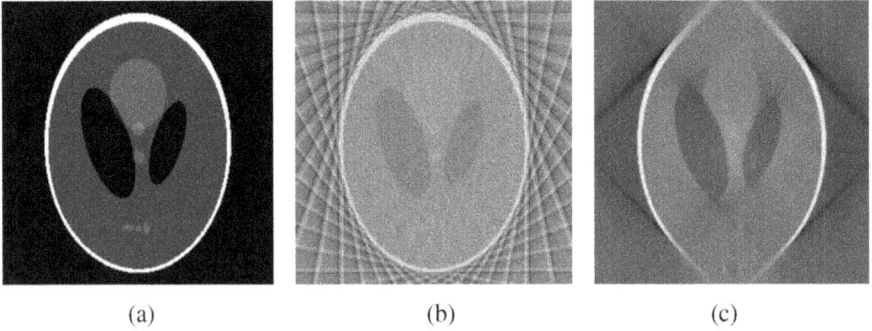

| (a) | (b) | (c) |

Figure 4.34. Reconstruction with incomplete data. (a) Original image. (b) Reconstruction of sparse views. (c) Reconstruction of limited views.

involved pixels according to p_{error}. Clearly, we need to iteratively correct the current images view-by-view or ray-by-ray. Each iteration can be expressed as

$$x^{k+1} = x^k + \frac{\lambda^k (p_i - A_i x^k)}{\|A_i\|} \frac{A_i}{\|A_i\|}, \tag{4.63}$$

where $\lambda(0 < \lambda < 2)$ is a relaxation factor. The formulation of each pixel is

$$x_j^{k+1} = x_j^k + \lambda^k \frac{p_i - A_i x^k}{\|A_i\|^2} a_{ij}. \tag{4.64}$$

ART has the following advantages:
1. ART is a line-by-line iterative algorithm that does not require the inverse of the matrix and is computationally insignificant.
2. In each iteration, only one row element of matrix A is used, which saves storage space.

3. It can directly solve the overdetermined equation without conversion to a regular equation. When the equations are compatible, an exact solution can be obtained.
4. The projection values and residuals are uniformly backprojected so that distortion is not caused by the concentration of errors.

However, ART also has a certain flaw. It treats each beam as irrelevant. In fact, each pixel contributes to all the rays, so it is not reasonable to use only one ray at a time to correct the image. To address this problem, the SART algorithm was proposed.

The SART algorithm. The simultaneous algebraic reconstruction technique (SART) (Andersen and Kak 1984) was proposed as an improved version of ART in 1984. A great drawback of ART is that only one ray is considered per iteration, ignoring the fact that each pixel contributes to all projections simultaneously. To solve this problem, SART was proposed. In the SART algorithm, instead of thinking of each ray independently, the whole projection is viewed as a correlative system. When we update a pixel, we will consider all the rays that intersect it and compute all the corrections along those rays. The average correction is used to update the pixel:

$$x_j^{k+1} = x_j^k + \lambda \frac{\sum_i \left(\frac{p_i - A_i x^k}{\sum_{n=1}^N a_{in}} \right)}{\sum_i a_{ij}}, \tag{4.65}$$

where the summation with respect to i is over all the rays intersecting the jth pixel. λ is the relaxation factor ($0 < \lambda < 2$). The term $\sum_{n=1}^N a_{in}$ is the actual physical length of the ith ray. Replacing the module with $\sum_{n=1}^N a_{in}$ will bring more uniformity. Mathematically, it can be shown that the SART algorithm must converge (Jiang and Wang 2003).

4.3.3 Statistical iterative reconstruction

Since the number of photons received by each detector element obeys a certain probability distribution, we can reconstruct an image based on the probabilistic model. The Bayes principle, which can be simplified to the maximum likelihood estimation, is often used to find the parameters of the probability model. In the field of CT reconstruction, observation data models are mixtures of Poisson and Gaussian distributions (in the ideal case, the Poisson distribution) (Barrett *et al* 1994, Chiao *et al* 1994). The Poisson distribution is considered physically accurate in modeling the emission and attenuation of x-ray photons. Based on the Poisson model, a classic statistical algorithm, the maximum likelihood expectation maximization (ML-EM) (Dempster *et al* 1977, Shepp and Vardi 1982, Lange *et al* 1984) was proposed for emission CT. As the name implies, the algorithm uses the EM algorithm to maximize the likelihood function. Let us now introduce the key details of the algorithm.

Because of statistical uncertainty, the number of photons received by the detector will be often more or less than a mean. This number of photons follows the Poisson distribution, since a linear combination of Poisson emission variables should still obey a Poisson distribution. The two most common forms of tomography, emission and transmission tomography, have different likelihood functions. Next we will introduce them separately.

In emission tomography, we assume that a variable c_{ij} represents the contribution of the jth pixel to the ith projection. Then, the ith projection is the summation of the contributions from all the involved pixels, i.e. $p_i = \sum_j c_{ij}$, where c_{ij} is a Poisson variable with an expectation $\lambda_{ij} = a_{ij}x_j$. Then, we can maximize the likelihood function of observed data to calculate an underlying emission image x. Let us formulate the likelihood function as follows:

$$\text{Prob}(p|x) = \prod_{i,j} e^{-\lambda_{ij}} \frac{\lambda_{ij}^{c_{ij}}}{c_{ij}!} = \prod_{i,j} e^{-a_{ij}x_j} \frac{(a_{ij}x_j)^{c_{ij}}}{c_{ij}!}. \tag{4.66}$$

Then, its log-likelihood function is straightforward:

$$\ln(\text{Prob}(p|x)) = \sum_{i,j}(c_{ij}\ln(a_{ij}x_j) - a_{ij}x_j) - \sum_{i,j}\ln(c_{ij}!). \tag{4.67}$$

Note that the last term does not involve x, and can be removed without affecting the image reconstruction. Hence, we have the equivalent objective function:

$$L = \sum_{i,j}(c_{ij}\ln(a_{ij}x_j) - a_{ij}x_j). \tag{4.68}$$

Generally, it is difficult to derive the unknown parameters directly that maximize the likelihood function. Hence, we use the EM algorithm to solve it iteratively. The EM algorithm is divided into two steps: expectation (E) and maximization (M).

Suppose that we have the kth iteration result x^k, and want to perform the next iteration. The difficulty in solving the problem is that c_{ij} is unknown. However, we can use x^k to construct the expectation of c_{ij}:

$$\mathbb{E}(c_{ij}|p_i, x^k) = a_{ij}x_j^k \frac{p_i}{A_i x^k}, \tag{4.69}$$

where $A_i x^k$ is the ith projection of x^k. Clearly, this formula can be transformed to $a_{ij}x_j/a_{ij}x_j^k = p_i/p_i^k$, which can be readily proved based on the property of the means of Poisson variables. Therefore, the objective function becomes

$$\mathbb{E}(L|p, x^k) = \sum_{i,j}\left(\frac{a_{ij}x_j^k}{A_i x^k}p_i\ln(a_{ij}x_j) - a_{ij}x_j\right). \tag{4.70}$$

Next, we must update the current iteration result to x^{k+1} so that the mean of the above objective function is maximized. The solution is to compute the partial derivatives with respect to x and set them to zero:

$$
\frac{\partial}{\partial x_j} \mathbb{E}(L|p,\, x^k) = \sum_i \left(\frac{a_{ij}x_j^k}{A_i x^k} p_i \frac{a_{ij}}{a_{ij}x_j} - a_{ij} \right)
$$

$$
= \frac{1}{x_j} \sum_i \frac{a_{ij}x_j^k}{A_i x^k} p_i - \sum_i a_{ij} \tag{4.71}
$$

$$
= 0.
$$

Finally, the closed-form solution for emission CT is

$$
x_j^{k+1} = \frac{x_j^k}{\sum_i a_{ij}} \sum_i a_{ij} \frac{p_i}{A_i x^k}. \tag{4.72}
$$

The derivation of the EM algorithm for transmission CT is similar but mathematically a little bit more involved, since the form of the likelihood function becomes more complex. For the ith x-ray, the parameter of the Poisson distribution is $\lambda_i = I_0 \exp(-A_i x)$. We define the observed measurement of the ith x-ray as I_i, and have the likelihood function as follows:

$$
\mathrm{Prob}(I|x) = \prod_i e^{-\lambda_i} \frac{\lambda_i^{I_i}}{I_i!} = \prod_i e^{-I_0 \exp(-A_i x)} \frac{[I_0 \exp(-A_i x)]^{I_i}}{I_i!}. \tag{4.73}
$$

As we have done in the case of emission CT, we compute the log-likelihood function:

$$
\ln(\mathrm{Prob}(I|x)) = \sum_i \left\{ -I_0 e^{-A_i x} - I_i A_i x + I_i \ln I_0 - \ln I_i! \right\}, \tag{4.74}
$$

After the removal of the terms that are not related to x, we obtain the following equivalent objective function to be maximized:

$$
L = -\sum_i \left\{ I_0 e^{-A_i x} + I_i A_i x \right\}. \tag{4.75}
$$

Equation (4.75) is difficult to optimize directly. Even if we compute partial derivatives with respect to x, we cannot have a closed-form solution. Hence, we need to find an indirect way. Let us consider each pixel x_j independently, and introduce two hidden variables u_{ij} and v_{ij} to represent the photon counts of entering and leaving that pixel, respectively, along the ith ray path. For the jth pixel, the number of photons leaving along the ith ray path should obey the binomial distribution with a parameter of $\exp(-a_{ij}x_j)$. Then, the complete likelihood is

$$
\mathrm{Prob} = \prod_i \prod_j \binom{u_{ij}}{v_{ij}} (\exp(-a_{ij}x_j))^{v_{ij}} (1 - \exp(-a_{ij}x_j))^{u_{ij}-v_{ij}}. \tag{4.76}
$$

It is easy to obtain the log-likelihood function

$$\ln(\text{Prob}) = \sum_i \sum_j \left\{ \ln\left(\frac{u_{ij}}{v_{ij}}\right) + v_{ij} \ln(e^{-a_{ij}x_j}) + (u_{ij} - v_{ij})\ln(1 - e^{-a_{ij}x_j}) \right\}. \quad (4.77)$$

Suppose that the variables u_{ij} and v_{ij} are known, we have the objective function:

$$L = \sum_i \sum_j \left\{ v_{ij} \ln(e^{-a_{ij}x_j}) + (u_{ij} - v_{ij})\ln(1 - e^{-a_{ij}x_j}) \right\}. \quad (4.78)$$

However, we need to find u_{ij} and v_{ij}. Based on the principles of the EM algorithm, we use the previous iterative solution to calculate the expectations of the hidden variables. Considering the observations and the kth iterative solution x^k, the expectations of u_{ij} and v_{ij} should be

$$M_{ij} = \mathbb{E}\left(u_{ij}|I_i, x^k\right) = I_0 e^{-\sum_{l \in S_{ij}} a_{il}x_l^k} + I_i - I_0 e^{-A_i x^k}$$

$$N_{ij} = \mathbb{E}\left(v_{ij}|I_i, x^k\right) = I_0 e^{-\sum_{l \in S_{ij} \cup \{j\}} a_{il}x_l^k} + I_i - I_0 e^{-A_i x^k}, \quad (4.79)$$

where S_{ij} is the set of pixels between the source and the jth pixel along the ith ray path. The proof of equation (4.79) can be found in Lange $et\ al$ (1984). After this, we must compute the partial derivatives of equation (4.78) with respect to x_j and set them to zero:

$$0 = \sum_i -N_{ij}a_{ij} + \sum_i (M_{ij} - N_{ij})\frac{a_{ij}e^{-a_{ij}x_j}}{1 - e^{-a_{ij}x_j}}$$

$$= \sum_i -N_{ij}a_{ij} + \sum_i (M_{ij} - N_{ij})\frac{a_{ij}}{e^{a_{ij}x_j} - 1}, \quad (4.80)$$

which is a transcendental function which can be solved for an approximate solution. Specifically, it can be solved with the following property obtained by the Taylor expansion:

$$\frac{1}{e^s - 1} = \frac{1}{s} - \frac{1}{2} + \frac{s}{12} + O(s^3) \approx \frac{1}{s} - \frac{1}{2}. \quad (4.81)$$

Substituting equation (4.81) into equation (4.80), an approximate solution can be obtained

$$x_j^{k+1} = \frac{\sum_i (M_{ij} - N_{ij})}{\frac{1}{2}\sum_i (M_{ij} + N_{ij})a_{ij}}. \quad (4.82)$$

The above two EM algorithms are classic for emission and transmission CT, respectively. Since the ML-EM formulation models projection data according to statistical characteristics, it can produce quite good results even if severe noise exists during data acquisition. However, the ML-EM algorithm still has limits. It requires a large number of calculations, and the convergence speed is slow. Therefore, many efforts were made for algorithmic acceleration. Here, we describe two notable acceleration techniques, ordered subsets (OS) (Hudson and Larkin 1994) and convex algorithms (Lange and Fessler 1995), which greatly improve the efficiency of the ML-EM-based reconstruction.

In the OS method, we divide the rays into a number of subsets S_1, S_2, ... , S_n. In each iteration, we use these subsets in turn to update an intermediate reconstruction. Taking emission CT as an example, for the kth iteration the updating process using the mth subset is expressed as

$$x_{j_{[m]}}^k = \frac{x_{j_{[m-1]}}^k}{\sum_{i \in S_m} a_{ij}} \sum_{i \in S_m} a_{ij} \frac{p_i}{A_i x_{[m-1]}^k}. \tag{4.83}$$

The OS technique greatly improves the convergence speed. Additionally, it accelerates not only the ML-EM algorithm (Hudson and Larkin 1994) but also the SART algorithm (Wang and Jiang 2004).

Lang et al proposed the convex algorithm in 1995 (Lange and Fessler 1995). It takes advantage of a property of convex functions and makes the ML algorithm converge faster. Taking transmission CT as an example, we rewrite the objective function equation (4.75) as follows:

$$L(x) = -\sum_i f_i(A_i x), \tag{4.84}$$

where

$$f_i(t) = I_0 e^{-t} + I_i t. \tag{4.85}$$

Equation (4.85) is a convex function. Hence, we have

$$L(x) = -\sum_i f_i \left(\sum_j \frac{a_{ij} x_j^k}{A_i x^k} \frac{x_j}{x_j^k} A_i x^k \right)$$

$$\geqslant -\sum_i \sum_j \frac{a_{ij} x_j^k}{A_i x^k} f_i \left(\frac{x_j}{x_j^k} A_i x^k \right) \tag{4.86}$$

$$= Q(x|x^k).$$

Evidently, $L(x) = Q(x|x^k)$ at $x = x^k$. Thus, when we maximize $Q(x|x^k)$, we optimize $L(x)$ as well. To maximize $Q(x|x^k)$, we set its partial derivatives to zero:

$$0 = \frac{\partial}{\partial x_j} Q(x|x^k)$$

$$= -\sum_i a_{ij} f_i' \left(\frac{x_j}{x_j^k} A_i x^k \right) \tag{4.87}$$

$$= -\sum_i a_{ij} \left[-I_0 e^{-\left(x_j / x_j^k \right) A_i x^k} + I_i \right].$$

Then, we solve equation (4.87) using Newton's method. We have

$$\frac{\partial^2}{\partial x_j^2}Q(x|x^k)|_{x=x^k} = -\sum_i \frac{a_{ij}}{x_j^k}A_i x^k f_i''\left(\frac{x_j}{x_j^k}A_i x^k\right)$$

$$= -\sum_i \frac{a_{ij}}{x_j^k}A_i x^k I_0 e^{-\left(x_j/x_j^k\right)A_i x^k} \tag{4.88}$$

$$= -\sum_i \frac{a_{ij}}{x_j^k}A_i x^k I_0 e^{-A_i x^k}$$

and

$$\frac{\partial}{\partial x_j}Q(x|x^k)|_{x=x^k} = \sum_i a_{ij}\left[I_0 e^{-A_i x^k} - I_i\right]. \tag{4.89}$$

Then, the iterative solution is

$$x_j^{k+1} = x_j^k - \frac{\dfrac{\partial}{\partial x_j}Q(x|x^k)|_{x=x^k}}{\dfrac{\partial^2}{\partial x_j^2}Q(x|x^k)|_{x=x^k}}$$

$$= x_j^k + \frac{x_j^k \sum_i a_{ij}\left[I_0 e^{-A_i x^k} - I_i\right]}{\sum_i a_{ij}A_i x^k I_0 e^{-A_i x^k}} \tag{4.90}$$

$$= x_j^k \frac{\sum_i a_{ij}\left[I_0 e^{-A_i x^k}\left(1 + A_i x^k\right) - I_i\right]}{\sum_i a_{ij}A_i x^k I_0 e^{-A_i x^k}}.$$

The convex algorithm guarantees convergence, and yet has a faster convergence rate than the EM algorithm. To further accelerate the image reconstruction, the OS and convex algorithms were combined with a substantial speed-up (Kamphuis and Beekman 1998).

In addition to the Poisson distribution, related studies were also performed on the Gaussian distribution; for example, the weighted least squares (WLS) method (Huesman *et al* 1977). The objective function of WLS is

$$L(x) = (Ax - p)^T\Sigma^{-1}(Ax - p), \tag{4.91}$$

where Σ is a diagonal matrix with the ith entry σ_i^2, which is an estimate of the variance of the ith projection p_i. The value of σ_i^2 is set usually to p_i. Various methods can be used to minimize equation (4.91), including the steepest descent, conjugate gradient, and other techniques.

The WLS method is sensitive to noise but it often leads to unsatisfactory results. Sauer and Bouman incorporated a smoothness penalty into WLS (Sauer and Bouman 1993). Fessler extended the method to positron emission tomography (PET) and proposed a penalized weighted least squares (PWLS) technique (Fessler 1994). Later, the PWLS approach was used and improved for emission CT reconstruction (Sukovic and Clinthorne 2000, Wang *et al* 2006). The objective function of PWLS is

$$L(\boldsymbol{x}) = \frac{1}{2}(A\boldsymbol{x} - \boldsymbol{p})^T \Sigma^{-1}(A\boldsymbol{x} - \boldsymbol{p}) + \beta R(\boldsymbol{x}), \tag{4.92}$$

where

$$R(\boldsymbol{x}) = \frac{1}{2}\boldsymbol{x}^T R\boldsymbol{x} = \frac{1}{2}\sum_{j}\sum_{m \in N_j} w_{jm}(x_j - x_m)^2, \tag{4.93}$$

where N_j is the set of eight neighbors of the jth pixel, and the weights w_{jm} equal 1 for horizontal and vertical neighbors, and $1/\sqrt{2}$ for diagonal neighbors. Equation (4.92) can be solved using the SOR algorithm. The procedure is given in algorithm 4.1. In algorithm 4.2, A_j represents the jth column of the system matrix and $\omega \in (0, 1]$. If ω is set to 1, the SOR algorithm becomes the Gauss–Seidel (GS) algorithm.

Algorithm 4.1. SOR for PWLS

Initialize: $\hat{x} = \mathrm{FBP}\{\boldsymbol{p}\}$, $\boldsymbol{r} = \boldsymbol{p} - A\hat{x}$, $s_j = A_j^T\Sigma^{-1}A_j$, $\forall j$, $d_j = s_j + \beta\sum_j\sum_{m \in N_j}w_{jm}$.
Output:
loop
$\quad \hat{x}_j^{\mathrm{old}} := \hat{x}_j$
$\quad \hat{x}_j^{\mathrm{new}} := \dfrac{A_j^T\Sigma^{-1}\boldsymbol{r} + s_j\hat{x}_j^{\mathrm{old}} + \beta\sum_{m \in N_j}w_{jm}\hat{x}_m}{d_j}$
$\quad \hat{x}_j := \max\left(0, (1 - \omega)\hat{x}_j^{\mathrm{old}} + \omega\hat{x}_j^{\mathrm{new}}\right)$
$\quad \boldsymbol{r} := \boldsymbol{r} + A_j(\hat{x}_j^{\mathrm{old}} - \hat{x}_j)$
end loop
return $\hat{x}$

PWLS is an earlier application of maximum posterior probability (MAP) for CT reconstruction. Because of the utilization of the prior information, such methods can achieve excellent performance. With technological advancements, people are eager to use less radiation for CT. In this case of low-dose CT scanning, the resultant images are easily contaminated by noise with compromised diagnostic performance. In this context, regularization iterative methods have been widely studied and translated into commercial CT scanners. In the next subsection, we will introduce them.

4.3.4 Regularized iterative reconstruction*

With the widespread use of CT in clinical diagnosis, there is a problem that cannot be ignored, namely x-ray radiation. According to statistics, patients who received more than 28 CT scans had a 12% higher oncogenic risk than the average, and children were more affected by radiation. There are two common ways to reduce the radiation dose of CT scans. First, reduce the number of x-ray photons by controlling the current or voltage of the tube, which is called low-dose. Second, reduce the number of rays by decreasing the scanning views, which is referred to as incomplete data. The reconstruction by FBP will be full of noise in the case of low-dose CT and the reconstruction with incomplete data by FBP will subject to severe artifacts.

In this book, we mainly discuss the problem of incomplete data. Of course, the algorithm mentioned later can also solve the problem of low-dose. There are two methods to reduce scan views. One is to collect continuous angle data at an angle range less than 180°, which is called limited-views, and the other is to evenly sample discontinuous angle data in a range of 180°, which is called sparse-views. We performed a set of experiments to simulate the reconstruction of sparse and limited views, respectively. In the case of the sparse-views, projection data of 18 angles is acquired at equal intervals in the range of 180°. In the limited-views case, the projection data are acquired uniformly in the range of 90°. The results are reconstructed by FBP. As shown in figure 4.34, the quality of the images reconstructed by FBP with incomplete data is severely degraded. In the case of sparse-views, the reconstruction is full of streak artifacts. In the reconstruction of limited-views data, there are severe artifacts in some directions. Both results are unsatisfactory and clinically useless.

So, is there a way to obtain an acceptable image when we use incomplete data? In statistical problems, the Bayesian probability is an effective method to obtain a satisfactory result when the sample set is so small that it cannot precisely reflect the data distribution (Bishop 2006). After obtaining the projection p, we can use the maximum posterior probability (MAP) to obtain the distribution of x. First, we need to develop a conditional probability. Bayesian law states that

$$\text{Prob}(x|p) = \frac{\text{Prob}(p|x)\text{Prob}(x)}{\text{Prob}(p)}, \tag{4.94}$$

and its log-likelihood function is

$$\ln(\text{Prob}(x|p)) = \ln(\text{Prob}(p|x)) + \ln(\text{Prob}(x)) - \ln(\text{Prob}(p)). \tag{4.95}$$

The third term has nothing to do with x. So, eliminating it will not affect the solution of this objective function. The Bayesian objective function then becomes

$$L = \ln(\text{Prob}(p|x)) + \ln(\text{Prob}(x)), \tag{4.96}$$

or

$$(\text{PosteriorFunction}) = (\text{LikelihoodFunction}) + \beta(\text{PriorFunction}). \tag{4.97}$$

The first term on the right is the maximum likelihood function and the second term is a prior function that reflects the distribution of x. Of course, we do not know the exact distribution of x, so it needs to be determined based on our understanding and experience of the image. In general, the choice of prior function is highly correlated with the quality of imaging. Therefore, we will introduce an important prior knowledge in the field of image processing, namely sparsity.

The sparsity of the signal means that the number of non-zero elements of the signal is very small. Usually, we use the l_0 norm to constrain the sparsity of the signal. Also, to avoid signal distortion in the optimization, an observation constraint is added into optimization function, such as

$$\min_{x} \ \|x\|_0, \quad \text{s. t. } Ax = p, \tag{4.98}$$

where

$$\|x\|_0 = \sum_{j} (x_j)^0.$$

In fact, the optimization of the l_0 norm is an NP-hard problem. It has been proven that relaxing the l_0 norm to an l_1 norm, which is convex and can easily be optimized, will have the same effect when the signal is sufficiently sparse (Donoho 2006). Then, the sparse constraint becomes

$$\min_{x} \ \|x\|_1, \quad \text{s. t. } Ax = p, \tag{4.99}$$

where

$$\|x\|_1 = \sum_{j} |x_j|.$$

Sparsity has been around for decades, but a critical finding comes from compressed sensing (CS) theory, first proposed by Donoho, Candes, and Tao in 2004 (Candes *et al* 2004, 2006, Donoho *et al* 2006). CS theory mathematically rigorously proved that a signal can be reconstructed perfectly if it can be represented sparsely with a certain sparse transform. CS theory breaks through the constraint of Nyquist's theorem and provides a powerful mathematical basis for sparsity. Currently, there are many excellent optimization algorithms based on CS.

Here, we will introduce a powerful sparse regularizer—total variation (TV). TV was proposed by Rudin *et al* in 1992 (Rudin *et al* 1992). For an image X, its TV can be formulated as

$$\|X\|_{TV} = \|\nabla X\|_1 = \sum_{i,j} \sqrt{(\nabla_i X)^2 + (\nabla_j X)^2}, \tag{4.100}$$

where $\nabla_i X$ and $\nabla_j X$ represent gradients along the row and column directions, respectively. In an image, the gradient of each pixel can be obtained by subtracting two adjacent pixels, such as $\nabla_i X = X(i + 1, j) - X(i, j)$ and $\nabla_j X = X(i, j + 1) - X(i, j)$. Through statistical experiments, Rudin *et al* found that the TV of the image would be small when the image is subjected to low noise. Therefore, the TV can be used as *a priori* knowledge of noise-free images and added to the objective function. It turns out that TV can perform well in denoising, in particular for images with the piecewise continuous property.

In 2005, Yu *et al* introduced the TV regularization into CT image reconstruction (Yu *et al* 2005). Then, TV was also used by other groups, for example Chen *et al* (2008) and Sidky *et al* (2006). TV minimization is not the most appropriate for faithful recovery of subtle details since this minimization promotes a piece-constant solution (Han *et al* 2009, Yu and Wang 2009, Yu *et al* 2009). Although TV minimization does smooth a noisy appearance, it may introduce blocky artifacts if it is emphasized too much. To overcome this weakness, high-order variation minimization (Yang *et al* 2010, 2012, Zhao *et al* 2015), low-rank regularization (Gao *et al* 2011), dictionary learning (Xu *et al* 2012, Tan *et al* 2015), and low-dimensional modeling (Cong *et al* 2017) can be utilized for better performance.

4.3.5 Model-based iterative reconstruction

The data model (re-projection) used in the above iterative reconstruction algorithms is idealized, that is, we do not consider all physical factors. Considering particular applications, we need realistically model physical conditions and interactions. Here we give a simple introduction, interested readers can see Nuyts *et al* (2013) for a review.

As discussed before, when performing the projection and backprojection operations, we view a continuous object as a discrete grid. Such implementation will lead to discrepancies between the measured projection data and the simulated re-projections. For every unit in the grid, an average attenuation is assumed, which will bring mismatches in areas corresponding to interfaces between tissues. These mismatches will introduce artifacts during reconstruction. The resultant artifacts will be more pronounced in the areas corresponding to higher contrasts and higher frequencies. Intuitively, using smaller units can alleviate this problem (Zbijewski and Beekman 2003). However, it will mean a larger grid and a higher computational cost. Sometimes, we will adopt the non-uniform discretization (Brankov *et al* 2004) and region-of-interest (ROI) techniques. With these techniques, we can focus on the regions of richer details.

The finite spatial resolution effects are caused by focal spot size, detector size, crosstalk and/or afterglow, as well as various motion effects, which make the projection data blur. The projector maps the pixel/voxel onto the detector, which can take into account these practical complications. The distance-driven method is a footprint based projector. Detector crosstalk will make the adjacent detector elements become correlated within the same view. The detector afterglow will make the signal of a certain view affect the next view(s). An intuitive modeling method is to use a convolution kernel when computing the signals within adjacent detector elements or views (Thibault *et al* 2007). Another method is to enlarge the detector elements into overlapped ones (Zeng *et al* 2009). As for the finite size of the focal spot, it means that the x-ray source is not a point source. It can be modeled by presenting the source as a combination of point sources.

The signals measured by the detectors are usually noisy. The flux of the x-ray tube fluctuates around the mean value. The number of photons and the energy of the photons are both random. The absorption or scattering within an object and a detector are not deterministic either. The rate of conversion of photons into photoelectrons is also stochastic. During the process in which photons are emitted, absorbed, scattered, converted into light photons, and then converted into electron signals, every step will bring noise into data. Hence, CT reconstruction based on a noise model is an important research topic. The researchers perform CT reconstruction by modeling different physical events, which leads to significantly improved results. The Poisson and Gaussian distributions we introduced earlier are the most commonly used noise models. We can use the standard Poisson model to formulate the likelihood function for pre-log data. Also, we can use the Gaussian model to process the post-log data using the weighted least square (WLS) method.

The energy spectrum of x-rays is important to image reconstruction. In practice, we often encounter the beam-hardening problem. The x-rays emitted by the tube contain a continuous spectrum. The attenuation of matter is energy-dependent. Usually, low-energy photons are more easily absorbed by an object. During the transmission of x-rays through the object, the mean energy of x-rays will gradually increase. In other words, the x-ray will become harder. This phenomenon will cause the well-known beam-hardening artifacts in a reconstructed CT image. Modeling the spectrum of x-rays is a prerequisite to rectify the beam-hardening artifacts. We usually build a composite data model to calculate line integrals with respect to different energies and then combine them into energy-integrating data. We also need to utilize the fact that different materials have different sensitivities to energy changes. Usually, we will model an object to be reconstructed in terms of basis materials (Alvarez and Macovski 1976).

Model-based iterative reconstruction usually increases the computational time. However, if the reconstruction is degraded by the severe noises and artifacts, such model-based reconstruction algorithms will be needed to improve image quality remarkably. In doing so, we need to find the best trade-off between image quality and computational cost.

4.4 CT scanner

4.4.1 CT scanning modes

In 1895, the German scientist Wilhelm Röntgen discovered x-rays that could penetrate objects and photographic film in a cathode ray tube experiment, producing an x-ray photograph of the hand of his wife Anna (see figure 4.35). Since then, a new form of radiology called x-ray radiology was created. X-rays were first used in medical diagnostics and were then widely used in the industrial field. X-rays have had a profound impact on human history and technological development. In 1901, Röntgen received the first Nobel Prize in Physics.

However, x-ray photography has its obvious disadvantage in that the image obtained by imaging the inside of the object is two-dimensional, and the information in the depth direction is overlapped. In order to solve this problem, tomography, a technique of imaging the cross-section of an object, has emerged.

In 1963, American physicist Allan McLeod Cormack first proposed an algebraic calculation method using the multi-directional projection of objects for tomography reconstruction (Cormack 1963), which basically solved the mathematical problem of tomography reconstruction. In 1972, British EMI engineer Godfrey Hounsfield developed the first clinical CT scanner, shown in figure 4.36. This was the first CT scanner, used to diagnose brain cysts in a female patient and produce the world's first CT image, which is shown in figure 4.37.

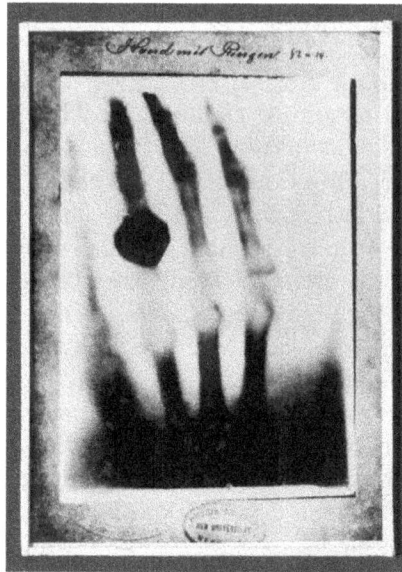

Figure 4.35. The first x-ray photograph. Reproduced with permission. Copyright the Science Museum, London. CC BY 4.0.

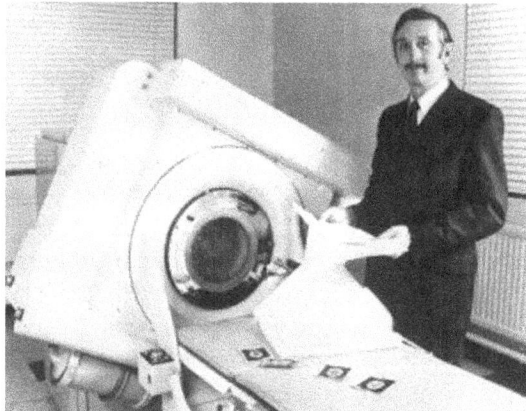

Figure 4.36. Hounsfield and the EMI head CT scanner. Downloaded from http://catalinaimaging.com/history-ct-scan/.

The advent of CT caused a sensation in the field of radiology. It is considered to be another epoch-making contribution to the diagnosis of radiology, after the discovery of x-ray by Röntgen. Considering Hounsfield and Cormack's pioneering contributions to CT development, the 1979 Nobel Prize in Physiology and Medicine was awarded to two scientists without specialized medical experience.

Since then, radiological diagnosis has entered the CT era. CT technology began to develop at high speed. In 1974, American scientist Robert Ledley developed whole-body CT. Electron beam CT (EBCT), developed by Douglans Boyd in 1983, was applied clinically. In 1985, slip-ring technology was developed. In 1989, spiral CT

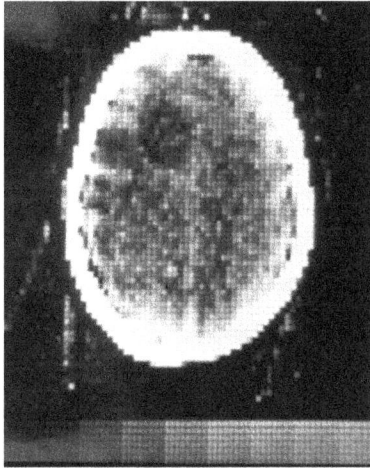

Figure 4.37. The first clinical CT image from Atkinson Morley's Hospital, October 1971. Downloaded from http://www.impactscan.org/CThistory.htm.

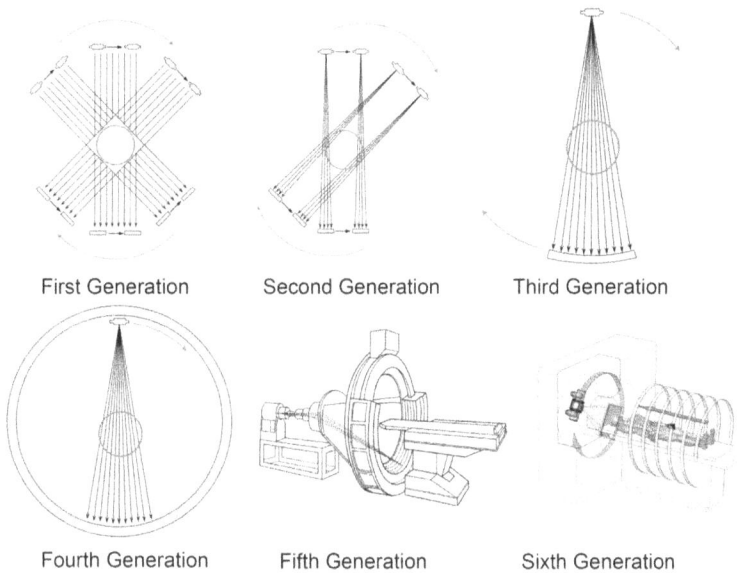

First Generation   Second Generation   Third Generation

Fourth Generation   Fifth Generation   Sixth Generation

Figure 4.38. Six generations of CT scanning modes.

(SCT) was successfully developed. In 1998, a multi-layer CT (MSCT) was success-fully developed. Since the advent of the technology in the early 1970s, CT machines can be roughly divided into six generations based on the timing and structural performance of their development. Figure 4.38 summarizes the main generations of CT scanning modes. The main features of each generation of CT machines are as described in the following.

The first-generation CT scanner. The first-generation CT machine, using a rotation/translation scan mode, is a head-specific machine. The x-ray tube is an oil-cooled fixed anode. This kind of machine sends out a pencil beam, and the number of detectors is generally 2–3. During the scan, the x-ray tube and the detector are rotated around the patient for linear rotation and synchronous linear motion. The x-ray tube rotates by 1° each time and simultaneously scans in a linear motion. Then, it rotates 1° and repeats the aforementioned scanning action until 180 parallel projection values within 180° are completed. The disadvantage of this CT machine is the long scan time. Generally, one section will take 3–5 min and the quality of the obtained image is poor.

The second-generation CT scanner. The second-generation CT machine still uses the rotation/translation scan mode. This kind of machine sends out a narrow fan beam of 5°–20° instead of the pencil beam, and the number of detectors is increased to 3–30. The rotation angle after the translational scan is increased from 1° to the angle of the fan beam. The scan time is shortened to 20–90 s. Compared with the first-generation CT machine, second-generation CT reduces the aperture of the detector, increases the matrix, and improves the accuracy of sampling, resulting in a significant improvement in image quality. The main disadvantage of this scan method is that since the detectors are arranged in a straight line, the measured values of the center and edge portions are not equal for the fan-beam. Therefore, correction is needed after scanning to avoid artifacts. Although the quality of the image obtained by second-generation scanners is improved, the artifacts caused by the physiological motion of the patient cannot be completely avoided.

The third-generation CT scanner. The third-generation CT machine changed the scan mode to rotation/rotation. The x-ray beam is a wide fan beam of 30°–45°, the number of detectors is increased to 300–800, and the scanning time is further shortened to 2–9 s or less. The detector or detector array in this manner is arranged in an arc-shape and without gaps with each other.

This arrangement makes the center and edge of the fan-beam have equal distance to the detector and does not require correction. The disadvantage of this type of scanning is that the sensitivity difference of each adjacent detector needs to be corrected during the scan. Otherwise, ringing artifacts will occur due to the synchronized scanning motion.

The fourth-generation CT scanner. The fourth-generation CT machine scans with only rotation of the tube. The fan angle of the x-ray beam is larger than that of the third-generation CT scanner, reaching 50°–90°. Therefore, the load of the x-ray tube is reduced, such that the scanning time is about 1–5 s. This type of CT machine has more detectors ranging from 600 to 1500, all distributed over a 360° circumference. There is no detector motion during the scan; only the tube is rotated 360° around the patient. The difference of scanning method between the fourth-generation and the third-generation is that the projection for each detector is equivalent to that obtained by taking the detector as the focus and rotating the x-ray tube to scan a fan around the object. This scanning method is also called inverse fan-beam scanning.

The fifth-generation CT scanner. The fifth-generation CT scanner, also known as electron beam CT, is significantly different in structure from previous generations of

CT machines. It consists of an electron beam x-ray tube, a set of 864 fixed detector arrays and a computer system for sampling, sorting, and data display. The biggest difference is the x-ray emitting part, which has an electron gun, a deflection coil and a semi-circular tungsten target in a vacuum. During the scan, the electron beam is accelerated in the axial direction of the x-ray tube, and then the electromagnetic coil focuses the electron beam and bombards the four tungsten targets by instantaneously deflecting the electron beam with the magnetic field. Since the detectors are arranged in two rows of $216°$ rings and four target faces are bombarded in one scan, eight slices can be obtained in one scan.

The sixth-generation CT scanner. The sixth-generation CT scanner is called helical CT or spiral CT, with the first embodiment in the fan-beam geometry. In the previous generation of CT scanners, the data acquisition is not a successive process because the gantry must be stopped after one slice is obtained. To accelerate the data acquisition process, a new technique called 'slip-ring' is introduced into this generation of CT, which makes the gantry rotate simultaneously with the movement of the patient couch through the bore of the scanner. The use of the slip-ring technique in the CT field greatly shortens the scan time. For example, it only takes about 30 s to scan the whole abdomen. With the slip-ring technology, the x-ray source traces a spiral/helical trajectory with respect to the patient, hence the name of this mode is spiral fan-beam CT.

The ideal of spiral fan-beam CT first appeared as a patent in 1987 (Mori 1986). Initial work began in the 1980s and continued in the 1990s (Bresler and Skrabacz 1989, Crawford and King 1990, Kalender *et al* 1990, Crawford 1991, Polacin *et al* 1992, Crawford and King 1993). Spiral fan-beam CT was really intended for faster scanning than what is possible in an incremental translation mode. Naturally, spiral CT fan-beam projections are not consistent on any place, causing longitudinal motion blurring. Initially, it appeared that an improved temporal resolution with a spiral scan must be achieved at the cost of a compromised longitudinal resolution. In contrast to this conventional wisdom of that time, it was theoretically and experimentally demonstrated that spiral fan-beam CT with overlapping reconstructions allows better longitudinal resolution than incremental CT (Kalender *et al* 1994, Wang and Vannier 1994, Kalender 1995). It has become clear now that spiral fan-beam CT is inherently superior to incremental CT in terms of all major image quality metrics.

Currently, the most commonly used form of spiral CT is spiral/helical cone-beam CT. The cone-beam probes a volumetric trunk of a body and acquires data rapidly, instead of passing through a narrow collimator like in the pencil beam and fan-beam cases. To take advantage of cone-beam data, a two-dimensional detector array must be used, which contains multiple detector rows. Basically, with a cone-beam of x-rays and an area detector, a larger number of CT image slices can be acquired in a very short time. As mentioned earlier, the first spiral/helical cone-beam CT algorithm was proposed in 1991 by Wang *et al* (Wang *et al* 1992, 1993), along with a description of the first spiral/helical cone-beam micro-CT prototype. Relative to fan-beam spiral CT, cone-beam spiral CT introduced major mathematical complexity due to the beam divergence and data truncation. While fan-beam spiral

CT reconstruction can be handled with longitudinal interpolation, cone-beam spiral CT demands significantly more complicated mathematical analysis. Note that the transition from fan-beam spiral CT to cone-beam spiral CT involved tremendous industrial and academic efforts (McCollough and Zink 1999, Miller *et al* 2008, Liang and *et al* 2010). In 1991, double-slice spiral CT was introduced. The double-layer spiral CT places two rows of detectors, which can obtain images of two slices in one scan. In 1998, engineers introduced a four-layer spiral CT, and spiral CT began to enter the multi-layer era. Almost every year, there is a new multi-slice spiral CT product. To date, the 64-slice spiral CT technology is very mature. The 64-slice spiral CT has a total of 64 rows of detectors with coverage of up to 4 cm and a thinnest layer of up to 0.64 mm. The spatial resolution on the z-axis is sub-millimeter, and the reconstructed three-dimensional image is subtler. Of course, 64-slice spiral CT also has certain problems. The detector range of the 64-slice spiral CT does not cover the entire organ, so a spiral scan must be performed on each organ. When performing coronary angiography, if the heart rate is fast or not uniform, imaging will easily fail. Therefore, the concept of the latter 64-slice CT has emerged, mainly around the temporal resolution of the multi-layer CT and the coverage of the detector. To improve the coverage of the detector, Toshiba introduced a CT scanner with 320 rows of detectors in 2008, which can obtain 320 layers of 0.5 mm images, and the detector coverage is 16 cm. In 2010, it was upgraded to 640-slice CT, keeping the number of detectors unchanged, and double-sampling by the dynamic bias of the z-axis during the scan. The 320-row CT can scan the main human organs in one circle without the need for a spiral scan. This not only reduces the radiation dose but also ensures that the scanning time of each point of the target organ on the z-axis is consistent. The problem of coronary angiography is well solved. The disadvantage of the 320-row CT is that the large width of the detector will introduce cone-beam artifacts, which degrades the image quality. In terms of increasing the rotational speed of the gantry and shortening the scanning time, Siemens produced a dual-source CT based on the idea of dual detectors and double tubes. The two sets of tubes and detectors are arranged at 90° to each other, and the tubes and detectors only need to be rotated a half circle. Therefore, the scanning speed and temporal resolution are improved. Dual-source CT acquisition is very fast, and it takes only 0.25 s to complete a heart scan. Of course, because the spiral scan is still being performed, the scanning time limit of the dual-source CT at each point of the z-axis will be different, which affects the accuracy of the image.

4.4.2 Detector technology

The detector is one of the core elements of the CT scanner. Detector technology has undergone significant improvements over the past few decades. Depending on the material process, detectors can be divided into solid detectors and gas detectors. Several representative detectors are now introduced in chronological order of development.

Cadmium tungstate crystal detector (CdWO$_4$). With their superior chemical properties, cadmium tungstate crystals became the preferred scintillator material for detectors in the 1970s. The cadmium tungstate crystal has a large x-ray absorption coefficient and a short radiation length. So the detectors can be arranged centrally, reducing the cost of the equipment. However, it also has some obvious shortcomings. First, it is vulnerable to moisture during use, which makes it unstable. Second, the afterglow effect cannot be handled.

Scintillation crystal detector (GOS). The surface of scintillation crystal detectors is covered by a reflective material and coupled to a row of photodiodes. The incident x-rays interact with the scintillator to produce secondary light, which then passes through the photodiode to generate an electrical signal. The photons generated by the scintillation process are oriented in all directions, so highly reflective material is needed to direct the emitted light toward the photodiode at the bottom of the detector. Due to the reflection and absorption of the scintillator, only a small fraction of the photons can reach the photodiode to generate an electrical signal. Therefore, in order to obtain a better image, there must be a high x-ray input energy. The scintillation crystal detector has a high absorption rate, luminous efficiency and photoelectric conversion efficiency. It also has some disadvantages. First, the light transmittance is poor. Only part of the photons excited by the x-rays can finally reach the photodiodes. Second, the poor uniformity of the z-axis affects the image quality. Third, it is necessary to avoid absorbing moisture during use.

High-pressure helium detector. High-pressure helium detectors were commonly used in the 1980s. They utilize the principle of inert gas ionization under x-ray irradiation. There are many sets of positive and negative plates inside the detector, which are separated by insulating material and filled with helium. When x-rays are injected into the detector under the action of high-voltage electricity, the plates collect the ions generated by the helium ionization and induce the corresponding current intensities. The gas detector has a lower absorption efficiency and lower photoelectric conversion efficiency than scintillation crystal detectors and generally requires higher incident intensity. However, due to the consistency of the internal environment, including pressure, density, purity, and temperature, it has high stability and tolerance, which makes frequent calibration unnecessary and it is less affected by temperature and humidity.

Solid-state rare earth ceramic detector. The most widely used rare earth ceramic detectors now have a higher light output rate than other detectors. Their photo-electric conversion rate is twice that of the cadmium tungstate crystal, and their x-ray utilization rate reaches 99%. The stability is excellent, making the images rarely produce ring artifacts. Scanners equipped with rare earth ceramic detectors can perform fast spiral scans because of the short afterglow time. Compared to ceramic detectors of the same period and similar structure, simultaneous and structurally similar cermet detectors have been eliminated by mainstream brand manufacturers due to the afterglow problem. The rare earth ceramic detector has the advantage of a high absorption rate, high luminous efficiency, short afterglow effect, high conversion rate, and high stability. However, there are two main disadvantages. First, the detector unit volume limits further improvement of the resolution. Second, the splicing gap between the adjacent detectors affects the x-ray detection efficiency.

Gemstone detector. Gemstone detectors are obtained by adding rare elements to the gem structure. This is the most revolutionary breakthrough in the CT industry in the 20 years since the invention of rare earth ceramic detectors. Their performance is greatly improved. They are fast, efficient, and stable because of the unique gem structure. Comparing the gemstone detector to other detectors, the initial responding speed of x-rays is increased by 150 times and the afterglow time is reduced by 10 times. Also, the gemstone detector material has better consistency with the photodiode response.

Photon counting detector. As mentioned previously, the attenuation coefficients of matter interacting with x-rays with different energies are different. The x-rays emitted by the tube contain a certain energy spectrum. Therefore, the results obtained by conventional detectors using the integral mode reflect the average attenuation characteristics of the x-rays. The latest photon counting detectors can analyze x-ray energy information. The core of the photon counting detectors is made up of some semiconductor materials, the most commonly used of which are cadmium telluride (CdTe) and cadmium zinc tellurium (CZT). The semiconductor subjected to an x-ray will generate an induced charge on the electrodes, thereby generating a pulse signal. The height of the pulse signal corresponds to the energy of the x-ray photons, and the number corresponds to the number of photons. By setting the electronic threshold, the incident x-ray energy can be discriminated to obtain the count values of different energies.

4.4.3 The latest progress in CT technology

4.4.3.1 Interior tomography

As mentioned above, spiral cone-beam CT is actually a reconstruction from longitudinal data truncation, which has been perfectly solved. Inspired by the success in the field of longitudinal data truncation, researchers became curious about the transverse data truncation, or 'interior problem' (Ye *et al* 2007). In the interior problem, we hope to reconstruct an internal region of interest (ROI) with the projection data of x-rays only through the ROI. It has been proved that the interior problem has no unique solution (Natterer 2001). A series of approximate solutions of the interior problem were studied. They can be summarized into two types: (i) approximate algorithms for image reconstruction over an ROI and (ii) lambda tomography algorithms for edge identification with an ROI, either of which only assumes that local projection data are available. The approximate local reconstruction uses some approximate algorithms and thus cannot obtain an exact reconstruction theoretically. Different from the approximate local reconstruction, the lambda tomography is able to reconstruct the gradient-like function, instead of the distribution function. The lambda tomography has a major mathematical advantage and can theoretically obtain exact fan-beam tomography in several circumstances. However, since the lambda tomography relies on the gradient of the image, it can only capture significant changes in an ROI (Ramm and Katsevich 1996). In a word, the reconstruction by the above traditional CT algorithms from truncated projection data cannot reach a satisfactory performance, and quantitative accuracy will be lost.

To obtain a stable solution for exact reconstruction of an interior problem, interior tomography has been developed since 2007. The key idea of interior tomography is that we need to additionally assume an appropriate prior knowledge for an exact reconstruction. Gel'fand–Graev theory gives the foundation of the correlation between image and projection domains in terms of a Hilbert transform (Gel'fand and Graev 1991). Truncated Hilbert transform data can be uniquely and stably inverted under various general prior knowledge on an ROI. Two common types of prior knowledge in interior tomography are a known sub-region in an ROI (Ye *et al* 2007, Kudo *et al* 2008) or a sparsity model of an ROI (Yu and Wang 2009, Yang *et al* 2010). In known sub-region-based interior tomography, we assume that a sub-region of an ROI is available. In many scenarios, a sub-region is indeed known in advance, such as air in airways, blood through vessels, or images from prior scans. The known sub-region is the inversion of a truncated Hilbert transform. Then, an excellent reconstruction can be obtained by projection onto convex sets (POCS) and singular value decomposition (SVD) methods iteratively from truncated data. However, there is usually no precise information of the sub-region. The sparsity-model-based interior tomography is a way to address this challenge. It is based on the assumption that an ROI is piecewise constant or piecewise polynomial. The total variation (TV) or high-order TV (HOT) have excellent performance in such situations and can solve the interior problem reliably.

Based on focus over an ROI, interior tomography has several advantages over traditional CT. First, interior tomography allows exact reconstruction from fewer data, which means it will require a lower radiation dose. Reducing the angular range of the beam will decrease the radiation dose effectively. Because of the narrow beam, the number of scattered photons becomes smaller, improving the contrast resolution. Second, reconstruction can be obtained with a narrow beam, which means that we can scan an object larger than the field of view. This improves the flexibility of CT, and the size of the object is no longer a concern. Interior tomography can be used for geo-science projects or scans of large patients. Third, interior tomography allows a smaller detector size, a faster frame rate and more imaging chains in a gantry space, all of which contribute to an accelerated data acquisition process. However, reconstruction with fewer data may carry the potential risk that diagnostically critical information may be hidden or lost. Even if sufficient views are available, interior tomography is not as stable as global reconstruction. This weakness can be handled with more prior knowledge, sparsely sampling more global data, or other means.

Interior tomography has been extended to other tomographic imaging modalities, such as single-photon emission computed tomography (SPECT), MRI, differential phase-contrast tomography, and spectral CT. Not only that, Wang *et al* also proposed a multimodality imaging strategy called omni-tomography following the interior tomography principle. As we know, multimodality imaging, including PET-CT, SPECT-CT, optical-CT, PET-MRI, etc, can achieve some specific tasks and such multimodality probes hold great potential for early screening, accurate diagnosis, prognostic value, and interventional guidance. Similarly, omni-tomography also has many potential clinical applications, for example, interior CT-MRI can be applied in cardiac and stoke imaging. More details about omni-tomography can be seen in (Wang *et al* 2012).

4.4.3.2 Multi-spectral CT

Spectral CT has many advantages over traditional CT by imaging x-rays under different energy bins. Traditional CT is actually a multi-energy hybrid imaging that is prone to metal artifacts and beam hardening when attenuated by dense objects. Spectral CT can effectively suppress these artifacts by single-energy imaging and de-artifact techniques. In addition, spectral CT can also increase the contrast in the image and improve the detection rate of the lesion. In traditional CT, the attenuation of some lesions is similar to that of organs and is difficult to distinguish in images. However, by using the difference in maximum attenuation between lesions and organs between specific energy bins and contrast agents, spectral CT can not only enhance the image contrast but also analyze the nature of the lesion. The appearance of spectral CT makes the diagnosis of lesions more abundant. After years of research, the technology to achieve spectral CT is improving.

Sequential scanning imaging technology. The CT imaging system will not change. Two scans are performed using x-rays of high kVp (such as 140 kVp) and low kVp (such as 80 kVp), respectively, while the data are collected. The two kinds of data are spatially matched in the image data to perform dual-energy subtraction.

Dual-tube dual-energy imaging technology. In the dual-energy CT, there are two sets of tubes that simultaneously produce x-rays with high kVp and low kVp x-rays, respectively. The two systems independently collect data and match them in the image space for dual-energy subtraction analysis.

Single-source instantaneous kVp switching technology. Data space energy spectrum analysis is achieved by using instantaneous switching of high and low dual energy (80 and 140 kVp) in a single tube (<0.5 ms energy time resolution) to generate fully matched dual-energy data in space and time.

Double-layer detector technology. Each layer of detectors only excites x-ray photons of a certain energy. The two detectors use a filter to shape the radiation to reduce the energy overlap between low-energy and high-energy rays and use separate detection to obtain high and low-energy projection data and perform dual-energy CT reconstruction.

Photon counting technology. Photon counting detectors can detect the energy of photons in x-rays. This is performed by dividing a wide energy bin into several energy bins and then analyzing the difference in energy information.

4.4.4 Practical applications

4.4.4.1 Medical applications

In 1972, Hounsfield invented the first CT scanner. The CT scanner at the time could only be used for scanning the head. Later, CT was developed to allow for full body scans. The CT during this period was still only used for lesion detection. With the advent of spiral CT and multi-slice spiral CT, CT has been widely used in the medical field. We will now introduce some of the latest applications of CT in the medical field.

CT perfusion imaging. CT perfusion imaging is one of the newly developed fields. It is different from previous CT morphological imaging and belongs to the category of functional imaging. It makes full use of the function of multi-slice spiral CT to show capillary staining. After injecting a contrast agent into a vein, continuous multi-layer scans of specific tissues or organs are performed to obtain the time density curve (TDC) of these areas. The TDC can be used to obtain parameters with different mathematical models, such as bleeding flow (BF), blood volume (BV), mean transit time (MTT), and transit time peak (TTP). These parameters are used to evaluate the function of the tissues or organs. Compared to traditional CT with its rough evaluation and unquantifiable nature, CT perfusion provides a more valuable imaging method for the evaluation of acute or chronic cerebral ischemia and differential diagnosis of benign or malignant tumors.

CT cardiac imaging. The moving organ has always been a blind spot in the clinical application of conventional CT machines. However, multi-slice spiral CT has provided a breakthrough. The latest 64-slice spiral CT can complete a whole heart scan in 5 s. The temporal resolution of each image has also been shortened to 0.4–0.25 s. The cardiac image obtained using 64-slice spiral CT clearly shows the details of structures, including soft plaques, hard plaques, and stents.

CT angiography. Multi-slice spiral CT has a fast scanning speed and high temporal resolution, which makes vascular imaging simple, convenient, safe, and non-invasive. CT angiography can partially or entirely replace traditional angiography and is another major means of non-invasive angiography.

Multi-slice spiral CT can also image and perform three-dimensional reconstruction of cerebrovascular vessels. Three-dimensional reconstruction of CT cerebrovascular vessels can obtain accurate and clear stereoscopic images and display cerebrovascular information in all directions. This can be used for surgical planning, preoperative positioning, etc, and has important guiding significance for cerebrovascular disease surgery.

Virtual endoscope. Virtual endoscopes include virtual angioscopes, virtual bronchoscopes, virtual colonoscopes, virtual gastroscopes, and virtual cholangioscopes. They have several advantages. First, they are non-invasive, safe, and painless for patients. Additionally, they can observe the parts that ordinary endoscopes cannot reach and adjust the transparency and color. Therefore, they can simultaneously observe the conditions inside and outside the cavity, which is more beneficial for observing lesions and provides more accurate and richer information for operations. Virtual endoscopy is a powerful complement and potential alternative to conventional endoscopy.

4.4.4.2 Industrial applications

The application of CT technology in industrial nondestructive testing began roughly in the mid to late 1970s. The initial research work was carried out with medical CT, which detected low-density workpieces such as petroleum cores, carbon composites, and light alloy structures. Because the radiation source used in medical CT has low energy and limited penetration, and the mechanical scanning system is specially designed for the human body, there are obvious limitations in detecting high-density and large-volume objects. Beginning in the early 1980s, the US military first proposed some specialized research programs to manufacture CT equipment for testing large rocket engines or small precision castings. After about 20 years of development, industrial CT research has become a specialized branch and has achieved rapid development in the last decade or so. Industrial CT has the characteristics of clear and intuitive images, high-density resolution, a wide dynamic range of detection signals, and digital images, so it has unique advantages in nondestructive testing.

Flaw detection. The industrial CT image corresponds to the material, geometry, composition, and density of the test piece. A plurality of two-dimensional CT image combinations actually reconstructs the three-dimensional object. The position, orientation, shape, and size of the defect can be obtained from the three-dimensional information. CT flaw detection provides a direct solution to spatial location, depth quantification, and comprehensive problems.

Dimensional measurement and assembly structure analysis. The three-dimensional spatial information obtained by industrial CT can also be used to measure the internal dimensions of complex structural parts and analyze assembled structures of the key components. This can be used to verify that the product size or assembly meets the design requirements.

Density distribution characterization. The density information provided by industrial CT images can be directly used to determine the physical density of uniform materials to verify that the product density meets the design requirements. Of course, establishing a correspondence between CT values and physical density requires a specific calibration technique.

4.4.4.3 Other applications

In terms of security, CT has now become a major security checkpoint for important entrances and exits (such as airports, ports, stations, customs, etc). As a security check, CT has the following advantages. (i) Multiple perspectives can resolve objects stacked in a complex background. (ii) Additional dual energy can quantify the exact atomic number. (iii) It has an extremely high resolution and can distinguish explosives from other low atomic numbers. (iv) It can detect the thickness of the substance and provide three-dimensional density information.

CT has been used in engineering for a long time. Geotomography is the result of applying CT technology to engineering. Currently, the field sources used in geotomography mainly include seismic waves, electromagnetic waves, sound waves, and ultrasonic waves, which result in elastic wave CT, electromagnetic wave CT, and resistivity CT. Its specific applications include geological engineering surveys, nondestructive testing of buildings, underground faults and karst detection, concrete quality testing, and engineering quality testing. It has achieved good results in these areas. In addition, the use of geotomography in the field of geophysics has created a new branch, namely seismic tomography. In recent years, the use of seismic tomography to study geological movements has produced many results.

In recent years, there have been some studies using CT in the agricultural field. Some of these studies use CT to evaluate the quality of agricultural and livestock products. Other studies use CT to analyze agricultural products under different conditions to improve the production and storage of agricultural products.

References

Alvarez R E and Macovski A 1976 Energy-selective reconstructions in x-ray computerised tomography *Phys. Med. Biol.* **21** 733

Andersen A H and Kak A C 1984 Simultaneous algebraic reconstruction technique (SART): a superior implementation of the ART algorithm *Ultrason. Imaging* **6** 81–94

Barrett H H and Swindell W 1996 *Radiological Imaging: The Theory of Image Formation, Detection, and Processing* vol 2 (Cambridge, MA: Academic)

Barrett H H, Wilson D W and Tsui B M 1994 Noise properties of the EM algorithm. I. Theory *Phys. Med. Biol.* **39** 833

Bishop C M 2006 *Pattern Recognition and Machine Learning* (Berlin: Springer)

Bracewell R N and Bracewell R N 1986 *The Fourier Transform and its Applications* vol 31999 (New York: McGraw-Hill)

Brankov J G, Yang Y and Wernick M N 2004 Tomographic image reconstruction based on a content-adaptive mesh model *IEEE Trans. Med. Imaging* **23** 202–12

Bresler Y and Skrabacz C J 1989 Optimal interpolation in helical scan 3D computerized tomography *Int. Conf. on Acoustics, Speech, and Signal Processing (IEEE)* pp 1472–5

Bushberg J T and Boone J M 2011 *The Essential Physics of Medical Imaging* (Philadelphia, PA: Lippincott, Williams and Wilkins)

Candes E, Romberg J and Tao T 2004 Robust uncertainty principles: exact signal reconstruction from highly incomplete frequency information, arXiv:math/0409186

Candes E J, Romberg J K and Tao T 2006 Stable signal recovery from incomplete and inaccurate measurements *Commun. Pure Appl. Math.* **59** 1207–23

Chen G-H, Tang J and Leng S 2008 Prior image constrained compressed sensing (PICCS): a method to accurately reconstruct dynamic CT images from highly undersampled projection data sets *Med. Phys.* **35** 660–3

Chiao P C, Rogers W L, Fessler J A, Clinthorne N H and Hero A O 1994 Model-based estimation with boundary side information or boundary regularization (cardiac emission CT) *IEEE Trans. Med. Imaging* **13** 227–34

Cong W, Wang G, Yang Q, Hsieh J, Li J and Lai R 2017 CT image reconstruction in a low dimensional manifold, arXiv:1704.04825

Cormack A M 1963 Representation of a function by its line integrals, with some radiological applications *J. Appl. Phys.* **34** 2722–7

Crawford C R 1991 Method for reducing skew image artifacts in helical projection imaging *US Patent* 5,046,003

Crawford C R and King K F 1990 Computed tomography scanning with simultaneous patient translation *Med. Phys.* **17** 967–82

Crawford C R and King K F 1993 Method for fan beam helical scanning using rebinning *US Patent* 5,216,601

Dempster A P, Laird N M and Rubin D B 1977 Maximum likelihood from incomplete data via the EM algorithm *J. Roy. Stat. Soc.* B **39** 1–22

De Man B and Basu S 2004 Distance-driven projection and backprojection in three dimensions *Phys. Med. Biol.* **49** 2463

Donoho D L 2006 For most large underdetermined systems of linear equations the minimal L1-norm solution is also the sparsest solution *Commun. Pure Appl. Math.* **59** 797–829

Donoho D L *et al* 2006 Compressed sensing *IEEE Trans. Inform. Theory* **52** 1289–306

Feldkamp L A, Davis L and Kress J W 1984 Practical cone-beam algorithm *Josa* A **1** 612–9

Fessler J A 1994 Penalized weighted least-squares image reconstruction for positron emission tomography *IEEE Trans. Med. Imaging* **13** 290–300

Gao H, Yu H, Osher S and Wang G 2011 Multi-energy CT based on a prior rank, intensity and sparsity model (PRISM) *Inverse Problems* **27** 115012

Gel'fand I M and Graev M I 1991 Crofton's function and inversion formulas in real integral geometry *Funct. Anal. Appl.* **25** 1–5

Gordon R, Bender R and Herman G T 1970 Algebraic reconstruction techniques (ART) for three-dimensional electron microscopy and x-ray photography *J. Theor. Biol.* **29** 471–81

Grangeat P 1991 Mathematical framework of cone beam 3D reconstruction via the first derivative of the radon transform *Mathematical Methods in Tomography* (Berlin: Springer), pp 66–97

Han W, Yu H and Wang G 2009 A general total variation minimization theorem for compressed sensing based interior tomography *J. Biomed. Imaging* **2009** 21

Hubbell J H and Seltzer S M 1995 Tables of x-ray mass attenuation coefficients and mass energy-absorption coefficients 1 keV to 20 meV for elements $z = 1$ to 92 and 48 additional substances of dosimetric interest *Technical Report* (Gaithersburg, MD: National Inst. of Standards and Technology) https://www.nist.gov/pml/x-ray-mass-attenuation-coefficients

Hudson H M and Larkin R S 1994 Accelerated image reconstruction using ordered subsets of projection data *IEEE Trans. Med. Imaging* **13** 601–9

Huesman R, Gullberg G, Greenberg W and Budinger T 1977 User manual: Donner algorithms for reconstruction tomography Lawrence Berkeley National Laboratory. LBNL Report #: PUB-214. Retrieved from https://escholarship.org/uc/item/1mz679x3

Jiang M and Wang G 2003 Convergence of the simultaneous algebraic reconstruction technique (SART) *IEEE Trans. Image Process.* **12** 957–61

Kak A C, Slaney M and Wang G 2002 Principles of computerized tomographic imaging *Med. Phys.* **29** 107

Kalender W A 1995 Thin-section three-dimensional spiral CT: is isotropic imaging possible? *Radiology* **197** 578–80

Kalender W A, Polacin A and Süss C 1994 A comparison of conventional and spiral CT: an experimental study on the detection of spherical lesions *J. Comput. Assist. Tomogr.* **18** 167–76

Kalender W A, Seissler W, Klotz E and Vock P 1990 Spiral volumetric CT with single-breath-hold technique, continuous transport, and continuous scanner rotation *Radiology* **176** 181–3

Kamphuis C and Beekman F J 1998 Accelerated iterative transmission CT reconstruction using an ordered subsets convex algorithm *IEEE Trans. Med. Imaging* **17** 1101–5

Karczmarz S 1937 Angenaherte Auflosung von Systemen linearer Glei-Chungen *Bull. Int. Acad. Pol. Sic. Let. Cl. Sci. Math. Nat.* **35** 355–7

Katsevich A 2002a Analysis of an exact inversion algorithm for spiral cone-beam CT *Phys. Med. Biol.* **47** 2583

Katsevich A 2002b Theoretically exact filtered backprojection-type inversion algorithm for spiral CT *SIAM J. Appl. Math.* **62** 2012–26

Katsevich A 2004 An improved exact filtered backprojection algorithm for spiral computed tomography *Adv. Appl. Math.* **32** 681–97

Kudo H, Courdurier M, Noo F and Defrise M 2008 Tiny *a priori* knowledge solves the interior problem in computed tomography *Phys. Med. Biol.* **53** 2207

Kudo H, Rodet T, Noo F and Defrise M 2004 Exact and approximate algorithms for helical cone-beam CT *Phys. Med. Biol.* **49** 2913

Lange K *et al* 1984 EM reconstruction algorithms for emission and transmission tomography *J. Comput. Assist. Tomogr.* **8** 306–16

Lange K and Fessler J A 1995 Globally convergent algorithms for maximum a posteriori transmission tomography *IEEE Trans. Image Process.* **4** 1430–8

Liang X *et al* 2010 A comparative evaluation of cone beam computed tomography (CBCT) and multi-slice CT (MSCT): Part I. On subjective image quality *Eur. J. Radiol.* **75** 265–9

McCollough C H and Zink F E 1999 Performance evaluation of a multi-slice CT system *Med. Phys.* **26** 2223–30

Miller J M *et al* 2008 Diagnostic performance of coronary angiography by 64-row CT *New Engl. J. Med.* **359** 2324–36

Mori I 1986 Computerized tomographic apparatus utilizing a radiation source *US Patent* 4,630,202

Natterer F 2001 *The Mathematics of Computerized Tomography* (Philadelphia, PA: SIAM)

Nuttall A 1981 Some windows with very good sidelobe behavior *IEEE Trans. Acoustics Speech Signal Process* **29** 84–91

Nuyts J, De Man B, Fessler J A, Zbijewski W and Beekman F J 2013 Modelling the physics in the iterative reconstruction for transmission computed tomography *Phys. Med. Biol.* **58** R63

Peters T M 1981 Algorithms for fast back-and re-projection in computed tomography *IEEE Trans. Nucl. Sci.* **28** 3641–7

Polacin A, Kalender W A and Marchal G 1992 Evaluation of section sensitivity profiles and image noise in spiral CT *Radiology* **185** 29–35

Radon J 1986 On the determination of functions from their integral values along certain manifolds *IEEE Trans. Med. Imaging* **5** 170–6

Ramachandran G and Lakshminarayanan A 1971 Three-dimensional reconstruction from radiographs and electron micrographs: application of convolutions instead of Fourier transforms *Proc. Natl Acad. Sci.* **68** 2236–40

Ramm A G and Katsevich A I 1996 *The Radon Transform and Local Tomography* (Boca Raton, FL: CRC)

Rudin L I, Osher S and Fatemi E 1992 Nonlinear total variation based noise removal algorithms *Physica* D **60** 259–68

Sauer K and Bouman C 1993 A local update strategy for iterative reconstruction from projections *IEEE Trans. Signal Process.* **41** 534–48

Shepp L A and Logan B F 1974 The Fourier reconstruction of a head section *IEEE Trans. Nucl. Sci.* **21** 21–43

Shepp L A and Vardi Y 1982 Maximum likelihood reconstruction for emission tomography *IEEE Trans. Med. Imaging* **1** 113–22

Sidky E Y, Kao C-M and Pan X 2006 Accurate image reconstruction from few-views and limited-angle data in divergent-beam CT *J. X-ray Sci. Technol.* **14** 119–39

Siddon R L 1985 Fast calculation of the exact radiological path for a three-dimensional ct array *Med. Phys.* **12** 252–5

Sukovic P and Clinthorne N H 2000 Penalized weighted least-squares image reconstruction for dual energy x-ray transmission tomography *IEEE Trans. Med. Imaging* **19** 1075–81

Tan S, Zhang Y, Wang G, Mou X, Cao G, Wu Z and Yu H 2015 Tensor-based dictionary learning for dynamic tomographic reconstruction *Phys. Med. Biol.* **60** 2803

Thaller B 2013 *The Dirac Equation* (Berlin: Springer)

Thibault J-B, Sauer K D, Bouman C A and Hsieh J 2007 A three-dimensional statistical approach to improved image quality for multislice helical ct *Med. Phys.* **34** 4526–44

Wang G and Jiang M 2004 Ordered-subset simultaneous algebraic reconstruction techniques (OS-SART) *J. X-ray Sci. Technol.* **12** 169–77

Wang G, Lin T-H, Cheng P-c and Shinozaki D M 1993 A general cone-beam reconstruction algorithm *IEEE Trans. Med. Imaging* **12** 486–96

Wang G, Lin T-H, Cheng P C, Shinozaki D M and Kim H-G 1992 Scanning cone-beam reconstruction algorithms for x-ray microtomography *Scanning Microscopy Instrumentation* vol 1556 (International Society for Optics and Photonics) pp 99–113

Wang G and Vannier M 1994 Longitudinal resolution in volumetric x-ray computerized tomography—analytical comparison between conventional and helical computerized tomography *Med. Phys.* **21** 429–33

Wang G, Ye Y and Yu H 2007 Approximate and exact cone-beam reconstruction with standard and non-standard spiral scanning *Phys. Med. Biol.* **52** R1

Wang G *et al* 2012 Towards omni-tomography—grand fusion of multiple modalities for simultaneous interior tomography *PloS One* **7** e39700

Wang J, Li T, Lu H and Liang Z 2006 Penalized weighted least-squares approach to sinogram noise reduction and image reconstruction for low-dose x-ray computed tomography *IEEE Trans. Med. Imaging* **25** 1272–83

Xu Q, Yu H, Mou X, Zhang L, Hsieh J and Wang G 2012 Low-dose x-ray CT reconstruction via dictionary learning *IEEE Trans. Med. Imaging* **31** 1682–97

Yang J, Cong W, Jiang M and Wang G 2012 Theoretical study on high order interior tomography *J. X-ray Sci. Technol.* **20** 423–36

Yang J, Yu H, Jiang M and Wang G 2010 High-order total variation minimization for interior tomography *Inverse Probl.* **26** 035013

Ye Y, Yu H, Wei Y and Wang G 2007 A general local reconstruction approach based on a truncated Hilbert transform *J. Biomed. Imaging* **2007** 2

Yu G, Li L, Gu J and Zhang L 2005 Total variation based iterative image reconstruction *Int. Workshop on Computer Vision for Biomedical Image Applications* (Berlin: Springer), pp 526–34

Yu H and Wang G 2009 Compressed sensing based interior tomography *Phys. Med. Biol.* **54** 2791

Yu H, Yang J, Jiang M and Wang G 2009 Supplemental analysis on compressed sensing based interior tomography *Phys. Med. Biol.* **54** N425

Zbijewski W and Beekman F J 2003 Characterization and suppression of edge and aliasing artefacts in iterative x-ray ct reconstruction *Phys. Med. Biol.* **49** 145

Zeng G L 2010 *Medical Image Reconstruction: A Conceptual Tutorial* (Berlin: Springer)

Zeng K, De Man B, Thibault J-B, Yu Z, Bouman C and Sauer K 2009 Spatial resolution enhancement in ct iterative reconstruction *2009 IEEE Nuclear Science Symp. Conf. Record (NSS/MIC)* pages 3748–51 IEEE

Zhao Z, Yang J and Jiang M 2015 A fast algorithm for high order total variation minimization based interior tomography *J. X-ray Sci. Technol.* **23** 349–64

IOP Publishing

Machine Learning for Tomographic Imaging

Ge Wang, Yi Zhang, Xiaojing Ye and Xuanqin Mou

Chapter 5

Deep CT reconstruction

5.1 Introduction

In the CT imaging process, image quality may be degraded by various factors. Since it is well known that radiation may cause potential cancerous or genetic diseases (Brenner and Hall 2007), the well-known ALARA (as low as reasonably achievable) principle is widely accepted for CT scanning. Reducing the x-ray flux toward each detector element or decreasing the number of projection views are two common methods to implement low-dose CT (LDCT) imaging. Decreasing the x-ray tube current and shortening the x-ray exposure time during a CT scan can efficiently reduce the radiation dose but also magnifies quantum noise in the projection domain, which may cause a reconstructed image to deteriorate with severe noise and artifacts. On the other hand, sparse sampling can accelerate the scanning procedure resulting in a lower radiation dose, but reconstructing the CT image becomes a question of solving an under-determined linear system of equations. In other words, sparse sampling leads to a system with the number of data points lower than the number of unknown pixels/voxels, which is a challenge for image reconstruction using traditional analytical or iterative algorithms, such as FBP, EM, or SART. Severe streak artifacts will appear in the reconstructed results.

In addition to low-dose sampling, other factors can cause image artifacts as well. For example, the scattering effect in CT images becomes more severe with increasing cone angles. Photon starvation happens when an object being scanned is over-sized or contains high-attenuation materials such as metallic implants. Beam hardening is often a serious problem in extracting quantitative energy-sensitive features. Furthermore, it is very difficult to avoid artifacts in the cardiac or pulmonary regions due to organ motion.

To deal with the above-indicated issues, extensive efforts have been made over the past few decades. The resultant methods can be categorized into three groups: data domain-based methods, image domain-based methods, and iterative reconstruction methods.

Data domain-based methods perform on either raw data or log-transformed data. The main advantage of this approach is that once the hardware parameters and scanning configuration are determined, the impacts on the scanning data can be well described. For example, structural adaptive filtering (Balda *et al* 2012), bilateral filtering (Manduca *et al* 2009), and penalized weighted least-squares (PWLS) algorithms (Wang *et al* 2006) were proposed to improve the sinograms directly. For example, different interpolation methods, including linear, polynomial, cubic interpolation, and their variants (Abdoli *et al* 2011, Gjesteby *et al* 2016), were used to fill in the parts of the sinogram corresponding to the metals. With similar ideas, several methods for image inpainting were introduced by treating sinogram interpolation as image inpainting (Zhang *et al* 2011, Gu *et al* 2006). As another example, based on the characteristics of beam hardening artifacts, Jian *et al* proposed a correction method, which applies several simple operations on sinogram data (Jian and Hongnian 2006).

Image domain-based methods do not need raw data from the scanner, which is more convenient for deployment on the current commercial CT systems. These methods can be typically applied to images reconstructed with analytical methods such as FBP. Several classical methods for image restoration, including the anisotropic diffusion filter, nonlocal means (NLM), dictionary learning, BM3D, etc, were developed for low-dose CT denoising (Ma *et al* 2011, Sheng *et al* 2014, Chen *et al* 2013a, Zhu *et al* 2012). Chen *et al* trained three discriminative dictionaries for the high-frequency bands with different orientations so that the artifacts caused by sparse sampling could be eliminated through a discriminative sparse representation operation in terms of these dictionaries (Chen *et al* 2014). Low-pass and radial adaptive filters can process specific regions to remove the metal artifacts (Bal *et al* 2005). Image-based weighted superposition of the results from different metal artifact reduction algorithms is also an efficient way to reduce metal artifacts (Watzke and Kalender 2004).

Iterative reconstruction methods were widely studied for noise and artifact reduction. According to the Bayesian rule, the reconstruction is equivalent to an unconstrained optimization problem for a specific energy function,

$$E(x) = L(x) + \lambda R(x), \ x \geqslant 0, \tag{5.1}$$

where x is an image to be reconstructed, $L(x)$ is the data fidelity term, $R(x)$ is the regularization term, and λ is the parameter to balance the two terms. The L_2-norm and penalized weighted least square (PWLS) methods are two widely used methods for the measurement of data fidelity. With different priors, various regularization terms were developed. As a popular example, the total variation (TV) discussed in a previous chapter assumes that the target image tends to be piecewise constant. This assumption is coherent with the structures of many simulated or physical phantoms and will achieve satisfactory reconstructed results. However, the structures in clinical images are complicated, and the piecewise constant assumption oversimplifies clinical images. As a result, the TV regularization often causes a notorious blocky effect in reconstructed images.

To outperform the performance of TV, many variants were developed. Using high-order derivatives or combining other priors are ways to compensate for the imperfectness of TV. Sidky *et al* evaluated the L_p-norm to substitute the original L_1-norm for TV minimization-based sparse data CT reconstruction (Sidky *et al* 2007). Niu *et al* combined the ideas of PWLS and total generalized variation (TGV), which is regarded as a second-order TV model, for sparse data CT (Niu *et al* 2014). Classical TV and high-order TV were combined in a weighted form to make a trade-off between the two norms (Zhang *et al* 2013). Fractional-order TV was introduced for sparse data and low-dose CT reconstruction (Zhang *et al* 2016c, 2014a). By adjusting the fractional order properly, more middle and low-frequency details can be preserved. To deal with the over-smoothing problem, an interesting approach is to modify the diffusion property of the regularization term. When introducing an anisotropic property to diffuse edges among neighboring pixels, various configurations are possible to give different edge-preserving TV-based models for sparse sampling, low-dose, and limited-angel CT reconstruction, respectively (Chen *et al* 2013b, Liu *et al* 2012, Tian *et al* 2011).

Before deep learning-based reconstruction, compressive sensing (CS) was widely used for solving ill-conditioned linear systems (Candès *et al* 2006, Donoho *et al* 2006). If the sampling pattern satisfies the restricted isometry property (RIP), aided by a proper sparsifying transform, the original signal can be accurately reconstructed at a much lower sampling rate than the Nyquist sampling rate. One of the most important steps for CS is to find a suitable sparsifying transform. TV is a widely used discrete gradient transform, which practically meets the requirement of CS. Chen *et al* added the TV constraint on the difference image between the target and prior images and formulated a prior image constrained compressed sensing (PICCS) framework for image reconstruction (Chen *et al* 2008). Beyond TV, many other regularization methods were proposed to improve iterative reconstruction results. In Rantala *et al* (2006), an unknown attenuation distribution represented by a wavelet expansion and a Besov space prior distribution were combined into the statistical framework for limited-angle tomography. To utilize the self-similarity within an image itself or even across images, NLM was introduced as the regularization term for different image reconstruction problems. Furthermore, the idea of nonlocal TV (NLTV) was adopted in several models for various imaging tasks (Kim *et al* 2016, Liu *et al* 2016, Zhang *et al* 2014b). Inspired by the robust principal component analysis (RPCA) (Gao *et al* 2011a), low-rank-based models were developed to deal with CT reconstruction, in particular high-dimensional reconstruction problems, such as 4DCT, spectral CT, cine CBCT, etc (Cai *et al* 2014, Gao *et al* 2011a, 2011b). Another popular sparse representation model is dictionary learning, which has been demonstrated to be powerful in signal processing. Xu *et al* proposed a low-dose CT reconstruction model by incorporating a redundant dictionary into an objective function for statistical iterative reconstruction (Xu *et al* 2012). Tan *et al* and Zhang *et al* extended this model into a tensor form for dynamic CT and spectral CT (Tan *et al* 2015, Zhang *et al* 2016b). Inspired by the work on image super-resolution, Lv *et al* constructed two dictionaries: a transitional dictionary for atom matching and a global dictionary for image updating, and proposed a dual dictionary-based reconstruction model for sparse data CT.

Although these methods achieved promising results, very few of them have been deployed on commercial CT scanners for several reasons, including computational complexity, hardware limitation, possible new artifacts, etc.

Recently, deep learning has achieved great successes in the field of image processing and computer vision, which provides a new powerful tool to deal with imaging problems. A brief sketch of the potential directions for deep learning-based medical imaging was drawn in Wang (2016) and McCann *et al* (2017). Several natural paths toward deep learning-based image reconstruction are illustrated in figure 5.1. Similar to the traditional methods, all the methods can be generally categorized into three groups: image domain processing, data domain processing, and deep reconstruction methods. The first two groups focus on filtering directly on a single domain representation (data or image). The third group denotes the methods with data as an input and the reconstructed image as an output without an explicit pseudo inversion operation, such as FBP. Referring to the data from RSNA'18, in the artificial intelligence (AI) based imaging studies, CT accounts for 43% of all imaging devices (the proportions are shown in figure 5.2). Meanwhile, deep learning-based methods have been embedded into commercial CT scanners to yield impressive reconstruction quality. Some representative results given by different companies, including Canon, GE, Philips and Siemens, are shown in figures 5.3, 5.4, 5.5, and 5.6.

In the following sections, we will briefly introduce the area of deep learning-based CT reconstruction. In doing so, we explain some numerical implementations of several methods in our online resource[1].

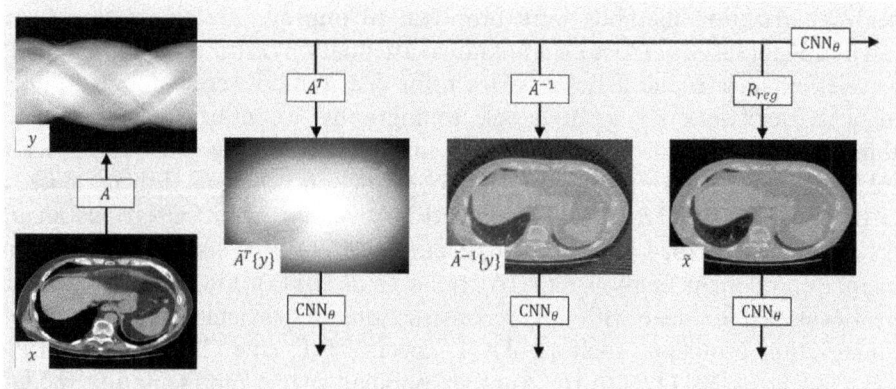

Figure 5.1. Natural paths toward deep learning-based image reconstruction, where x is an image, y is the corresponding projection data, A denotes the system matrix, A^T denotes the transpose of A, $\tilde{A}^{-1}$ is the pseudo-inverse of A, R_{reg} is a regularization term, and CNN_θ represents the neural network characterized by a vector Θ. Reproduced with permission from McCann *et al* (2017). Copyright 2017 IEEE.

---

[1] http://www.fully3d.org/list-60-1.html.

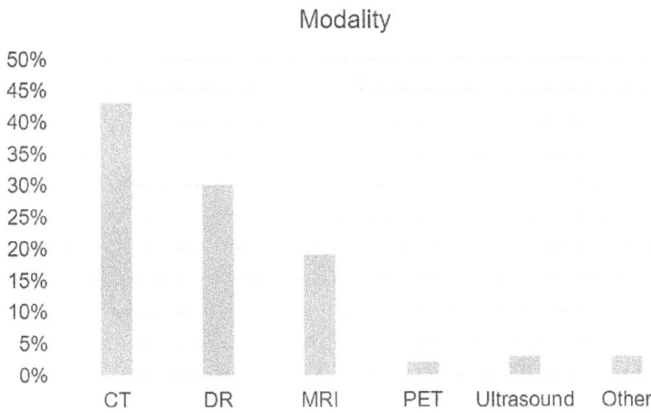

Figure 5.2. Deep learning engagement of different imaging modalities in RSNA'18.

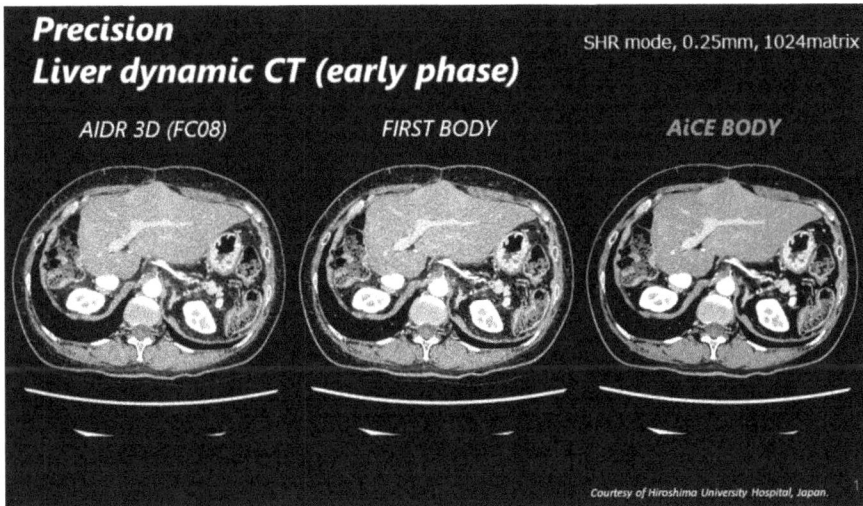

Figure 5.3. Deep learning denoising by the Canon CT scanner. The results are provided by Canon.

5.2 Image domain processing

For image domain processing, let x denote an image produced by a scanner. Due to sampling and dose limitations or physical defects, x usually contains noise and artifacts, which may heavily affect the diagnostic performance. Image domain processing methods estimate $\hat{x} = g(x)$ as close as x^* corresponding to the ground truth (with neither noise nor artifacts). Since it is rather difficult to exactly determine the distribution of the noise and the extent of artifacts in the image domain, traditional image processing methods cannot achieve a satisfactory performance for CT noise/artifact reduction. However, learning-based methods show a great potential to overcome this hurdle. Once a sufficiently large number of samples with decent quality are supplied, current deep learning techniques can effectively deal with this problem. Recently, deep learning has proven powerful in this

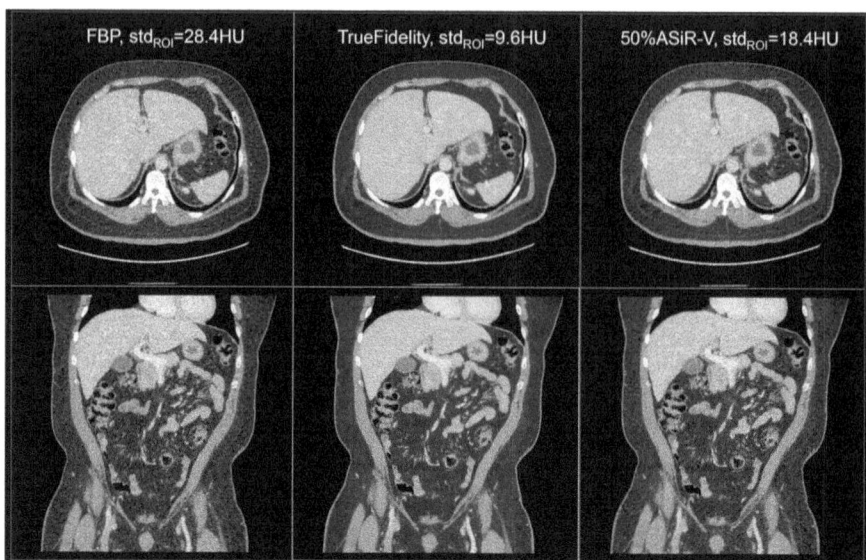

Figure 5.4. Deep learning denoising by GE CT scanner. The results are provided by GE.

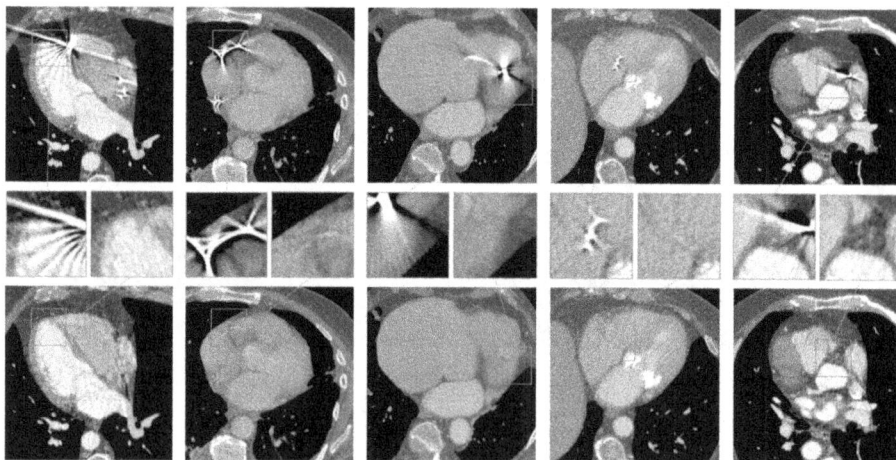

Figure 5.5. Pacemaker artifact removal using a deep learning technique. The top row shows the images with strong artifacts. The bottom row gives the results processed by a deep learning method. The results are provided by Philips.

situation, which can be formulated to estimate the mapping function $g(\cdot)$. The optimization problem can be written as minimizing the following objective function:

$$\hat{g} = \arg\min_{g} \|g(x) - x^*\|_2^2 . \tag{5.2}$$

Inspired by the studies on image restoration with deep learning, two representative neural networks, stacked denoising auto-encoders, and convolutional neural

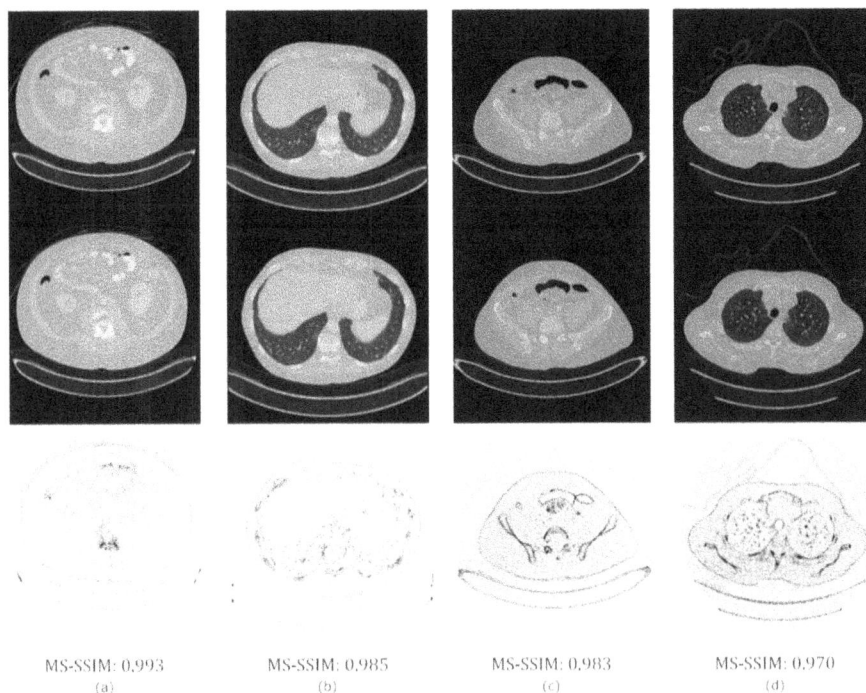

MS-SSIM: 0.993   MS-SSIM: 0.985   MS-SSIM: 0.983   MS-SSIM: 0.970
(a)   (b)   (c)   (d)

Figure 5.6. Image reconstruction via deep learning . The ground-truth images are on the top, deep learning reconstructed images in the middle, and the difference images are at the bottom. The results are provided by Siemens. Reproduced with permission from Whiteley and Gregor (2019). Copyright 2019 SPIE.

networks, were first adapted for low-dose CT denoising (Ma *et al* 2016) and limited-angle artifact reduction (Zhang *et al* 2016a), respectively.

Stacked sparse denoising auto-encoders (SSDAs) were first proposed for unsupervised learning from noisy samples. By ignoring the operation of splitting the trained symmetrical network into two parts to extract the features, SSDAs predict the noise free samples using noisy ones. In Ma *et al* (2016) and Liu and Zhang (2018), SSDAs were used as a supervised learning tool, whose samples (noisy and clean patches) were well paired. However, associated with fully connected layers there are two drawbacks for SSDAs: (i) the sizes of inputs and outputs are fixed, and the network needs to be retrained if the image size changes; and (ii) the number of parameters is huge, and the depth of the network will be limited.

The famous super-resolution convolutional neural network (SRCNN) (Dong *et al* 2015), which was designed for image super-resolution and only has three convolutional layers, was adopted (Zhang *et al* 2016a) for limited-angle CT. Similar ideas were used for low-dose CT and metal artifact reduction (Chen *et al* 2017b, Gjesteby *et al* 2017). In Chen *et al* (2017b), an input image of any size is allowed for CT image denoising, thanks to the convolution operation. In this first peer-reviewed journal paper on deep-learning-based low-dose CT denoising, paired image patches were extracted to represent local structures and boost the number of training samples. The network architecture is illustrated in figure 5.7.

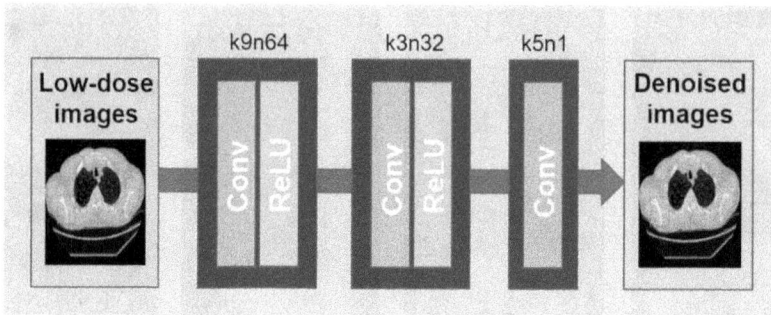

Figure 5.7. The network architecture of Chen *et al* (2017b).

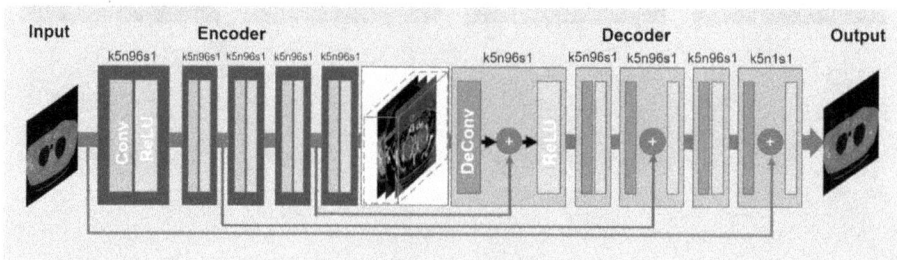

Figure 5.8. The network architecture of RED-CNN featured by symmetric matching of convolutional and deconvolutional layers (Chen *et al* 2017a).

5.2.1 RED-CNN

Structural details in images are important for clinical diagnosis. Some famous network architectures for computer vision, such as SRCNN, may lose details due to the convolutional and pooling operations. Meanwhile, the representation ability of these methods is limited due to the shallow depth of the architectures, which initially treated LDCT image denoising as a low-level task without intention to extract features. This is in sharp contrast to high-level tasks such as detection or classification, in which pooling and other layers are widely adopted to circumvent image details and extract topological structures. To maintain more image details, more advanced technologies of deep learning were developed in the subsequent investigations.

Inspired by the idea of auto-encoders and as a major improvement for deep learning-based CT denoising, the deconvolution layers and shortcut connections were introduced into the CNN model, which is referred to as the residual encoder–decoder convolutional neural network (RED-CNN) (Chen *et al* 2017a). The top-level flowchart of RED-CNN is shown in figure 5.8. This network is composed of ten layers, including five convolutional and five deconvolutional layers which are symmetrically arranged. Shortcut connections are utilized to joint the matching convolutional and deconvolutional pairs. A rectified linear unit (ReLU) is used behind each layer.

To train RED-CNN, overlapped patches are extracted effectively and efficiently. Traditional stacked auto-encoders usually consist of fully connected layers, resulting in a large model with a huge number of parameters. Instead of fully connected layers, RED-CNN adopts a chain of convolutional layers, which essentially act as multi-resolution noise filters. Moreover, due to its high likelihood of losing details of input images, pooling operations are discarded. As a result, there are only two types of elements in our encoder part: convolutional layers and ReLU units. Although the pooling layers are removed, successive convolutions will still obliterate details of input images. In reference to the recent results on image segmentation (Noh *et al* 2015, Drozdzal *et al* 2016), deconvolution layers are used in RED-CNN to retain details. Similarly, there are only two types of elements in the decoder network: deconvolution and ReLU. Since the encoders and decoders appear in pairs in the architecture of auto-encoders, the convolutional and deconvolutional layers were symmetrically configured. To ensure the input and output of the network have the same sizes, the convolutional and deconvolutional layers should use the same kernel size. As illustrated in figure 5.8, the first convolution layer connects to the last deconvolutional layer, the third convolution layer corresponds to the eighth deconvolutional layer, and the fifth convolution layer connects to the sixth deconvolutional layer. In summary, one of the most important features of RED-CNN is the symmetrical architecture featured by paired convolution and deconvolutional layers. Note that deepening the network may not be always good, and an aggressive network depth causes two issues. First, the problem becomes too complicated to preserve structural details well, even with the deconvolutional layers. Second, it will aggravate the gradient vanishing problem and make the network difficult to train. To handle these issues, the residual compensation mechanism (He *et al* 2016) was introduced into RED-CNN. Different from the U-Net, which concatenates features directly, a residual mapping is adopted as demonstrated in figure 5.9. By defining the input and output of the residual block as I and O,

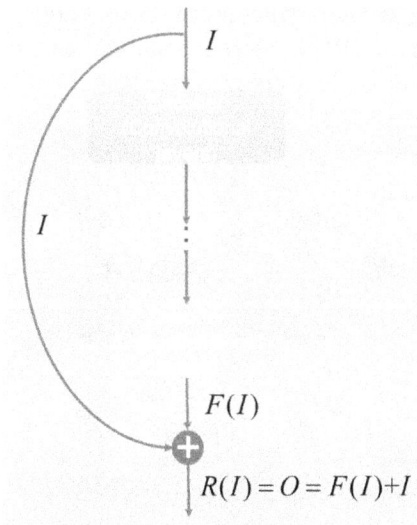

Figure 5.9. Residual mapping structure so that the network learns a residual mapping instead of the direct mapping.

respectively, the final output can be rewritten as $R(I) = O = F(I) + I$, where $F(I)$ denotes the output of the decoder. As a result, the problem is transformed from learning a direct mapping to a residual mapping. There are two benefits associated with the residual mapping. First, as stated in He *et al* (2016), a residual mapping is helpful to avoid gradient vanishing and make the training procedure easier than direct mapping, since it is easier to optimize the residual mapping than to optimize the direct mapping. Second, since the residual mapping directly transfers the input into the computation of the output, it helps to avoid the structural loss at the stage of feature extraction and preserves subtle features, which are clinically important for low-dose CT imaging.

Typical results of RED-CNN are shown in figure 5.10. It can be noticed that the results of RED-CNN represent considerable improvements in terms of both noise suppression and contrast retention. In the magnified region, a lesion can be better identified by RED-CNN than revealed by the traditional methods. However, it also can be noticed that the main difference between the reference image and the result of RED-CNN is that the mottle-like details in the normal-dose CT (NDCT) image are smoothed by RED-CNN. The reason for this lies mainly in the use of the mean squared error (MSE) as the loss function, as suggested by Johnson *et al* (2016) and Ledig *et al* (2017) that this per-pixel MSE is often associated with over-smoothed edges and compromised details.

To further enhance the denoising performance, Zhang *et al* (2018) replaced convolution layers in the encoder of RED-CNN with a dense block (Huang *et al* 2017). This modification has been proven successful in reducing the number of parameters and enhancing the feature conservation.

5.2.2 AAPM-Net

The well-known U-Net (Ronneberger *et al* 2015) for medical image segmentation was first adopted for image reconstruction to suppress the streak artifacts caused by sparse sampling (Jin *et al* 2017). In the resultant network FBPConvNet, three properties of U-Net are utilized for image reconstruction from sparse data: (a) multi-

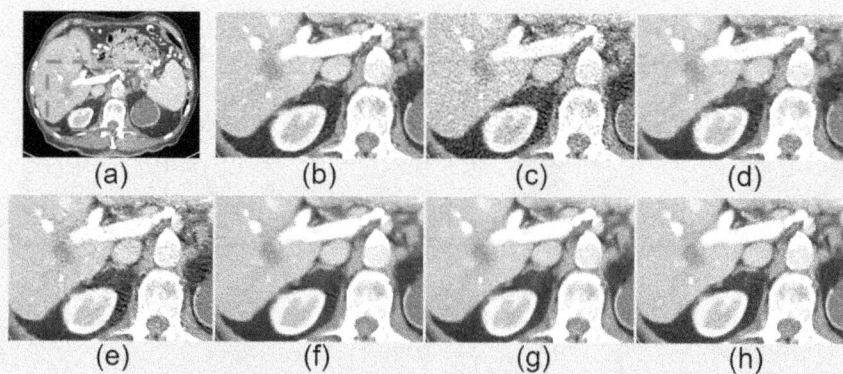

Figure 5.10. The results of an abdominal scan with different methods. (a) The reference image; (b) the zoomed part of (a) in the dashed line box; (c) a low-dose image; (d) TV-POCS (Sidky *et al* 2006); (e) K-SVD (Chen *et al* 2013a); (f) BM3D (Sheng *et al* 2014); (g) AAPM-Net (Kang *et al* 2017); and (h) RED-CNN (Chen *et al* 2017a).

level decomposition, similar to the mechanism of wavelets, is implemented as a dyadic scale-based pooling to produce multi-level features; (b) multi-channel filtering is implemented by stacking multiple feature maps at each layer to enhance the expression ability of the network; and (c) residual learning addresses the problem of gradient vanishing. With the help of these properties, FBPConvNet achieved a better performance than the classical TV-based method.

For LDCT imaging, it is difficult for existing image denoising methods to obtain satisfactory results, as image noise is usually combined with streak artifacts caused by beam hardening and photon starvation. In Kang *et al* (2017), Kang *et al* proposed a new CNN-based AAPM-Net, which employs the U-Net as the backbone network. First, a contourlet transform is used as a directional local analysis and to facilitate network training. Specifically, for a given high-pass filter $H_1(z)$ and low-pass filter $H_0(z)$, a number of pyramids can be obtained using the filter banks. More specifically, the kth level pyramid is expressed as

$$H_n^{eq}(z) = \begin{cases} H_1(z^{2n-1})\Pi_{j=0}^{n-2}H_0(z^{2j}), & 1 \leqslant n < 2^k \\ \Pi_{j=0}^{n-1}H_0(z^{2j}), & n = 2^k \end{cases}. \qquad (5.3)$$

The directional wavelet transform is applied to the high-pass subbands to decompose them into different directional components. Aided by this operation, the noise and artifacts in the LDCT image can be extracted into the high-frequency components. Then, several new CNN techniques were introduced into the architecture of AAPM-Net as illustrated in figure 5.11. An input LDCT image is initially decomposed into four levels using a contourlet transform. Since the noise and artifacts are mainly contained in the high-frequency components, the low-frequency part is directly passed to the output, which forms a residual structure. Meanwhile, shortcuts are also used in internal modules to alleviate the difficulty in training the deep network. Another important technique utilized in this network is the

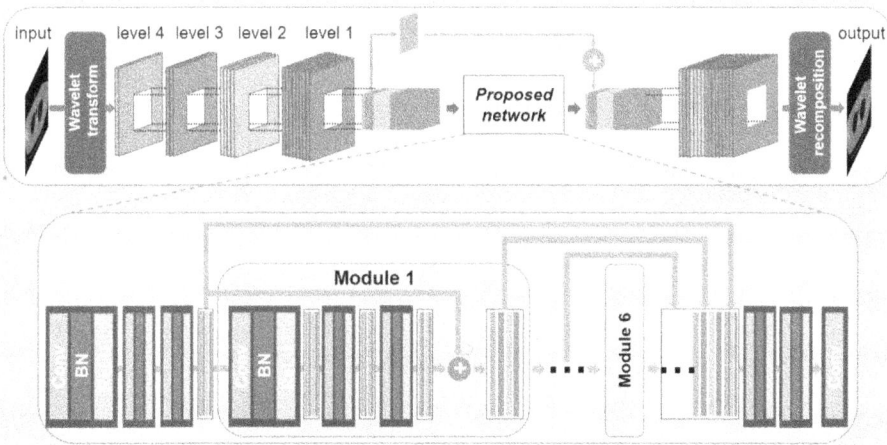

Figure 5.11. The network architecture of AAPM-Net. Reproduced with permission from Kang *et al* (2017). Copyright 2017 John Wiley and Sons.

contracting patch (Ronneberger *et al* 2015). Since the pooling and up-sampling operations may cause loss of details, high-resolution features are directly stacked with the up-sampled output to provide more information.

The results demonstrated that this method gave a performance comparable to the current commercial model-based iterative reconstruction (MBIR) methods. The same architecture was also applied for interior tomography and obtained competitive results (Han *et al* 2017).

Ye *et al* studied the connection between deep learning and signal processing theory by extending the idea of convolution framelets to interpret a deep neural network in terms of perfect reconstruction (Ye *et al* 2018b). Based on an iterative low-dose CT denoising algorithm functionally similar to framelet-based image denoising (Kang *et al* 2018), two variants of U-Net were designed (Han and Ye 2018) for sparse data CT, which are called dual frame U-Net and tight frame U-Net, respectively. To meet the condition for the nonlocal basis, these two methods use short connections to pass low-frequency or high-frequency components of images to the relevant U-Net layers.

5.2.3 WGAN-VGG

Recently, the generative adversarial network (GAN) has drawn rapidly increasing attention in the field of machine learning (Goodfellow *et al* 2014). In GAN, a discriminator network is trained to maximize the discrepancy between the distributions of real and synthetic data, which can significantly improve the performance of traditional networks; and also a generative network is trained to generate data as realistically as possible.

Wolterink *et al* were the first who applied GAN for 3D low-dose cardiac CT denoising (Wolterink *et al* 2017). A seven-layer CNN was adopted as the generator to predict the normal-dose CT image. A discriminator network consisting of nine convolutional layers and one fully connected layer was used to help improve the performance. The method proposed in Wolterink *et al* (2017) has two main drawbacks. First, GAN suffers from a considerable difficulty in training. Second, as revealed by Johnson *et al* (2016) and Ledig *et al* (2017), this per-pixel MSE is subject to over-smoothing and blurry features.

To tackle these two problems, Yang *et al* replaced GAN with WGAN (Yang *et al* 2018), which facilitates the training process with Wasserstein distance. Furthermore, since the MSE-based loss function is associated with structural blurring, the perceptual loss, which originated from the work for image super-resolution (Ledig *et al* 2017) and was implemented with the VGG network, was used to sense the perceptual image quality.

In the classical GAN (Goodfellow *et al* 2014), the generator G and discriminator D are trained by optimizing the following objective function:

$$\min_G \max_D L_{GAN}(D, G) = E_{x \sim P_r}[\log D(x)] + E_{z \sim P_z}[\log(1 - D(G(z)))], \quad (5.4)$$

where $E(\cdot)$ denotes the expectation operator, and P_r and P_z are the real and noisy data distributions, respectively. The generator G maps a sample drawn from the naive distribution to a synthetic sample, which comes from a data distribution denoted by P_g. Since D is trained to be the optimal discriminator for a fixed G, the minimization of G is equivalent to minimizing the Jensen–Shannon (JS) divergence between P_r and P_g, which may lead to gradient vanishing for G. As a solution, Arjovsky *et al* (2017) proposed using the Wasserstein distance instead of the JS divergence for GAN, which is referred to as WGAN, since the Wasserstein distance is continuously differentiable everywhere under certain mild assumptions while neither the Kullback–Leibler (KL) nor JS divergence is. Subsequently, WGAN was further coupled with the gradient penalty to accelerate the convergence (Gulrajani *et al* 2017).

The introduction of the perceptual loss is also important for two reasons. As mentioned in the first chapter, in the human vision system (HVS) features are basic elements for human visual perception, which are compared structurally instead of being compared pixel-wise. Based on this insight, a pre-trained VGG (Simonyan and Zisserman 2014) network is employed to measure differences in a proper feature space to simulate the inner-workings of the HVS. Second, LDCT images approximately distribute over a low-dimensional manifold. The pixel-wise MSE can only measure the superficial differences in the naive Euclidean space, which is not consistent with the way human perception works and may cause unpleasant artifacts. After being transformed into the feature space, images are projected onto a specific manifold, and the geodesic distance is readily calculated to compare images with respect to their intrinsic structures. As a result, the perceptual loss can enhance the diagnostic performance of the network outcomes with superior structural details.

The architecture of WGAN-VGG is illustrated in figure 5.12. While the last three fully connected layers of the VGG network were discarded, the first 16 CNN layers were kept as the feature extractor. The VGG network was pre-trained with natural images from ImageNet, and the Euclidean distance in the feature space was calculated as the perceptual loss. In figure 5.13, it can be observed that the results of WGAN-VGG enjoy patterns similar to the original normal-dose CT counterparts; for example, see the mottle-like patterns.

In another LDCT denoising study (Shan *et al* 2018), the ideas of conveying path and convolutional encoder–decoder (CPCE) were combined for 3D low-dose CT

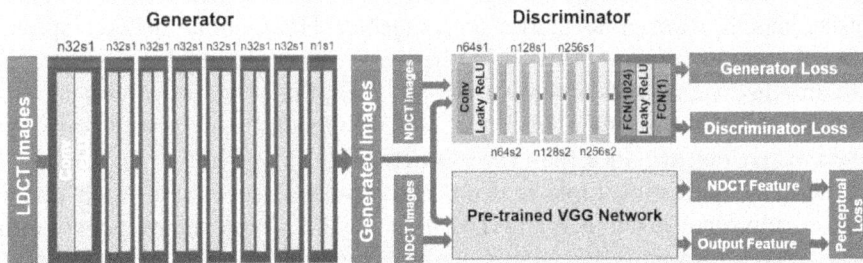

Figure 5.12. Network architecture of WGAN-VGG featuring the Wasserstein distance and the perceptual loss (Yang *et al* 2018).

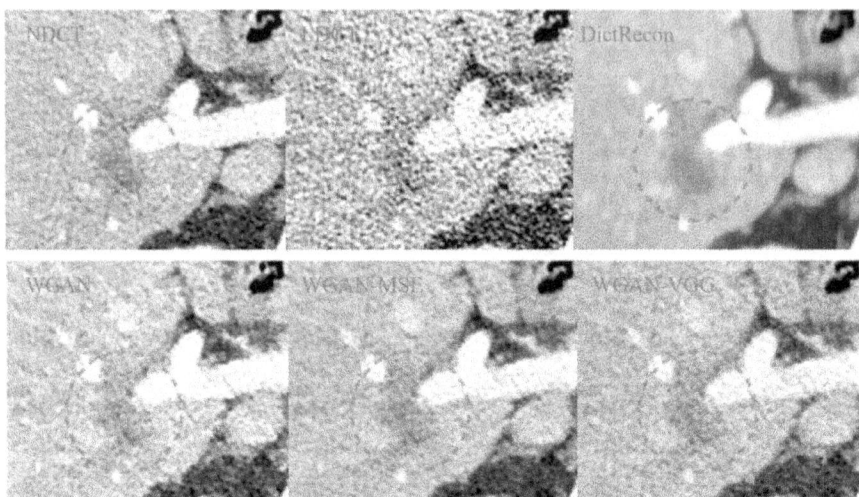

Figure 5.13. Results reconstructed using different methods, including DicRecon (Xu *et al* 2012), WGAN (with no other additive losses), and WGAN-MSE (with MSE loss in the WGAN framework). The regions in the dashed red circles enclose a lesion. Reproduced with permission from Yang *et al* (2018). Copyright 2018 IEEE.

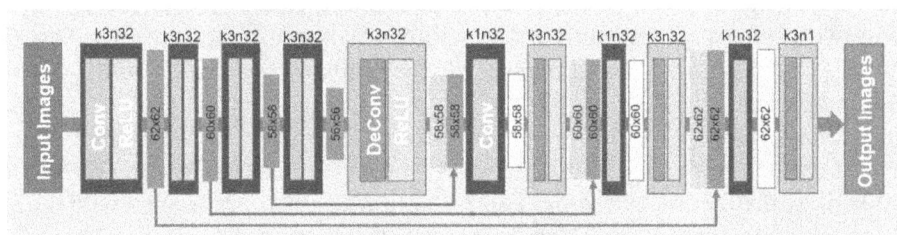

Figure 5.14. Network architecture of CPCE (Shan *et al* 2018).

denoising. The conveying path, originally proposed in the U-Net, transfers the previous feature maps in the encoders to the decoders as the input by concatenating the feature maps from the two sides of the conveying path. Figure 5.14 shows the characteristics of the conveying path. This operation can preserve high-resolution details very well. This strategy was previously utilized in DenseNet and achieved a promising classification performance on ImageNet (Huang *et al* 2017). Inspired by the idea of transfer learning, after a 2D network was pre-trained the weights were used to initialize the weights in a 3D network. Then, the 3D network was fine-tuned with 3D scratches. Using this technique, the 3D network can rapidly converge to achieve a better performance than the 2D network.

Yet another interesting topic is deep learning-based super-resolution imaging. Such an example was given in Wang (2016), in which a simple neural network maps a blurry image at an early iteration stage to a high-quality one at a late iteration stage. Due to the limitation to accessing the paired normal- and low-dose images in multi-phase coronary CT angiography (CTA), traditional supervised learning-based methods cannot obtain satisfactory results. As a result, a cycle-consistent adversarial denoising network (CycleGAN) (Zhu *et al* 2017), which can effectively deal with unsupervised learning, was employed to learn the transform between two different dose levels (Kang *et al* 2019). To efficiently constrain the mapping learned with GAN and maintain features exhibited in target images, CycleGAN was also adopted as the backbone of the proposed network for CT super-resolution reconstruction (You *et al* 2019) with promising results, shown in figure 5.15.

For dual-energy CT image decomposition, with double entries and double exits, a butterfly network was proposed to link paired dual-energy images and their corresponding decomposed images (Zhang *et al* 2019). The crossover architecture helps information exchange synergistically.

For metal artifact reduction, Zhang and Yu trained a five-layer CNN to predict a CNN prior image (Zhang and Yu 2018). After forward projection, the prior image was transformed to a prior sinogram. Finally, the prior sinogram was combined with the original sinogram and the metal trace to correct the sinogram.

Finally, let us look at a high-level comparative study recently published in *Nature Machine Intelligence* (Shan *et al* 2019). With CPCE as a basic module, Shan *et al* proposed a modularized adaptive processing neural network (MAP-NN), which is a series of conveying-link-oriented network encoder–decoder (CLONE) units, for LDCT progressive denoising. The main novelty of this work is to involve the radiologists in the progressive loop so that the denoising process can be guided by domain experts to an optimal extent in a task-specific fashion. This approach can be

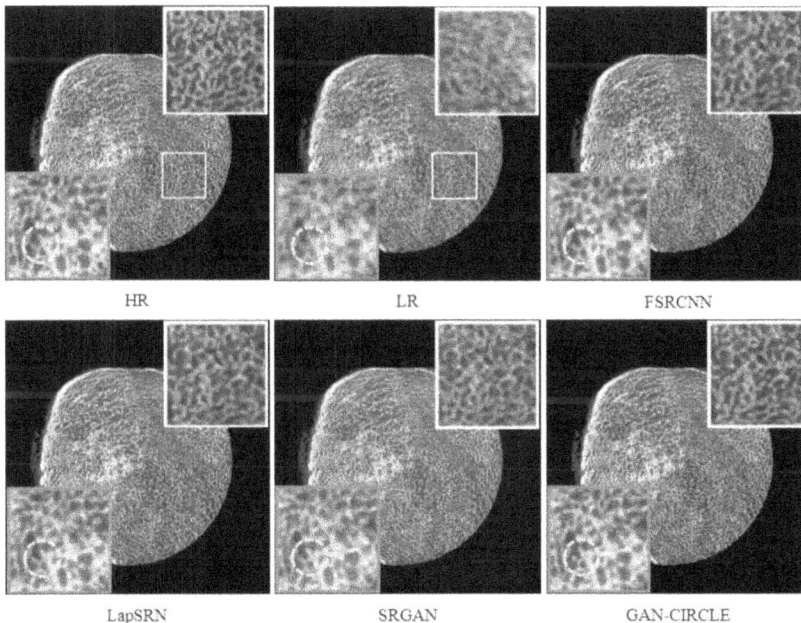

Figure 5.15. CT super-resolution results with different methods, including FSRCNN (Dong *et al* 2016), LapSRN (Lai *et al* 2017), SRGAN (Ledig *et al* 2017), and GAN-CIRCLE (You *et al* 2019), respectively.

also applied on NDCT images for denoising beyond the reference image, which is a unique merit different from other end-to-end methods that directly map LDCT images to the NDCT counterparts. A reader study was performed to compare the results obtained using the proposed method and three typical iterative reconstruction methods used for three brands of CT scanners (GE, Siemens, and Philips). The results are given in figure 5.16. In figure 5.16, A, B, and C, respectively, are denoted the results from one of the three types of commercial scanners in a randomized order to protect the identities of the scanners and in a double blind fashion to eliminate any bias. The results show a competitive performance of the deep learning-based method compared to the commercial iterative reconstruction methods tested in this study across vendors and across body regions.

5.3 Data domain and hybrid processing

For data domain processing, let y denote the measurement obtained from the scanner. Under practical conditions, y usually contains data corruption (noise and biases), which may significantly degrade the image quality. Data domain processing methods estimate $\bar{y} = f(y)$ as closely as feasible to recover the ideal measurement y^* corresponding to the perfect data acquisition system:

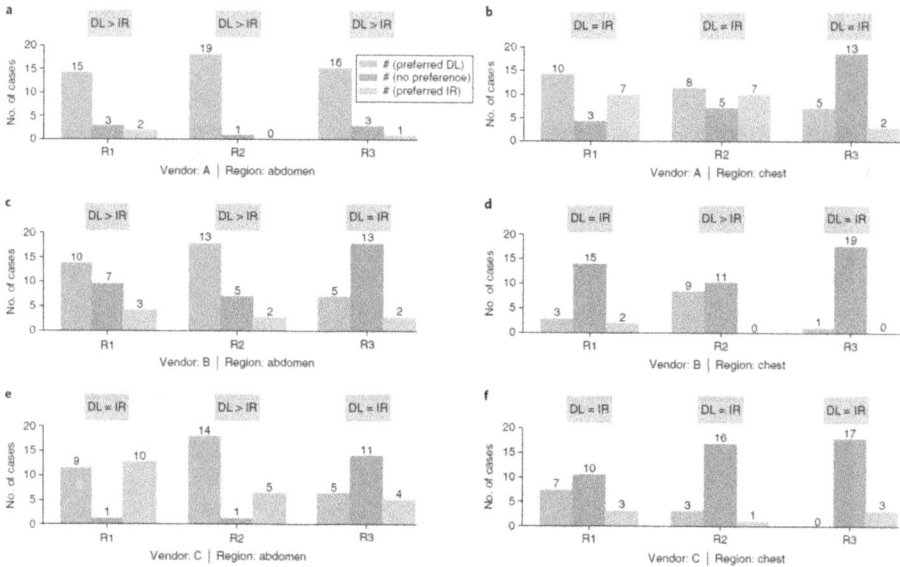

Figure 5.16. Comparison between the best deep learning reconstruction and best iterative reconstruction algorithms in this study for the abdominal and chest regions across three major vendors (A, B, and C) and three experienced readers (R1, R2, and R3). (a)–(f) Histograms showing the number of cases per class in 20 cases for the abdomen from vendor A (a), the chest from vendor A (b), the abdomen from vendor B (c), the chest from vendor B (d), the abdomen from vendor C (e) and the chest from vendor C (f). The text in the gray box above each plot for each reader gives the results evaluated by the sign test at a 5% significant level. Reprinted by permission from *Nature Machine Intelligence* (Shan *et al* 2019). Copyright 2019 Macmillan Publishers Ltd.

$$\hat{f} = \arg\min_{f} \|f(\bar{y}) - y^*\|_2^2.$$ (5.5)

For traditional methods, the main idea is to design filters according to the statistical characteristic of noise and biases in the data domain, which are usually handcrafted and mathematically involved. Over the past few decades, extensive efforts have focused on developing different filters for image feature extraction, in particular in the field of computer vision.

As an emerging direction, deep learning-based methods are being developed to deal with the corrupted raw data or sinogram data.

Claus *et al* reported preliminary results by adopting a three-layer fully connected network to recover contaminated parts in the sinogram (Claus *et al* 2017). In Park *et al* (2018), to simplify the learning process, a patient's implant type-specific learning model (hip prosthesis) was considered. Then, U-Net was utilized to learn the mapping, aided by the data inconsistency.

Compared to the aforementioned image domain methods, the deep imaging efforts in the data domain are relatively limited. The main reason is that raw data or sinogram data are not widely available. In fact, any side effects caused by operations in the data domain will introduce image artifacts, being typically diffused over the whole field of view. A synergistic strategy is to combine image domain and data domain methods to have dual-domain learning capabilities. Han *et al* concatenated two networks for highly sparse data CT (Han *et al* 2018). After being reconstructed by FBP from an original under-sampled sinogram, a noisy image was first fed into an image domain CNN to suppress the artifacts. After that, a forward projection operation was applied to obtain the corresponding sinogram. Then, a sinogram domain CNN was used to further filter the sinogram. The final reconstruction will be obtained by applying FBP to the improved sinogram.

To address the limited-angle CT reconstruction problem, a GAN framework was proposed (Anirudh *et al* 2018). Inspired by the success of 1D CNNs in natural language processing, the proposed generator network treats a sinogram as a 'sequence', which models projections from successive views, and captures the relationship across different views through an attention mechanism.

In Lee *et al* (2019), the authors compared deep learning models in the image, sinogram, and hybrid domains for sparse data CT. Three diagrams are given in figure 5.17 for the three types of models. The experimental results suggested that the hybrid domain deep learning approach was able to reconstruct images most similar to the original fully sampling references. Zhao *et al* reached a similar conclusion, that the hybrid domain learning method showed better performance than the single domain scheme. Also, they used GAN in their dual-domain learning method (Zhao *et al* 2018).

Due to the approximation to the physical process from a nonlinear integral model to a linear one, a reconstructed image usually suffers from poor quantification of attenuation coefficients and significant beam-hardening artifacts. Different from other learning methods for image reconstruction in this category, Cong and Wang fixed the gap via deep learning (Cong and Wang 2017). They utilized a multi-layer

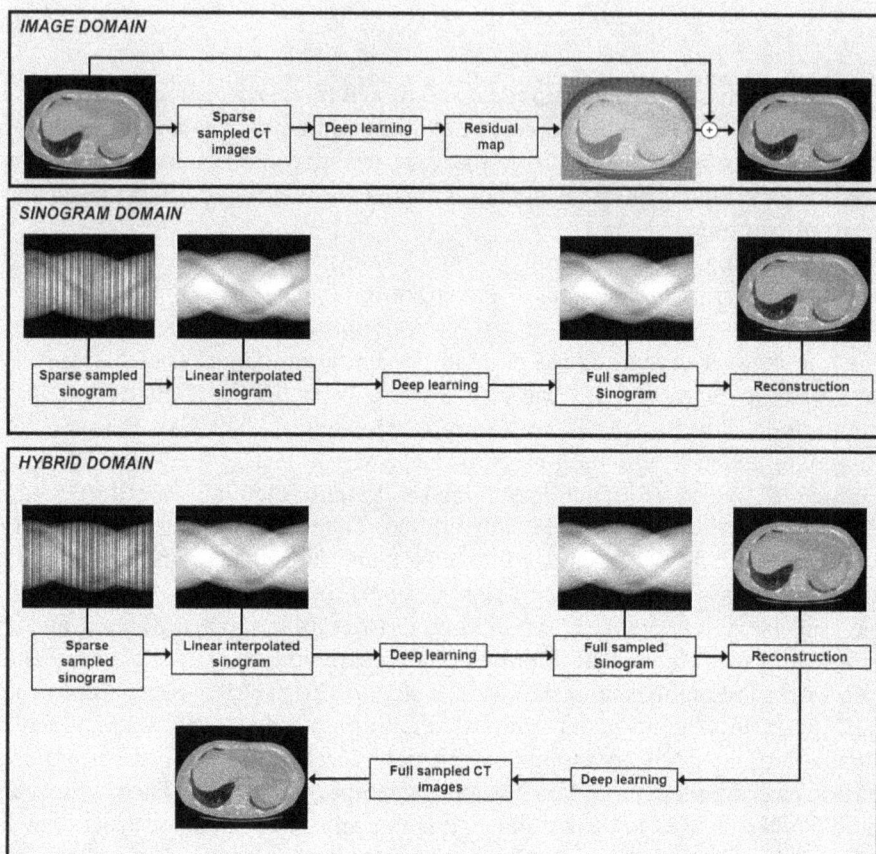

Figure 5.17. Diagrams for deep learning models in the image, sinogram, and hybrid domains, respectively. Reproduced with permission from Lee *et al* (2019). Copyright 2019 John Wiley and Sons.

perceptron (MLP) to learn an x-ray path-based mapping from big data, which can efficiently correct measured projection data to accurately match the linear integral model and eventually realize monochromatic imaging without significant beam-hardening artifacts for the purpose of therapeutic planning.

5.4 Iterative reconstruction combined with deep learning

In the previous two sections, the networks directly process information in either the data or image domain. Although the results are encouraging, these methods inherently take little advantage of traditional algorithms that were developed through physical modeling and mathematical derivation. This lamentable omission might make the network-based reconstruction results sub-optimal. To compensate for this weakness, we can design reconstruction networks in reference to established reconstruction algorithms.

The first method is to mix the iterative reconstruction algorithms and deep learning techniques for improved imaging performance. A direct way is to impose the network trained in a single domain into the framework of regularized iterative reconstruction. The traditional iterative reconstruction model can be formulated in the following unconstrained optimization form:

$$E(x) = \frac{\lambda}{2}\|Ax - y\|_2^2 + R(x), \tag{5.6}$$

where the first term is for the L_2 data fidelity, which enforces the consistency between measured data y and a vectorized target image x, A is the system matrix of $I \times J$, I and J denote the numbers of image pixels and measurements, respectively, R is the regularization term as the prior information, and λ is the parameter to balance the trade-off between fidelity and regularization. Previous studies were focused on the development of different regularization terms, such as TV, nonlocal means, low rank, and dictionary learning-based sparsity.

Deep learning was first introduced into iterative reconstruction in Wang *et al* (2016) to obtain a mapping function as a regularization term for MRI reconstruction:

$$E(x) = \|C(F^H y; \Theta) - x\|_2^2 + \frac{\lambda}{2}\|F_P x - y\|_2^2, \tag{5.7}$$

where F indicates the Fourier encoding matrix normalized as $F^H F = E$ (where E denotes the identity matrix) and H denotes the Hermitian transpose operation. $C(F^H y; \Theta)$ is an image-to-image mapping function using SRCNN with the parameter vector Θ. The network is trained in the image domain, whose input and label are the zero-filled and ground truth images, respectively.

In the field of CT reconstruction, several studies were presented to map data to images in light of iterative reconstruction algorithms. Different from Wang *et al* (2016), Wu *et al* adopted the k-sparse auto-encoders (KSAE) instead of a convolutional network to model the data manifold in a latent space (Wu *et al* 2017). Due to the large number of parameters, the depth of KSAE is limited.

In Wang *et al* (2018), RLNet (Zhang *et al* 2017) was trained to estimate an initial residual image, which contained most noise and artifacts. Then, wavelet decomposition was applied on the estimated residual image, and the corresponding high-frequency part was treated as the noise and artifacts to be subtracted from the original input image. Finally, the previous two steps were embedded in the penalized weight least square (PWLS)-based iterative reconstruction framework as follows:

$$\hat{x} = \arg \min_x \lambda \|x - I\text{RLNet}(x)\|_2^2 + (y - Ax)^T \Sigma^{-1}(y - Ax), \tag{5.8}$$

where Σ^{-1} is a diagonal weighting matrix which determines the impact of each measurement according to noise variance, and IRLNet($\cdot$) is the learned map corresponding to RLNet. This model couples the network for post-processing with the traditional iterative reconstruction method in an effective way.

Another relevant idea was represented in Chen *et al* (2018) for cone-beam CT (CBCT) reconstruction. Taking blurring into account, the PWLS model can be expressed as follows:

$$\hat{x} = \arg\min_{x}(y - AGx)^T\Sigma^{-1}(y - AGx) + \lambda R(x), \tag{5.9}$$

where G is the estimated blurring kernel. By introducing an auxiliary variable $u = Gx$, equation (5.9) can be reformulated as

$$(\hat{u}, \hat{x}) = \arg\min_{u,x}(y - Au)^T\Sigma^{-1}(y - Au) + \lambda R(x) + \beta\|u - Gx\|_2^2, \tag{5.10}$$

where β is a penalty parameter. Half quadratic splitting (Geman and Yang 1995) is used for equation (5.10). Then, we have the following joint optimization problem:

$$\hat{u} = \arg\min_{u}(y - Au)^T\Sigma^{-1}(y - Au) + \beta\|u - Gx\|_2^2, \tag{5.11}$$

and

$$\hat{x} = \arg\min_{x} R(x) + \mu\|u - Gx\|_2^2, \tag{5.12}$$

where $\mu = \beta/\lambda$. Equation (5.11) is convex and can be solved via conjugate gradient search, Gauss–Seidel or other methods. Equation (5.12) can be rewritten as follows:

$$(\hat{z}, \hat{x}) = \arg\min_{z,x} R(z) + \mu\|u - Gx\|_2^2 + \eta\|z - x\|_2^2. \tag{5.13}$$

Equation (5.13) can be further split as

$$\hat{x} = \arg\min_{x}\|u - Gx\|_2^2 + \gamma\|z - x\|_2^2, \tag{5.14}$$

and

$$\hat{z} = \arg\min_{z} R(z) + \eta\|z - x\|_2^2, \tag{5.15}$$

where $\gamma = \eta/\mu$. Equation (5.14) can be solved via optimization in a similar way by which we solve equation (5.11). Equation (5.15) is the standard image denoising model. Many classical methods were proposed for this denoising problem, such as TV, nonlocal means, and BM3D. A general way is to solve equation (5.15) with state-of-the-art methods, such as plug-and-play (Venkatakrishnan *et al* 2013) and RED (Romano *et al* 2017).

In the well-known plug-and-play framework (Venkatakrishnan *et al* 2013), the augmented Lagrangian method was used to transfer the following regularization-based PWLS model:

$$\hat{x} = \arg\min_{x}(y - Ax)^T\Sigma^{-1}(y - Ax) + \lambda R(x), \tag{5.16}$$

to

$$(\hat{x}, \hat{v}, \hat{u}) = \arg\min_{x,v,u}(y - Ax)^T\Sigma^{-1}(y - Ax) + \lambda R(v) + \gamma\|x - v + u\|_2^2, \quad (5.17)$$

where v is the splitting variable and u is a scaled dual variable. The alternating direction method of multipliers (ADMM) (Boyd *et al* 2011) can be applied to equation (5.17), and variables are updated in an alternating fashion as follows:

$$\hat{x} = \arg\min_{x}(y - Ax)^T\Sigma^{-1}(y - Ax) + \gamma\|x - v + u\|_2^2, \quad (5.18)$$

$$\hat{v} = \arg\min_{v} \lambda R(v) + \gamma\|x - v + u\|_2^2, \quad (5.19)$$

$$\hat{u} = \arg\min_{u} \gamma\|x - v + u\|_2^2. \quad (5.20)$$

Equations (5.18) and (5.20) have closed-form solutions. Similar to equation (5.15), equation (5.19) can be treated as an image restoration problem, which was dealt with DnCNN (Zhang *et al* 2017).

Like the classical TV-POCS method (Sidky *et al* 2006), the constrained optimization problem can be separated into two parts: one for data fidelity, the other for regularization, and both steps can be performed alternatively. Based on this framework, Ma *et al* (2018) and Wu *et al* (2018) adopted conjugate gradient (CG) and separable quadratic surrogate with ordered subsets (OSSQS) techniques, respectively, to enforce the data fidelity and model regularization terms with different CNNs. The main steps of the algorithm (Ma *et al* 2018) are presented in algorithm 5.1.

In algorithm 5.1, x_0 is the initial input, β is the balancing parameter, and $C(\cdot, \Theta)$ is the network with the parameter vector Θ. Different optimization algorithms (such as CG, OSSQS, and SART) and various network architectures can be used to implement the the data fidelity and regularization terms, respectively. Since the forward propagation of a neural network is efficient, the main computational cost depends on the optimization algorithms.

Algorithm 5.1 Main steps of the algorithm (Ma *et al* 2018).

Input: A, y, x_0, β
Output: x
Procedure:
1. Set $i = 0$;
2. Compute $\hat{x}_i = \mathrm{CG}(A, x_{i-1}, y)$;
3. Compute $x_i^* = \frac{\hat{x}_i + \beta C(\hat{x}_i, \Theta)}{1 + \beta}$;
4. If the stop criterion is met, set $x = x_i^*$; else set $i = i + 1$, and go to step 2.

Another approach is not to replace the regularization term in equation (5.6). The initial idea was proposed in Gregor and LeCun (2010) for sparse coding. The classical iterative shrinkage and thresholding algorithm (ISTA) (Daubechies *et al* 2004) and coordinate descent algorithm (CoD) (Li and Osher 2009) were unfolded to networks with specific architectures and fixed depths to predict a close approximation of the sparse code produced by the original algorithms. Along this direction, many efforts were made to unfold the optimization procedure into a trainable neural network (Chen and Pock 2016, Yang *et al* 2016). In the CT field, two representative studies, LEARN and 3pADMM, were proposed in Chen *et al* (2018b) and He *et al* (2018), respectively.

5.4.1 LEARN

The learned experts' assessment-based reconstruction network (LEARN) is designed to tackle the few-view CT reconstruction problem, which unfolds the numerical scheme of the 'fields of experts' (FoE)-based (Roth and Black 2009) iterative reconstruction algorithm up to a fixed number of iterations for data-driven training and constructs a residual CNN in each block. The FoE model is a framework for learning expressive image priors that capture features of natural scenes and can be used for a variety of machine vision tasks. The regularization term in LEARN, referred to as FoE, is described as

$$R(x) = \sum_{k=1}^{K} \phi_k(G_k x), \qquad (5.21)$$

where K is the number of regularizers, G_k is a transform matrix of size $N_f \times N_f$, which can be treated as a convolutional operator for a CT image x, and $\phi_k(\cdot)$ denotes a potential function. Both G_k and $\phi_k(\cdot)$ can be learned from a training set. With the FoE model, the optimization problem for few-view CT reconstruction is described as follows:

$$E(x) = \frac{\lambda}{2}\|Ax - y\|_2^2 + \sum_{k=1}^{K} \phi_k(G_k x), \quad s.t. \ x \geqslant 0, \qquad (5.22)$$

where A denotes a system matrix, y is measured projection data, and λ is a weighting parameter. If the regularization term is differentiable and convex, equation (5.22) can be optimized using a gradient descent scheme:

$$x^{t+1} = x^t - \alpha \cdot \eta(x^t) = x^t - \alpha\frac{\partial E}{\partial x}, \qquad (5.23)$$

where α is the step size, t is the iteration number, and

$$\eta(x^t) = \lambda A^T(Ax^t - y) + \sum_{k=1}^{K} (G_k)^T \gamma_k(G_k x^t), \qquad (5.24)$$

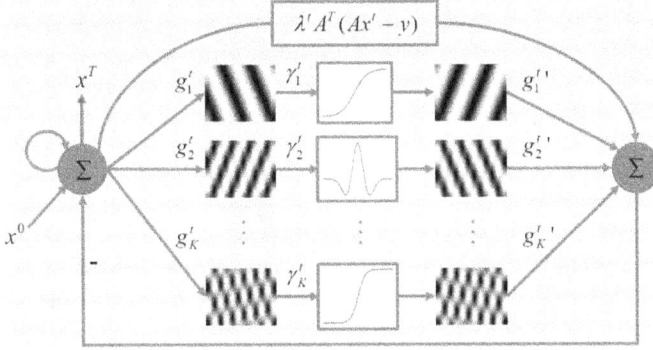

Figure 5.18. Network architecture corresponding to equation (5.25).

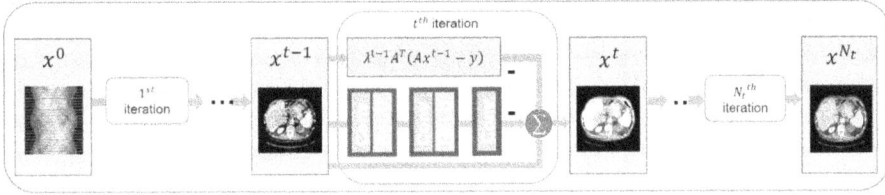

Figure 5.19. Overall network architecture of LEARN (Chen *et al* 2018b).

where $\gamma(\cdot) = \phi'(\cdot)$, the superscript T denotes the transpose operation, and A^T can be seen as the backprojection operator. Letting equation (5.23) be iteration-independent, equation (5.24) becomes

$$x^{t+1} = x^t - \left(\lambda^t A^T (Ax^t - y) + \sum_{k=1}^{K} (G_k^t)^T \gamma_k^t (G_k^t x^t) \right), \qquad (5.25)$$

where α disappeared in equation (5.25) because $\eta(x^t)$ can be freely scaled. The key to solve equation (5.25) is to choose the specific transforms G_k^t and γ_k^t. In fact, $\sum_{k=1}^{K}(G_k^t)^T \gamma_k^t (G_k^t x^t)$ can be interpreted as a CNN. In particular, G_k^t and $(G_k^t)^T$ can be seen as two convolutional layers. Then, equation (5.25) can be represented as a recurrent network as shown in figure 5.18. That is, the transform matrices G_k^t and $(G_k^t)^T$ can be replaced by the corresponding convolutional kernels g_k^t and $g_k^{t\prime}$, respectively. To show the feasibility of this network-based implementation, $\sum_{k=1}^{K}(G_k^t)^T \gamma_k^t (G_k^t x^t)$ is implemented as a three-layer CNN. With a fixed number of iterations, the network architecture in figure 5.18 can be unfolded into a deep CNN, resulting in LEARN shown in figure 5.19. It can be seen that there are N_t blocks (corresponding to the specified number of iterations) in LEARN, and each block has the same structure illustrated in the cyan box in figure 5.19. The data fidelity term $\lambda^t A^T (Ax^t - y)$, which keeps the coherence with the original data, is well preserved.

The LEARN model can be trained by minimizing the loss function in the form of accumulated mean squared errors (MSE) or another alternative loss function.

Different from most of the previous deep leaning-based methods, which treat the network architectures as a black-box, LEARN was directly motivated by the numerical scheme for solving the optimization problem based on the forward imaging model. The main merits of LEARN are threefold: (i) due to the forward propagation, the reconstruction is much faster than traditional iterative reconstruction methods, such as TV and dictionary learning; (ii) all the regularization terms and parameters in this framework are learned from a training dataset, and are more relevant or task-specific; and (iii) the regularization terms and parameters can be different for various iterations (blocks) to improve the accuracy and efficiency of the reconstruction process.

5.4.2 3pADMM

Currently, most model-based iterative reconstruction algorithms only consider the sinogram or image domain prior for CT image reconstruction. However, according to the work of Liu *et al* (2017), it is promising to reconstruct a high-quality CT image by synergizing prior information in the sinogram and image domains. The general dual-domain optimization problem for CT imaging is expressed as follows:

$$\min_{x,y} \frac{1}{2}\|\hat{y} - y\|^2_{\Sigma_y^{-1}} + \frac{1}{2}\|Ax - y\|^2_{\Sigma_x^{-1}} + \lambda R_y(y) + \gamma R_x(x), \qquad (5.26)$$

where $\hat{y}$ denotes real measurements, y is the measurements to be recovered, x represents an image to be reconstructed, A is the system matrix, Σ_x^{-1} and Σ_y^{-1} are the two diagonal weighted matrices for the image and sinogram, respectively, calculated according to Ma *et al* (2012), and λ and γ are two regularization parameters for R_x and R_y, respectively. The ADMM algorithm (Boyd *et al* 2011) can be used to solve equation (5.26). By introducing an auxiliary variable z, equation (5.26) can be transformed into the following optimization problem:

$$\min_{x,y,z} \frac{1}{2}\|\hat{y} - y\|^2_{\Sigma_y^{-1}} + \frac{1}{2}\|Ax - y\|^2_{\Sigma_x^{-1}} + \lambda R_y(y) + \gamma R_x(z), \quad s.t. \quad z = x. \quad (5.27)$$

Then, its augmented Lagrangian function can be expressed as follows:

$$L(x, y, z, \alpha) = \frac{1}{2}\|\hat{y} - y\|^2_{\Sigma_y^{-1}} + \frac{1}{2}\|Ax - y\|^2_{\Sigma_x^{-1}} + \lambda R_y(y)$$
$$+ \gamma R_x(z) + \langle \alpha, x - z \rangle + \frac{\rho}{2}\|x - z\|^2_2 , \qquad (5.28)$$

where α is the Lagrangian multiplier and ρ is a penalty parameter. Letting $\beta = \alpha/\rho$, equation (5.28) can be divided into the following sub-problems:

$$\begin{cases} \min_{y} \dfrac{1}{2}\left\| y - \dfrac{\Sigma_y^{-1}}{\Sigma_y^{-1} + \Sigma_x^{-1}}\hat{y} - \dfrac{\Sigma_x^{-1}}{\Sigma_y^{-1} + \Sigma_x^{-1}}Ax \right\|_{\Sigma_y^{-1}+\Sigma_x^{-1}}^{2} + \lambda R_y(y), \\[2ex] \min_{x} \dfrac{1}{2}\|Ax - y\|_{\Sigma_x^{-1}}^{2} + \dfrac{\rho}{2}\|x + \beta - z\|_2^2, \\[2ex] \min_{z} \gamma R_x(z) + \dfrac{\rho}{2}\|x + \beta - z\|_2^2, \\[2ex] \max_{\beta} \langle \beta, x - z \rangle. \end{cases} \qquad (5.29)$$

These sub-problems have the following solutions:

$$\begin{cases} y^{(n)} = [I - l_{ry}(\Sigma_y^{-1} + \Sigma_x^{-1})]y^{(n-1)} + l_{ry}(\Sigma_y^{-1}\hat{y} + \Sigma_x^{-1}Ax^{(n-1)}) \\ \qquad\quad - \lambda \nabla R_y(y^{(n-1)}), \\[1ex] x^{(n)} = (1 - l_{rx}\rho)x^{(n-1)} + l_{rx}\rho(z^{(n-1)} - \beta^{(n-1)}) \\ \qquad\quad - l_{rx}A^T\Sigma_x^{-1}(Ax^{(n-1)} - y^{(n)}), \\[1ex] z^{(n)} = (1 - l_{rz}\rho)z^{(n-1)} + l_{rz}\rho(x^{(n)} + \beta^{(n-1)}) - \gamma \nabla R_x(z^{(n-1)}), \\[1ex] \beta^{(n)} = \beta^{(n-1)} + \eta(x^{(n)} - z^{(n)}), \end{cases} \qquad (5.30)$$

where I denotes the identity matrix, the superscript T represents the transpose operation, l_{rx}, l_{ry}, and l_{rz} are the step sizes, η denotes the update rate, and ∇ is a gradient operator.

However, there are too many parameters that need to be chosen manually in equation (5.30). As a result, a parameterized plug-and-play ADMM (3pADMM) was proposed (He *et al* 2018), which modifies the ADMM iteration equation (5.30) as follows. First, the universal parameters are allowed to be different in each iteration to improve the reconstruction quality. Specifically, $\tilde{\Sigma}_x^{(n)} = l_{ry}\Sigma_x^{-1}$, $\tilde{\Sigma}_y^{(n)} = l_{ry}\Sigma_y^{-1}$, $\tilde{\lambda}^{(n)} = \lambda$, $\theta^{(n)} = l_{rx}\rho$, $\tilde{\Sigma}^{(n)} = l_{rx}\Sigma_x^{-1}$, $\tilde{\theta}^{(n)} = l_{rz}\rho$, $\tilde{\gamma}^{(n)} = \gamma$, and $\tilde{\eta}^{(n)} = \eta$. Second, since it is challenging to choose any specific form of the regularization term R_x, the residual CNNs $\mathrm{Res}^{(n)}(\cdot)$, shown in figure 5.20, are used to represent the corresponding gradient ∇R_x. To reduce the number of parameters and ensure that only one residual image is produced, the first 'Conv+BN+ReLU' contains L filters

Figure 5.20. The structure of $\mathrm{Res}^{(n)}(\cdot)$, which consists of several residual units as its building blocks. Reproduced with permission from He *et al* (2018). Copyright 2018 IEEE.

of size $w_f \times w_f$ and the second one only has one filter. Third, to simplify the problem a quadratic penalty is adopted for y, i.e. $R_y = \frac{1}{2}\sum_j \sum_{m \in N_j} \omega_{jm}(y_j - y_m)^2$. Then, $\nabla R_y = y - Dy$, where D represents a filtering operation corresponding to the weight value ω (Wang *et al* 2006) (the strategy of constructing ∇R_x can also be used to construct ∇R_y). With these modifications, the 3pADMM procedure is represented as follows:

$$
\begin{cases}
y^{(n)} = (I - \tilde{\Sigma}_y^{(n)} - \tilde{\Sigma}_x^{(n)} - \lambda^{(n)}I)y^{(n-1)} + \tilde{\Sigma}_y^{(n)}\hat{y} + \tilde{\Sigma}_x^{(n)}Ax^{(n-1)} \\
\qquad + \tilde{\lambda}^{(n)}Dy^{(n-1)}, \\
x^{(n)} = (1 - \theta^{(n)})x^{(n-1)} + \theta^{(n)}(z^{(n-1)} - \beta^{(n-1)}) - A^T\tilde{\Sigma}^{(n)}(Ax^{(n-1)} - y^{(n)}), \\
z^{(n)} = (1 - \tilde{\theta}^{(n)})z^{(n-1)} + \tilde{\theta}^{(n)}(x^{(n)} + \beta^{(n-1)}) - \tilde{\gamma}^{(n)}\text{Res}^{(n)}(z^{(n-1)}), \\
\beta^{(n)} = \beta^{(n-1)} + \tilde{\eta}^{(n)}(x^{(n)} - z^{(n)}).
\end{cases}
\tag{5.31}
$$

Similar to LEARN, 3pADMM can also be unfolded into a deep reconstruction neural network, which includes four basic blocks. The overall architecture is shown in figure 5.21, and the four basic blocks are shown in figure 5.22. The parameters of 3pADMM can be optimized by minimizing MSE between reconstructed CT images and high-quality reference images.

5.4.3 Learned primal–dual reconstruction

The methods mentioned above in this subsection are all based on a specific numerical scheme, such as ADMM and half quadratic splitting. Based on the work of Putzky and Welling (2017), Adler and Oktem proposed learning a broader class of schemes instead of restricting the attention to specific schemes (Adler and Öktem 2018). A more general objective function for regularization-based signal recovery can be expressed as

$$
\arg \min_x E(x) \approx \arg \min_x [L(T(x), y) + \lambda R(x)],
\tag{5.32}
$$

where T is a forward operator, which is equal to the system matrix A in CT imaging. L is a suitable transform, which can be seen as the norm of data fidelity.

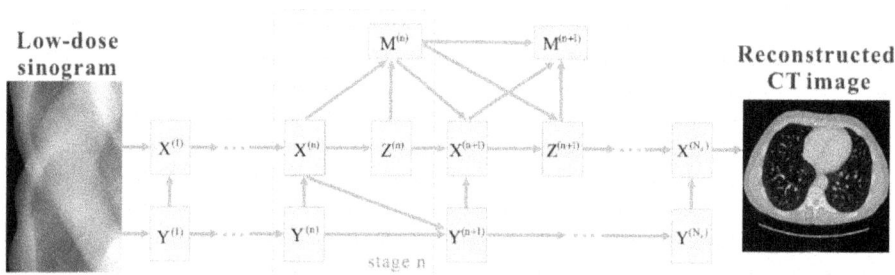

Figure 5.21. Overall architecture of the 3pADMM network. Reproduced with permission from He *et al* (2018). Copyright 2018 IEEE.

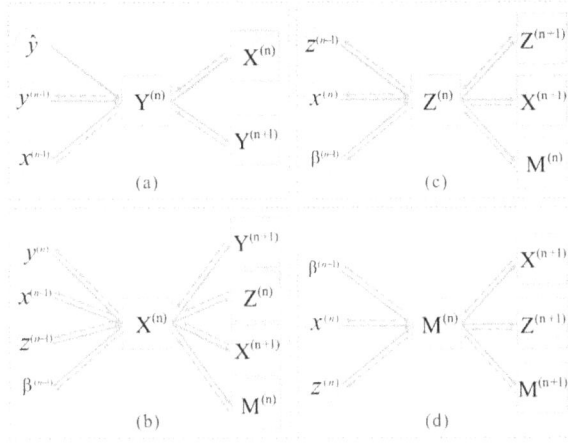

Figure 5.22. Data flows of the four basic blocks including (a) the sinogram restoration block, (b) the image reconstruction block, (c) the residual denoising block, and (d) the multiplier update block. The solid lines indicate forward processes, while the dotted lines indicate backward procedures. Reproduced with permission from He *et al* (2018). Copyright 2018 IEEE.

Under the assumption that the right-hand side of equation (5.32) is differentiable and convex, a simple gradient descent method can be applied to find the minimum of equation (5.32):

$$x_k = x_{k-1} - \sigma(\nabla L[T(\cdot), y](x_{k-1}) + \lambda \nabla R(x_{k-1})), \tag{5.33}$$

where σ is the step size. By introducing a parameterized updating operator Λ_Θ and a persistent memory term s that represents the information from earlier iterations, equation (5.33) can be transformed into the following two steps:

$$(s_k, \Delta x_k) \leftarrow \Lambda_\Theta(x_{k-1}, s_{k-1}, \nabla L[T(\cdot), y](x_{k-1}), \nabla R(x_{k-1})) \tag{5.34}$$

and

$$x_k \leftarrow x_{k-1} + \Delta x_k. \tag{5.35}$$

A three-layer CNN was trained to be Λ_Θ, which is called the learned partial gradient decent (LPGD) method. Any forms of L, T, and R can be used, showing the flexibility of this approach.

LPGD assumes that the right part of equation (5.32) is differentiable. This assumption cannot be always satisfied, since many regularizers of interest are not differentiable. Utilization of a smooth approximation is a proper alternative, but this method will introduce more parameters and lead to non-exact solutions. To circumvent this obstacle, the same groups employed a primal–dual hybrid gradient (PDHG) algorithm for non-smooth convex optimization (Chambolle and Pock 2011). A more general version of equation (5.32) is given as

$$\arg\min_x E(x) \approx \arg\min_x[F(J(x)) + G(x)], \tag{5.36}$$

where J is a proper operator, and F and G are functions on the dual/primal spaces, respectively. PDHG can solve equation (5.36) by performing the steps in algorithm 5.2 iteratively:

$$h_{k+1} \leftarrow \text{prox}_{\sigma F^*}(h_k + \sigma J(\bar{x}_k)), \tag{5.37}$$

$$x_{k+1} \leftarrow \text{prox}_{\tau G}(x_k - \tau[\partial J(x_k)]^*(h_{k+1})), \tag{5.38}$$

and

$$\bar{x}_{k+1} \leftarrow x_{k+1} + \gamma(x_{k+1} - x_k), \tag{5.39}$$

where h is the dual variable, F^* is the Fenchel conjugate of F, and $[\partial J(x_k)]^*$ is the adjoint of the derivative of J at the point x_k. In reference to Venkatakrishnan *et al* (2013) and Romano *et al* (2017), the proximal operators can be substituted by other operators, such as BM3D and nonlocal means. In CT imaging, the mapping function J and its derivative's adjoint are typically projection and backprojection.

Algorithm 5.2. Learned PDHG.

Initialize: x_0, h_0
for $k = 1, 2, \ldots, N$ **do**
  $h_k \leftarrow \Gamma_{\theta^d}(h_{k-1} + \sigma J(\bar{x}_{k-1}))$,
  $x_k \leftarrow \Lambda_{\theta^p}(x_{k-1} - \tau[\partial J(x_{k-1})]^*(h_k))$,
  $\bar{x}_k \leftarrow x_k + \gamma(x_k - x_{k-1})$,
end for
Return x_N

To unfold PDHG into a neural network, the number of iterations needs to be specified. In each iteration, the proximal operators of the primal and dual spaces are replaced by several convolutional layers. Algorithm 5.2 describes a variant of the PDHG algorithm with N iterations in which the proximal operators are obtained through training. Γ_{θ^d} and Λ_{θ^p} are, respectively, the dual and primal proximal operators with learned parameters θ^d and θ^p. Other parameters including σ, τ, and γ are also learned from training data.

Learned PDHG has a better performance than traditional methods, but there is still potential for improvement. The following ideas can help improve the perform-ance of the neural network:

- Referring to Adler and Öktem (2017) and Putzky and Welling (2017), the primal and dual spaces can be extended to increase the efficiency of each iteration using the information from previous iterations (memory). Then, the primal and dual variables become $x = [x^{(1)}, x^{(2)}, \ldots, x^{(N_{\text{primal}})}]$ and $h = [h^{(1)}, h^{(2)}, \ldots, h^{(N_{\text{dual}})}]$, respectively.

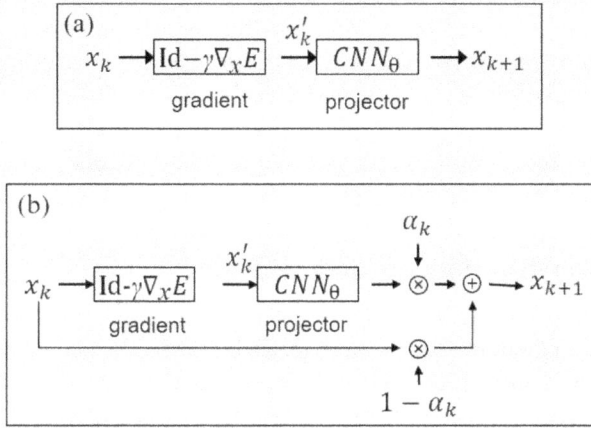

Figure 5.23. (a) Block diagram of projected gradient descent using a CNN as the projector. (b) Block diagram of the relaxed projected gradient descent proposed in Gupta *et al* (2018). Reproduced with permission from Gupta *et al* (2018). Copyright 2018 IEEE.

- Instead of updating the variables in the fixed form $h_k + \sigma J(\bar{x}_k)$, let the network learn how to combine h_k and $J(\bar{x}_k)$ through training.
- Instead of hard-coding the over-relaxation $\bar{x}_{k+1} = x_{k+1} + \theta(x_{k+1} - x_k)$, the network can be allowed to find a more appropriate point for the forward operator.
- Let the proximal operators in each iteration differ, which increases the number of parameters but notably improves the reconstruction quality.

The above updates can produce a more desirable learned primal–dual network, which is shown in algorithm 5.3, with θ_k^d and θ_k^p being the learned parameters obtained in the kth iteration. Similar ideas were also mentioned by Kelly *et al* (2017), which can be seen as a simplified version of this method.

Algorithm 5.3. Learned primal–dual.

Initialize: x_0, h_0
for $k = 1, 2, \ldots, N$ **do**
   $h_k = \Gamma_{\theta_k^d}(h_{k-1}, J(x_{k-1}^{(2)}), y)$,
   $x_k = \Lambda_{\theta_k^p}(x_{k-1}, [\partial J(x_{k-1}^{(1)})]^*(h_k^{(1)}))$,
end for
Return $x_N^{(1)}$

In Gupta *et al* (2018), the projected gradient descent (PGD) method was modified by replacing the projector with a CNN. The block diagram of PGD using a CNN as the projector is given in figure 5.23(a), where E is the data fidelity term and γ is the step size. A relaxed PGD was also proposed to improve PGD to maintain the consistency between a reconstructed image and measured data as given in figure 5.23(b). Also, a hybrid reconstruction framework was proposed to combine model-based sparse regularization with data-driven deep learning (Bubba *et al* 2019). The decomposition into visible and invisible parts is achieved using a shearlet transform that allows one to resolve wavefront sets in the phase space. U-Net was employed to infer the invisible part, and a traditional regularization-based model was adopted to recover the visible part.

5.5 Direct reconstruction via deep learning

Before deep learning had drawn much attention, neural networks were applied for CT reconstruction. The main efforts were made based on the classical FBP and least-squares methods (Cierniak 2008, 2009, Kerr and Bartlett 1995, Paschalis *et al* 2004, Rodriguez *et al* 2001), but due to the limitations of hardware and techniques, the networks usually had only one hidden layer. The small scale of networks were the bottleneck of performance. With the rapid development of deep learning, the depths of the networks become deeper and the performance is significantly improved.

In Würfl *et al* (2018), the authors rewrote the Radon transform with the operator notation as

$$x = A_{bp}^T Cy, \tag{5.40}$$

where A_{bp}^T is the adjoint of the Radon transform, which can be seen as the backprojection operator, and C denotes the convolution of the projection data with the discrete filter. With equation (5.40), it is straightforward to construct a network by specifying layers corresponding to A_{bp}^T and C. The architecture of the proposed network for parallel beam geometry is shown in figure 5.24(a). For the situations of fan-beam and cone-beam, the modified networks are given in figures 5.24(b) and (c). In figure 5.24(b), W_{cos} denotes the pixel-wise independent weighting of the projection data with cosine weights. In figure 5.24(c), $W_{cos_{2D}}$ and W_{red} denote the two-dimensional cosine weighting of the projection data and Parker weights (Parker 1982). Hammernik *et al* cascaded this network with a variational network (Hammernik *et al* 2017). The first part plays the role of FBP and corrects the intensity inhomogeneities. The second part is used to eliminate the streak artifacts as a nonlinear filter.

Ye *et al* proposed to backproject each view separately to form a set of back projections at first (Ye *et al* 2018a) and then feed these projections into a network to fuse these data to the final reconstructed result. The single backprojection can decode the appropriate location from the original sinogram and the network was trained to map these back projections to the reconstruction.

He and Ma presented a similar idea in He and Ma (2018). To diminish the scale of parameters in the network, two networks were cascaded. The first one was used to reconstruct the initial image and the second one was utilized to refine the imaging quality. Following the idea of FBP, two fully connected layers are included in the first segments, respectively, corresponding to the filtering operation and back-projection. A residual CNN was used as the second segments.

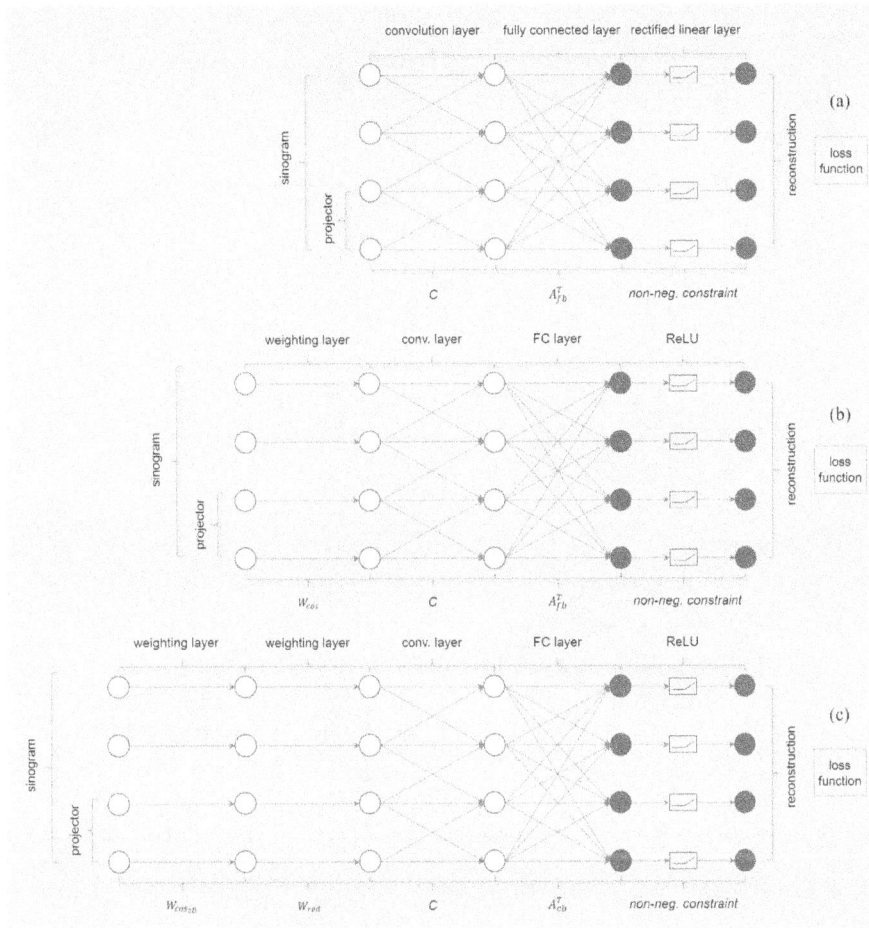

Figure 5.24. The network architectures proposed in Würfl *et al* (2018). (a) Parallel-beam architecture, (b) fan-beam architecture, and (c) cone-beam architecture. Reproduced with permission from Würfl *et al* (2018). Copyright 2018 IEEE.

With the rapid development of deep learning in the field of medicine, more and more works are being published in high-impact journals, such as *Science*, *Nature*, and its series. In 2018, a group from Massachusetts General Hospital and Harvard University published an image reconstruction work (AUTOMAP) in *Nature* (Zhu *et al* 2018). They proposed a general deep learning framework for image reconstruction. A very simple network architecture was proposed, shown in figure 5.25. The first three layers are fully connected layers, which transform the data from the sensor domain (projection domain for CT) into the image domain, as illustrated in figure 5.26. The fully connected layers are followed by two convolutional layers and one deconvolutional layer, which perform the role in noise and artifact reduction. The idea is essentially similar to He and Ma (2018) and the main difference lies in that AUTOMAP adopted three fully connected layers to implement domain

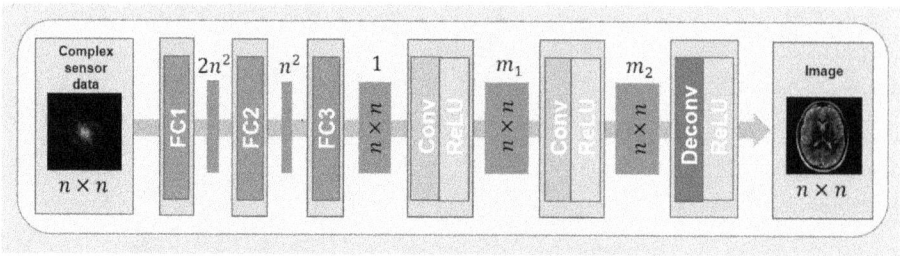

Figure 5.25. The network architecture of AUTOMAP (Zhu *et al* 2018).

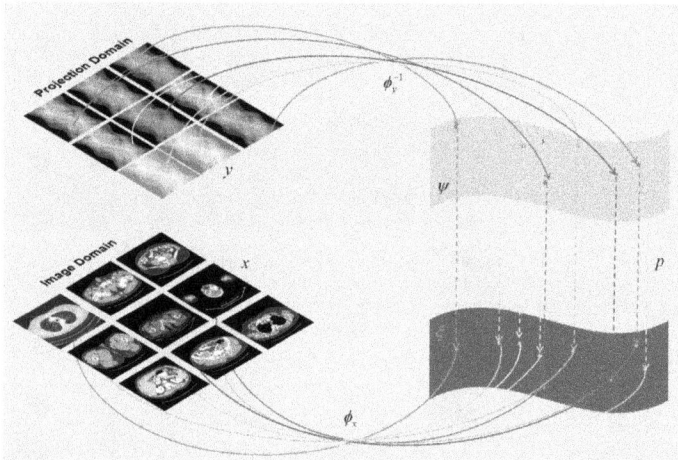

Figure 5.26. A mapping between the projection domain and image domain is determined via supervised learning of projection (top) and image (bottom) domain pairs. Reproduced with permission from *Nature* (Zhu *et al* 2018). Copyright 2018 Macmillan Publishers Ltd.

transform, which can be generalized for different imaging modalities, such as CT and MRI, but other methods focus on the concrete form of FBP for CT. However, due to the utilization of three fully connected layers, the number of parameters for AUTOMAP is too huge to be implemented in practice. Although AUTOMAP has this flaw, it will be not vital due to the rapid development of hardware and software. Li *et al* (2019) constructed a network to simulate the main steps of FBP and reduce the complexity of AUTOMAP by a factor of *N*. Xie *et al* (2019) further reduced the complexity and adopted the transfer learning technique by training the network only with samples from ImageNet. Fu and De Bruno proposed a scalable learning method for CT reconstruction, which recursively decomposes the Radon transform into hierarchical steps involving local transforms (Fu and De Bruno 2019). This strategy can efficiently improve the learning process.

References

Abdoli M, De Jong J R, Pruim J, Dierckx R A and Zaidi H 2011 Reduction of artefacts caused by hip implants in CT-based attenuation-corrected PET images using 2-D interpolation of a virtual sinogram on an irregular grid *Eur. J. Nucl. Med. Mol. Imaging* **38** 2257–68

Adler J and Öktem O 2017 Solving ill-posed inverse problems using iterative deep neural networks *Inverse Problems* **33** 124007

Adler J and Öktem O 2018 Learned primal–dual reconstruction *IEEE Trans. Med. Imaging* **37** 1322–32

Anirudh R, Kim H, Thiagarajan J J, Aditya Mohan K, Champley K and Bremer T 2018 Lose the views: limited angle CT reconstruction via implicit sinogram completion *Proc. of the IEEE Conf. on Computer Vision and Pattern Recognition* pp 6343–52

Arjovsky M, Chintala S and Bottou L 2017 Wasserstein generative adversarial networks *Int. Conf. on Machine Learning* pp 214–23

Bal M, Celik H, Subramanyan K, Eck K and Spies L 2005 A radial adaptive filter for metal artifact reduction *Medical Imaging 2005: Image Processing* vol 5747 (Bellingham, WA: International Society for Optics and Photonics) pp 2075–83

Balda M, Hornegger J and Heismann B 2012 Ray contribution masks for structure adaptive sinogram filtering *IEEE Trans. Med. Imaging* **31** 1228–39

Boyd S *et al* 2011 Distributed optimization and statistical learning via the alternating direction method of multipliers *Found. Trends Mach. Learn.* **3** 1–122

Brenner D J and Hall E J 2007 Computed tomography—an increasing source of radiation exposure *New Engl. J. Med.* **357** 2277–84

Bubba T A, Kutyniok G, Lassas M, Maerz M, Samek W, Siltanen S and Srinivasan V 2019 Learning the invisible: a hybrid deep learning-shearlet framework for limited angle computed tomography *Inverse Problems* **35** 064002

Cai J-F, Jia X, Gao H, Jiang S B, Shen Z and Zhao H 2014 Cine cone beam CT reconstruction using low-rank matrix factorization: algorithm and a proof-of-principle study *IEEE Trans. Med. Imaging* **33** 1581–91

Candès E, Romberg J and Tao T 2006 Robust uncertainty principles: exact signal reconstruction from highly incomplete frequency information *IEEE Trans. Inform. Theory* **52** 489–509

Chambolle A and Pock T 2011 A first-order primal–dual algorithm for convex problems with applications to imaging *J. Math. Imaging Vis.* **40** 120–45

Chen B, Xiang K, Gong Z, Wang J and Tan S 2018 Statistical iterative CBCT reconstruction based on neural network *IEEE Trans. Med. Imaging* **37** 1511–21

Chen G-H, Tang J and Leng S 2008 Prior image constrained compressed sensing (PICCS): a method to accurately reconstruct dynamic CT images from highly undersampled projection data sets *Med. Phys.* **35** 660–3

Chen H, Zhang Y, Chen Y, Zhang J, Zhang W, Sun H, Lv Y, Liao P, Zhou J and Wang G 2018b LEARN: learned experts' assessment-based reconstruction network for sparse-data CT *IEEE Trans. Med. Imaging* **37** 1333–47

Chen H, Zhang Y, Kalra M K, Lin F, Chen Y, Liao P, Zhou J and Wang G 2017a Low-dose CT with a residual encoder–decoder convolutional neural network *IEEE Trans. Med. Imaging* **36** 2524–35

Chen H, Zhang Y, Zhang W, Liao P, Li K, Zhou J and Wang G 2017b Low-dose CT via convolutional neural network *Biomed. Optics Exp.* **8** 679–94

Chen Y and Pock T 2016 Trainable nonlinear reaction diffusion: a flexible framework for fast and effective image restoration *IEEE Trans. Pattern Anal. Mach. Intell.* **39** 1256–72

Chen Y, Shi L, Feng Q, Yang J, Shu H, Luo L, Coatrieux J-L and Chen W 2014 Artifact suppressed dictionary learning for low-dose CT image processing *IEEE Trans. Med. Imaging* **33** 2271–92

Chen Y, Yin X, Shi L, Shu H, Luo L, Coatrieux J-L and Toumoulin C 2013a Improving abdomen tumor low-dose CT images using a fast dictionary learning based processing *Phys. Med. Biol.* **58** 5803–20

Chen Z, Jin X, Li L and Wang G 2013b A limited-angle CT reconstruction method based on anisotropic TV minimization *Phys. Med. Biol.* **58** 2119–41

Cierniak R 2008 A 2D approach to tomographic image reconstruction using a Hopfield-type neural network *Artif. Intell. Med.* **43** 113–25

Cierniak R 2009 New neural network algorithm for image reconstruction from fan-beam projections *Neurocomputing* **72** 3238–44

Claus B E, Jin Y, Gjesteby L A, Wang G and De Man B 2017 Metal-artifact reduction using deep-learning based sinogram completion: initial results *Proc. 14th Int. Meeting Fully Three-Dimensional Image Reconstruction Radiol. Nucl. Med.* pp 631–4

Cong W and Wang G 2017 Monochromatic CT image reconstruction from current-integrating data via deep learning, arXiv:1710.03784

Daubechies I, Defrise M and De Mol C 2004 An iterative thresholding algorithm for linear inverse problems with a sparsity constraint *Commun. Pure Appl. Math.* **57** 1413–57

Dong C, Loy C C, He K and Tang X 2015 Image super-resolution using deep convolutional networks *IEEE Trans. Pattern Anal. Mach. Intell.* **38** 295–307

Dong C, Loy C C and Tang X 2016 Accelerating the super-resolution convolutional neural network *European Conf. on Computer Vision* (Berlin: Springer) pp 391–407

Donoho D L *et al* 2006 Compressed sensing *IEEE Trans. Inform. Theory* **52** 1289–306

Drozdzal M, Vorontsov E, Chartrand G, Kadoury S and Pal C 2016 *The Importance of Skip Connections in Biomedical Image Segmentation (Proc. Int. Workshop on Deep Learning Medical Image Analysis)* (Berlin: Springer) pp 179–87

Fu L and De Man B 2019 A hierarchical approach to deep learning and its application to tomographic reconstruction *Proc. 14th Int. Meeting Fully Three-Dimensional Image Reconstruction Radiol. Nucl. Med.* **11072** 1107202

Gao H, Cai J-F, Shen Z and Zhao H 2011a Robust principal component analysis-based four-dimensional computed tomography *Phys. Med. Biol.* **56** 3181–98

Gao H, Yu H, Osher S and Wang G 2011b Multi-energy CT based on a prior rank, intensity and sparsity model (PRISM) *Inverse Problems* **27** 115012

Geman D and Yang C 1995 Nonlinear image recovery with half-quadratic regularization *IEEE Trans. Image Process.* **4** 932–46

Gjesteby L, De Man B, Jin Y, Paganetti H, Verburg J, Giantsoudi D and Wang G 2016 Metal artifact reduction in CT: where are we after four decades? *IEEE Access* **4** 5826–49

Gjesteby L, Yang Q, Xi Y, Zhou Y, Zhang J and Wang G 2017 Deep learning methods to guide CT image reconstruction and reduce metal artifacts *Proc. SPIE* **10132** 101322W

Goodfellow I, Pouget-Abadie J, Mirza M, Xu B, Warde-Farley D, Ozair S, Courville A and Bengio Y 2014 Generative adversarial nets *Advances in Neural Information Processing Systems* (San Francisco, CA: Morgan Kaufmann) pp 2672–80

Gregor K and LeCun Y 2010 Learning fast approximations of sparse coding *Proc. of the 27th Int. Conf. on Machine Learning* (Madison, WI: Omnipress), pp 399–406

Gu J, Zhang L, Yu G, Xing Y and Chen Z 2006 X-ray CT metal artifacts reduction through curvature based sinogram inpainting *J. X-ray Sci. Technol.* **14** 73–82

Gulrajani I, Ahmed F, Arjovsky M, Dumoulin V and Courville A C 2017 Improved training of Wasserstein GANs *Advances in Neural Information Processing Systems* (San Francisco, CA: Morgan Kaufmann) pp 5767–77

Gupta H, Jin K H, Nguyen H Q, McCann M T and Unser M 2018 CNN-based projected gradient descent for consistent CT image reconstruction *IEEE Trans. Med. Imaging* **37** 1440–53

Hammernik K, Würfl T, Pock T and Maier A 2017 A deep learning architecture for limited-angle computed tomography reconstruction *Bildverarbeitung für die Medizin 2017* (Berlin: Springer) pp 92–7

Han Y, Gu J and Ye J C 2017 Deep learning interior tomography for region-of-interest reconstruction, arXiv:1712.10248

Han Y, Kang J and Ye J C 2018 Deep learning reconstruction for 9-view dual energy CT baggage scanner, arXiv:1801.01258

Han Y and Ye J C 2018 Framing U-net via deep convolutional framelets: application to sparse-view CT *IEEE Trans. Med. Imaging* **37** 1418–29

He J and Ma J 2018 Radon inversion via deep learning, arXiv:1808.03015

He J, Yang Y, Wang Y, Zeng D, Bian Z, Zhang H, Sun J, Xu Z and Ma J 2018 Optimizing a parameterized plug-and-play ADMM for iterative low-dose CT reconstruction *IEEE Trans. Med. Imaging* **38** 371–82

He K, Zhang X, Ren S and Sun J 2016 Deep residual learning for image recognition *Proc. of the IEEE Conf. on Computer Vision and Pattern Recognition* pp 770–8

Huang G, Liu Z, Van Der Maaten L and Weinberger K Q 2017 Densely connected convolutional networks *Proc. of the IEEE Conf. on Computer Vision and Pattern Recognition* pp 4700–8

Jian F and Hongnian L 2006 Beam-hardening correction method based on original sinogram for X-CT *Nucl. Instrum. Meth. A* **556** 379–85

Jin K H, McCann M T, Froustey E and Unser M 2017 Deep convolutional neural network for inverse problems in imaging *IEEE Trans. Image Process.* **26** 4509–22

Jin P, Bouman C A and Sauer K D 2015 A model-based image reconstruction algorithm with simultaneous beam hardening correction for x-ray CT *IEEE Trans. Comput. Imaging* **1** 200–16

Johnson J, Alahi A and Fei-Fei L 2016 Perceptual losses for real-time style transfer and super-resolution *European Conference on Computer Vision* (Berlin: Springer) pp 694–711

Kang E, Chang W, Yoo J and Ye J C 2018 Deep convolutional framelet denosing for low-dose CT via wavelet residual network *IEEE Trans. Med. Imaging* **37** 1358–69

Kang E, Koo H J, Yang D H, Seo J B and Ye J C 2019 Cycle-consistent adversarial denoising network for multiphase coronary CT angiography *Med. Phys.* **46** 550–62

Kang E, Min J and Ye J C 2017 A deep convolutional neural network using directional wavelets for low-dose x-ray CT reconstruction *Med. Phys.* **44** e360–75

Kelly B, Matthews T P and Anastasio M A 2017 Deep learning-guided image reconstruction from incomplete data, arXiv:1709.00584

Kerr J P and Bartlett E B 1995 A statistically tailored neural network approach to tomographic image reconstruction *Med. Phys.* **22** 601–10

Kim H, Chen J, Wang A, Chuang C, Held M and Pouliot J 2016 Non-local total-variation (NLTV) minimization combined with reweighted L1-norm for compressed sensing CT reconstruction *Phys. Med. Biol.* **61** 6878–91

Lai W-S, Huang J-B, Ahuja N and Yang M-H 2017 Deep Laplacian pyramid networks for fast and accurate super-resolution *Proc. of the IEEE Conf. on Computer Vision and Pattern Recognition* pp 624–32

Ledig C *et al* 2017 Photo-realistic single image super-resolution using a generative adversarial network *Proc. of the IEEE Conf. on Computer Vision and Pattern Recognition* pp 4681–90

Lee D, Choi S and Kim H-J 2019 High quality imaging from sparsely sampled computed tomography data with deep learning and wavelet transform in various domains *Med. Phys.* **46** 104–15

Li Y, Li K, Zhang C, Montoya J and Chen G-H 2019 Learning to reconstruct computed tomography (CT) images directly from sinogram data under a variety of data acquisition conditions *IEEE Trans. Med. Imaging* **38** 2469–81

Li Y and Osher S 2009 Coordinate descent optimization for L1 minimization with application to compressed sensing; a greedy algorithm *Inverse Problems Imaging* **3** 487–503

Liu J, Ding H, Molloi S, Zhang X and Gao H 2016 TICMR: total image constrained material reconstruction via nonlocal total variation regularization for spectral CT *IEEE Trans. Med. Imaging* **35** 2578–86

Liu J *et al* 2017 Discriminative feature representation to improve projection data inconsistency for low dose CT imaging *IEEE Trans. Med. Imaging* **36** 2499–509

Liu Y, Ma J, Fan Y and Liang Z 2012 Adaptive-weighted total variation minimization for sparse data toward low-dose x-ray computed tomography image reconstruction *Phys. Med. Biol.* **57** 7923–56

Liu Y and Zhang Y 2018 Low-dose CT restoration via stacked sparse denoising autoencoders *Neurocomputing* **284** 80–9

Ma G, Shen C and Jia X 2018 Low dose CT reconstruction assisted by an image manifold prior, arXiv:1810.12255

Ma J, Huang J, Feng Q, Zhang H, Lu H, Liang Z and Chen W 2011 Low-dose computed tomography image restoration using previous normal-dose scan *Med. Phys.* **38** 5713–31

Ma J, Liang Z, Fan Y, Liu Y, Huang J, Chen W and Lu H 2012 Variance analysis of x-ray CT sinograms in the presence of electronic noise background *Med. Phys.* **39** 4051–65

Ma Z, Zhang Y, Zhang W, Wang Y, Lin F, He K, Li X, Pu Y and Zhou J 2016 Noise reduction in low-dose CT with stacked sparse denoising autoencoders *2016 IEEE Nuclear Science Symp., Medical Imaging Conf. and Room-Temperature Semiconductor Detector Workshop (NSS/MIC/RTSD) (IEEE)* pp 1–2

Manduca A, Yu L, Trzasko J D, Khaylova N, Kofler J M, McCollough C M and Fletcher J G 2009 Projection space denoising with bilateral filtering and CT noise modeling for dose reduction in CT *Med. Phys.* **36** 4911–19

McCann M T, Jin K H and Unser M 2017 Convolutional neural networks for inverse problems in imaging: a review *IEEE Signal Process. Mag.* **34** 85–95

Niu S, Gao Y, Bian Z, Huang J, Chen W, Yu G, Liang Z and Ma J 2014 Sparse-view x-ray CT reconstruction via total generalized variation regularization *Phys. Med. Biol.* **59** 2997–3017

Noh H, Hong S and Han B 2015 Learning deconvolution network for semantic segmentation *Proc. of the IEEE Int. Conf. on Computer Vision* pp 1520–8

Park H S, Lee S M, Kim H P, Seo J K and Chung Y E 2018 CT sinogram-consistency learning for metal-induced beam hardening correction *Med. Phys.* **45** 5376–84

Parker D L 1982 Optimal short scan convolution reconstruction for fan beam CT *Med. Phys.* **9** 254–7

Paschalis P, Giokaris N, Karabarbounis A, Loudos G, Maintas D, Papanicolas C, Spanoudaki V, Tsoumpas C and Stiliaris E 2004 Tomographic image reconstruction using artificial neural networks *Nucl. Instrum. Methods Phys. Res. Sect. A* **527** 211–15

Putzky P and Welling M 2017 Recurrent inference machines for solving inverse problems, arXiv:1706.04008

Rantala M, Vanska S, Jarvenpaa S, Kalke M, Lassas M, Moberg J and Siltanen S 2006 Wavelet-based reconstruction for limited-angle x-ray tomography *IEEE Trans. Med. Imaging* **25** 210–17

Rodriguez A F, Blass W E, Missimer J H and Leenders K L 2001 Artificial neural network radon inversion for image reconstruction *Med. Phys.* **28** 508–14

Romano Y, Elad M and Milanfar P 2017 The little engine that could: regularization by denoising (RED) *SIAM J. Imaging Sci.* **10** 1804–44

Ronneberger O, Fischer P and Brox T 2015 U-Net: convolutional networks for biomedical image segmentation *Int. Conf. on Medical Image Computing and Computer-assisted Intervention* pp 234–41

Roth S and Black M J 2009 Fields of experts *Int. J. Comput. Vis.* **82** 205

Shan H, Padole A, Homayounieh F, Kruger U, Khera R D, Nitiwarangkul C, Kalra M K and Wang G 2019 Competitive performance of a modularized deep neural network compared to commercial algorithms for low-dose CT image reconstruction *Nat. Mach. Intell.* **1** 269–76

Shan H, Zhang Y, Yang Q, Kruger U, Kalra M K, Sun L, Cong W and Wang G 2018 3-D convolutional encoder-decoder network for low-dose CT via transfer learning from a 2-D trained network *IEEE Trans. Med. Imaging* **37** 1522–34

Sheng K, Gou S, Wu J and Qi S X 2014 Denoised and texture enhanced MVCT to improve soft tissue conspicuity *Med. Phys.* **41** 101916

Sidky E Y, Chartrand R and Pan X 2007 Image reconstruction from few views by non-convex optimization *IEEE Nuclear Science Symposium Conference Record* **5** 3526–30

Sidky E Y, Kao C-M and Pan X 2006 Accurate image reconstruction from few-views and limited-angle data in divergent-beam CT *J. X-ray Sci. Technol.* **14** 119–39

Simonyan K and Zisserman A 2014 Very deep convolutional networks for large-scale image recognition, arXiv:1409.1556

Tan S, Zhang Y, Wang G, Mou X, Cao G, Wu Z and Yu H 2015 Tensor-based dictionary learning for dynamic tomographic reconstruction *Phys. Med. Biol.* **60** 2803–18

Tian Z, Jia X, Yuan K, Pan T and Jiang S B 2011 Low-dose CT reconstruction via edge-preserving total variation regularization *Phys. Med. Biol.* **56** 5949–67

Venkatakrishnan S V, Bouman C A and Wohlberg B 2013 Plug-and-play priors for model based reconstruction *2013 IEEE Global Conf. on Signal and Information Processing* pp 945–8

Wang G 2016 A perspective on deep imaging *IEEE Access* **4** 8914–24

Wang J, Li T, Lu H and Liang Z 2006 Penalized weighted least-squares approach to sinogram noise reduction and image reconstruction for low-dose x-ray computed tomography *IEEE Trans. Med. Imaging* **25** 1272–83

Wang S, Su Z, Ying L, Peng X, Zhu S, Liang F, Feng D and Liang D 2016 Accelerating magnetic resonance imaging via deep learning *2016 IEEE 13th Int. Symp. on Biomedical Imaging (ISBI)* pp 514–17

Wang Y *et al* 2018 Iterative quality enhancement via residual-artifact learning networks for low-dose CT *Phys. Med. Biol.* **63** 215004

Watzke O and Kalender W A 2004 A pragmatic approach to metal artifact reduction in CT: merging of metal artifact reduced images *Eur. Radiol.* **14** 849–56

Wolterink J M, Leiner T, Viergever M A and Išgum I 2017 Generative adversarial networks for noise reduction in low-dose CT *IEEE Trans. Med. Imaging* **36** 2536–45

Wu D, Kim K, El Fakhri G and Li Q 2017 Iterative low-dose CT reconstruction with priors trained by artificial neural network *IEEE Trans. Med. Imaging* **36** 2479–86

Wu D, Kim K and Li Q 2018 Computationally efficient cascaded training for deep unrolled network in CT imaging, arXiv:1810.03999

Würfl T, Hoffmann M, Christlein V, Breininger K, Huang Y, Unberath M and Maier A K 2018 Deep learning computed tomography: learning projection-domain weights from image domain in limited angle problems *IEEE Trans. Med. Imaging* **37** 1454–63

Xie H, Shan H, Cong W, Zhang X, Liu S, Ning R and Wang G 2019 Dual network architecture for few-view CT–trained on ImageNet data and transferred for medical imaging, arXiv:1907.01262

Xu Q, Yu H, Mou X, Zhang L, Hsieh J and Wang G 2012 Low-dose x-ray CT reconstruction via dictionary learning *IEEE Trans. Med. Imaging* **31** 1682–97

Yang Q, Yan P, Zhang Y, Yu H, Shi Y, Mou X, Kalra M K, Zhang Y, Sun L and Wang G 2018 Low-dose CT image denoising using a generative adversarial network with Wasserstein distance and perceptual loss *IEEE Trans. Med. Imaging* **37** 1348–57

Yang Y, Sun J, Li H and Xu Z 2016 Deep ADMM-Net for compressive sensing MRI *Advances in Neural Information Processing Systems* (San Francisco, CA: Morgan Kaufmann) pp 10–18

Ye D H, Buzzard G T, Ruby M and Bouman C A 2018a Deep back projection for sparse-view CT reconstruction *2018 IEEE Global Conf. on Signal and Information Processing* pp 1–5

Ye J C, Han Y and Cha E 2018b Deep convolutional framelets: a general deep learning framework for inverse problems *SIAM J. Imaging Sci.* **11** 991–1048

You C *et al* 2019 CT super-resolution GAN constrained by the identical, residual, and cycle learning ensemble (GAN-CIRCLE) *IEEE Trans. Med. Imaging* https://doi.org/10.1109/TMI.2019.2922960

Zhang H, Li L, Qiao K, Wang L, Yan B, Li L and Hu G 2016a Image prediction for limited-angle tomography via deep learning with convolutional neural network, arXiv:1607.08707

Zhang K, Zuo W, Chen Y, Meng D and Zhang L 2017 Beyond a Gaussian denoiser: residual learning of deep CNN for image denoising *IEEE Trans. Image Process.* **26** 3142–55

Zhang W, Zhang H, Wang L, Wang X, Hu X, Cai A, Li L, Niu T and Yan B 2019 Image domain dual material decomposition for dual-energy CT using butterfly network *Med. Phys.* **46** 2037–51

Zhang Y, Mou X, Wang G and Yu H 2016b Tensor-based dictionary learning for spectral CT reconstruction *IEEE Trans. Med. Imaging* **36** 142–54

Zhang Y, Pu Y-F, Hu J-R, Liu Y and Zhou J-L 2011 A new CT metal artifacts reduction algorithm based on fractional-order sinogram inpainting *J. X-ray Sci. Technol.* **19** 373–84

Zhang Y, Wang Y, Zhang W, Lin F, Pu Y and Zhou J 2016c Statistical iterative reconstruction using adaptive fractional order regularization *Biomed. Optics Exp.* **7** 1015–29

Zhang Y and Yu H 2018 Convolutional neural network based metal artifact reduction in x-ray computed tomography *IEEE Trans. Med. Imaging* **37** 1370–81

Zhang Y, Zhang W, Lei Y and Zhou J 2014a Few-view image reconstruction with fractional-order total variation *J. Opt. Soc. Am.* A **31** 981–95

Zhang Y, Zhang W and Zhou J 2014b Accurate sparse-projection image reconstruction via nonlocal TV regularization *Sci. World J.* **2014** 458496

Zhang Y, Zhang W-H, Chen H, Yang M-L, Li T-Y and Zhou J-L 2013 Few-view image reconstruction combining total variation and a high-order norm *Int. J. Imaging Syst. Technol.* **23** 249–55

Zhang Z, Liang X, Dong X, Xie Y and Cao G 2018 A sparse-view CT reconstruction method based on combination of DenseNet and deconvolution *IEEE Trans. Med. Imaging* **37** 1407–17

Zhao J, Chen Z, Zhang L and Jin X 2018 Unsupervised learnable sinogram inpainting network (SIN) for limited angle CT reconstruction, arXiv:1811.03911

Zhu B, Liu J Z, Cauley S F, Rosen B R and Rosen M S 2018 Image reconstruction by domain-transform manifold learning *Nature* **555** 487–92

Zhu J-Y, Park T, Isola P and Efros A A 2017 Unpaired image-to-image translation using cycle-consistent adversarial networks *Proc. of the IEEE Int. Conf. on Computer Vision* pp 2223–32

Zhu Y, Zhao M, Zhao Y, Li H and Zhang P 2012 Noise reduction with low dose CT data based on a modified ROF model *Opt. Express* **20** 17987–8004

Part III

Magnetic resonance imaging

IOP Publishing

Machine Learning for Tomographic Imaging

Ge Wang, Yi Zhang, Xiaojing Ye and Xuanqin Mou

Chapter 6

Classical methods for MRI reconstruction

Magnetic resonance imaging (MRI) is a relatively new non-invasive medical imaging technology. In MRI, a main magnetic field (also known as the background field) is applied so that the nuclei in the human body are magnetized and aligned with the main field. With a specifically tuned radio-frequency pulse sequence, the nuclei become stimulated and spin out of the equilibrium state. After that, the perturbed spins of nuclei gradually realign with the main field and return to the equilibrium state, known as relaxation. During this relaxation phase, MRI sensors can detect the energy released from the nuclei, which characterizes the tissue types, from which extremely valuable anatomical and physiological information can be extracted.

A major advantage of MRI over CT and nuclear imaging is that MRI provides great soft tissue contrast without using harmful ionizing radiation. The acquisition parameters of MRI can be tuned flexibly to optimize image quality for specific clinical tasks. MRI contrast agents are also well tolerated and less likely to cause allergic reactions or alter kidney function compared to the x-ray counterparts. The numerous advantages of MRI make it widely used in disease detection, diagnosis, and treatment monitoring in modern medical practice.

However, MRI also has drawbacks and limitations. MRI equipment is expensive, and requires greater technological expertise to maintain and operate. MRI scanning takes a long time for high image quality, which causes patient discomfort and motion artifacts. Therefore, a significant amount of MRI research efforts have been devoted to accelerating MRI scan processes in the past few decades. One of the main state-of-the-art approaches for rapid MRI is to reduce the number of samples during data acquisition, employ proper reconstruction methods to remove aliasing artifacts caused by undersampling, and recover high-quality images for clinical use.

In this chapter, we provide an overview of the physical principles of MRI, which serves as the background of the MRI data acquisition and image formation. Then, we introduce several well-known reconstruction methods developed in the past two decades. These include algorithms for total variation regularized image reconstruction in the compressed sensing MRI context, and several parallel MRI methods which have been implemented on commercial MRI scanners. This chapter sets up the foundation of the MRI image reconstruction problem, and prepares the reader for the developments of more recent deep-learning-based approaches for MRI reconstruction in the next chapter.

6.1 The basic physics of MRI

MRI is based on nuclear magnetic resonance (NMR), which was discovered by Bloch in his seminal work (Bloch 1946, Packard 1946). The application of NMR to medical imaging was realized by Lauterbur (1974). The mathematical theory for fast MRI scanning using gradient variation was presented by Mansfield (1977). In 2003, Lauterbur and Mansfield shared the Nobel Prize in Medicine or Physiology for their fundamental contribution to MRI.

To understand MRI, we first review the basic principle of NMR. Physicists found that every nucleus that has a nonzero spin angular moment $\vec{J}$ (also called spin for short) that is associated with a magnetic moment $\vec{\mu}$. These two three-dimensional vectors have the same orientation and are related by

$$\vec{\mu} = \gamma \vec{J}, \tag{6.1}$$

where γ is the gyromagnetic ratio characterized by the specific nucleus. The values of γ for different nuclei are known and are usually given in the form of $\gamma/(2\pi)$. For example, $\gamma/(2\pi)$ is 42.57 MHz/T for $_1^1$H (hydrogen), 10.71 MHz/T for $_6^{13}$C (carbon), etc, where T stands for tesla, the unit to measure the strength of a magnetic field, and Hz stands for hertz, the unit of frequency defined as one cycle per second.

If the nuclei are placed in an external constant magnetic field (i.e. the main field) $\vec{B}$ (often with strength 1 T, 1.5 T, or 3 T in MRI), whose direction is set as the z direction for convenience, i.e. $\vec{B} = (0, 0, B_0)^{\mathsf{T}} \in \mathbb{R}^3$, then $\vec{\mu}$ and $\vec{B}$ interact and yield a precession motion with torque $\vec{\tau}$:

$$\vec{\tau} = \vec{\mu} \times \vec{B}, \tag{6.2}$$

where $\times$ is the cross product between two vectors in $\mathbb{R}^3$. On the other hand, the torque $\vec{\mu}$ and the spin $\vec{J}$ are related by

$$\frac{\mathrm{d}\vec{J}}{\mathrm{d}t} = \vec{\tau}. \tag{6.3}$$

Combining (6.1), (6.2), and (6.3) yields

$$\frac{\mathrm{d}\vec{\mu}}{\mathrm{d}t} = \vec{\mu} \times (\gamma \vec{B}). \tag{6.4}$$

The ordinary differential equation (6.4) with an initial value $\vec{\mu}(0)$ at $t = 0$ has the solution

$$\vec{\mu}(t) = R(t)\vec{\mu}(0), \quad \text{where } R(t) \triangleq \begin{bmatrix} \cos(\omega_0 t) & \sin(\omega_0 t) & 0 \\ -\sin(\omega_0 t) & \cos(\omega_0 t) & 0 \\ 0 & 0 & 1 \end{bmatrix} \quad (6.5)$$

and $\omega_0 \triangleq \gamma B_0$. In other words, $\vec{\mu}(t)$ is constant along the z-axis, and rotates around the z-axis (clock-wise if viewing from the positive direction of z) with frequency ω_0. In physics, it is often the convention to write the x–y component, i.e. the transverse component $\vec{\mu}_{tr}$, of $\vec{\mu}$ as a complex value as

$$\vec{\mu}_{tr}(t) = \vec{\mu}_{tr}(0)e^{i\omega_0 t}. \quad (6.6)$$

In addition to the torque $\vec{\mu}$, the magnetic field $\vec{B}$ also generates a potential (energy) $E = -\vec{\mu} \cdot \vec{B} = -\mu_z(0)B_0$ for a magnetic moment. Although a classic spin may have any energy E, it is not the case at the atomic and subatomic levels where particles exhibit quantization. Quantum mechanics, which is necessary to accurately describe physical phenomena in these situations, indicates that the nuclei can only have a finite number of energy states. More precisely, the energy levels of a spin can only be

$$E = -m\gamma\hbar B_0, \quad \text{where } m = -j, -j+1, \dots, j-1, j, \quad (6.7)$$

and $\hbar = h/(2\pi)$ where h, approximately 6.626×10^{-34} J s (Joule seconds), is the Planck constant. In (6.7), j is the spin quantum number which is a characteristic of the specific nucleus and takes values such as 0, 1/2, 1, 3/2,.... For example, the nuclei of $_1^1\text{H}$ and $_6^{13}\text{C}$ both have $j = 1/2$, and the energy values of their nuclei can only be $E_\pm = \pm\gamma\hbar B_0/2$. This phenomenon of nuclei occupying quantized energy states is called the Zeeman effect.

An important quantity associated with quantized energy states is the gap between the spin up energy E_+ and spin down energy E_- (for $j = 1/2$, other j are similar but yield more energy states and gaps):

$$\Delta E = E_+ - E_- = \hbar\gamma B_0. \quad (6.8)$$

Taking $_1^1\text{H}$ (proton) as an example, the proton in the state E_- can switch to the state E_+ by absorbing a photon with energy ΔE, and return to E_- by emitting such a photon. As the energy of a photon with frequency ω (or wavelength $\lambda = c/\omega$, where c is the speed of light) is $\hbar\omega$, we know from (6.8) that only photons with the specific frequency $\omega_{RF} = \gamma B_0$, called the Larmor frequency, can stimulate the energy state change of a proton. Notice that the Larmor frequency ω_{RF} is exactly the same as the precessing frequency ω_0 in (6.5). Therefore, under $B_0 = 1$ and 1.5 T, respectively, ω_{RF} of the protons is about 42.6 and 63.85 MHz.

In the remainder of this section, we will only consider the case of hydrogen (proton) $_1^1\text{H}$ spin as these are the nuclei to be observed in MRI. This is because the goal of MRI is to visualize the human body, such as the brain, muscle, fat, bone marrow, cerebrospinal fluid (CSF), etc, which contain massive amounts of hydrogen nuclei.

However, even under $\vec{B}$, thermal agitation makes a fraction of hydrogen occupy the higher energy state E_+ whose spins in the z direction are anti-parallel to $\vec{B}$. The ratios of nuclei in the higher energy state E_+ and lower energy state E_- are

$$\frac{e^{-E_+/(k_b T)}}{e^{-E_-/(k_b T)} + e^{-E_+/(k_b T)}} \quad \text{and} \quad \frac{e^{-E_-/(k_b T)}}{e^{-E_-/(k_b T)} + e^{-E_+/(k_b T)}}, \tag{6.9}$$

where $k_b \approx 1.38 \times 10^{-23}$ J/K is the Boltzmann constant, and T is the temperature in kelvins. For example, at room temperature $T \approx 300$ K and with $\vec{B}$ at 1.5 T, the two ratios in (6.9) for hydrogen $_1^1\text{H}$ differ only by about 0.5×10^{-6}. Therefore, there is only slightly higher probability for a nucleus to occupy the lower energy state E_- under 1.5 T. Nevertheless, due to the presence of a massive amount of hydrogen spins in human body tissues, the magnetization of the nuclei is still readily detectable. More precisely, a voxel or region of interest in an MR image volume contains a large number of hydrogen spins, such that its net magnetization is

$$\vec{M}_0 = \sum_{i \in \text{voxel}} \vec{\mu}_i, \tag{6.10}$$

where the sum is taken over all hydrogen spins in that voxel. Due to the complete randomness of $\vec{\mu}_i$ in the transverse plane (the x–y plane) and the bias to E_- state in (6.9) in the longitudinal direction (z direction), the net magnetization $\vec{M}_0$ is aligned with the external magnetic field B, i.e. $\vec{M}_0 = (0, 0, M_0)^\mathsf{T}$. Taking the same sum of (6.4) over all these spins in the voxel as in (6.10), we obtain the macroscopic magnetization precess:

$$\frac{\mathrm{d}\vec{M}_0}{\mathrm{d}t} = \vec{M}_0 \times (\gamma \vec{B}), \tag{6.11}$$

for which $\vec{M}_0 = (0, 0, M_0)^\mathsf{T}$ is an obvious solution. This matches the description for processional motion of a spinning top in classical mechanics. The net magnetization $\vec{M}_0$ is the baseline equilibrium state of the spins due to the main magnetic field $\vec{B} = (0, 0, B_0)^\mathsf{T}$. To detect the different types of tissues, additional RF fields need to be applied to stimulate the spins so that the spins can be rotated away from this equilibrium and then gradually return, during which the RF signals they emit can be detected.

The RF excitation field can in principle be realized with two coils positioned along the x- and y-axes, called the quadrature transmitter, which generate a magnetic field $\vec{B}_1$ with the same Larmor frequency ω_0, as follows:

$$\vec{B}_1(t) = (B_1 \cos(\omega_0 t), -B_1 \sin(\omega_0 t), 0)^\mathsf{T}. \tag{6.12}$$

The transverse component of $\vec{B}_1(t)$ can simply be written as $B_1 e^{-i\omega_0 t}$. Therefore, the dynamics of the net magnetization vector $\vec{M}$ under the magnetic fields $\vec{B}$ and $\vec{B}_1(t)$ is given by the Bloch equation:

$$\frac{d\vec{M}}{dt} = \vec{M} \times \gamma(\vec{B} + \vec{B}_1(t)). \tag{6.13}$$

If we use the rotating frame (x', y', z) instead of the static frame (x, y, z), where $(x', y', z) = R(t)(x, y, z)^\top$ and $R(t)$ is the transverse rotation matrix in (6.5), then $\vec{B}_1(t)$ is always pointing in the x' direction, and the effect of $\vec{B}_1(t)$ is to make $\vec{M}$ further precess about $\vec{B}_1$ with precession frequency $\omega_1 = \gamma B_1$. After time t, there is an angle $\alpha = \gamma B_1 t$, called the flip angle, between the z direction and $\vec{M}(t)$. By varying B_1 and t, one can obtain any flip angle α. Among commonly used flip angles, 90° and 180°, for which $\vec{M}$ is put in the transverse plane and along the negative z direction, respectively, are the most important ones. We illustrate the aforementioned magnetic fields and net magnetization in the 3D stationary (left) and rotating (right) coordinate systems in figure 6.1.

After the desired flip angle α is reached, the RF field is switched off and all the perturbed spins enter the relaxation phase, during which the net transverse component M_{tr} of $\vec{M}$ gradually reduces to 0, and the longitudinal component M_{lo} returns to M_0:

$$M_{tr} = M_0 e^{-t/T_2} \sin \alpha, \tag{6.14a}$$

$$M_{lo} = M_0 e^{-t/T_1} \cos \alpha + M_0(1 - e^{-t/T_1}), \tag{6.14b}$$

where the rates for the two components M_{tr} and M_{lg} are denoted by T_2 and T_1, respectively. The values of T_1 and T_2 strongly depend on the tissue type, and hence they are extremely important quantities to differentiate body tissue types in MRI. For example, under 1.5 T, T_1 is about 200 ms for fat and 3000 ms for CSF. In contrast, T_2 is 100 and 2000 ms for these two tissue types, respectively. In addition, T_1 also depends on the strength of the external magnet field $\vec{B}$. For each fixed tissue type, there is always $T_1 > T_2$, where the former is usually at the scale of 10^3 ms and the latter at 10^2 ms. The relaxation in (6.14a) is called spin–spin relaxation, and (6.14b) is called spin–lattice relaxation. Note that, in an actual imaging process, the decay rate of M_{tr} is determined not only by the intrinsic tissue properties but also by

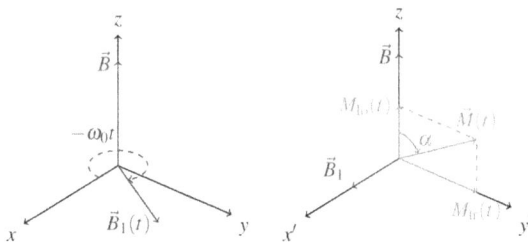

Figure 6.1. Left: the main magnetic field $\vec{B}$ and the transverse excitation field $\vec{B}_1(t)$ in (6.12) with rotation angle $-\omega_0 t$ in the stationary coordinate system. Right: the net magnetization $\vec{M}(t)$ of tissues in a voxel with flip angle α (and its transverse component $M_{tr}(t)$ and longitudinal component $M_{lo}(t)$) in the rotating coordinate system where $\vec{B}_1$ is always pointing in the x' direction.

the inhomogeneities in the magnetic field (as a result, M_{tr} decreases according to T_2^* which is much smaller than T_2).

In a typical spin-echo MRI process, an RF pulse first brings the net magnetization $\vec{M}$ to a flip angle $\alpha = 90°$, which will gradually relax according to (6.14a) (with T_2 replaced by T_2^*) and (6.14b). In the relaxation phase at $t = \text{TE}/2$, where TE is called the echo time, another RF pulse is applied. This second RF pulse flips the dephased in-plane magnetic moments 180° from one side of the transverse plane to the other side of the same plane. This will allow re-phasing of the involved spins, since the inhomogeneities of the magnetic field are assumed to be constant so that with the re-phasing effect we can reduce the T_2^* effect to the T_2 effect when we collect RF signals during a short time window around $t = \text{TE}$. Then, the nuclei continue to relax until $t = \text{TR}$, where TR stands for the repetition time and is much larger than TE, the same excitation process started with $\alpha = 90°$ is repeated.

At time $t = \text{TR}$, the transverse magnetization has vanished, and the longitudinal magnetization becomes $M_{lo} = M_0(1 - e^{-\text{TR}/T_1})$, which is flipped to the transverse plane by the new excitation with flip angle $\alpha = 90°$. With this spin-echo pulse sequence, the magnitude of the transverse component, denoted by $s(t)$, is the signal to be detected:

$$s(t) = M_0(1 - e^{-\text{TR}/T_1})e^{-t/T_2}, \tag{6.15}$$

where the signal detection takes place in a short period around $t = \text{TE}$ in this new phase of excitation.

Up to this point, we have explained the fundamental principles of NMR, including how the nuclei can be stimulated by external magnetic fields and what signals are measured. However, the signals acquired by NMR do not contain any spatial information, i.e. the locations of different tissues, which is necessary for MRI. To encode the spatial information in the detected signal, Lauterbur developed a sophisticated modification of the background magnetic field by imposing a series of linear magnetic field gradients to the field $\vec{B} = (0, 0, B_0)^\top$. For instance, if a linear magnetic field with gradient G_z along the z direction (i.e. $\vec{G} = (0, 0, G_z z)^\top$) is imposed, the magnetic field becomes $(0, 0, B_0 + G_z z)^\top$ which varies for different slices parallel to the x–y plane. Therefore, we can generate the RF field with frequency $\gamma(B_0 + G_z z)$ so that only the nuclei in the slice of z (i.e. the plane with the normal direction $(0, 0, 1)$ and through the point $(0, 0, z)$) respond to the RF excitation. A similar strategy can be employed by imposing a gradient field in an arbitrary direction (x, y, z) rather than $(0, 0, 1)$. By imposing the proper gradient field, the slice location information can be encoded in the detected RF signal. Furthermore, in the same spirit the pixel location information in the selected slice can be encoded as well, which are often called phase encoding and frequency encoding, respectively.

Finally, we need to decode the tissue information contained in the detected signal to reconstruct the cross-section. This can be done in multiple ways. As a primary example, we can collect RF signals using the well-known spin-echo sequence. Suppose that a magnetic field $\vec{G}$ (we will see how to select this $\vec{G}$ soon) is applied.

Then, the net magnetization of the voxel at location $\vec{r}$ has angular frequency $\omega(\vec{r}) = \gamma \vec{G} \cdot \vec{r}$, with its magnitude given in (6.15). The detected signal is the integral of such signals over the entire space $\mathbb{R}^3$ in the time window [TE, TE + Δt], where Δt is a very short time period during which the signal is measured. During this short time period, $e^{-t/T_2} \approx e^{-TE/T_2}$, and the detected signal is approximately

$$\hat{u}(\vec{k}) = \int_{\mathbb{R}^3} \rho(\vec{r})(1 - e^{-TR/T_1(\vec{r})})e^{-TE/T_2(\vec{r})}e^{-2\pi i \vec{k} \cdot \vec{r}} d\vec{r}, \tag{6.16}$$

where $\rho(\vec{r})$ is the net spin density (also called proton density), $T_1(\vec{r})$ and $T_2(\vec{r})$ are the T_1 and T_2 parameter values of the tissue in the voxel located at $\vec{r}$, respectively, and $\vec{k}$ is defined by

$$\vec{k} = \frac{\gamma \Delta t}{2\pi} \vec{G}. \tag{6.17}$$

As we can see, the detected signal $\hat{u}(\vec{k})$ in (6.16) is the Fourier transform of $u(\vec{r})$, which is the MR image of interest, defined by

$$u(\vec{r}) = \rho(\vec{r})(1 - e^{-TR/T_1(\vec{r})})e^{-TE/T_2(\vec{r})}. \tag{6.18}$$

Therefore, by choosing $\vec{G}$ for every desired $\vec{k}$ point according to (6.17), we can obtain the Fourier spectrum $\hat{u}(\vec{k})$ for the reconstruction of the underlying image $u(\vec{r})$. The Fourier space where $\hat{u}(\vec{k})$ are measured is also called the k-space. In practice, there are rules for k-space sampling trajectories suggesting the sampling order of $\vec{k}$ with which $\hat{u}(\vec{k})$ can be scanned efficiently. The image $u(\vec{r})$ can be computed by taking an inverse Fourier transform of the k-space data $\hat{u}(\vec{k})$.

From (6.18), we know that the image value at voxel $\vec{r}$ is not the 'pure' proton density $\rho(\vec{r})$. Specifically, it also depends on $T_1(\vec{r})$, $T_2(\vec{r})$, TR, and TE. This is in fact advantageous because we can tune TR and TE to see different contrasts that exhibit different emphases on $\rho(\vec{r})$, $T_1(\vec{r})$, and $T_2(\vec{r})$. For example, if we choose a short TR, then tissues with different T_1 will have significantly different values of $1 - e^{-TR/T_1}$: it is close to 0 for tissues with small T_1 time and 1 for those with large T_1 time. The resultant image is called T_1 weighted. Similarly, if we choose a long TE, then the image is called T_2 weighted. If we choose a long TR but short TE, then $u(\vec{r}) \approx \rho(\vec{r})$ according to (6.18), and the proton density (PD) weighted (or ρ weighted) image is obtained. In figure 6.2, we show examples of T_1, T_2, and ρ weighted brain images obtained from BrainWeb (Cocosco *et al* 1997). The rich tissue contrast and flexibility by tuning parameters of the pulse sequence make MRI a rather attractive medical imaging technology, with great applicability in real-world clinical tasks.

In summary, we have provided a brief overview of the basic principle of MRI, including its physical foundation in NMR, the excitation and gradient fields, and the signal and image formation. For more comprehensive treatment of this subject, we refer to Suetens (2017).

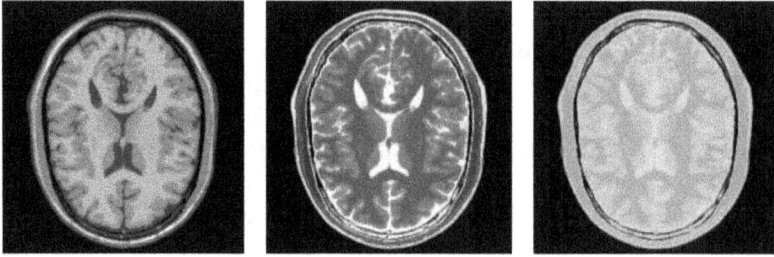

Figure 6.2. From left to right: the T_1, T_2, and ρ weighted images of the same brain generated by the data downloaded from the BrainWeb site at https://brainweb.bic.mni.mcgill.ca/brainweb/ (Cocosco *et al* 1997).

6.2 Fast sampling and image reconstruction

A main problem with MRI is that it requires a long time to collect all $\hat{u}$ in (6.16) to completely cover the k-space. A prolonged imaging process can cause patient discomfort, and even worse, introduce the notorious motion artifacts into MR images. However, it is difficult for humans, in particular children, the elderly, and patients with certain health issues, to remain still for an extended time during an MRI scan, which often takes 30–60 min. In addition, the heart beat, breathing, and blood flow are inevitable and can also introduce motion artifacts. The consequence of motion is that the proton density is not constant at $\vec{r}$, which disturbs the k-space signal $\hat{u}$, and degrades the quality of the image u after inverse Fourier transform.

In the past few decades, great efforts have been devoted to accelerating MRI scanning. One of the main state-of-the-art methods for rapid MRI is to reduce the number of k-space samples $\hat{u}$. That is, the MRI scanner only acquires $\hat{u}(\vec{k})$ at a limited number of $\vec{k}$ locations rather than the full set of k-space required by a standard inverse Fourier transform to obtain $u(\vec{r})$. Then, a reconstruction algorithm needs to be properly designed to recover a high-quality image u from the incomplete k-space data.

For ease of presentation and derivation in the remainder of this chapter, we focus on the two-dimensional (2D) discrete case where an MR image u can be regarded as a 2D array of size $\sqrt{n} \times \sqrt{n}$, where n is the number of pixels in the image u. For example, a square image u of $n = 128^2 = 16{,}384$ pixels can be treated as a 2D array with 128×128 pixels. Extension from a 2D image to 3D images is straightforward. Let $u_{i,j} \in \mathbb{R}$ denote the value of u at pixel (i, j), called the intensity value, for $i, j = 1, \ldots, \sqrt{n}$. In the formulation of numerical algorithms for MRI reconstruction, it is convenient and common to stack the columns of u vertically to form a column vector u of dimension n, and $u_i \in \mathbb{R}$ is the intensity value at a pixel indexed by i for $i = 1, \ldots, n$. For example, $u_1 = u_{1,1}$, $u_{\sqrt{n}+1} = u_{1,2}, \ldots, u_n = u_{\sqrt{n},\sqrt{n}}$, etc. The relation between the array and vector forms of u can be visualized in (6.19):

$$
\begin{bmatrix}
u_{1,1} & u_{1,2} & \cdots & u_{1,\sqrt{n}} \\
u_{2,1} & u_{2,2} & \cdots & u_{2,\sqrt{n}} \\
\vdots & \vdots & \ddots & \vdots \\
u_{\sqrt{n},1} & u_{\sqrt{n},2} & \cdots & u_{\sqrt{n},\sqrt{n}}
\end{bmatrix}
\iff
\begin{bmatrix}
u_{1,1} \\
u_{2,1} \\
\vdots \\
u_{\sqrt{n},1} \\
u_{1,2} \\
\vdots \\
u_{\sqrt{n},\sqrt{n}}
\end{bmatrix}
=
\begin{bmatrix}
u_1 \\
u_2 \\
\vdots \\
u_{\sqrt{n}} \\
u_{\sqrt{n}+1} \\
\vdots \\
u_n
\end{bmatrix}.
\tag{6.19}
$$

We will use these two notations interchangeably, often the latter one. The corresponding k-space data $\hat{u}$ can be regarded as either a 2D array or an n-dimensional vector, with its indices for the specific $\vec{k}$ values instead of the spatial locations $\vec{r}$. It is also worth noting that although u is real-valued in the ideal case, noise and perturbation in the k-space data and the numerical reconstruction algorithm may render u complex-valued in practice.

Now, we consider what is meant by partial k-space sampling mathematically. Let $\mathcal{F} \in \mathbb{C}^{n \times n}$ be the 2D discrete Fourier transform (DFT) matrix, with which the image u and the full k-space data $\hat{u}$ are related by $\hat{u} = \mathcal{F}u$. If only partial k-space data are acquired, then it is equivalent to left-multiplying a matrix $P \in \mathbb{R}^{m \times n}$ to $\mathcal{F}u$, where m is the number of Fourier coefficients sampled. More precisely, each row of P has one entry as 1 indicating the index of the sampled Fourier coefficient (location of $\vec{k}$) and the others as 0. Hence, $P\mathcal{F}u \in \mathbb{C}^m$ is the vector of the sampled, partial Fourier coefficients. The detected signal, which may involve noise during the k-space data acquisition, is expressed as

$$
f = P\mathcal{F}u + e,
\tag{6.20}
$$

where e represents the noise in data acquisition (the noise distribution may be available or estimated). The goal of image reconstruction is thus to solve u from the problem (6.20) provided the partial Fourier data f.

However, even if f contains no noise, i.e. $e = 0$, it is still highly non-trivial to recover u from the partial Fourier data f in (6.20). First of all, one can immediately see that $P\mathcal{F}u = f$ with $P\mathcal{F} \in \mathbb{C}^{m \times n}$ and $m < n$ is an underdetermined system which admits infinitely many solutions. If one were to pursue the 'minimal-norm' solution as is common practice in many other inverse problems, i.e.

$$
\underset{u}{\text{minimize}} \ \|u\| \quad \text{subject to} \ P\mathcal{F}u = f,
\tag{6.21}
$$

where $\|u\| = \left(\sum_{i=1}^{n} |u_i|^2\right)^{1/2}$ is the standard Euclidean norm of u, then we can first substitute the objective $\|u\|$ by $(1/2) \cdot \|u\|^2$, which does not alter the solution (6.21), and form the associated Lagrangian as

$$
L(u; \lambda) = \frac{1}{2}\|u\|^2 + \lambda^{\top}(P\mathcal{F}u - f).
\tag{6.22}
$$

Then, the Karush–Kuhn–Tucker (KKT) condition (see appendix A) implies the existence of λ such that the following two equations hold simultaneously:

$$u + \mathcal{F}^\mathsf{T} P^\mathsf{T} \lambda = 0, \tag{6.23a}$$

$$P\mathcal{F}u - f = 0. \tag{6.23b}$$

Left-multiplying $P\mathcal{F}$ to (6.23a), using the facts that $\mathcal{F}\mathcal{F}^\mathsf{T} = I_n$ and $PP^\mathsf{T} = I_m$, where I_n stands for the n-by-n identity matrix, and replacing $P\mathcal{F}u$ by f according to (6.23b), we obtain $\lambda = -f$. Plugging this back into (6.23a) yields the minimal-norm solution

$$u = \mathcal{F}^\mathsf{T} P^\mathsf{T} f. \tag{6.24}$$

Note that the solution (6.24) is essentially setting the missing Fourier coefficients to 0 to obtain a pseudo full k-space dataset $P^\mathsf{T} f \in \mathbb{C}^n$, and then applying the inverse Fourier transform $\mathcal{F}^{-1} = \mathcal{F}^\mathsf{T}$ to reconstruct u. Therefore, the solution (6.24) is also called the *zero-filled* reconstruction (also known as the zero-filling reconstruction).

As the inverse problem (6.20) is ill-posed, it is necessary to regularize the solution to recover u from f specifically. The zero-filled solution (6.24) resulting from the ℓ_2-norm regularization (6.21), however, appears to be practically inappropriate since it exhibits severe aliasing artifacts (see figure 6.4 as an example). Therefore, it is crucial to design a suitable regularization term by incorporating proper prior knowledge into the image reconstruction.

6.2.1 Compressed sensing MRI

Using proper regularization to remedy the ill-posedness of an inverse problems has a long history. The design of a regularization term should be based upon the prior information we know or the property we prefer to have in the underlying image to be reconstructed. Hence, regularization can either be based on physical properties or inspired by common sense, or both. As we will see in the next chapter, regularization can also be designed by a data-driven approach in deep-learning.

In the late 1990s and the early 2000s, the idea of compressed sensing (also called compressive sensing, or CS for short) (Candes *et al* 2006, Donoho 2006) became widely accepted in the signal processing community, since it provides a rigorous mathematical justification of regularization for important types of inverse problems. Such regularization is primarily on the sparsity of the underlying signal.

A digital signal x, regarded as a vector in $\mathbb{R}^n$, is called sparse if the majority of its components are zeros. In other words, the number of nonzero components, denoted by p, is much smaller than n. We call p (or p/n) the sparsity level of x. However, even if x is known to be sparse, the locations and values of the nonzero components are still unknown. Nevertheless, it would not be economical to acquire or store all components of a sparse x one by one since most of them are just zeros. CS theory implies that if a signal x is sufficiently sparse and a proper sensing matrix $A \in \mathbb{R}^{m\times n}$ is used, where $p < m \ll n$, then it is possible to encode x into the vector $b = Ax \in \mathbb{R}^m$ without any significant loss of information. The acquisition and storage of b are much less resource-consuming, and we can recover $x \in \mathbb{R}^n$ from b in the decoding phase by solving the following minimization problem:

$$\underset{x}{\text{minimize}} \; \|x\|_0, \quad \text{subject to } Ax = b, \tag{6.25}$$

where $\|x\|_0$ counts the number of nonzero components in x. The minimization problem in (6.25) is known to be NP-hard, since the objective function is nonconvex and discontinuous, and hence lacks robust numerical solvers with guaranteed optimality. However, it is shown that x can also be accurately recovered by solving the problem (6.25) with the ℓ_0 norm replaced by the continuous (but not differentiable) convex relaxation, the ℓ_1 norm, i.e. $\|x\|_1 = \sum_{i=1}^{n} |x_i|$, if x is sufficiently sparse and the matrix A has the restricted isometry property (RIP) (Candes $et\ al$ 2006). With this relaxation, problem (6.25) becomes convex for which a number of efficient convex optimization algorithms can be applied.

The CS formulation (6.25) is quite similar to the one we have in (6.21). However, there are two problems remaining: (i) is the MR image u to be reconstructed really sparse and (ii) does the sensing matrix $P\mathcal{F}$ (6.20) satisfy the RIP required by the CS theory? Regarding problem (i), we know that a medical image u itself is unlikely to be sparse, but $x = \Psi u$ may be sparse if a proper sparsifying transform Ψ is employed. For example, we can set $\Psi \in \mathbb{R}^{n \times n}$ to an orthogonal wavelet transform matrix (such that $\Psi^\top \Psi = \Psi \Psi^\top = I_n$), then $x = \Psi u \in \mathbb{R}^n$ is the wavelet coefficients of u which is often sparse. Therefore, we can instead solve for x from the following ℓ_1 minimization problem

$$\underset{x}{\text{minimize}} \; \|x\|_1, \quad \text{subject to } P\mathcal{F}\Psi^\top x = f, \tag{6.26}$$

and set $u = \Psi^\top x$ to obtain the reconstructed image. The constraint in (6.26) is deduced from the one in (6.21) with $u = \Psi^{-1}x = \Psi^\top x$ under the noiseless data assumption. Problem (ii) is a much more tedious issue. However, it is found that, with randomized k-space sampling P, the resulting sensing matrix $A = P\mathcal{F}\Psi^\top$ is robust enough to reconstruct x and hence u faithfully in CS-MRI in practice (Lustig $et\ al$ 2007).

The idea of employing a sparsifying matrix and solving ℓ_1 minimization in CS is closely connected to the total variation regularization developed for image processing initially and adapted for image reconstruction subsequently. These two concepts merged and generated numerous encouraging results, in particular for image reconstruction in MRI, which we discuss in the next section.

6.2.2 Total variation regularization

In the past decade, total variation (TV) has been widely recognized as a good choice for robust and effective regularization in image reconstruction. The use of TV as regularization in image reconstruction was first proposed in the seminal work (Rudin $et\ al$ 1992). Total variation is a concept in the theory of functional analysis and partial differential equations (PDE). Briefly speaking, it is a regularity measure to quantify the oscillations (variations) of functions which are possibly discontinuous and not (weakly) differentiable in the classical sense. For an L^1 function u (i.e. $|u|$ is an integrable function in the Lebesgue sense) defined on a domain $\Omega \subset \mathbb{R}^2$, its TV semi-norm is defined by

$$\text{TV}(u) \triangleq \sup\left\{ -\int_\Omega u \, \text{div} \, p \, \mathrm{d}x \colon p \in C_0^\infty(\Omega; \mathbb{R}^2), \, |p(x)| \leqslant 1, \, \forall \, x \in \Omega \right\}. \quad (6.27)$$

Furthermore, the set of L^1 functions on Ω with finite TV (6.27) is called the space of bounded variations (BV), denoted by BV(Ω). The characteristics of TV (6.27) are outside the scope of this book, and we only briefly discuss its relevance to image reconstruction and its discrete counterpart for numerical computation here.

First of all, a 2D square grayscale image u can be considered as a function defined on $\Omega = [0, 1]^2 \in \mathbb{R}^2$ in the continuous setting, where $u(x) \in \mathbb{R}$ is the intensity value at $x \in \Omega$. Ideally, we have a clear image if the objects, also called targets or regions (such as white/gray matters, CSF, and fat in a brain image) in the image have sharp boundaries, and within each object the intensity values are nearly constant. See the Shepp–Logan phantom in figure 6.4 as an example, where different objects are shown with distinct intensity levels. The images with such properties actually look like 'cartoons', but natural and medical images may be far more complex than cartoons. However, let us be content with 'cartoon-like' images for now, as they at least show clear objects with low noise levels. In the context of functional analysis and PDEs, the functions representing such images are not continuous due to the jump of intensity values at object boundaries. But the jumps should be controlled to a certain level such that the boundaries are not too irregular, and the intensities should be quite similar or even identical inside each target. Such functions are perfectly characterized by the BV space BV(Ω) with a small TV norm for regular boundary contours and low noise.

Discretizing an image represented by the function u on $\Omega = [0, 1]^2$ requires applying a $\sqrt{n} \times \sqrt{n}$ mesh grid to the square $[0, 1]^2$, and the value of the discretized image u at any one of the n pixels, each having an area of $(1/\sqrt{n}) \times (1/\sqrt{n})$, is the integral (or average) of the continuous image u over that pixel area. In the discrete case, the TV norm defined in (6.27) reduces to the following form, which turns out to also be the ℓ_1 norm of $\|D_i u\|$ since the optimal $_{pi}$ is obviously $D_i u / \|D_i u\|$:

$$\text{TV}(u) = \max\left\{ \sum_{i=1}^n p_i^\top D_i u \colon p_i \in \mathbb{R}^2, \, \|p_i\| \leqslant 1, \, \forall \, i \right\} = \sum_{i=1}^n \|D_i u\|, \quad (6.28)$$

where $D_i \in \mathbb{R}^{2 \times n}$ is binary and has only two nonzero entries (1 and -1) corresponding to the forward finite difference approximations to partial derivatives along the coordinate axes. For example, the forward difference of u at the first pixel $u_{1,1} = u_1$ is $(u_{2,1} - u_{1,1}, u_{1,2} - u_{1,1})^\top = (u_2 - u_1, u_{\sqrt{n}+1} - u_1)^\top \in \mathbb{R}^2$ (see (6.19) for the conversion between the array and vector forms of u) if the x and y directions are defined as downward and rightward, respectively. Therefore, $D_1 \in \mathbb{R}^{2 \times n}$ is the matrix with $[D_1]_{1,1} = -1, [D_1]_{1,2} = 1, [D_1]_{2,1} = -1, [D_1]_{2,\sqrt{n}+1} = 1$, and the other entries as 0. Hence, $D_1 u = (u_2 - u_1, u_{\sqrt{n}+1} - u_1)^\top \in \mathbb{R}^2$ is the discretized gradient of u at pixel indexed by

$i = 1$ using forward differences to approximate partial derivatives. The other $D_i \in \mathbb{R}^{2 \times n}$ for $i = 2, \ldots, n$ can be similarly formulated.

The forward differences at the border pixels, such as $u_{\sqrt{n},j}$ and $u_{i,\sqrt{n}}$ for $i, j = 1, \ldots, \sqrt{n}$, need slight more care since they require pixel values such as $u_{\sqrt{n}+1,j}$ and $u_{i,\sqrt{n}+1}$, which are outside the range of the 2D array u. There are three common ways to address this issue, which are related to the boundary condition imposed on the image u, with the latter two illustrated in figure 6.3:

- *Zero-padding*: This is to set $u_{\sqrt{n}+1,j} = u_{i,\sqrt{n}+1} = 0$ for all $i, j = 1, \ldots, \sqrt{n}$.
- *Symmetric extension*: This is to assume $u_{\sqrt{n}+1,j} = u_{\sqrt{n},j}$ and $u_{i,\sqrt{n}+1} = u_{i,\sqrt{n}}$ for $i, j = 1, \ldots, \sqrt{n}$. In other words, the pixel values near the image boundary are symmetric about the boundary.
- *Periodic extension*: This is to let $u_{\sqrt{n}+1,j} = u_{1,j}$ and $u_{i,\sqrt{n}+1} = u_{i,1}$ for $i, j = 1, \ldots, \sqrt{n}$. That is equivalent to concatenating the same 2D array u to each side of the image u.

Since the main features are often inside an MR image and the intensities are mostly zero near the image boundaries (see the brain images in figure 6.2), the three boundary conditions above yield the same result. However, it is often convenient to use the periodic extension in MRI, since it is coherent with the Fourier transform as we will see later. The symmetric extension is advantageous when non-trivial features are truncated at image boundaries such as region of interest images acquired in local MRI and many natural images. Zero-padding is simple but does not have properties as nice as the other two, and may introduce artifacts if the pixel values near boundaries are nonzero. In this chapter, without loss of generality, we only use the periodic extension as the boundary condition unless otherwise noted.

Now, we have clearly defined the discrete TV norm of an image u in (6.28), and intend to use it as the regularization term for MRI reconstruction (6.20). More precisely, we want to reconstruct an image u by solving the following minimization problem:

$$\underset{u}{\text{minimize }} \mu \, \mathrm{TV}(u) + \frac{1}{2} \| P \mathcal{F} u - f \|^2. \tag{6.29}$$

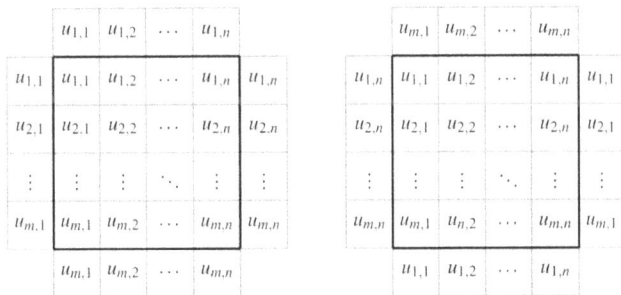

Figure 6.3. Illustrations of symmetric (left) and periodic (right) extension for the boundary conditions of a 2D image u of size $m \times n$. Each small box represents a pixel. The center $m \times n$ array (inside thick boundary lines) is the original image u, the values outside u are extrapolated.

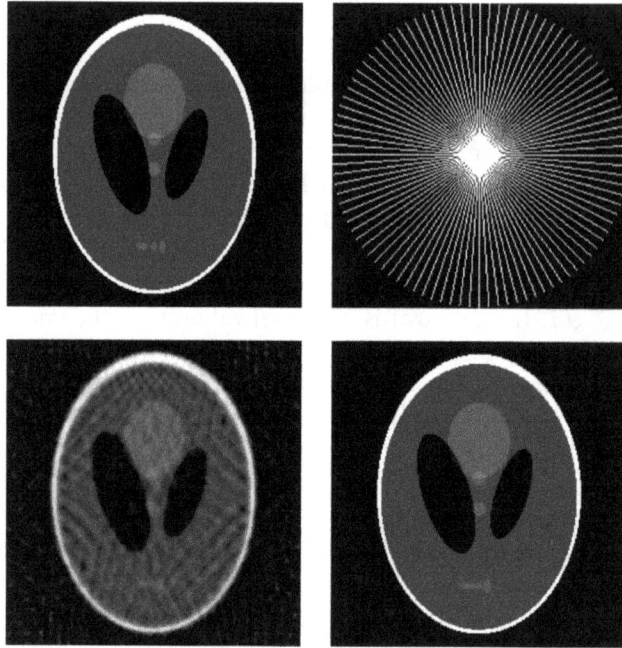

Figure 6.4. An example of CS-MRI reconstruction by the PDHG algorithm. Upper left: the Shepp–Logan phantom image. Upper right: synthesized radial k-space undersampling pattern (mask), where white pixels indicate sampled k-space locations. Lower left: zero-filled reconstruction, which exhibits severe aliasing artifacts. Lower right: reconstruction from the undersampled k-space data using TV regularization by the PDHG algorithm, which appears to be faithful to the original image obtained by full k-space data.

Note that here we use this unconstrained minimization since the constrained ones like (6.21) are only suitable for noiseless data f. In this formulation, $\mu > 0$ is called the weight parameter to balance the regularization term $\text{TV}(u)$ and the data fidelity (also called data consistency) term $(1/2) \cdot \|P\mathcal{F}u - f\|^2$. The weight parameter μ is chosen based on our confidence about the data f: if f is noiseless or the noise level is low, then we use smaller μ, so that the data fidelity term gains priority in the minimization; otherwise we should use a larger μ. From the practical point of view, larger μ helps to suppress noise and artifacts, but it can also smear fine structures in the reconstructed image u. Therefore, μ needs to be properly chosen for an optimal balance between the two terms.

Before we move on to the computation of (6.29), we would like summarize some properties of the TV regularization. TV is widely recognized as a good candidate of regularization for image reconstruction for several reasons: (i) the continuous version of (6.29) admits a solution since TV (6.27) is closed and lower semi-continuous, which is an important property in the PDE theory if one approaches the image reconstruction problem from the geometric analysis point of view; (ii) the discrete version of (6.29) is convex for which numerical optimization algorithms are efficient and have optimality guarantees; (iii) compared to other regularization methods in the literature, TV is simple, effective, and robust in preserving object boundaries and suppressing noise; and (iv) as shown in (6.28), the discrete TV behaves as the ℓ_1 norm of the (magnitude of) gradient $\|D_i u\|$, which is closely related to the theory of CS in (6.26). In other words, we tend to recover an image with a 'sparse gradient' via (6.29), which means that the intensity values are nearly constant within each target/region and only jump at the boundary between two different adjacent targets/regions (e.g. white and gray matter in the brain), resulting a clear delineation of object contours.

However, TV regularized image reconstruction (6.29) has several known drawbacks. The main issue with TV is that the reconstructed images could look like 'oil paintings' since they tend to be 'piecewise constant'—a side effect of enforcing sparse gradients prompted by (6.28). This oil painting effect is undesirable in clinical applications, since it may smear extremely important subtle features in images, such as fine details and textures. This issue with TV regularization may be (partially) alleviated by tuning the weight parameter μ in (6.29). However, a proper weight parameter yielding desired image quality in reconstruction may not exist for a specific problem due to the nature of TV; and even if it exists, it may be very difficult to find in practice.

6.2.3 ADMM and primal–dual

In this subsection, we introduce two state-of-the-art algorithms for solving the TV regularized image reconstruction problem (6.29). Since TV (6.28) is convex in u due to the convexity of the norm, and the data fidelity in (6.29) is a quadratic function of u and hence also convex, the minimization (6.29) falls into the class of convex optimization, for which highly efficient algorithms are available.

There have been numerous attempts to solve (6.29) with high efficiency numerically. In the past decade, convex optimization techniques have been employed, through which algorithms developed to solve (6.29) have become very mature. Among these algorithms, the alternating direction method of multipliers (ADMM) (Boyd *et al* 2011, Goldstein and Osher 2009, Tai and Wu 2009) and the primal–dual hybrid gradient method (PDHG) (Chambolle and Pock 2011, Esser *et al* 2010, Zhu and Chan 2008) are widely considered as the most robust and efficient.

The derivation of ADMM starts with variable splitting. Variable splitting is a useful trick to handle composite optimization such as the discrete TV (6.28) where Du appears in another non-differentiable but simple function—the norm $\|\cdot\|$. Recall that the TV regularized minimization problem (6.29) can be written as

$$\underset{u}{\text{minimize}} \left\{ \mu \sum_{i=1}^{n} \|D_i u\| + \frac{1}{2} \|P\mathcal{F}u - f\|^2 \right\}. \tag{6.30}$$

The difficulty in (7.34) is that the norm in the first term is not differentiable and hence gradient-based algorithms cannot be directly applied. To overcome this issue, we introduce auxiliary variables $w_i \in \mathbb{R}^2$ and require $w_i = D_i u$ for $i = 1,\ldots, n$. Then, we can rewrite (7.34) as an equivalent, constrained optimization problem as

$$\underset{w,u}{\text{minimize}} \left\{ \mu \sum_{i=1}^{n} \|w_i\| + \frac{1}{2} \|P\mathcal{F}u - f\|^2 \right\}, \text{ subject to } w_i = D_i u, \text{ for all } i. \quad (6.31)$$

Although the new formulation (7.35) still has the non-differentiable norm on w_i, it is now possible to apply alternating minimizations to w and u such that they can be alternately updated easily. This treatment follows the 'divide-and-conquer' strategy, since the original problem is decomposed into smaller and much simpler subproblems, as we will show below.

To solve the constrained minimization (7.35), we first formulate its associated augmented Lagrangian:

$$L(u, w; \lambda) = \mu \sum_{i=1}^{n} \|w_i\| + \frac{1}{2} \|Au - f\|^2 + \langle \lambda, w - Du \rangle + \frac{\rho}{2} \|w - Du\|^2, \quad (6.32)$$

where λ is the Lagrangian multiplier, or multiplier for short, such that $\lambda_i \in \mathbb{R}^2$ corresponds to the constraint $w_i = D_i u$ for $i = 1, \ldots, n$. The last two terms in (7.36) are interpreted as $\langle \lambda, w - Du \rangle = \sum_{i=1}^{n} \lambda_i^{\top}(w_i - D_i u)$ and $\|w - Du\|^2 = \sum_{i=1}^{n} \|w_i - D_i u\|^2$. The penalty parameter $\rho > 0$ (7.36) is user-chosen, which may affect the speed of convergence in practice.

The ADMM algorithm is designed to solve (7.35) with the augmented Lagrangian (7.36) by the following iterative procedure:

$$w_i^{(k+1)} = \arg \min_{w_i} \left\{ \mu \|w_i\| + \langle \lambda_i^{(k)}, w_i - D_i u^{(k)} \rangle + \frac{\rho}{2} \|w_i - D_i u^{(k)}\|^2 \right\}, \forall i, \quad (6.33a)$$

$$u^{(k+1)} = \arg \min_{u} \left\{ \frac{1}{2} \|P\mathcal{F}u - f\|^2 + \langle \lambda^{(k)}, w^{(k+1)} - Du \rangle + \frac{\rho}{2} \|w^{(k+1)} - Du\|^2 \right\} \quad (6.33b)$$

$$\lambda^{(k+1)} = \lambda^{(k)} + \rho(w^{(k+1)} - Du^{(k+1)}). \quad (6.33c)$$

That is, with an initial guess $(u^{(0)}, w^{(0)}, \lambda^{(0)})$ (in practice only $u^{(0)}$ and $\lambda^{(0)}$ are needed here), ADMM repeats the cycle of the three steps (7.37) for the iteration index $k = 1, 2, \ldots$, until a stopping condition is satisfied, and outputs the final u as the reconstruction.

For ADMM to run efficiently, the subproblems in (7.37) must be easy to solve. This is the case for the MRI image reconstruction (7.34). As we can see, the λ subproblem (7.37c) involves a straightforward computation only. For the w subproblem (7.37a), we can combine the last two terms by completing the square, and reformulate it as

$$w_i^{(k+1)} = \arg \min_{w_i} \left\{ \mu \|w_i\| + \frac{\rho}{2} \|w_i - D_i u^{(k)} + \rho^{-1} \lambda_i^{(k)}\|^2 + \text{const} \right\}, \quad (6.34)$$

where 'const' is a constant dependent on $\lambda_i^{(k)}$ but not on w_i, and hence makes no difference in the minimization (6.34). Solving (6.34) amounts to finding the solution of the minimization problem in the following form:

$$\underset{x}{\text{minimize}} \left\{ \|x\| + \frac{1}{2\alpha} \|x - b\|^2 \right\},$$ (6.35)

for some given $\alpha > 0$ and $b \in \mathbb{R}^2$, which has the close-form solution, called soft shrinkage, defined by

$$\text{shrink}_\alpha(b) = \max\{0, \|b\| - \alpha\} \cdot \frac{b}{\|b\|}.$$ (6.36)

Hence, (6.34) has a closed form solution, $\text{shrink}_{\rho^{-1}\mu}(D_i u^{(k)} - \rho^{-1}\lambda_i^{(k)})$, which is easy to compute according to (6.36). For the u subproblem (7.37b), we first realize that the objective function is convex and differentiable. Therefore, we can take the gradient of the objective function in (7.37b) with respect to u, set it to zero, and solve for u to obtain $u^{(k+1)}$. After setting the gradient to zero, we obtain the so-called normal equation of u given by

$$(\rho D^\mathsf{T} D + \mathcal{F}^\mathsf{T} P^\mathsf{T} P \mathcal{F})u = \rho D^\mathsf{T}(w^{(k+1)} + \rho^{-1}\lambda^{(k)}) + \mathcal{F}^\mathsf{T} P^\mathsf{T} f.$$ (6.37)

Note that the right-hand side is known and independent of u. The normal equation (6.37) is a large linear system of size n and can be challenging to solve in general. However, for the TV regularized MRI image reconstruction (7.34), the finite difference operator D can be diagonalized by the DFT matrix $\mathcal{F}$ if the periodic boundary condition is used (this is why we mostly use the periodic boundary condition in MRI reconstruction). That is, $D = \mathcal{F}\hat{D}\mathcal{F}^\mathsf{T}$ for some diagonal matrix $\hat{D}$. Then, (6.37) has a solution in the closed form

$$u = \mathcal{F}^\mathsf{T}[(\rho\hat{D}^\mathsf{T}\hat{D} + P^\mathsf{T}P)^{-1}(\rho D^\mathsf{T}(w^{(k+1)} + \rho^{-1}\lambda^{(k)}) + \mathcal{F}^\mathsf{T} P^\mathsf{T} f)],$$ (6.38)

where $\rho\hat{D}^\mathsf{T}\hat{D} + P^\mathsf{T}P$ is diagonal and hence trivial to invert. Therefore, the main computation cost in (6.38) is the Fourier transforms $\mathcal{F}$ and $\mathcal{F}^\mathsf{T}$, which is nearly linear in problem size n at $O(n \log n)$. However, it should be noted that the closed form (6.38) exists because the sensing matrix is $P\mathcal{F}$ in (6.20). For more complex MRI reconstruction, such as parallel MRI, this matrix is not that simple due to the involvement of coil sensitivities. We will discuss this issue in more detail in section 6.3.

We have shown that the computations in ADMM (7.37) are all easy for the TV regularized MR image reconstruction problem (7.34): the w subproblem has a closed form solution by soft shrinkage (6.36), the u problem mainly involves two Fourier transforms in (6.38), and the λ problem is a straightforward computation.

Convergence of an optimization algorithm is of both theoretical and practical importance. The ADMM algorithm (7.36) is known to be convergent due to the convexity of the constrained problem (7.35) and its two-block (i.e. u and w) structure. The overall computational cost depends on the per-iteration complexity (which we have shown to be low for ADMM (7.36) above) and the number of iterations. In most MRI reconstruction problems that can be formulated as (7.34), ADMM (7.37) often outputs satisfactory reconstruction in 50–100 iterations, which is quite fast considering its low per-iteration cost. Therefore, ADMM is particularly suitable for convex optimization with linear constraints, and its application to TV regularized MR image reconstruction (7.35) is one of the many successful examples.

The PDHG algorithm (Chambolle and Pock 2011, Esser *et al* 2010, Zhu and Chan 2008) is an alternative method to solve (6.29). Although PDHG was originally developed to tackle this TV regularized image reconstruction problem, it is shown to fall into the class of general primal–dual algorithms, which have much broader applications in scientific computation. PDHG starts with the original max-type formulation of TV in (6.28), and formulates the TV regularized image reconstruction problem (6.29) as a saddle-point problem:

$$\min_{u} \max_{\|p_i\| \leqslant 1} \left\{ \mu \sum_{i=1}^{n} \langle p_i, D_i u \rangle + \frac{1}{2} \|P \mathcal{F} u - f\|^2 \right\}. \tag{6.39}$$

To present the PDHG in a slightly more general form, we first introduce the indicator function J^* as follows:

$$J^*(p_i) = \begin{cases} 0 & \text{if } \|p_i\| \leqslant 1, \\ +\infty & \text{otherwise.} \end{cases} \tag{6.40}$$

This indicator function J^* is the Fenchel-dual of $J(u) = \|q\|$ for $q \in \mathbb{R}^2$, defined by $J^*(p) = \max_p \{\langle p, q \rangle - J(q)\}$. Then, the saddle-point problem (6.41) can be written as

$$\min_{u} \max_{p} \left\{ \mu \sum_{i=1}^{n} \langle p_i, D_i u \rangle - J^*(p_i) + \frac{1}{2} \|P \mathcal{F} u - f\|^2 \right\}, \tag{6.41}$$

and the PDHG algorithm (with an additional extragradient step to ensure convergence) is given by

$$p_i^{(k+1)} = \arg \min_{p_i} \left\{ J^*(p_i) - \mu \langle p_i, D_i \hat{u}^{(k)} \rangle + \frac{1}{2\beta_k} \|p_i - p_i^{(k)}\|^2 \right\} \tag{6.42a}$$

$$u^{(k+1)} = \arg \min_{u} \left\{ \frac{1}{2} \|P \mathcal{F} u - f\|^2 + \mu \langle p^{(k+1)}, Du \rangle + \frac{1}{2\alpha_k} \|u - u^{(k)}\|^2 \right\} \tag{6.42b}$$

$$\hat{u}^{(k+1)} = u^{(k+1)} + \theta_k (u^{(k+1)} - u^{(k)}), \tag{6.42c}$$

where α_k and β_k are the primal and dual step sizes, respectively, and $\theta_k \in (0, 1]$ is the extragradient parameter. Again, starting from an initial guess $(p^{(0)}, u^{(0)}, \hat{u}^{(0)})$, the steps in (6.42) are repeated until a stopping criterion is met, and the final u is output as the reconstruction.

Similar to ADMM, the subproblems in (6.42) must be easy to solve for PDHG to run efficiently. Based on our derivation of subproblem solutions for ADMM, we can readily derive the solutions for the PDHG iteration. First of all, the $\hat{u}$ subproblem (6.42c) is again a straightforward computation. The p subproblem (6.42a) is equivalent to

$$p_i^{(k+1)} = \arg\min_{\|p_i\|\leqslant 1}\left\{\frac{1}{2\beta_k}\|p_i - p_i^{(k)} - \mu\beta_k D_i\hat{u}^{(k)}\|^2 + \text{const}\right\}, \tag{6.43}$$

where 'const' is a constant dependent on $\hat{u}_i^{(k)}$ but not p_i, and hence does not affect the solution since the minimization (6.43) is in p_i. The minimization (6.43) amounts to solving the projection problem in the form

$$\underset{\|x\|\leqslant\alpha}{\text{minimize}}\|x - b\|^2, \tag{6.44}$$

for some given $b \in \mathbb{R}^2$, which has a closed form solution, called the (Euclidean) projection, defined by

$$\Pi_\alpha(b) = \frac{b}{\max\{\alpha, \|b\|\}}. \tag{6.45}$$

Therefore, the p subproblem (6.42a), or equivalently (6.43), has a closed form solution $\Pi_1(p_i^{(k)} + \mu\beta_k D_i\hat{u}^{(k)})$. The u subproblem (6.42b) can be solved in a similar fashion as (7.37b), which yields a normal equation

$$(\mathcal{F}^{\mathsf{T}}P^{\mathsf{T}}P\mathcal{F} + \alpha_k^{-1}I_n)u = \mathcal{F}^{\mathsf{T}}P^{\mathsf{T}}f - \mu D^{\mathsf{T}}p^{(k+1)} + \alpha_k^{-1}u^{(k)}. \tag{6.46}$$

The solution of this normal equation again mainly involves two Fourier transforms, i.e. $\mathcal{F}$ and $\mathcal{F}^{\mathsf{T}}$:

$$u = \mathcal{F}^{\mathsf{T}}\left[\left(P^{\mathsf{T}}P + \alpha_k^{-1}I_n\right)^{-1}\left(P^{\mathsf{T}}f - \mu\mathcal{F}D^{\mathsf{T}}p^{(k+1)} + \alpha_k^{-1}\mathcal{F}u^{(k)}\right)\right], \tag{6.47}$$

where the matrix $P^{\mathsf{T}}P + \alpha_k^{-1}I_n$ is diagonal and hence trivial to invert.

In summary, the PDHG steps in (6.42) are also easy to compute, with a similar complexity as ADMM (Esser *et al* 2010, Zhu and Chan 2008). The convergence of PDHG has been established under the conditions on the primal and dual step sizes. More specifically, PDHG (6.42) is proved to converge if $\alpha_k\beta_k < 1/\|D^{\mathsf{T}}D\|$, which is 1/8 for 2D image reconstruction problems using a forward finite difference to approximate partial derivatives (Chambolle and Pock 2011). In figure 6.4, we show the reconstruction result by PDHG on the synthetic Shepp–Logan phantom image. The download link of the computer code and the implementation details of this example are provided in appendix B.

Both ADMM and PDHG fall into the class of more general primal–dual algorithms for solving saddle-point problems such as (6.41). The MRI reconstruction problem formulated in (6.29) is one of many successful applications of primal–dual algorithms. However, the MRI technology never ceases to advance, and there are more powerful and complex improvements that allow for both fast scanning and high-contrast imaging. One of such major advancements in the recent two decades is parallel MRI, which we introduce next.

6.3 Parallel MRI*

Parallel MRI (pMRI), is a state-of-the-art MRI technology that employs multiple sensor coils in an MRI scanner to acquire data simultaneously. The coils are placed over the patient body region to be scanned. Each of these coils acquires a set of k-space data as we discussed earlier. The datasets collected by these coils are heterogeneous and complementary due to the differences in their spatial locations and sensitivities. Hence, there may be considerable amount of redundancy in these datasets, with which we can improve the reconstruction quality and/or further reduce the sampling rates in k-space.

In this section, we focus on two methods that have been used in commercial MRI devices: GRAPPA and SENSE. The two methods deal with the MRI reconstruction in quite different ways: GRAPPA directly works in the k-space to infer a pseudo full k-space from the sampled partial one, whereas SENSE performs image reconstruction in the image domain by incorporating the coil sensitivities and the special k-space undersampling pattern. Methods such as GRAPPA are thus called k-space methods, while those based on the principle of SENSE are called image-space methods. Furthermore, we extend the TV regularized image reconstruction methods for single-coil CS-MRI discussed in section 6.2.2 to the pMRI case here. Note that the TV regularized reconstruction falls into the class of image-space methods since TV is a regularization directly imposed to the underlying image.

In contrast to k-space methods, image-space methods are based on the relation between the images acquired by different sensor coils and the associated coil sensitivities, which are simple to interpret visually. Consider imaging a subject (2D array u of size $\sqrt{n} \times \sqrt{n}$) in a J-coil pMRI system, then the sensitivity $s_j \in \mathbb{C}^n$ and the image $u_j \in \mathbb{C}^n$ obtained by the jth coil relate to u point-wise (pixel-wise) as

$$[u_j]_i = [s_j]_i[u]_i \tag{6.48}$$

for all pixels $i = 1,\ldots, n$ and coil indices $j = 1,\ldots, J$, where $[u]_i$ stands for the ith component of a vectorized image $u \in \mathbb{C}^n$. For example, the jth coil in an eight-coil (also called eight-channel) pMRI system can obtain a set of k-space data whose inverse Fourier transform is $u_j \in \mathbb{C}^n$ for an image u. An example of these images and the corresponding coil sensitivities are shown in figure 6.5.

In this section, we assume that there are J sensor coils each acquiring a k-space dataset using the same k-space mask. We denote the two axes in the k-space by k_x and k_y, respectively, and assume that undersampling the k-space is realized by skipping along the k_y-axis only. That is, for each sampled k_y value, all k_x coefficients are acquired. We use R to denote the reduction factor, which is effectively the reciprocal of the sampling ratio. For example, if we sample every other k_y line, then $R = 2$; if we skip three lines after each k_y, then $R = 4$, etc. These correspond to 50% and 25% of the full dataset in the k-space, respectively. An example k-space mask with $R = 4$ is shown in the middle panel of figure 6.8.

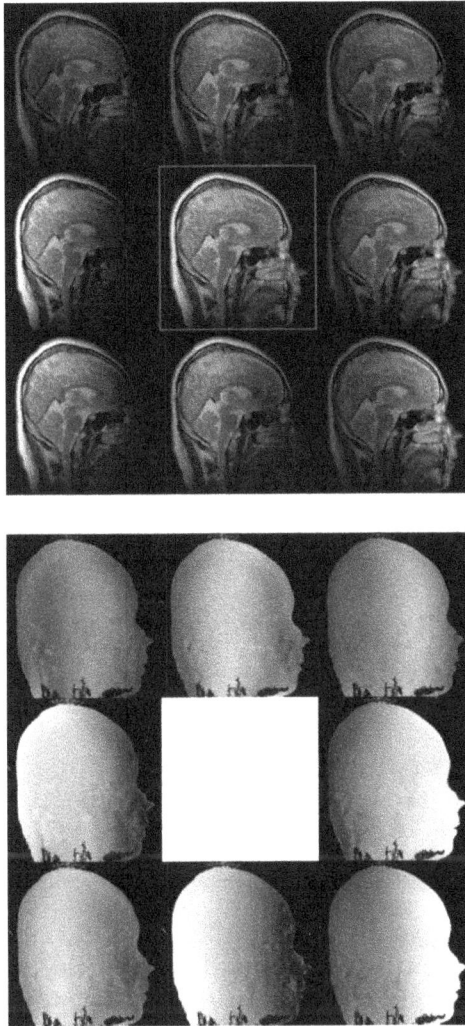

Figure 6.5. Top: brain images $u_1, \ldots, u_8$ (the surrounding eight images) obtained by an eight-coil pMRI system. Bottom: the corresponding coil sensitivities $s_1, \ldots, s_8$ during the pMRI scan. Each u_j is the inverse Fourier transforms of the k-space data obtained by the jth sensor coil. These u_j relate to the image u (center image in the top figure) via the coil sensitivities s_j according to (6.48).

6.3.1 GRAPPA

Generalized autocalibrating partially parallel acquisitions (GRAPPA) is a k-space data interpolation method for pMRI reconstruction (Griswold *et al* 2002). The idea of GRAPPA is to interpolate the missing Fourier coefficients using the sampled ones to create a pseudo full dataset $f_j \in \mathbb{C}^n$ for each coil j. Then, the inverse Fourier transform is applied to f_j, resulting an image $u_j = \mathcal{F}^{\mathsf{T}} f_j \in \mathbb{C}^n$ with suppressed artifacts, for $j = 1, \ldots, J$. The final reconstruction u is obtained by the root of summed squares of u_j as

$$u_i = \left\{ \sum_{j=1}^{J} |(u_j)_i|^2 \right\}^{1/2}, \quad \text{for } i = 1, \ldots, n, \tag{6.49}$$

where $(u_j)_i \in \mathbb{C}$ is the intensity value of u_j at pixel i.

The key of GRAPPA is the interpolation of k-space data to obtain f_j. For convenience, we interpret f_j as a 2D complex-valued array of size $\sqrt{n} \times \sqrt{n}$ in this subsection. For each y, the k-space line $\{(x, y): x = 1, \ldots, \sqrt{n}\}$ is either fully sampled or not sampled at all. To interpolate the missing k-space lines, GRAPPA assumes that the missing $[f_j]_{x,y} \in \mathbb{C}$ is a weighted sum of the Fourier coefficients in the adjacent, sampled k-space lines from all coils. For example, if $R = 2$ and only the nearest four sampled k-space lines are used, then such a weighted sum can be expressed as

$$[f_j]_{x,y} = \sum_{j'=1}^{J} \sum_{b \in B} w_{j'b} [f_{j'}]_{x,y+b}, \quad j = 1, \ldots, J, \tag{6.50}$$

where $B = \{\pm 1, \pm 3\}$. The weight parameters $W \triangleq \{w_{j'b}: j' = 1, \ldots, J, b \in B\}$ need to be determined in advance, which we will explain soon. The choice B corresponds to an 'interpolation window' of size 1. One can use a wider window by setting B to $\{\pm 1, \pm 3, \pm 5\}$ or even $\{\pm 1, \pm 3, \pm 5, \pm 7\}$ for improved accuracy. Wider windows mean that more sampled k-space data can be used for interpolation, and hence they require more weight parameters $w_{j'b}$ and higher computational cost. This interpolation procedure is illustrated in figure 6.6.

To obtain the weight parameters W in (6.50), several additional k-space lines are also acquired. It is common practice to sample the low-frequency k-space lines, i.e. $\{(x, y): x = 1, \ldots, \sqrt{n}\}$ for small values of y. The reason is that these Fourier coefficients are important as they provide the background information, and hence they can be directly used rather than interpolated in the creation of a pseudo full dataset in the k-space. Moreover, their magnitudes are considerably larger than the high-frequency coefficients and hence have a potentially higher signal-to-noise ratio. These additional sampled k-space lines are called auto-calibration signals (ACS). An example k-space sampling pattern (mask) of size 256×256 with reduction factor $R = 4$ along the k_y direction for GRAPPA is shown in figure 6.7, where k_x and k_y are in the rightward and upward directions, respectively, and the 16 low-frequency lines within the range $k_y = \pm 8$ are the acquired ACS. The weights are then determined by the least squares as

$$W = \underset{W}{\arg\min} \sum_{j=1}^{J} \sum_{y \in C} \sum_{x=1}^{\sqrt{n}} \left\| [f_j]_{x,y} - \sum_{j'=1}^{J} \sum_{b \in B} w_{j'b} [f_{j'}]_{x,y+b} \right\|^2, \tag{6.51}$$

where y is summed over the set of ACS lines indexed in C (e.g. $C = \{y \in \mathbb{N} \mid -8 \leqslant y \leqslant 8\}$ in figure 6.7).

To summarize the steps in GRAPPA, we first sample equally spaced k-space lines in the k_y direction and several additional ACS lines. Then, the ACS lines are used to

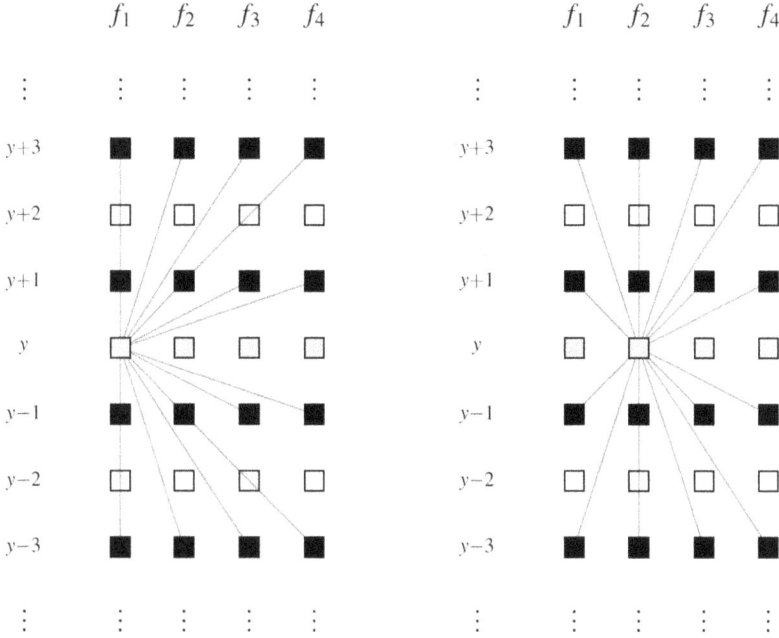

Figure 6.6. Illustration of k-space interpolation by GRAPPA (Griswold *et al* 2002) with reduction factor $R = 2$ in a four-coil pMRI system. A column with f_j labeled on top represents the k-space data acquired by the jth coil. k_x and k_y are in the inward and upward directions, respectively. Solid black (empty white, respectively) boxes indicate sampled (unsampled) k-space data. Each unsampled $[f_j]_{x,y}$ is then interpolated (labeled as a gray box subsequently) by a weighted sum of sampled $[f_{j'}]_{x,y+b}$ (connected to $[f_j]_{x,y}$ by lines) for $j' = 1, 2, 3, 4$ and $b = \pm1, \pm3$ as shown in (6.50). The weights are obtained using the ACS as given by (6.51). Illustrations show how $[f_1]_{x,y}$ (left) and $[f_2]_{x,y}$ (right) are interpolated.

determine the interpolation weights W in (6.51). With the estimated weights W, we interpolate all missing Fourier coefficients in the k-space according to (6.50). After interpolation, we have J full k-space datasets $\{f_j: j = 1,\ldots, J\}$, and hence their corresponding $\{u_j: j = 1,\ldots, J\}$ through the inverse Fourier transform $\mathcal{F}^{\mathsf{T}}$ and the final reconstruction u by (6.49).

6.3.2 SENSE

Sensitivity encoding (SENSE) (Pruessmann and *et al* 1999) is an image domain 'unfolding' algorithm developed specifically for pMRI reconstruction. In the Cartesian-type sampled k-space (i.e. sampled locations k_x and k_y are on the Cartesian grid of the k-space), an undersampling pattern along the k_y direction results in a folded image with a reduced field-of-view (FOV) along the y-axis. For example, with a reduction factor $R = 2$, i.e. skipping every other k-space line in the sampling process, the inverse Fourier transform of the undersampled k-space data produces an image composed of two identical half-sized components, where each component is the sum of the upper and lower halves of the ground truth image. The same phenomenon occurs generally such as for $R = 3, 4 \ldots$, where the image is a

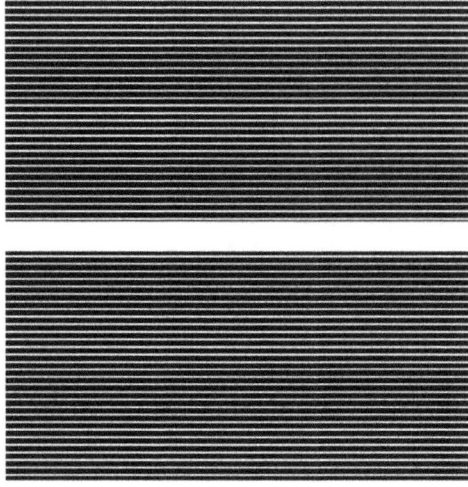

Figure 6.7. An example k-space sampling pattern (mask) of size 256×256 with reduction factor $R = 4$ along the k_y direction for GRAPPA, where k_x and k_y are in the rightward and upward directions, respectively. The 16 low-frequency lines within the range $k_y = \pm 8$ are the ACS used to estimate the weight in (6.51).

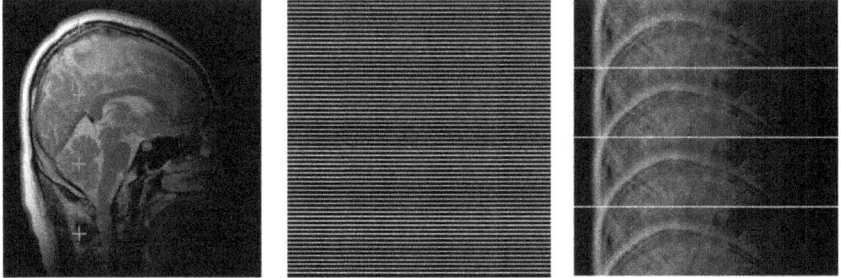

Figure 6.8. Illustration of reduced FOV and SENSE with reduction factor $R = 4$. Left: image u_j obtained from full k-space acquired at a selected coil j. Middle: k-space undersampling along the k_y direction with reduction factor $R = 4$. Right: The folded image v_j, which is composed of four replicas, obtained by the inverse Fourier transform of the undersampled k-space, where the intensity value at the red marker in v_j is the sum of the intensity values at the four green markers in u_j on the left. This folding is obtained for every coil j in SENSE.

repetition of R components, and each one has a reduced FOV of size $(\sqrt{n}/R) \times \sqrt{n}$ and is the superposition of the R portions of the ground truth image. Figure 6.8 shows this folding effect with $R = 4$.

Assume that the coil sensitivities $\{s_j \in \mathbb{C}^n : j = 1, \ldots, J\}$ are known (we treat them as 2D arrays of size $\sqrt{n} \times \sqrt{n}$ in this subsection for convenience), we can relate the folded image v_j with a reduced FOV for each coil j to the ground truth image u to be recovered as follows:

$$[v_j]_{i,m} = \sum_{m' \in F(m)} [s_j]_{i,m'}[u]_{i,m'}, \quad \text{for } i = 1, \ldots, \sqrt{n} \text{ and } m = 1, \ldots, \sqrt{n}/R, \quad (6.52)$$

where $F(m) \triangleq \{m + r\sqrt{n}/R: r = 0, \ldots, R - 1\}$. Here, $[u]_{i,m} \in \mathbb{C}$ is the intensity value of the 2D image u at pixel location (i, m). To understand (6.52), we first realize that the intensity value of the folded image v_j at (i, m) in the reduced FOV is the sum of intensity values of $[u_j]$ at the folded locations, denoted by $F(m)$. That is,

$$[v_j]_{i,m} = \sum_{m' \in F(m)} [u_j]_{i,m'}, \quad \text{for } i = 1, \ldots, \sqrt{n} \text{ and } m = 1, \ldots, \sqrt{n}/R. \tag{6.53}$$

Note that $|F(m)| = R$, i.e. the folding number is the reduction factor. Furthermore, the unfolded image u_j in coil j relates to the ground truth image via the coil sensitivity s_j as

$$[u_j]_{i,m} = [s_j]_{i,m}[u]_{i,m}, \quad \text{for } i, m = 1, \ldots, \sqrt{n}, \tag{6.54}$$

for all $j = 1, \ldots, J$. Combining (6.53) and (6.54) yields (6.52).

To reconstruct u, we first realize that there are in total of nJ/R equations in (6.52). If $J > R$, i.e. the number of coils is larger than the reduction factor, (6.52) is potentially an overdetermined linear system since $nJ/R > n$, the number of unknowns in u. Then, we can reconstruct the image u by solving the least squares problem induced by (6.52) as follows:

$$u = \arg\min_u \sum_{j=1}^{J} \sum_{i=1}^{\sqrt{n}} \sum_{m=1}^{\sqrt{n}/R} \left\| [v_j]_{i,m} - \sum_{m' \in F(m)} [s_j]_{i,m'} u_{i,m'} \right\|^2. \tag{6.55}$$

If the system (6.52) is overdetermined, then (6.55) yields a unique solution u, which is the desired reconstruction from pMRI data.

Because of the broad availability of SENSE, this technique has become the most used parallel imaging method in the clinical routine. Many diagnostic applications already benefit from the enhanced image quality with SENSE. The main issue with SENSE is the availability of sensitivity maps. In contrast, GRAPPA is particularly beneficial in areas where accurate coil sensitivity maps may be difficult to obtain. In inhomogeneous regions with low spin density such as the lung and the abdomen, it can be difficult to determine precise coil sensitivity information. In these regions, the image quality with SENSE might suffer from inaccurate sensitivity maps. In contrast, the GRAPPA algorithm provides decent image quality since the sensitivity information is extracted from the k-space itself.

6.3.3 TV regularized pMRI reconstruction

The idea of TV regularized image reconstruction can also be applied to reconstruct an image from undersampled pMRI data if the coil sensitivity maps are known. In this subsection, we assume that these sensitivity maps, denoted by $\{s_j \in \mathbb{R}^n: j = 1, \ldots, J\}$, are given or estimated in advance. As we have seen before, it is advantageous to use a random k-space sampling pattern for improved data incoherence, which in turn helps to recover the image more faithfully according to the CS theory. This is still the case for pMRI reconstruction.

Through the derivation of SENSE in (6.54), we know that the partial k-space data f_j acquired by coil j relates to the original ground truth image u by

$$f_j = P\mathcal{F}u_j = P\mathcal{F}S_j u, \tag{6.56}$$

where $u_j = S_j u \in \mathbb{C}^n$ and $S_j = \text{diag}(s_j) \in \mathbb{C}^{n \times n}$. Therefore, it is straightforward to form a TV regularized minimization of u as follows,

$$\underset{u}{\text{minimize}} \left\{ \mu \, \text{TV}(u) + \sum_{j=1}^{J} \left\| P\mathcal{F}S_j u - f_j \right\|^2 \right\}. \tag{6.57}$$

Unfortunately, despite the fact that the above formulation looks similar to (6.29), the problem (6.57) cannot be easily tackled by ADMM or PDHG due to the presence of S_j, which makes the closed form solution for (7.37b) in ADMM and (6.42b) in PDHG unavailable (Ye $et\ al$ 2011). In this case, one needs to employ iterative algorithms, such as the conjugate gradient method, to solve the u subproblem, in each iteration of (7.37) and (6.42). However, this is obviously inefficient since inner iterations are introduced into ADMM and PDHG, and the overall cost of these methods can increase significantly since the subproblem in each iteration has much higher computation complexity.

The TV regularized inverse problems with complicated sensing matrices such as the one in (6.57) are very common in real-world applications. These problems can be presented in a general form as

$$\underset{u}{\text{minimize}} \{ G(Ku) + H(u) \}, \tag{6.58}$$

where $K: \mathbb{R}^n \to \mathbb{R}^m$ is a matrix, and G and H are both close and convex functions. Moreover, G is often assumed to be simple but possibly non-differentiable, and H is differentiable with a Lipschitz gradient but not simple. Then, (6.57) is a special case of (6.58) with $K = D$, $G(w) = \sum_{i=1}^{n} \|w_i\|$ for $w_i \in \mathbb{R}^2$, $1 \leqslant i \leqslant n$, $H(u)$ being the data fidelity term given by

$$H(u) = (1/2) \cdot \|Au - f\|^2, \tag{6.59}$$

where $A = [P\mathcal{F}S_1; \cdots ; P\mathcal{F}S_J] \in \mathbb{C}^{(mJ) \times n}$ and $f = [f_1; \cdots ; f_J] \in \mathbb{C}^{mJ}$. Due to the generality of the abstract form (6.58), a large amount of effort has been devoted to solving such problems efficiently in the optimization community in the past decade.

A straightforward approach to solving (6.58) with general H is to linearize H in the u subproblem of ADMM. To this end, recall that if ADMM is applied to (6.58), then the u subproblem (7.37b) becomes

$$u^{(k+1)} = \arg\min_u \left\{ H(u) + \frac{\rho}{2} \|Ku - z^{(k)}\|^2 \right\}, \tag{6.60}$$

where $z^{(k)} \triangleq w^{(k+1)} - \rho^{-1}\lambda^{(k)}$ is known. Instead of solving (6.60) exactly with inner iterations, we can approximate $H(u)$ by its first-order Taylor expansion plus a quadratic penalty term on $u - u^{(k)}$ with penalty parameter $(2\alpha)^{-1}$, and update $u^{(k+1)}$ simply by

$$u^{(k+1)} = \arg\min_u \left\{ H(u^{(k)}) + \langle \nabla H(u^{(k)}, u - u^{(k)}) \rangle + \frac{1}{2\alpha} \|u - u^{(k)}\|^2 \right.$$
$$\left. + \frac{\rho}{2} \|Ku - z^{(k)}\|^2 \right\} \qquad (6.61)$$
$$= \arg\min_u \left\{ \frac{1}{2\alpha} \|u - (u^{(k)} - \alpha \nabla H(u^{(k)}))\|^2 + \frac{\rho}{2} \|Ku - z^{(k)}\|^2 \right\},$$

where we completed the square and discarded the constant term not involving u to obtain the second equality. If $K = D$ is the discrete gradient operator as in TV regularized MR image reconstruction (6.57), then K can be diagonalized by the DFT matrix and hence the problem (6.61) again has a closed form solution. Therefore, using (6.61) in place of (7.37b) still ensures low per-iteration computational cost (the main cost is usually in the computation of $\nabla H(u^{(k)})$ since H involves the large dataset f). This approach is called the linearized ADMM. The linearized ADMM is shown to be convergent if $\alpha \leqslant 1/L_H$ where L_H is the Lipschitz constant of ∇H. However, this requirement on α (which behaves as the step size) is often too restrictive if L_H is large, and the linearized ADMM takes an excessive amount of iterations to converge in practice. To overcome this issue, one can employ the Barzilai–Borwein step size policy (Barzilai and Borwein 1988) and implement an efficient line search strategy to significantly improve practical performance and guarantee theoretical convergence simultaneously (Chen *et al* 2013, Ye *et al* 2011). Alternatively, one can also integrate Nesterov's accelerated gradient scheme (Nesterov 1983, 2013) into the linearized ADMM or PDHG algorithms to obtain improved theoretical and practical convergence rates (Chen *et al* 2014, Ouyang *et al* 2015). These algorithms can solve the general composite optimization problem (6.58) and hence can be applied to many other problems beyond CS-MRI.

References

Barzilai J and Borwein J M 1988 Two-point step size gradient methods *IMA J. Numer. Anal.* **8** 141–8

Bloch F 1946 Nuclear induction *Phys. Rev.* **70** 460

Boyd S, Parikh N, Chu E, Peleato B and Eckstein J 2011 Distributed optimization and statistical learning via the alternating direction method of multipliers *Found. Trends Mach. Learn.* **3** 1–122

Candes E J, Romberg J and Tao T 2006 Robust uncertainty principles: exact signal reconstruction from highly incomplete frequency information *IEEE Trans. Inform. Theory* **52** 489–509

Chambolle A and Pock T 2011 A first-order primal–dual algorithm for convex problems with applications to imaging *J. Math. Imaging Vis.* **40** 120–45

Chen Y, Hager W W, Yashtini M, Ye X and Zhang H 2013 Bregman operator splitting with variable stepsize for total variation image reconstruction *Comput. Optim. Appl.* **54** 317–42

Chen Y, Lan G and Ouyang Y 2014 Optimal primal–dual methods for a class of saddle point problems *SIAM J. Optim.* **24** 1779–814

Cocosco C A, Kollokian V, Kwan R K-S, Pike G B and Evans A C 1997 Brainweb: online interface to a 3D MRI simulated brain database *NeuroImage* Citeseer

Donoho D L 2006 Compressed sensing *IEEE Trans. Inform. Theory* **52** 1289–306

Esser E, Zhang X and Chan T F 2010 A general framework for a class of first order primal–dual algorithms for convex optimization in imaging science *SIAM J. Imaging Sci.* **3** 1015–46

Goldstein T and Osher S 2009 The split Bregman method for $L1$-regularized problems *SIAM J. Imaging Sci.* **2** 323–43

Griswold M A, Jakob P M, Heidemann R M, Nittka M, Jellus V, Wang J, Kiefer B and Haase A 2002 Generalized autocalibrating partially parallel acquisitions (GRAPPA) *Magn. Reson. Med.* **47** 1202–10

Lauterbur P 1974 Image formation by induced local interactions: examples employing nuclear magnetic resonance *Nature* **246** 469

Lustig M, Donoho D and Pauly J M 2007 Sparse MRI: the application of compressed sensing for rapid MR imaging *Magn. Reson. Med.* **58** 1182–95

Mansfield P 1977 Multi-planar image formation using NMR spin echoes *J. Phys. C: Solid State Phys.* **10** L55

Nesterov Y 1983 A method for unconstrained convex minimization problem with the rate of convergence $O(1/k^2)$ *Sov. Math. Dokl.* **27** 372–6

Nesterov Y 2013 Gradient methods for minimizing composite functions *Math. Programm.* **140** 125–61

Ouyang Y, Chen Y, Lan G and Pasiliao E Jr. 2015 An accelerated linearized alternating direction method of multipliers *SIAM J. Imaging Sci.* **8** 644–81

Packard F B W H M 1946 The nuclear induction experiment *Phys. Rev.* **70**

Pruessmann K P *et al* 1999 Sense: sensitivity encoding for fast MRI *Magn. Reson. Med.* **42** 952–62

Rudin L I, Osher S and Fatemi E 1992 Nonlinear total variation based noise removal algorithms *Physica* D **60** 259–68

Suetens P 2017 *Fundamentals of Medical Imaging* (Cambridge: Cambridge University Press)

Wu C and Tai X-C 2010 Augmented Lagrangian method, dual methods, and split Bregman iteration for ROF, vectorial TV, and high order models *SIAM J. Imaging Sci.* **3** 300–39

Ye X, Chen Y and Huang F 2011 Computational acceleration for MR image reconstruction in partially parallel imaging *IEEE Trans. Med. Imaging* **30** 1055–63

Zhu M and Chan T 2008 An efficient primal–dual hybrid gradient algorithm for total variation image restoration *UCLA CAM Report* pp 8–34

IOP Publishing

Machine Learning for Tomographic Imaging

Ge Wang, Yi Zhang, Xiaojing Ye and Xuanqin Mou

Chapter 7

Deep-learning-based MRI reconstruction

This chapter is dedicated to deep-learning-based MRI reconstruction methods. Although deep-learning-based tomographic reconstruction research emerged only a few years ago, it has proven to be very promising with a great number of successful instances on clinical datasets including MRI datasets. Since this field is still undergoing very fast development, and there are currently many research papers coming out every month, our intention in this chapter is to focus on a few representative deep-learning-based methods. Our goal is to guide the reader to learn the basics of deep-learning-based MRI reconstruction methods, after which they can easily explore other related possibilities in this field.

In section 7.1, we introduce several deep-learning methods, including ISTA-Net, ADMM-Net, and Var-Net, and their connections to the classical MRI reconstruction methods in the previous chapter. To be concrete, we study their network structures and derive all key steps to compute the network parameters in detail. We also showcase several deep-learning-based methods that leverage the generic network architectures and models to show how these well developed network structures become instrumental in building powerful deep MRI reconstruction networks in section 7.2. In section 7.3, we provide an overview of several deep-learning-based approaches for recent and advanced MRI techniques, such as dynamic MRI, magnetic resonance fingerprinting (MRF), and synergized pulsing-imaging network (SPIN). We also discuss several miscellaneous topics related to MRI reconstruction in section 7.4, such as calculus with complex variables, activation functions for complex-valued inputs and outputs, optimal k-space sampling strategies, etc. In section 7.5, we provide more references for the reader for further reading.

7.1 Structured deep MRI reconstruction networks

In this section, we introduce several deep neural networks specifically designed for MRI image reconstruction. As we learned in the previous chapter, rapid MR imaging can be realized using the CS-MRI strategy which samples the k-space at a sub-Nyquist rate to reduce imaging time, followed by a proper reconstruction algorithm to recover high-quality images from the undersampled k-space data. A majority of these reconstruction algorithms are formulated as iterative procedures to solve an optimization problem, where the objective function is typically composed of two terms: a data fidelity term that quantifies the faithfulness of the reconstruction to the undersampled k-space data and a regularization term that imposes proper prior information of the image to be reconstructed. The data fidelity term is derived from the definitive physical principles that describe the formation of the k-space data. The regularization term, on the other hand, is often crafted subjectively to reflect our understanding of or preference for the underlying image.

Among many regularization terms proposed in the literature, the ℓ_1-type regularizations, such as the total variation (TV) introduced in the previous chapter, are shown to be robust in suppressing artifacts and noise while preserving image edges, as they encourage sparse gradients in reconstructed images. In addition to ℓ_1-type regularization, non-convex ℓ_p type regularization with $0 \leqslant p < 1$ is also used and sometimes shows improved performance over ℓ_1 regularization. However, ℓ_p regularization requires a non-convex optimization, which is computationally challenging due to the lack of robust and efficient numerical algorithms with guaranteed global optimality and comprehensive convergence analysis.

Despite the concise and elegant formulation of image reconstruction with regularization, the resulting optimization problem needs to be solved with an iterative procedure which generates a sequence of estimates $u^{(k)}$ successively which gradually approaches the solution. However, there are still three main issues that hinder the application of these methods in real-world MRI reconstruction problems:

1. *Choice of regularization.* There are numerous regularization terms proposed in the literature. Although many of them have proven robust in practice, they are still overly simplified and cannot capture the subtle details in medical images which are critical in diagnosis and treatment. For example, TV regularization is known for its 'staircase' (also known as 'oil painting') issue due to its enforcement of sparse gradients, such that the reconstructed images tend to be piece-wise constant. Due to this issue, fine structures are often degraded in the images reconstructed with TV regularization, which are not suitable for medical applications.

2. *Parameter tuning.* To achieve the desired balance between artifacts/noise reduction and faithful structural reconstruction, the parameters of a reconstruction model and its associated optimization algorithm need to be carefully tuned. Unfortunately, the image quality is often very sensitive to these parameters, and the optimal parameters are also shown to be highly dependent on the specific acquisition settings and imaging datasets. In many cases, it is simply impossible to tune the parameters to meet the clinical standard due to the intrinsic properties of the selected regularization which cannot capture the fine structures in real images.

3. *Reconstruction time.* Despite the efficiency of optimization algorithms being continuously being improved, these algorithms, even for convex problems, often require many iterations to converge. As a consequence, the reconstruction

process can take an excessive amount of time and hence is not suitable for the emerging advanced MRI technologies that require online reconstruction results within sub-seconds.

The issues with regularized reconstruction models mentioned above inspired a new class of deep reconstruction networks. The architectures of these networks mimic the iterative optimization algorithms. In particular, the data fidelity term that describes the image formation is retained in these networks, since it is based on the well-established physical principles, which are known and do not need to be relearned. The regularization, which is often manually designed and overly simplified in the classical reconstruction methods, is replaced by deep neural networks. More specifically, these reconstruction networks are mostly composed of a small number of phases, where each phase mimics one iteration of a classical, optimization-based reconstruction algorithm. The terms corresponding to the manually designed regularization in the classical methods are parameterized by multilayer perceptrons whose parameters are to be learned in the offline training process. After training, these networks work as fast feedforward mappings with extremely low computational cost for online MRI image reconstruction.

7.1.1 ISTA-Net

ISTA-Net (Zhang and Ghanem 2018) is a deep neural network architecture for image reconstruction motivated by the iterative shrinkage thresholding algorithm (ISTA). To understand the architecture of ISTA-Net, we first have a quick overview of the original ISTA to see how it works.

ISTA is a simple numerical algorithm for solving a class of unconstrained convex optimization problems in the following form:

$$\underset{x}{\text{minimize}}\, f(x), \quad \text{where } f(x) = g(x) + h(x), \tag{7.1}$$

where the objective function f consists of a convex, differentiable component h with L_h-Lipschitz continuous gradient ∇h, i.e. $\|\nabla h(x) - \nabla h(y)\| \leqslant L_h \|x - y\|$ for all $x, y \in \mathbb{R}^n$, and a convex but possibly non-differentiable component g that is simple. Here, we call a function g simple if its proximity operator prox_g defined below has a closed form or is easy to compute for any given b:

$$\text{prox}_g(b) \triangleq \arg\min_x \left\{ g(x) + \frac{1}{2}\|x - b\|^2 \right\}. \tag{7.2}$$

Note that ISTA still works even if g is not simple, but the subproblem in the form (7.2) needs to be solved with some iterative procedure (i.e. requires inner iterations) in every iteration of ISTA, which can make the overall computational cost of ISTA high.

The terms shrinkage and threshold in the name of ISTA are due to a specific and widely used choice of $g(x) \triangleq \mu\|x\|_1$ for a fixed weighting parameter $\mu > 0$. In this case, the proximity operator prox_g reduces to the shrinkage operator shrink_μ defined in the previous chapter. That is, the ith component of $\text{prox}_g(b) \in \mathbb{R}^n$ is

$$[\text{prox}_g(b)]_i = \text{sign}(b_i) \cdot \max\{|b_i| - \mu, 0\}. \tag{7.3}$$

Therefore, $\text{prox}_g(b)$ 'shrinks' the magnitude of each component of its argument b by μ; if the magnitude is smaller than μ then it becomes 0 after the shrinkage.

To solve (7.1), ISTA first approximates $h(u)$ by its first-order Taylor expansion at the previous iterate $u^{(k-1)}$ plus a quadratic penalty term with weight $1/(2\alpha)$ as follows:

$$h(u) \approx h(u^{(k-1)}) + \langle \nabla h(u^{(k-1)}), u - u^{(k-1)} \rangle + \frac{1}{2\alpha}\|u - u^{(k-1)}\|^2$$
$$= \frac{1}{2\alpha}\|u - (u^{(k-1)} - \alpha\nabla h(u^{(k-1)}))\|^2 + \text{const}, \tag{7.4}$$

where we completed the square to obtain the equality above, and the term 'const' represents a constant independent of u. Substituting this approximation (7.4) back into (7.1), the new iteration $u^{(k)}$ is then obtained by solving

$$u^{(k)} = \arg\min_u \left\{ g(u) + \frac{1}{2\alpha}\|u - (u^{(k-1)} - \alpha\nabla h(u^{(k-1)}))\|^2 \right\}, \tag{7.5}$$

where the constant term is omitted since it does not affect the result $u^{(k)}$ in (7.5). ISTA thus iterates the update rule (7.5) for $k = 1, 2, \ldots$, until a stopping criterion is met.

The update rule (7.5) can be rewritten as an equivalent, two-step scheme by introducing an auxiliary variable $r^{(k)} = u^{(k-1)} - \alpha\nabla h(u^{(k-1)})$ and using the definition of the proximity operator in (7.2):

$$r^{(k)} = u^{(k-1)} - \alpha\nabla h(u^{(k-1)}), \tag{7.6a}$$

$$u^{(k)} = \text{prox}_{\alpha g}(r^{(k)}). \tag{7.6b}$$

This two-step scheme is the architecture that motivates the ISTA-Net below.

Now, let us come back to the MR image reconstruction problem. Recall that the single-coil CS-MRI reconstruction given in the previous chapter can be written as an ℓ_1 regularized minimization problem as follows:

$$\underset{u}{\text{minimize}} \left\{ \mu\|\Psi u\|_1 + \frac{1}{2}\|P\mathcal{F}u - f\|^2 \right\}, \tag{7.7}$$

where $\Psi \in \mathbb{R}^{n\times n}$ is a sparsifying operator such that Ψu becomes a sparse vector, and $\mu > 0$ is the weight parameter of this ℓ_1 regularization term. Therefore, we can recast the CS-MRI reconstruction (7.7) into a similar form as (7.1):

$$\underset{u}{\text{minimize}} \{g(\Psi u) + h(u)\}, \tag{7.8}$$

where $g(\cdot) = \mu\|\cdot\|_1$, and $h(u) = (1/2) \cdot \|P\mathcal{F}u - f\|^2$ is the data fidelity term. Note that ISTA (and hence ISTA-Net below) applies as long as h has a Lipschitz continuous gradient ∇h, which allows for a much broader range of applications in addition to the single-coil CS-MRI problem (7.7). For example, it can be applied to the pMRI problems where the data fidelity term involves coil sensitivities.

Although (7.8) does not exactly match the ISTA (7.1) due to the presence of Ψ, this can be easily resolved using an orthogonal sparsifying operator Ψ and setting $x = \Psi u$ as the unknown for (7.1). A typical choice of such a Ψ is an orthogonal 2D wavelet transform. In this case, we just need to solve x from the exact form of (7.1) with $g(x) = \mu\|x\|_1$ and $\tilde{h}(x) \triangleq h(\Psi^\top x)$ as the data fidelity, and recover $u = \Psi^\top x$ using the output x of ISTA. Integrating this change of variables into the scheme (7.6), we obtain a slightly modified version of ISTA for (7.7) as follows:

$$r^{(k)} = u^{(k-1)} - \alpha \nabla h(u^{(k-1)}), \qquad (7.9a)$$

$$u^{(k)} = \Psi^\top \mathrm{prox}_{\alpha g}(\Psi r^{(k)}) = \Psi^\top \mathrm{shrink}_\theta(\Psi r^{(k)}), \qquad (7.9b)$$

where $\theta = \alpha\mu$ combines the two parameters, and (7.9b) involves shrinkage due to the choice $g(x) = \mu\|x\|_1$.

Now, we take a closer look at this CS-MRI version of ISTA (7.9). The gradient ∇h in (7.9a) is due to the data fidelity h in (7.8), which establishes the relation between the k-space data and the image u to be reconstructed based on the physical principles of MRI introduced in the previous chapter. Therefore, we do not need to 'learn' this part in the reconstruction. On the other hand, the use of the sparsifying transform Ψ and ℓ_1 regularization is rather artificial, as it merely reflects our heuristic expectation about the image, i.e. the wavelet coefficients of the image 'should' be sparse. If there is a sufficient amount of training data, it is likely that we can learn a better representation of this regularization using a deep-learning technique.

According to this idea, ISTA-Net is proposed to replace the transform Ψ and $\Psi^\top$ by multilayer convolutional neural networks (CNN), while keeping the $\mathrm{prox}_{\alpha g}$, i.e. the shrinkage due to the ℓ_1 norm, as it seems robust in suppressing noises. To this end, ISTA-Net follows the scheme of ISTA (7.9), and constructs a deep neural network of K phases, where K corresponds to the total number of iterations of (7.9). However, while the original ISTA (7.9) requires K to be tens, hundreds, or even thousands to converge, ISTA-Net sets a fixed K (less than 20, often just 10) and hopes that these well-learned K phases are capable of recovering high-quality images.

Now the kth phase of ISTA-Net is to mimic the two steps in the kth iteration of ISTA given in (7.9). Given the output $u^{(k-1)}$ of the previous, $(k-1)$th phase, the update of $r^{(k)}$ follows (7.9a) closely by the physical principles incorporated in the data fidelity term h and its gradient ∇h. Only the parameter α, which behaves as the step size in ISTA, is set to α_k, which is to be learned during the training process in ISTA-Net. After $r^{(k)}$ is updated, it is passed to (7.9b) with Ψ and $\Psi^\top$ replaced by two multilayer CNNs

$H^{(k)}$ and $\tilde{H}^{(k)}$, respectively, and the shrinkage parameter θ is replaced by $\theta^{(k)}$, which is to be learned as well. Namely, $u^{(k)}$ is updated by

$$u^{(k)} = \tilde{H}^{(k)}(\text{shrink}_{\theta_k}(H^{(k)}(r^{(k)}))). \tag{7.10}$$

In ISTA-Net (Zhang and Ghanem 2018), both $H^{(k)}$ and $\tilde{H}^{(k)}$ are set to simple two-layer CNNs as follows:

$$H^{(k)}(r) = w_2^{(k)} * \sigma(w_1^{(k)} * r) \quad \text{and} \quad \tilde{H}^{(k)}(\tilde{r}) = \tilde{w}_2^{(k)} * \sigma(\tilde{w}_1^{(k)} * \tilde{r}), \tag{7.11}$$

where $w_1^{(k)}$, $w_2^{(k)}$, $\tilde{w}_1^{(k)}$, and $\tilde{w}_2^{(k)}$ are convolutional kernels in the kth phase to be learned, and σ is a component-wise activation function such as ReLU.

In summary, ISTA-Net is a deep neural network with K phases where K is prescribed. Each phase of ISTA-Net mimics one iteration (7.9) of ISTA and updates $r^{(k)}$ and $u^{(k)}$ by

$$r^{(k)} = u^{(k-1)} - \alpha_k \nabla h(u^{(k-1)}), \tag{7.12a}$$

$$u^{(k)} = \tilde{H}^{(k)} S_{\theta_k}(H^{(k)} r^{(k)}), \tag{7.12b}$$

where we have omitted excessive parentheses for simplicity of notation, i.e. $H^{(k)} r^{(k)}$ stands for $H^{(k)}(r^{(k)})$, etc, and we denote shrink_{θ_k} by S_{θ_k} for short. The K phases are concatenated in order, where the kth phase accepts the output $u^{(k-1)}$ of the previous phase, updates $r^{(k)}$ using (7.12a) with α_k, and finally outputs $u^{(k)}$ using (7.12b). In the kth phase, the parameters to be learned are: α_k, θ_k, and $w_1^{(k)}$, $w_2^{(k)}$ in $H^{(k)}$ and $\tilde{w}_1^{(k)}$ and $\tilde{w}_2^{(k)}$ in $\tilde{H}^{(k)}$, respectively. In the first phase, the input is the initial guess $u^{(0)}$, which can be set to the zero-filled reconstruction $\mathcal{F}^{\mathsf{T}} P^{\mathsf{T}} f$. The output of the last phase, $u^{(K)}$, is used in the loss function that measures its squared discrepancy to the corresponding ground truth image u_{true} obtained from the full k-space data,

$$L_{\text{dis}}(\Theta; f, u_{\text{true}}) = \frac{1}{2}\|u^{(K)}(f; \Theta) - u_{\text{true}}\|^2, \tag{7.13}$$

where (f, u_{true}) is a training pair with f being the partial k-space data (including the undersampling mask P) and u_{true} the ground truth reference image obtained by full k-space, and $\Theta = \{\alpha_k, \theta_k, w_1^{(k)}, w_2^{(k)}, \tilde{w}_1^{(k)}, \tilde{w}_2^{(k)} \mid k = 1,\ldots, K\}$. The structure of the ISTA-Net can be visualized in figure 7.1, where each arrow indicates a mapping from its input to the output with the required network parameters labeled next to it.

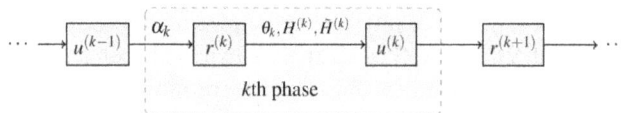

Figure 7.1. Architecture of ISTA-Net (7.12). The kth phase updates $r^{(k)}$ and $u^{(k)}$. The dependences of each variable on other variables are shown as incoming arrows, and the network parameters used for updating are labeled next to the corresponding arrows.

In addition, since $H^{(k)}$ and $\tilde{H}^{(k)}$ in (7.11) are replacing Ψ and Ψ^{T}, respectively, they are expected to satisfy $\tilde{H}^{(k)}H^{(k)} = I$, the identity mapping. To make this constraint approximately satisfied, the mismatch between $\tilde{H}^{(k)}(H^{(k)}(u_{\text{true}}))$ and u_{true} can be integrated into the following loss function, despite that it is much weaker than $\tilde{H}^{(k)}H^{(k)} = I$:

$$L_{\text{id}}(\Theta; u_{\text{true}}) = \frac{1}{2}\sum_{k=1}^{K}\|\tilde{H}^{(k)}(H^{(k)}(u_{\text{true}})) - u_{\text{true}}\|^2. \tag{7.14}$$

The loss function for a particular training pair (f, u_{true}) is thus the sum of the losses in (7.13) and (7.14) with a balancing parameter $\gamma > 0$ (set to 0.01 in ISTA-Net (Zhang and Ghanem 2018)):

$$L(\Theta; f, u_{\text{true}}) = L_{\text{dis}}(\Theta; f, u_{\text{true}}) + \gamma L_{\text{id}}(\Theta; u_{\text{true}}), \tag{7.15}$$

and the total loss function during training is the sum of $L(\Theta; f, u_{\text{true}})$ in (7.15) over all training pairs of form (f, u_{true}) in the training dataset.

The optimal parameter Θ can be solved by minimizing the loss function (7.15), which can be accomplished using the stochastic gradient descent (SGD) method. The key in the implementation of SGD is the computation of the gradient of (7.15) with respect to each network parameter, i.e. $\alpha_k, \theta_k, w_1^{(k)}, w_2^{(k)}, \tilde{w}_1^{(k)}, \tilde{w}_2^{(k)}$ for $k = 1, \dots, K$. To this end, we need to consult figure 7.1 and see how the backpropagation should be performed. More specifically, we first need to compute the gradient of L defined in (7.15) with respect to the main variables $u^{(k)}$ and $r^{(k)}$. Then, we compute the gradients of $u^{(k)}$ with respect to its parameters, i.e. $\theta_k, w_1^{(k)}, w_2^{(k)}, \tilde{w}_1^{(k)}, \tilde{w}_2^{(k)}$, and the gradient of $r^{(k)}$ with respect to $\alpha^{(k)}$. Finally, the gradients of L with respect to these network parameters can be built by multiplying the involved partial derivatives according to the chain rule. We provided the details as follows.

We first check the gradients of L defined in (7.15) with respect to $u^{(k)}$ and $r^{(k)}$. Note that L takes $u^{(k)}$ and $r^{(k)}$, which are vectors in $\mathbb{R}^n$ (in fact both are complex vectors in MRI applications, however, we pretend that they are real, and provide details about how to handle the complex version in section 7.4.1), and output scalars, we know the gradients of L with respect to $u^{(k)}$ and $r^{(k)}$ are both vectors in $\mathbb{R}^n$ as well. We use partial derivatives to indicate spatial dependences and compute the gradients here. First, we have

$$\frac{\partial L}{\partial r^{(k)}} = \frac{\partial L}{\partial u^{(k)}}\frac{\partial u^{(k)}}{\partial r^{(k)}}, \tag{7.16}$$

due to the fact that $u^{(k)}$ is a function of $r^{(k)}$, as shown in figure 7.1. The gradient $\partial u^{(k)}/\partial r^{(k)}$ in (7.16) is straightforward to compute due to the relation between $r^{(k)}$ and $u^{(k)}$ in (7.12b) and the chain rule,

$$\frac{\partial u^{(k)}}{\partial r^{(k)}} = \nabla\tilde{H}^{(k)}(s_k) \cdot S'_{\theta_k}(h_k) \cdot \nabla H^{(k)}(r^{(k)}), \tag{7.17}$$

where the notations are simplified using the following definitions:

$$h_k \triangleq H^{(k)}r^{(k)} \quad \text{and} \quad s_k \triangleq S_{\theta_k}(h_k). \tag{7.18}$$

Substituting (7.17) into (7.16), we see that $\partial L/\partial r^{(k)}$ can be obtained once we have $\partial L/\partial u^{(k)}$. The gradient $\partial L/\partial u^{(k)}$ can also be computed by the chain rule:

$$\frac{\partial L}{\partial u^{(k)}} = \frac{\partial L}{\partial r^{(k+1)}}\frac{\partial r^{(k+1)}}{\partial u^{(k)}}, \tag{7.19}$$

where $\partial r^{(k+1)}/\partial u^{(k)}$ is obtained by (7.12a) for $k \leftarrow k+1$ as

$$\frac{\partial r^{(k+1)}}{\partial u^{(k)}} = I - \alpha_{k+1}\nabla^2 h(u^{(k)}). \tag{7.20}$$

Hence, we can obtain $\partial L/\partial u^{(k)}$ once $\partial L/\partial r^{(k+1)}$ is computed. Therefore, we can compute the gradients of L with respect to $u^{(k)}$ and $r^{(k)}$ for all k in the order from left to right using (7.16), (7.17), (7.19), and (7.20), starting from $\partial L/\partial u^{(K)} = u^{(K)} - u_{\text{true}}$, as follows:

$$\frac{\partial L}{\partial u^{(K)}} \rightarrow \frac{\partial L}{\partial r^{(K)}} \rightarrow \cdots \rightarrow \frac{\partial L}{\partial r^{(k+1)}} \rightarrow \frac{\partial L}{\partial u^{(k)}} \rightarrow \frac{\partial L}{\partial r^{(k)}} \rightarrow \cdots \rightarrow \frac{\partial L}{\partial u^{(0)}}. \tag{7.21}$$

That is, we first compute $\partial L/\partial u^{(K)} = u^{(K)} - u_{\text{true}}$ according to the definition of L in (7.15), use it to compute $\partial L/\partial r^{(K)}$ according to (7.16) and (7.17), and then $\partial L/\partial u^{(K-1)}$ according to (7.19) and (7.20), and so on. Note that this order (7.21) simply has a reverse order of figure 7.1. This is the effect of backpropagation.

Now, we compute the gradients of $r^{(k)}$ and $u^{(k)}$ with respect to the network parameters used in (7.12a) and (7.12b), respectively. The derivative of $r^{(k)}$ with respect to α_k is straightforward due to (7.12a):

$$\frac{\partial r^{(k)}}{\partial \alpha_k} = -\nabla h(u^{(k)}). \tag{7.22}$$

The gradient of $u^{(k)}$ with respect to $w_j^{(k)}$ in the jth layer of the CNN $H^{(k)}$ defined in (7.11) can be obtained by applying the chain rule to (7.12b):

$$\frac{\partial u^{(k)}}{\partial w_j^{(k)}} = \nabla\tilde{H}^{(k)}(s_k) \cdot S'_{\theta_k}(h_k) \cdot \frac{\partial h_k}{\partial w_j^{(k)}} \tag{7.23}$$

for $j = 1, 2$, where h_k the is output of $H^{(k)}$ given the input $r^{(k)}$ and s_k is the output of S_{θ_k} given the input h_k defined in (7.18). The partial derivative $\partial h_k/\partial w_j^{(k)}$ is standard as in the backpropagation of CNN, as explained in part I. Similarly, the gradient of $u^{(k)}$ with respect to $\tilde{w}_j^{(k)}$ in the jth layer of the CNN $\tilde{H}^{(k)}$ defined in (7.11) can be obtained since $u^{(k)}$ and s_k are the output and input of $\tilde{H}^{(k)}$, respectively. The gradient of $u^{(k)}$ with respect to θ_k is slightly different:

$$\frac{\partial u^{(k)}}{\partial \theta_k} = \nabla\tilde{H}^{(k)}(s_k) \cdot \frac{\partial S_{\theta_k}(h_k)}{\partial \theta_k}. \tag{7.24}$$

In this case, we will need to treat $S_{\theta_k}(h_k) \in \mathbb{R}^n$ as a function of θ_k for given h_k, i.e. $S.(h_k): \theta_k \mapsto S_{\theta_k}(h_k)$ defined by

$$[S_{\theta_k}(h_k)]_i = \begin{cases} -\theta_k + [h_k]_i & \text{if } 0 < \theta_k < h_k, \\ \theta_k - [h_k]_i & \text{if } 0 < \theta_k < -h_k, \\ 0 & \text{otherwise.} \end{cases} \tag{7.25}$$

Hence, the derivative of $S_{\theta_k}(h_k)$ with respect θ_k is

$$\left[\frac{\partial S_{\theta_k}(h_k)}{\partial \theta_k} \right]_i = \begin{cases} -1 & \text{if } 0 < \theta_k < h_k, \\ 1 & \text{if } 0 < \theta_k < -h_k, \\ 0 & \text{otherwise.} \end{cases} \tag{7.26}$$

With all the partial derivatives obtained above, we can apply the chain rule to compute the gradient of L with respect to each of the network parameters. For example,

$$\frac{\partial L}{\partial \alpha_k} = \frac{\partial L}{\partial r^{(k)}} \frac{\partial r^{(k)}}{\partial \alpha_k}, \tag{7.27}$$

where $\partial L / \partial r^{(k)}$ is obtained by (7.16) and (7.17) following the backpropagation process in figure 7.1, and $\partial r^{(k)} / \partial \alpha_k$ is obtained by (7.22). The partial derivatives with respect to the other parameters can be similarly computed as follows:

$$\frac{\partial L}{\partial \theta_k} = \frac{\partial L}{\partial u^{(k)}} \frac{\partial u^{(k)}}{\partial \theta_k}, \quad \frac{\partial L}{\partial w_j^{(k)}} = \frac{\partial L}{\partial u^{(k)}} \frac{\partial u^{(k)}}{\partial w_j^{(k)}}, \quad \frac{\partial L}{\partial \tilde{w}_j^{(k)}} = \frac{\partial L}{\partial u^{(k)}} \frac{\partial u^{(k)}}{\partial \tilde{w}_j^{(k)}}, \tag{7.28}$$

where $\partial L / \partial u^{(k)}$ is obtained by (7.19) and (7.20) following the backpropagation in figure 7.1, and the partial derivatives of $u^{(k)}$ with respect to θ_k, $w_j^{(k)}$, and $\tilde{w}_j^{(k)}$ are obtained similarly, as explained above.

With these gradients of L with respect to the network parameters, we can employ a stochastic gradient descent method and find the optimal parameters Θ that minimize (7.15) over the entire training dataset. With the optimal Θ, ISTA-Net works as a feedforward mapping, which takes a new, partial k-space data f, sets $u^{(0)} = \mathcal{F}^T P^T f$, and outputs a reconstructed image $u^{(K)}$. This feedforward mapping can be computed very fast since all operations in (7.12) are explicit given Θ.

The performance of ISTA-Net can be further improved by modifying the structure of u in (7.10) similar to the residual network (ResNet). The resulting architecture, called ISTA-Net$^+$, still keeps (7.12a) for $r^{(k)}$ as ISTA-Net, but modifies the update (7.12b) by first rewriting $u^{(k)}$ as

$$u^{(k)} = r^{(k)} + e^{(k)}, \tag{7.29}$$

where $e^{(k)}$ is expected to be a function of $u^{(k)}$ that captures the difference between $u^{(k)}$ and $r^{(k)}$. More precisely, ISTA-Net$^+$ specifies that $e^{(k)}$ is obtained by the composition of two mappings, $G^{(k)}$ and $D^{(k)}$, of $u^{(k)}$ as follows:

$$e^{(k)} = (G^{(k)} \circ D^{(k)})(u^{(k)}), \tag{7.30}$$

where both $G^{(k)}$ and $D^{(k)}$ are multilayer CNNs to be learned. On the other hand, the sparsifying operator Ψ in ISTA is replaced by the composition of two CNNs, the same $D^{(k)}$ as in (7.30) and some $H^{(k)}$. Note that we reuse the notation $H^{(k)}$ here, but it is not related to and can be different from the one in ISTA-Net (7.11). If $H^{(k)} \circ D^{(k)}$ is still approximately orthogonal as Ψ, then the u step (7.9b) with Ψ replaced by $H^{(k)} \circ D^{(k)}$ is equivalent to

$$
\begin{aligned}
u^{(k)} &= \arg\min_u \left\{ \theta \| (H^{(k)} \circ D^{(k)}) u \|_1 + \frac{1}{2} \| u - r^{(k)} \|^2 \right\} \\
&= \arg\min_u \left\{ \theta \| (H^{(k)} \circ D^{(k)}) u \|_1 + \frac{1}{2} \| (H^{(k)} \circ D^{(k)}) u - (H^{(k)} \circ D^{(k)}) r^{(k)} \|^2 \right\}.
\end{aligned}
\tag{7.31}
$$

Therefore, $(H^{(k)} \circ D^{(k)})(u^{(k)}) = S_\theta((H^{(k)} \circ D^{(k)})(r^{(k)}))$ due to (7.31). If we have a left inverse $\tilde{H}^{(k)}$ of $H^{(k)}$ such that $\tilde{H}^{(k)} \circ H^{(k)} = I$, then

$$
\begin{aligned}
e^{(k)} &= G^{(k)} \circ D^{(k)}(u^{(k)}) = (G^{(k)} \circ \tilde{H}^{(k)} \circ H^{(k)} \circ D^{(k)})(u^{(k)}) \\
&= (G^{(k)} \circ \tilde{H}^{(k)})(S_\theta((H^{(k)} \circ D^{(k)})(r^{(k)}))).
\end{aligned}
\tag{7.32}
$$

Combining (7.32) and (7.29) yields the update rule of $u^{(k)}$ as

$$
u^{(k)} = r^{(k)} + (G^{(k)} \circ \tilde{H}^{(k)})(S_{\theta_k}((H^{(k)} \circ D^{(k)})(r^{(k)}))),
\tag{7.33}
$$

where the shrink threshold θ_k is also to be learned together with the convolution kernels in $G^{(k)}$, $\tilde{H}^{(k)}$, $H^{(k)}$, and $D^{(k)}$. In ISTA-Net$^+$, $\tilde{H}^{(k)}$ and $H^{(k)}$ have the same symmetric structure as that of ISTA-Net. In contrast, both $G^{(k)}$ and $D^{(k)}$ are implemented as simple linear convolutions in Zhang and Ghanem (2018). The architecture of ISTA-Net$^+$ is illustrated in figure 7.2.

Similar to ISTA-Net, the loss function for each training pair in ISTA-Net$^+$ also consists of two parts (7.13) and (7.14), but the data flow to obtain $u^{(K)}$ differs from that of ISTA-Net due to the new structure (7.33) for the u step in each phase k, and the parameter Θ to be learned in ISTA-Net$^+$ contains α_k, θ_k, as well as all convolutions in $G^{(k)}$, $\tilde{H}^{(k)}$, $H^{(k)}$, and $D^{(k)}$ in each phase k. Figure 7.3 shows an example reconstruction result by ISTA-Net$^+$ (Zhang and Ghanem 2018) on an MR brain image.

ISTA-Net and ISTA-Net$^+$ are typical examples of optimization-inspired deep reconstruction architectures. The data fidelity term in the traditional optimization-based reconstruction models is kept, since it is established by the physical principles of MRI and hence does not need to be relearned. The regularization term, on the

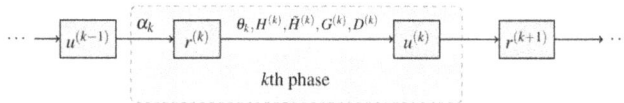

Figure 7.2. Architecture of ISTA-Net$^+$ (7.12a) and (7.33). The kth phase updates $r^{(k)}$ and $u^{(k)}$. The dependences of each variable on other variables are shown as incoming arrows, and the network parameters used for update are labeled next to the corresponding arrows.

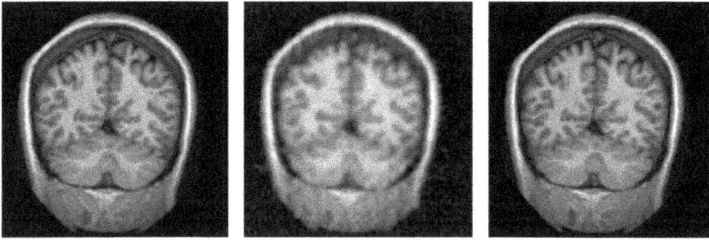

Figure 7.3. An example reconstruction result by ISTA-Net$^+$ (Zhang and Ghanem 2018) on a brain MR image with a synthesized radial mask of sampling ratio 20%. Left: ground truth reference image reconstructed from full k-space data. Middle: zero-filled reconstruction from undersampled k-space data. Right: reconstruction from undersampled k-space data by ISTA-Net$^+$.

other hand, is hand-crafted and may not work effectively for a specific image reconstruction task and/or imaging protocol. Therefore, this regularization term is replaced by deep neural networks. ISTA-Net and ISTA-Net$^+$ contain K phases where each phase mimics one iteration of ISTA. However, unlike the optimization algorithm ISTA, the iteration number K is prescribed in ISTA-Net and ISTA-Net$^+$ and does not need to be large. This is because K corresponds to the depth of ISTA-Net and ISTA-Net$^+$, which can be too difficult and expensive to train for larger K. Moreover, it was shown empirically that K can be merely around 10, and larger K does not incur significant improvement in reconstruction quality.

7.1.2 ADMM-Net

In the previous chapter, we introduced the alternating minimization method of multipliers, or ADMM for short, for single-coil MR image reconstruction. ADMM is a numerical algorithm particularly effective for convex optimization problems with linear equality constraints. Combined with the variable splitting technique, ADMM has been very successful in solving image reconstruction problems with total-variation regularization.

Compared to ISTA, ADMM can handle problems where the primal variable (i.e. the variable to be solved in the optimization problem) consists of two blocks related by a linear equality constraint[1]. In addition, there is a dual variable, i.e. the Lagrangian multiplier, associated with the equality constraint. In each iteration, ADMM updates the two blocks of the primal variables in order, one at each time with the other one fixed, and then the dual variable using the updated primal variable. ADMM yields more complex iterations due to the multiple-variable structure than ISTA. The deep neural network architecture, called ADMM-Net, inspired by the algorithmic design of ADMM, is the focus of this subsection.

---

[1] Multi-block ADMM has also gained attention in the optimization community over recent years, but it is beyond the scope of this chapter.

As before, to build ADMM-Net, we first recall the original ADMM applied to the single-coil MR image reconstruction problem with TV regularization formulated by

$$\underset{u}{\text{minimize}} \left\{ g(Du) + \frac{1}{2}\|P\mathcal{F}u - f\|^2 \right\}. \tag{7.34}$$

The first term in (7.34) is the TV regularization, written as a composite function $g(Du) \triangleq \mu\sum_{i=1}^{n}\|D_iu\|$ with weight parameter $\mu > 0$. The second term is the data fidelity to ensure the Fourier transform of u is close to the Fourier data f at the acquired k-space locations determined by the mask P. To apply ADMM, we first use variable splitting by introducing an auxiliary variable w such that $w = Du$, and rewrite (7.34) as the following equivalent problem:

$$\underset{w,u}{\text{minimize}} \left\{ g(w) + \frac{1}{2}\|P\mathcal{F}u - f\|^2 \right\}, \text{ subject to } w = Du. \tag{7.35}$$

Then, we formulate its associated augmented Lagrangian:

$$L(u, w; \lambda) = g(w) + \frac{1}{2}\|P\mathcal{F}u - f\|^2 + \langle \lambda, w - Du \rangle + \frac{\rho}{2}\|w - Du\|^2, \tag{7.36}$$

with Lagrangian multiplier λ. ADMM is then applied to solve (7.35) with the augmented Lagrangian (7.36). In each iteration of ADMM, the primal variables w and u are updated in order, and then the dual variable λ is updated. All of the subproblems in ADMM have closed form solutions for (7.35), and hence in each iteration the variables are updated as follows:

$$w^{(k)} = S_\theta(Du^{(k-1)} - \lambda^{(k-1)}), \tag{7.37a}$$

$$u^{(k)} = \mathcal{F}^{\mathsf{T}}(P^{\mathsf{T}}P + \rho\mathcal{F}D^{\mathsf{T}}D\mathcal{F}^{\mathsf{T}})^{-1}(P^{\mathsf{T}}f + \rho\mathcal{F}D^{\mathsf{T}}(w^{(k)} + \lambda^{(k-1)})), \tag{7.37b}$$

$$\lambda^{(k)} = \lambda^{(k-1)} + (w^{(k)} - Du^{(k)}), \tag{7.37c}$$

where $\theta = \mu/\rho$. Given an initial guess $(w^{(0)}, u^{(0)}, \lambda^{(0)})$, ADMM repeats the cycle of the three steps (7.37) for iteration $k = 1, 2, \ldots$, until a stopping criterion is satisfied.

ADMM-Net (Sun et al 2016) is a deep reconstruction network architecture inspired by the ADMM scheme (7.37). Similar to the case of ISTA-Net, each phase of ADMM-Net mimics one iteration of ADMM (7.37). Note that except for the match between $P\mathcal{F}u$ and f, other components in the reconstruction model (7.34) and the augmented Lagrangian (7.36), such as the TV regularization, the weight parameter μ, and the penalty parameter ρ, are all chosen manually in ADMM. These manually chosen regularizations and parameters may not work optimally for a specific task, and hence the quality of reconstructed images can be low due to inappropriate parameter setting and the computation time can be long because of the relatively slow convergence rate of ADMM.

To overcome these issues, ADMM-Net follows the same scheme of ADMM, but learns the parameters and filters from training data rather than setting them manually. More specifically, ADMM-Net sets a fixed iteration number K. The

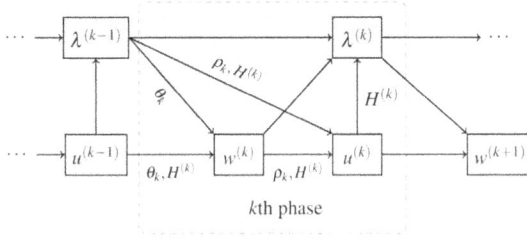

Figure 7.4. Architecture of ADMM-Net (7.38). The kth phase updates $w^{(k)}$, $u^{(k)}$, and $\lambda^{(k)}$. The dependences of each variable on other variables are shown as incoming arrows, and the network parameters used for update are labeled next to the corresponding arrows.

kth phase of ADMM-Net mimics the kth iteration of ADMM (7.37), but ADMM-Net replaces the gradient operator D by a parameterized filter (convolution) $H^{(k)}$ and the fixed parameters θ and ρ by θ_k and ρ_k to be learned through training. Therefore, the kth phase of ADMM-Net is formulated as

$$w^{(k)} = S_{\theta_k}(H^{(k)}u^{(k-1)} - \lambda^{(k-1)}),\qquad(7.38a)$$

$$u^{(k)} = \mathcal{F}^{\mathsf{T}}(P^{\mathsf{T}}P + \rho_k\mathcal{F}H^{(k)\mathsf{T}}H^{(k)}\mathcal{F}^{\mathsf{T}})^{-1}(P^{\mathsf{T}}f + \rho_k\mathcal{F}H^{(k)\mathsf{T}}(w^{(k)} + \lambda^{(k-1)})),\qquad(7.38b)$$

$$\lambda^{(k)} = \lambda^{(k-1)} + (w^{(k)} - H^{(k)}u^{(k)}),\qquad(7.38c)$$

where S_θ is the shrinkage by $\theta > 0$ as in (7.12b). In ADMM-Net (Sun *et al* 2016), $H^{(k)}$ is set to a linear combination of a set of given filters $\{B_l\}$ with coefficients $\gamma^{(k)} = (\cdots, \gamma_l^{(k)}, \cdots) \in \mathbb{R}^{|\{B_l\}|}$, i.e. $H^{(k)} = \sum_l \gamma_l^{(k)}B_l$. Therefore, $H^{(k)}$ is completely determined by the coefficients $\gamma^{(k)}$ in the kth phase.

The complete structure of ADMM-Net consists of K phases described above. The first phase accepts $(u^{(0)}, \lambda^{(0)})$ as input, and the last phase outputs $u^{(K)}$. The initial $\lambda^{(0)}$ is often set to 0, and $u^{(0)}$ can be set to the zero-filled reconstruction from partial k-space data f. The output $u^{(K)}$ is a function of the input f and network parameters $\Theta = \{\theta_k, \rho_k, \gamma^{(k)} \mid k = 1, \dots, K\}$. The architecture of ADMM-Net is shown in figure 7.4. As usual, the loss function can be set to the squared error of $u^{(K)}$ from the reference image u_{true} obtained from full k-space data for each training pair (f, u_{true}) (note that f also contains the information of the k-space sampling mask P necessary in (7.38b)):

$$L(\Theta; f, u_{\text{true}}) = \frac{1}{2}\|u^{(K)}(f; \Theta) - u_{\text{true}}\|^2.\qquad(7.39)$$

The total loss function is the sum of the loss in (7.39) above over all training pairs (f, u_{true}) in the given training dataset. Then, the total loss function is minimized using the (stochastic) gradient descent method, and the minimizer Θ is the learned network parameters.

To implement SGD, we again need the gradient of L in (7.39) with respect to each parameter in Θ. To this end, we first need to compute the gradient of L with respect to the main variables, which are also the outputs of each layer, i.e. $\partial L/\partial w^{(k)}$, $\partial L/\partial u^{(k)}$, $\partial L/\partial \lambda^{(k)}$. For $\partial L/\partial w^{(k)}$, we apply the chain rule to obtain

$$\frac{\partial L}{\partial w^{(k)}} = \frac{\partial L}{\partial \lambda^{(k)}}\frac{\partial \lambda^{(k)}}{\partial w^{(k)}} + \frac{\partial L}{\partial u^{(k)}}\frac{\partial u^{(k)}}{\partial w^{(k)}} = \frac{\partial L}{\partial \lambda^{(k)}}I + \frac{\partial L}{\partial u^{(k)}}\mathcal{F}^{\mathsf{T}}M_k^{-1}\mathcal{F}H^{(k)\mathsf{T}}, \qquad (7.40)$$

where $M_k \triangleq \rho_k^{-1}P^{\mathsf{T}}P + \mathcal{F}H^{(k)}H^{(k)}\mathcal{F}^{\mathsf{T}}$, and $\partial \lambda^{(k)}/\partial w^{(k)}$ and $\partial u^{(k)}/\partial w^{(k)}$ above are obtained by taking gradients of (7.38c) and (7.38b) with respect to $w^{(k)}$, respectively. Similarly, we can compute $\partial L/\partial u^{(k)}$:

$$\frac{\partial L}{\partial u^{(k)}} = \frac{\partial L}{\partial \lambda^{(k)}}\frac{\partial \lambda^{(k)}}{\partial u^{(k)}} + \frac{\partial L}{\partial w^{(k+1)}}\frac{\partial w^{(k+1)}}{\partial u^{(k)}} = -\frac{\partial L}{\partial \lambda^{(k)}}H^{(k)} + \frac{\partial L}{\partial w^{(k+1)}}S'_{\theta_{k+1}}(h_{k+1})H^{(k+1)}, \quad (7.41)$$

where $h_k \triangleq H^{(k)}u^{(k-1)} - \lambda^{(k-1)}$, and $\partial \lambda^{(k)}/\partial u^{(k)}$ and $\partial w^{(k+1)}/\partial u^{(k)}$ are obtained by taking gradients of (7.38c) and (7.38a) with respect to $u^{(k)}$, respectively. Finally, we compute $\partial L/\partial \lambda^{(k)}$:

$$\begin{aligned}\frac{\partial L}{\partial \lambda^{(k)}} &= \frac{\partial L}{\partial \lambda^{(k+1)}}\frac{\partial \lambda^{(k+1)}}{\partial \lambda^{(k)}} + \frac{\partial L}{\partial u^{(k+1)}}\frac{\partial u^{(k+1)}}{\partial \lambda^{(k)}} + \frac{\partial L}{\partial w^{(k+1)}}\frac{\partial w^{(k+1)}}{\partial \lambda^{(k)}} \\ &= \frac{\partial L}{\partial \lambda^{(k+1)}}I + \frac{\partial L}{\partial u^{(k+1)}}\mathcal{F}^{\mathsf{T}}M_{k+1}^{-1}\mathcal{F}H^{(k+1)\mathsf{T}} - \frac{\partial L}{\partial w^{(k+1)}}S'_{\theta_{k+1}}(h_{k+1}).\end{aligned} \qquad (7.42)$$

From (7.40), (7.41), and (7.42), and the fact that L immediately depends on $u^{(K)}$ in (7.39), we can see that the order of computing the $\partial L/\partial w^{(k)}$, $\partial L/\partial u^{(k)}$, $\partial L/\partial \lambda^{(k)}$ in backpropagation is as follows:

$$\frac{\partial L}{\partial u^{(K)}} \rightarrow \frac{\partial L}{\partial w^{(K)}} \rightarrow \frac{\partial L}{\partial \lambda^{(K-1)}} \rightarrow \cdots \rightarrow \frac{\partial L}{\partial \lambda^{(k)}} \rightarrow \frac{\partial L}{\partial u^{(k)}} \rightarrow \frac{\partial L}{\partial w^{(k)}} \rightarrow \cdots \rightarrow \frac{\partial L}{\partial w^{(1)}}, \qquad (7.43)$$

where we set $\partial L/\partial \lambda^{(K)} = 0$ since the output phase K does not compute $\lambda^{(K)}$.

Next, we need to compute the gradient of $w^{(k)}$, $u^{(k)}$, and $\lambda^{(k)}$ with respect to the network parameters $\{\theta_k, \rho_k, \gamma^{(k)}\}$ in (7.38), respectively. We first work on θ_k since it only appears in the computation of $w^{(k)}$ as in (7.38a). Similarly as in ISTA-Net, we need to treat $w^{(k)} = S_{\theta_k}(h_k)$ as a function of θ_k for given $h_k = H^{(k)}u^{(k-1)} - \lambda^{(k-1)}$ as in (7.25), then the gradient $\partial w^{(k)}/\partial \theta_k$ can be obtained by taking derivative of $S_{\theta_k}(h_k)$ with respect to θ_k as in (7.26). For ρ_k, it appears in the computation of $u^{(k)}$ in (7.38b) only. Note that the term involving ρ_k on the right-hand side of (7.38b) is

$$\mathcal{F}^{\mathsf{T}}M_{k+1}^{-1}(\rho_k)\mathcal{F}H^{(k)\mathsf{T}}(w^{(k)} + \lambda^{(k-1)}), \qquad (7.44)$$

where we treat M_{k+1} as a function of ρ_k for fixed $H^{(k)}$:

$$M_{k+1}(\rho_k) = \rho_k^{-1}P^{\mathsf{T}}P + \mathcal{F}H^{(k)\mathsf{T}}H^{(k)}\mathcal{F}^{\mathsf{T}}. \qquad (7.45)$$

Therefore, the derivative of $M_{k+1}^{-1}(\rho_k)$ with respect to ρ_k is

$$\frac{\partial M_{k+1}^{-1}(\rho_k)}{\partial \rho_k} = \frac{1}{\rho_k^2}M_{k+1}^{-1}(\rho_k)P^{\mathsf{T}}PM_{k+1}^{-1}(\rho_k) \qquad (7.46)$$

and hence the gradient of $u^{(k)}$ with respect to ρ_k is

$$\frac{\partial u^{(k)}}{\partial \rho_k} = \frac{1}{\rho_k^2} \mathcal{F}^\top M_{k+1}^{-1}(\rho_k) P^\top P M_{k+1}^{-1}(\rho_k) \mathcal{F} H^{(k)\top}(w^{(k)} + \lambda^{(k-1)}). \tag{7.47}$$

Note that the inverse M_{k+1}^{-1} can be difficult to compute in general. We will discuss how to handle this soon after the computation of $\partial u^{(k)}/\partial \gamma_l^{(k)}$, where this inverse will appear again. Lastly, $H^{(k)}$, or equivalently, $\gamma^{(k)}$ since $H^{(k)} = \sum_l \gamma_l^{(k)} B_l$ for a fixed set of filters $\{B_l\}$, appears in all of the three updates in (7.38), and therefore we need $\partial w^{(k)}/\partial \gamma_l^{(k)}$, $\partial u^{(k)}/\partial \gamma_l^{(k)}$, $\partial \lambda^{(k)}/\partial \gamma_l^{(k)}$ for all l. The first and last are straightforward to compute:

$$\frac{\partial w^{(k)}}{\partial \gamma_l^{(k)}} = S'_{\theta_k}(h_k) B_l u^{(k-1)} \quad \text{and} \quad \frac{\partial \lambda^{(k)}}{\partial \gamma_l^{(k)}} = -B_l u^{(k)}. \tag{7.48}$$

For $\partial u^{(k)}/\partial \gamma_l^{(k)}$, we first treat $M_{k+1}(\gamma^{(k)})$ as a function of $\gamma^{(k)}$ for fixed ρ_k, i.e.

$$M_{k+1}(\gamma^{(k)}) = \rho_k^{-1} P^\top P + \sum_{l,l'} \gamma_l^{(k)} \gamma_{l'}^{(k)} \mathcal{F} B_l^\top B_{l'} \mathcal{F}^\top. \tag{7.49}$$

Therefore, (7.38b) can be written as

$$u^{(k)} = \mathcal{F}^\top M_{k+1}^{-1}(\gamma^{(k)})(\rho_k^{-1} P^\top f + \mathcal{F} H^{(k)\top}(w^{(k)} + \lambda^{(k-1)})). \tag{7.50}$$

Furthermore, we apply the product rule and chain rule to obtain

$$\frac{\partial u^{(k)}}{\partial \gamma_l^{(k)}} = \mathcal{F}^\top M_{k+1}^{-1}(\gamma^{(k)}) \left[\frac{\partial M_{k+1}^{-1}(\gamma^{(k)})}{\partial \gamma_l^{(k)}} M_{k+1}^{-1}(\gamma^{(k)}) f_k + \mathcal{F} B_l^\top (w^{(k)} + \lambda^{(k-1)}) \right], \tag{7.51}$$

where $f_k \triangleq \mathcal{F} H^{(k)\top}(w^{(k)} + \lambda^{(k-1)})$, and $\partial M_{k+1}(\gamma^{(k)})/\partial \gamma_l^{(k)} = \sum_{l'} \gamma_{l'}^{(k)} \mathcal{F} B_l^\top B_{l'} \mathcal{F}^\top$.

To address the issue with M_{k+1}^{-1}, we can choose an orthonormal basis $\{B_l\}$ such that $B_l^\top B_{l'} = \delta_{ll'}$, then $H^{(k)\top} H^{(k)} = \|\gamma^{(k)}\|^2 I$, with which there is

$$M_{k+1}(\rho_k, \gamma^{(k)}) = \rho_k^{-1} P^\top P + \|\gamma^{(k)}\|^2 I, \tag{7.52}$$

which is diagonal and hence trivial to invert.

Now, we have all the ingredients to compute the gradients of L with respect to Θ. The gradient with respect to the main variables $w^{(k)}$, $u^{(k)}$, and $\lambda^{(k)}$ are given by (7.40), (7.41), and (7.42), respectively. On the other hand, the gradient of the main variables with respect to the network parameters $\Theta = \{\theta_k, \rho_k, \gamma^{(k)}\}$ can be computed as explained above. Assembling them according to the chain rule yields the gradient of L with respect to Θ.

The version of ADMM-Net (7.38) we presented above is a modification of the original one proposed in Sun *et al* (2016). In (7.38), the scheme closely follows the ADMM (7.37). In the original ADMM-Net (Sun *et al* 2016), however, the filter $H^{(k)}$ in (7.38a) and (7.38c) can be different from the one in (7.38b). Moreover, the shrinkage in (7.38a) is replaced by a piece-wise linear function (PLF) determined by a set of control points and the associated function values. More specifically, let $\{p_0, \ldots, p_{N_c}\}$ be a set of $N_c + 1$ control points on $\mathbb{R}$. In Sun *et al* (2016), these control points are simply chosen as uniform mesh grid points on the interval $[-1, 1]$, i.e. $p_0 = -1$ and $p_{N_c} = 1$, and $p_l - p_{l-1} = 2/N_c$ for $l = 1, \ldots, N_c$. Then, the PLF $S(h; \{p_l, q_l^{(k)}\})$ in $[-1, 1]$ is completely determined by the values $\{q_l^{(k)}\}$ at the corresponding control points $\{p_l\}$. Outside the interval $[-1, 1]$, the PLF $S(h; \{p_l, q_l^{(k)}\})$ is set to have slope 1 and concatenates with its part in $[-1, 1]$ at p_0 and p_{N_c} to form a continuous function. Then, instead of learning θ_k in the shrinkage operation S_{θ_k} in (7.38a), the original ADMM-Net learns the values $\{q_l^{(k)}\}$ as part of the network parameters. In figure 7.5, we show an example reconstruction result obtained by ADMM-Net for a brain image using synthesized radial mask with sampling ratio 20%.

7.1.3 Variational reconstruction network

The ISTA-Net/ISTA-Net$^+$ and ADMM-Net introduced previously were both inspired by the numerical optimization algorithms with great successes in solving image reconstruction problems with regularization. However, to overcome the issue of generic regularization terms such as ℓ_1 and TV, they parameterize the regularization using the combination of linear filters and nonlinear neural networks, and learn the parameters by the data-driven approach as in deep learning.

If we step back and look at the original image reconstruction model, we can see both ISTA and ADMM are designed to solve a composite optimization problem whose general form is given by

$$\underset{u}{\text{minimize}}\, f(u), \quad \text{where } f(u) \triangleq g(Du) + h(u). \tag{7.53}$$

Such problems are also known as the variational models in the context of partial differential equations (PDE), where the objective function f is also called the energy or

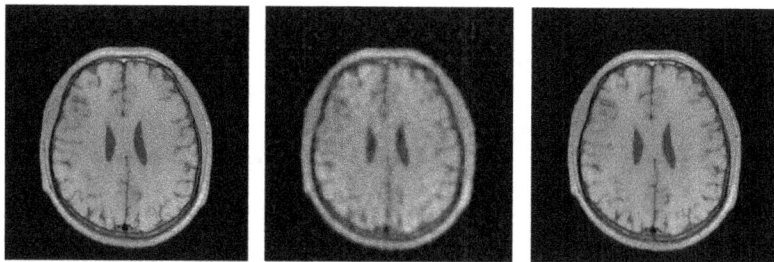

Figure 7.5. An example reconstruction result by ADMM-Net (Sun *et al* 2016) on a brain MR image with a synthesized radial mask of sampling ratio 20%. Left: a ground truth reference image reconstructed from full *k*-space data. Middle: zero-filled reconstruction from undersampled *k*-space data. Right: reconstruction from undersampled *k*-space data by ADMM-Net.

potential function. To find the minimizer of (7.53), u is considered time-dependent, and its gradient flow follows the negative first variation δf of the objective function f:

$$\frac{\partial u}{\partial t} = -\delta f(u), \tag{7.54}$$

where δf is essentially the gradient of f with respect to u. The time discretization of (7.54) is the well-known gradient descent method in numerical optimization:

$$u^{(k)} = u^{(k-1)} - \alpha_k(D^\mathsf{T}\nabla g(Du^{(k-1)}) + \nabla h(u^{(k-1)})), \tag{7.55}$$

where α_k is the step size in iteration k. Note that above we adopted a slight abuse of notation ∇g, since in image reconstruction g often represents the ℓ_1 norm or similar which is not differentiable. Hence, it is more rigorous to interpret ∇g as a subgradient of g, and the updating rule (7.55) is the subgradient descent. Nevertheless, this term will be replaced by a parameterized function to be learned in training, and thus its differentiability is not an important issue in the following derivation of the variational reconstruction network.

The variational network (Hammernik *et al* 2018) was inspired by this concise updating rule (7.55). Similar as ISTA-Net and ISTA-Net$^+$, h is kept in its original form since it represents the relation between the acquired data and the image to be reconstructed, which is governed by the physical principles described in the previous chapter. The terms g and D, on the other hand, are due to the manually designed regularization term in (7.53), which can be learned from the training data. Therefore, the variational network (Var-Net) sets a fixed iteration number K, and mimics each iteration of (7.55) by one phase in the reconstruction network. The kth phase of Var-Net is built as

$$u^{(k)} = u^{(k-1)} - H^{(k)\mathsf{T}}\phi_k(H^{(k)}u^{(k-1)}) - \alpha_k\nabla h(u^{(k-1)}), \tag{7.56}$$

where α_k, $H^{(k)}$, and ϕ_k are all to be learned from data. In particular, $H^{(k)}$ is a convolution to replace the gradient operation D in (7.55), and ϕ_k is a parameterized function to replace ∇g. The step size α_k appeared in the second term of the right-hand side in (7.55) but not in (7.56) since it has been absorbed into ϕ_k, which is to be learned together.

In Var-Net (Hammernik *et al* 2018), ϕ_k in (7.56) is represented as a linear combination of Gaussian functions. First of all, ϕ_k is to be applied to $H^{(k)}u^{(k-1)} \in \mathbb{R}^n$ component-wise, and hence it is sufficient to describe the component-wise operation of ϕ_k using a univariate function. To this end, we first determine a set of $N_c + 1$ control points $\{p_l: l = 0,\dots, N_c\}$ uniformly spaced on a prescribed interval $[-I, I]$ such that $-I = p_0 < p_1 < \cdots < p_{N_c} = I$ and $p_l - p_{l-1} = 2I/N_c$ for $l = 1,\dots, N_c$. For each point p_l, the Gaussian function with a prescribed standard deviation σ is given by

$$B_l(x) = \mathrm{e}^{-(x-p_l)^2/(2\sigma^2)}. \tag{7.57}$$

Then, ϕ_k is set to a linear combination of $B_l(x)$ with coefficients $\gamma_l^{(k)}$ to be determined:

$$\phi_k(x) = \sum_{l=0}^{N_c} \gamma_l^{(k)} B_l(x). \tag{7.58}$$

One can also design other basis functions, instead of (7.57) or even parametrize ϕ_k as a generic neural network. To be consistent with the original version of Var-Net in Hammernik *et al* (2018), we stick to the choice (7.58). For $H^{(k)}$, it is a convolution operation applied to $u^{(k-1)}$, and hence it suffices to determine the convolution kernel. This is a very simplified case of convolution layers of CNNs, and we omit the details here. It is worth noting that the convolution $H^{(k)}$ is applied to $u^{(k-1)}$ which can be complex-valued. On the other hand, $H^{(k)}u^{(k-1)}$ needs to be real-valued to be fed into ϕ_k defined in (7.58). This is a prevalent issue in deep-learning-based MRI reconstruction, for example, the commonly used activation functions such as ReLU requires real-valued inputs, just like ϕ_k here, but $u^{(k)}$ is complex-valued due to the k-space data and Fourier transforms in the $h(u)$ term. We will discuss this issue specifically in section 7.4.1. For ϕ_k here, Var-Net (Hammernik *et al* 2018) considers $H^{(k)}$ to contain two convolutions, one for the real part and the other one for the imaginary part of $u^{(k-1)}$, respectively. The two convolved results are then summed to obtain $H^{(k)}u^{(k-1)} \in \mathbb{R}^n$.

Now we can see that Var-Net consists of K phases, where each phase operates as (7.56). In particular, the first phase accepts $u^{(0)}$ as the input such as the zero-filled reconstruction from partial k-space data f. The last, Kth phase outputs $u^{(K)}$, which is used in the loss function to compare with the reference image u_{true} reconstructed from full k-space data:

$$L(\Theta; f, u_{\text{true}}) = \frac{1}{2}\|u^{(K)}(f; \Theta) - u_{\text{true}}\|^2, \tag{7.59}$$

where the network parameter $\Theta \triangleq \{\alpha_k, \gamma^{(k)}, H^{(k)} \mid k = 1, \ldots, K\}$. The total loss function is then the sum of (7.59) over all training pairs of form (f, u_{true}). The architecture of Var-Net is presented in figure 7.6.

To find the optimal parameter Θ that minimizes the loss function (7.59), we can again use the SGD algorithm. The key component of the implementation of SGD is again the computation of the gradients of L with respect to each parameter in Θ. As we have showed for ISTA-Net and ADMM-Net before, the gradients can be obtained by finding the gradient of L with respect to the main variable $u^{(k)}$ and then

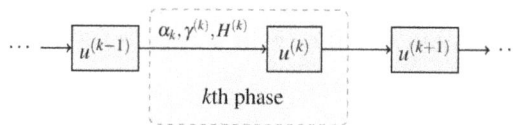

Figure 7.6. Architecture of Var-Net (7.56). The kth phase updates $u^{(k)}$. The dependences of each variable on other variables are shown as incoming arrows, and the network parameters used for update are labeled next to the corresponding arrows.

the gradient of $u^{(k)}$ with respect to these parameters. Due to the single updating rule in (7.56), the gradient $\partial L / \partial u^{(k)}$ is straightforward to compute:

$$\frac{\partial L}{\partial u^{(k)}} = \frac{\partial L}{\partial u^{(k+1)}} \frac{\partial u^{(k+1)}}{\partial u^{(k)}} = \frac{\partial L}{\partial u^{(k+1)}} (I - H^{(k+1)\top} \nabla \phi_k(h_{k+1}) H^{(k+1)} - \alpha_k \nabla^2 h(u^{(k)})), \quad (7.60)$$

where $h_k \triangleq H^{(k)} u^{(k-1)}$. Therefore, the backpropagation implies that these gradients can be computed in the following order:

$$\frac{\partial L}{\partial u^{(K)}} \rightarrow \frac{\partial L}{\partial u^{(K-1)}} \rightarrow \cdots \rightarrow \frac{\partial L}{\partial u^{(k)}} \rightarrow \frac{\partial L}{\partial u^{(1)}}. \quad (7.61)$$

The computation of gradients of $u^{(k)}$ with respect to the parameters α_k, $\gamma_l^{(k)}$, and $H^{(k)}$ are straightforward by observing (7.56) and (7.58). We leave these to the reader as an exercise. In figure 7.7, we show an example reconstruction result obtained by Var-Net (Hammernik *et al* 2018) for a PD-weighted MR image using Cartesian mask of variable density along phase encoding direction and sampling ratio 31.6%.

7.2 Leveraging generic network structures

7.2.1 Cascaded CNNs

The ISTA-Net, ADMM-Net, and Var-Net are all originated from the variational model consisting of the data fidelity term and the regularization term such as ℓ_1 and TV. In the past decade, an alternative regularization approach has also proved promising by learning redundant dictionaries and using a sparse representation to suppress image noise and artifacts. The dictionary can be either a generic, fixed set of redundant basis, such as an overcomplete discrete cosine transform (DCT) basis, or a learned one from a given dataset. A learned dictionary can even be adaptively constructed online during a reconstruction process. However, since we only need to motivate the reconstruction network of cascaded CNNs using dictionary-based image reconstruction, we consider the dictionary here to be fixed.

Dictionary-based regularization exploits sparse representations of images, or more specifically image patches, using the redundant basis. With a suitable set of redundant bases, any image patch is expected to be closely approximated by a small

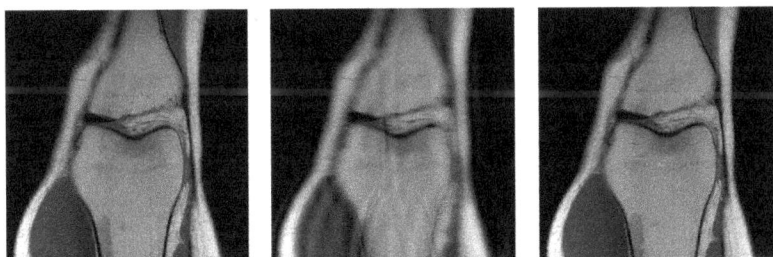

Figure 7.7. An example reconstruction result by Var-Net (Hammernik *et al* 2018) on a PD-weighted MR image with a Cartesian mask of variable density and sampling ratio 31.6%. Left: ground truth reference image reconstructed from full *k*-space data. Middle: zero-filled reconstruction from undersampled *k*-space data. Right: reconstruction from underdamped *k*-space data by Var-Net.

number of bases, also called *atoms*, of the dictionary. For example, consider a set of 2D MR images of size $\sqrt{n} \times \sqrt{n}$, and image patches of size $\sqrt{n'} \times \sqrt{n'}$ extracted from these images, e.g. $n = 256^2$ and $n' = 8^2$. For a redundant dictionary $D \in \mathbb{R}^{n' \times L}$ ($L \gg n'$) with L atoms $d_1, \ldots, d_L \in \mathbb{R}^{n'}$, every image patch written as a vector in $\mathbb{R}^{n'}$ can be approximated by only a few atoms in D. Following this idea, a dictionary-learning-based image reconstruction problem can be written as a minimization problem as follows:

$$\underset{u, \gamma}{\text{minimize}} \ \mu \sum_j \left(\frac{1}{2} \|Q_j u - D\gamma_j\|^2 + \nu\|\gamma_j\|_1 \right) + \frac{1}{2} \|P\mathcal{F}u - f\|^2, \tag{7.62}$$

where $Q_j : \mathbb{R}^n \to \mathbb{R}^{n'}$ extracts the jth patch of an image u, and $\gamma_j \in \mathbb{R}^L$ is the combination coefficients to represent the patch $Q_j u$ using only a few atoms in the dictionary D, and the sum of j in (7.62) runs over all patches that cover the image u. The sparsity of γ_j is enforced by the ℓ_1 norm (or ℓ_p norm for $0 \leqslant p < 1$) of γ_j with weight parameter $\nu > 0$. Let us assume that we use n/n' non-overlapping patches to cover the entire image u, since using overlapping patches does not yield significant difficulty in dictionary-based image reconstruction and we here only need to inspire the idea of the cascaded CNNs. Then, the solution of (7.62) can be obtained by alternately minimizing the objective function in (7.62) with respect to γ and u as follows:

$$\gamma_j^{(k)} = \arg\min_{\gamma_j} \left\{ \frac{1}{2} \|Q_j u^{(k-1)} - D\gamma_j\|^2 + \nu\|\gamma_j\|_1 \right\}, \quad \forall j \tag{7.63a}$$

$$u^{(k)} = \arg\min_u \left\{ \frac{\mu}{2} \|u - u_D^{(k)}\|^2 + \frac{1}{2} \|P\mathcal{F}u - f\|^2 \right\}, \tag{7.63b}$$

where $u_D^{(k)} \triangleq \sum_j Q_j^\top D\gamma_j^{(k)} \in \mathbb{R}^n$ is the image assembled with its jth patch set to $D\gamma_j^{(k)} \in \mathbb{R}^{n'}$. We used $\sum_j Q_j^\top Q_j = I$ to simplify the objective function (7.62) and obtained (7.63b) since the patches are non-overlapping and cover the entire image u. In this case, the u subproblem (7.63b) can be easily solved: the gradient of the objective function in (7.63b) with respect to u yields the normal equation $(I + \mu\mathcal{F}^\top P^\top P\mathcal{F})u^{(k)} = u_D^{(k-1)} + \mu\mathcal{F}^\top Pf$ which has the closed form solution

$$u^{(k)} = \mathcal{F}^\top(\mu I + P^\top P)^{-1}(\mu\mathcal{F}u_D^{(k)} + P^\top f) \tag{7.64}$$

or equivalently

$$u^{(k)} = \mathcal{F}^\top f_D^{(k)} \text{ where}$$

$$\left[f_D^{(k)} \right]_i \triangleq \begin{cases} [\mathcal{F}u_D^{(k)}]_i & \text{if location } i \text{ is not sampled} \\ \dfrac{\mu[\mathcal{F}u_D^{(k)}]_i + [P^\top f]_i}{\mu + 1} & \text{if location } i \text{ is sampled} \end{cases} \tag{7.65}$$

That is, $u^{(k)}$ is the inverse Fourier transform of $f_D^{(k)}$, where $f_D^{(k)}$ takes the k-space of $u_D^{(k)}$ if the k-space location i is missing in the data f, or takes the weighted sum of the k-space of $u_D^{(k)}$ and f with weights $\mu/(\mu + 1)$ and $1/(\mu + 1)$ otherwise. Nevertheless, the mapping from $u_D^{(k)}$ to $u^{(k)}$ in (7.64) is linear with a scalar parameter $\mu > 0$. We denote this mapping by $\mathrm{DC}(\cdot; \mu)$, as it performs a data consistency operation.

Note that the γ subproblem (7.63a) can be solved by ℓ_1 minimizations in parallel for all j values. Let $\Phi(u^{(k-1)})$ denote the composition of the mapping from $u^{(k-1)}$ to $\{\gamma_j^{(k)}\}$ in (7.63a) and the assembling from $\{\gamma_j^{(k)}\}$ to $u_D^{(k)} = \sum_j Q_j^\top D\gamma_j^{(k)}$, i.e. $u_D^{(k)} = \Phi(u^{(k-1)})$, then the iterations in (7.63) can be written in a more abstract form as

$$u^{(k)} = \mathrm{DC}(\Phi(u^{(k-1)}); \mu). \tag{7.66}$$

The regularization by the dictionary-based sparse representation in (7.62) often yields better preserved fine structures in reconstructed images than the counterparts through the ℓ_1 or TV minimization. A major reason is that the ℓ_1 or TV are generic and too simple to account for subtle details in images. On the other hand, dictionaries, in particular those trained using high-quality diverse images in a given domain, include most relevant fine structures as prior information, and hence are able to recover images more faithfully.

It is evident from (7.63a), updating $\gamma^{(k)}$ is to sparsely represent $u^{(k-1)}$ so that fine structures can be captured and noise can be removed. Therefore, the sparse representation using a dictionary can be regarded as a trained denoiser. On the other hand, deep learning also provides well developed deep neural networks that can produce excellent denoising results. In particular, trained convolutional neural networks (CNNs) are particularly suitable for image denoising. Therefore, we may replace the dictionary denoising operation Φ in (7.66) by a CNN with parameters to be learned from data. More specifically, let $\Phi(\cdot; \theta_k)$ denote a generic (or slightly modified) CNN with network parameter θ_k. Then, for a fixed iteration number K, we can construct a network of K cascaded CNNs by following (7.66):

$$u^{(k)} = \mathrm{DC}(\Phi(u^{(k-1)}; \theta_k); \mu_k). \tag{7.67}$$

Namely, the network of cascaded CNNs is constructed by concatenating K phases, where the kth phase consists of a CNN $\Phi(\cdot; \theta_k)$ with parameter θ_k and a simple, linear DC layer $\mathrm{DC}(\cdot; \mu_k)$ with parameter μ_k. Therefore, the network structure is conceptually very similar to the generic CNNs except that it is deeper and has an additional DC layers in between. Using the standard least squares loss function

$$L(\Theta; f) = \frac{1}{2}\|u^{(K)}(f; \Theta) - u_{\mathrm{true}}\|^2, \tag{7.68}$$

where $\Theta = \{\theta_k, \mu_k | k = 1, \ldots, K\}$, we can the obtain gradients of L with respect to θ_k and μ_k in a straightforward manner as in the CNN training. Note that the gradient of the loss function L in (7.68) with respect to μ_k is very simple: denote $M(\mu) \triangleq I + \mu^{-1} P^{\mathrm{T}} P$ which is diagonal, then we have $\partial M(\mu)/\partial \mu = -\mu^{-2} P^{\mathrm{T}} P$ and

$$\frac{\partial \mathrm{DC}(u; \mu_k)}{\partial \mu_k} = -\frac{1}{\mu_k^2} \mathcal{F}^{\mathrm{T}} M^{-1}(\mu_k) P^{\mathrm{T}} P M^{-1}(\mu_k) \mathcal{F} u \tag{7.69}$$

since the DC layer defined in (7.64) is a linear mapping.

7.2.2 GAN-based reconstruction networks

Generative adversarial networks, or GAN for short, are a class of generative models that train the generative network (generator) and the adversarial network (discriminator) simultaneously to reach optimum at an equilibrium (Goodfellow *et al* 2014). In the original GAN, the generator is trained to be a push-forward mapping that turns easy-to-generate random variables (such as Gaussian noise) into images similar to those sample images in the training set (in the sense that both lie in the same low-dimensional latent manifold of real images in $\mathbb{R}^n$), whereas the discriminator is trained to distinguish the generated images from the real ones by assigning 0 to the former and 1 to the latter. At an equilibrium, the generator produces images that look so realistic such that the discriminator cannot tell the difference but considers them equally likely to be real or fake.

The idea of GAN has motivated several network architectures for CS-MRI reconstruction. Existing works based on GAN replace the generator with a reconstruction network that maps an undersampled k-space (or its zero-filled reconstruction) to a reconstructed image of decent quality, and set the discriminator similar to the standard GAN to distinguish reconstructed images from the reference images reconstructed from full k-space data in the training set. The structures of the generator (i.e. the reconstruction network, we still call it a generator as in the GAN context for convenience here) and the discriminator are all manually designed and vary substantially, except that the input and output dimensions need to be consistent with the problem set-up. For example, if the generator G performs a mapping from a zero-filled image to a desired reconstruction, and D assigns each input image a scalar-valued probability of being real or reconstructed, then $G: \mathbb{R}^n \to \mathbb{R}^n$ and $D: \mathbb{R}^n \to [0, 1]$. After all, the key to the training of the generator G_θ and discriminator D_η, which are deep neural networks with parameters denoted by θ and η, respectively, is their objective (loss) functions. Then, with the optimal θ, G_θ is a feedforward network that can directly compute a reconstructed image from any input data (either partial k-space or the zero-filled reconstruction).

We briefly introduce two GAN-based CS-MRI reconstruction methods in this subsection. As discussed above, the main feature of these methods lies in the setting of the loss functions for the generator and the discriminator, respectively. In Mardani *et al* (2019), the optimization with respect to the generator is set to minimize

$$\alpha \mathbb{E}_f[\|f - A G_\theta(u_a)\|^2] + \beta \mathbb{E}_{(f, u_{\mathrm{true}})}[\|u_{\mathrm{true}} - G_\theta(u_a)\|^2] + \mathbb{E}_f[(1 - D_\eta(G_\theta(u_a)))^2] \tag{7.70}$$

with respect to θ, where (f, u_{true}) represents a training pair with u_{true} being the reference image from full k-space data, and f the undersampled k-space using mask (undersampling pattern) P, $u_a \triangleq \mathcal{F}^{\mathsf{T}} P^{\mathsf{T}} f$ is the zero-filled reconstruction from f, and A denotes the mapping from the reconstructed image to the partial k-space data (e.g. $A = P\mathcal{F}$ in single-coil CS-MRI). Here, the expectations are implemented as the averages over the training set $\{(f^{(i)}, u_{\text{true}}^{(i)}): 1 \leqslant i \leqslant N_{\text{train}}\}$, where N_{train} is the number of training pairs. For example, the first term in (7.70) is implemented as

$$\mathbb{E}_f[\|f - AG_\theta(u_a)\|^2] = \frac{1}{N_{\text{train}}} \sum_{i=1}^{N_{\text{train}}} \|f^{(i)} - AG_\theta(u_a^{(i)})\|^2. \tag{7.71}$$

The other terms in (7.70) are implemented similarly. The optimization with respect to the discriminator is set to minimize

$$\gamma \mathbb{E}_{u_{\text{true}}}[(1 - D_\eta(u_{\text{true}}))^2] + \mathbb{E}_f[(D_\eta(G_\theta(u_a)))^2] \tag{7.72}$$

with respect to η. In (7.70) and (7.72), $\alpha, \beta, \gamma > 0$ are weighting parameters to balance the terms in the minimizations. The goal of (7.70) is to train G_θ such that it maps a zero-filled image u_a to a reconstructed image $G_\theta(u_a)$ such that $G_\theta(u_a)$ is consistent with the partial k-space data f, close to the reference image u_{true}, and can trick the discriminator to assign it a high probability of being real, according to the first, second, and third terms in (7.70), respectively. On the other hand, (7.72) is to train D_η such that it assigns 1 to a real reference image u_{true} but 0 to a reconstructed image $G_\theta(u_a)$. The two networks G_θ and D_η are updated alternately using the SGD algorithm until convergence.

In Yang *et al* (2018), the generator is designed for de-aliasing, i.e. as a mapping from a zero-filled image with severe aliasing artifacts to an artifact-free reconstructed image. The optimization with respect to the generator parameter θ is to minimize

$$\begin{aligned} &\alpha \mathbb{E}_f[\|f - AG_\theta(u_a)\|^2] + \beta \mathbb{E}_{(f, u_{\text{true}})}[\|u_{\text{true}} - G_\theta(u_a)\|^2] \\ &+ \gamma \mathbb{E}_{(f, u_{\text{true}})}[\|F_{\text{VGG}}(u_{\text{true}}) - F_{\text{VGG}}(u_a)\|^2] + \mathbb{E}_f[-\log D_\eta(G_\theta(u_a))] \end{aligned} \tag{7.73}$$

and that with respect to the discriminator parameter η is to maximize

$$\mathbb{E}_f[-\log D_\eta(G_\theta(u_a))] + \mathbb{E}_{u_{\text{true}}}[\log D_\eta(u_{\text{true}})]. \tag{7.74}$$

The first two terms in (7.73) are the same as those in (7.70) (the first term was set to $\mathbb{E}_{(f, u_{\text{true}})}[\|\mathcal{F}u_{\text{true}} - \mathcal{F}G_\theta(u_a)\|^2]$ in Yang *et al* (2018) but this is identical to the second term $\mathbb{E}_{(f, u_{\text{true}})}[\|u_{\text{true}} - G_\theta(u_a)\|^2]$ due to the orthogonality of $\mathcal{F}$). The third term in (7.73) employs the pre-trained deep convolutional network from the Visual Geometry Group (VGG) at Oxford University, UK (Simonyan and Zisserman 2014), F_{VGG}, to construct a perceptual loss measure. The last term in (7.73) and the two terms in (7.74) related to the discriminator D_η are set to the logarithmic form as in the standard GAN (Goodfellow *et al* 2014), rather than the least-square losses used in (7.70) and (7.72). The parameters θ and η are again updated alternately until convergence, and then G_θ with the final θ is a feedforward network that maps any zero-filled image u_a to an artifact-free reconstruction $G_\theta(u_a)$.

7.3 Methods for advanced MRI technologies

7.3.1 Dynamic MRI

Dynamic MRI is an advanced MR imaging technique that exploits both spatial and temporal profiles of anatomy, which has extensive applications such as perfusion studies and cardiovascular imaging. A main feature of dynamic MRI is that there are a series of images acquired at a sequence of T time points $t = 1, \ldots, T$, so that the motion of the anatomy over time can be tracked to provide more diagnostic information in contrast to the standard static MRI.

Dynamic MRI data contain redundancy information between timeframes, which suggests great potential for aggressive undersampling and fast imaging. The main question in dynamic MRI is how to effectively incorporate organ/patient motions and temporal correlations into the reconstruction process. Therefore, the temporal redundancy and image correlations are the key features to be exploited in a reconstruction algorithm.

Let $u = [u_1, \ldots, u_T] \in \mathbb{C}^{n \times T}$ be a sequence of images in a dynamic MRI scan, where n is the number of pixels in each frame of u_t for $t = 1, \ldots, T$. Similar to CS-MRI, the k-space of u is highly undersampled for fast imaging. The undersampling pattern in the k-space is often set to be identical for different timeframes. The goal is to reconstruct u from the undersampled dynamic MRI data $f = [f_1, \ldots, f_T] \in \mathbb{C}^{m \times T}$ where $f_t \in \mathbb{C}^m$ is the undersampled k-space data with m sampled Fourier coefficients of image u_t.

Recurrent neural networks (RNN) are a class of network architectures to exploit temporal dynamics in streaming data. In Qin *et al* (2019), a deep reconstruction network was proposed to leverage the convolutional RNN (CRNN) architecture to recover dynamic MR images from undersampled k-space data. Similar to the network of cascaded CNNs (Schlemper *et al* 2018), the CRNN component plays the role of a de-aliasing operator, followed by the DC layer that incorporates sampled k-space data in each iteration. Therefore, the complete architecture is the same as the previous cascade network, and the difference is in the CRNN which replaces the CNN with a data sharing (DS) layer to exploit the correlation between timeframes in dynamic MRI.

Let $\Phi_{\mathrm{CR}}(\cdot; \theta_k)$ denote the (residual) CRNN component with all necessary parameters denoted by θ_k in the kth phase, then the update of $u^{(k)}$ can be formulated as

$$u^{(k)} = \mathrm{DC}(u^{(k-1)} + \Phi_{\mathrm{CR}}(u^{(k-1)}; \theta_k); \mu_k), \tag{7.75}$$

where the DC layer $\mathrm{DC}(\cdot; \mu_k)$ is the same linear DC operator as in (7.67). The key of (7.75) is the design of the CRNN component Φ_{CR}, which we will explain in detail below.

The main goal of the CRNN component $\Phi_{\mathrm{CR}}(\cdot; \theta_k)$ is to exploit the temporal relations between adjacent timeframes. The CRNN component $\Phi_{\mathrm{CR}}(\cdot; \theta_k)$ concatenates three subnetworks in order: the bidirectional CRNN (BCRNN-t-i), the

CRNN (CRNN-i), and a standard CNN. Here, BCRNN-t-i indicates that the network establishes connections between adjacent timeframes (t) and iterations (i), i.e. $u_t^{(k)}$ is connected to both $u_{t-1}^{(k)}$ and $u_t^{(k-1)}$ for all $k = 1, \ldots, K$ and $t = 1, \ldots, T$. Similarly, CRNN-i means that connections are only established between adjacent iterations, i.e. $u_t^{(k)}$ and $u_t^{(k-1)}$ for all k. Finally, CNN is applied for each specific $u_t^{(k)}$ without time and iteration connections.

In Qin *et al* (2019), the input $u^{(k-1)} \in \mathbb{C}^{n \times T}$ of the CRNN component $\Phi_{\mathrm{CR}}(\cdot; \theta_k)$ first passes through the BCRNN-t-i layers, then the CRNN-i layers, and finally the CNN layers. The output $h_{l,t}^{(k)}$ of the lth layer of BCRNN-t-i in the tth time frame of the kth phase is obtained by

$$\overrightarrow{h}_{l,t}^{(k)} = \sigma\left(W_l h_{l-1,t}^{(k)} + W_t \overrightarrow{h}_{l,t-1}^{(k)} + W_k h_{l,t}^{(k-1)} + \overrightarrow{b}_l \right), \tag{7.76a}$$

$$\overleftarrow{h}_{l,t}^{(k)} = \sigma\left(W_l h_{l-1,t}^{(k)} + W_t \overleftarrow{h}_{l,t+1}^{(k)} + W_k h_{l,t}^{(k-1)} + \overleftarrow{b}_l \right), \tag{7.76b}$$

$$h_{l,t}^{(k)} = \overleftarrow{h}_{l,t}^{(k)} + \overrightarrow{h}_{l,t}^{(k)}, \tag{7.76c}$$

where $h_{0,t}^{(k)}, \overleftarrow{h}_{0,t}^{(k)}$, and $\overrightarrow{h}_{0,t}^{(k)}$ are set to the input $u_t^{(k-1)}$, and W_l, W_t, and W_k represent the convolutions applied to the inputs from the previous layer $h_{l-1}^{(k)}$, the adjacent timeframes $\overrightarrow{h}_{l,t-1}^{(k)}$ and $\overleftarrow{h}_{l,t+1}^{(k)}$, and the previous iterate $h_{l,t}^{(k-1)}$, respectively. After passing through all the BCRNN-t-i layers, the output is fed to the CRNN-i layers. The output of the lth CRNN-i layer is obtained by

$$h_{l,t}^{(k)} = \sigma\left(W_l h_{l-1,t}^{(k)} + W_k h_{l,t}^{(k-1)} + b_l \right) \tag{7.77}$$

for each $t = 1, \ldots, T$ independently. Finally, after passing through all the CRNN-i layers, the output is fed to the CNN layers each of which is expressed by

$$h_{l,t}^{(k)} = \sigma\left(W_l h_{l-1,t}^{(k)} + \tilde{b}_l \right). \tag{7.78}$$

After passing through all these CNN layers, we obtain $\Phi_{\mathrm{CR}}(u^{(k-1)}; \theta_k) \in \mathbb{C}^{n \times T}$, the final output of the CRNN component in the kth phase. All the learnable parameters, namely, W_l, W_t, W_k, $\overrightarrow{b}_l$, $\overleftarrow{b}_l$, b_l, and $\tilde{b}_l$, are collectively called θ_k, although they are all shared by all the K phases in Qin *et al* (2019). This $\Phi_{\mathrm{CR}}(u^{(k-1)}; \theta_k)$ is used as the residual and added to $u^{(k-1)}$, which will pass the linear DC layer with parameter $\mu_k > 0$ to generate $u^{(k)}$, as shown in (7.75). The loss function is set to the standard squared error of the reconstructed $u^{(K)}$ from the reference image u_{true} as

$$L(\Theta; f, u_{\mathrm{true}}) = \frac{1}{2}\|u^{(K)}(f; \Theta) - u_{\mathrm{true}}\|^2, \tag{7.79}$$

where $(f, u_{\text{true}}) \in \mathbb{C}^{m \times T} \times \mathbb{C}^{n \times T}$ is a pair in the training dataset, and $\Theta = \{\theta_k \mid k = 1, \dots, K\}$ is the set of network parameters to be learned during training. The total loss function is the sum of (7.79) over all training pairs. The gradient of L in (7.79) with respect to Θ can be carried out in a straightforward manner according to the direct updating rule (7.75) and the formulation of the BCRNN-t-i layers (7.76), CRNN-t layers (7.77), and CNN layers (7.78). Figure 7.8 shows an example of reconstructed MR images after the first to tenth phases, respectively, by the CRNN Qin *et al* (2019). More details about the dataset and experimental set-up can be found in Qin *et al* (2019).

7.3.2 MR fingerprinting

Magnetic resonance fingerprinting (MRF) (Ma *et al* 2013) is a recently developed MR imaging technique that quantitatively measures tissue composition by matching signal dynamic behaviors to specific types of body tissues, in reference to a given dictionary. In MRF, the magnetic excitation and data acquisition settings are prescribed and fixed, such as the main magnetic field B_0, the number of readout times $T \in \mathbb{N}$, the flip angles $\{\alpha_t \mid t = 1, \dots, T\}$, the repetition times $\{\text{TR}_t \mid t = 1, \dots, T\}$, etc. Under such prescribed settings, the tissue in a pixel (voxel in 3D), characterized by its net magnetization parameters $\eta \triangleq (\text{T1, T2, PD}, \dots) \in \mathbb{R}^p$, induces an informative temporal signal $x \in \mathbb{C}^T$ governed by the Bloch equation with η as the parameter, where p is the prescribed number of nucleus parameters to infer in MRF, e.g. if T1, T2, PD are the quantities of interests, then $p = 3$.

To infer $\eta \in \mathbb{R}^p$ from a temporal signal $x \in \mathbb{C}^T$, the original MRF reconstruction scheme (Ma *et al* 2013) relies on a dictionary (i.e. a look-up table) of temporal signal and parameter pairs, denoted by $D = \{(d_l, \eta_l) \in \mathbb{C}^T \times \mathbb{R}^p \mid l = 1, \dots, |D|\}$. The size $|D|$ of the dictionary D is determined by the number of configurations of η. For example, if $p = 3$ and $\eta = (\text{T1, T2, PD})$, then η enumerates all possible combinations of T1, T2, and PD which form a subset of $\mathbb{R}^3$. Other parameters, such as susceptibility and chemical shift, can be added as well. Then, given an observed

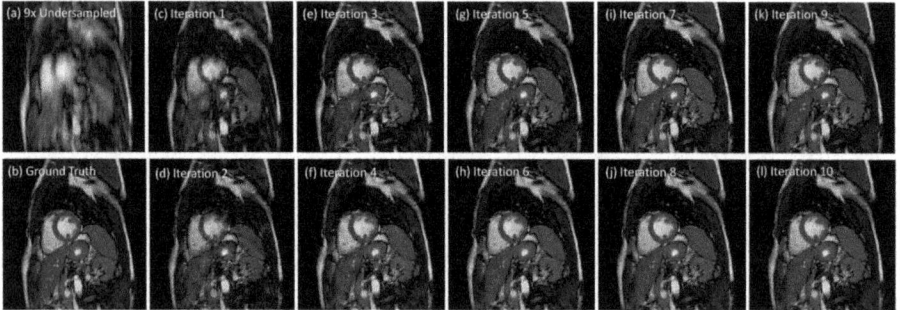

Figure 7.8. An example reconstruction of dynamic MR image by CRNN (Qin *et al* 2019). (a) Zero-filling reconstruction from undersampled *k*-space data with acceleration factor 9 (11% sampling ratio). (b) Ground truth reference image from full *k*-space data, (c)–(l) results after the first to tenth phases of CRNN reconstruction network using the undersampled *k*-space data. Reproduced with permission from (Qin *et al* 2019). Copyright 2019 IEEE.

$x \in \mathbb{C}^T$, one can identify its corresponding magnetization parameter $\eta_{l^*} \in \mathbb{R}^p$ by finding the atom d_{l^*} closest to x in D, where $l^* = \arg\min_{l=1,\ldots,|D|}\|x - d_l\|^2$. This process is an analog to fingerprint identification in forensic science, and hence is called MR fingerprinting.

To formulate the inverse problem of MRF, we first introduce an abstract mapping $B: \mathbb{R}^p \to \mathbb{C}^T$ to denote the solution of the Bloch equation with parameter η. That is, $x = B(\eta) \in \mathbb{C}^T$ is the solution of the Bloch equation with parameters η. As mentioned above, the mapping B also depends on the prescribed excitation and acquisition parameters (e.g. α_t, TR$_t$, etc), which we omitted for simplicity of notation since they are fixed under a specific setting of MRF scan and reconstruction. Let $u \in \mathbb{C}^{n \times T}$ be the magnetization image sequence acquired at the T readout times, where n is the number of pixels of u at each time t. Furthermore, we use $u_{:,t} \in \mathbb{C}^n$ to denote the tth readout of the whole image (reshaped into a vector of dimension n as usual), and $u_{i,:} \in \mathbb{C}^T$ to denote the magnetization response sequence at pixel i. Let $\eta_i \in \mathbb{R}^p$ be the net magnetization parameter at pixel i, then we can formulate the relation between data $u_{i,:}$ and η_i as follows:

$$u_{i,:} = B(\eta_i) \in \mathbb{C}^T, \quad \text{for } i = 1,\ldots, n. \tag{7.80}$$

On the other hand, the image u is related to the undersampled k-space data $f = \{f_t \in \mathbb{C}^{m_t} \mid t = 1,\ldots, T\}$ as

$$f_t = P_t \mathcal{F} u_{:,t} \in \mathbb{C}^{m_t}, \quad \text{for } t = 1,\ldots, T, \tag{7.81}$$

where $P_t \in \mathbb{R}^{m_t \times n}$ is the k-space undersampling mask at the tth readout time. For ease of presentation, we ignored the k-space noise in f in (7.81) but they should also be taken into consideration in practice. The inverse problem of MRF is to recover magnetization parameters $\eta = \{\eta_i \mid i = 1,\ldots, n\}$ from undersampled k-space data f, which are related by coupling the temporal evolution (7.80) and spatial information (7.81) of u.

Most of the existing MRF reconstruction methods tackle the inverse problem of MRF by solving (7.81) and then (7.80) in order. To solve u from f in (7.81), one can employ any CS-MRI reconstruction algorithm we introduced earlier. Then, the parameter η_i at each pixel i can be obtained by an 'inverse mapping' Φ of B such at $\eta_i = \Phi(u_{i,:})$ for $i = 1,\ldots, n$ point-wise. The flowchart in figure 7.9 shows this strategy.

In the original MRF (Ma et al 2013), the image reconstruction of (7.81) is obtained by simple zero-filling $u_{:,t} = \mathcal{F}^T P_t^T f_t$, and the inverse Bloch equation mapping Φ of (7.80) is obtained by dictionary matching point-wise as mentioned above:

$$\Phi(u_{i,:}) = \eta_{l_i}, \quad \text{where } l_i = \arg\min_{1 \leqslant l \leqslant |D|} \|u_{i,:} - d_l\|^2 \tag{7.82}$$

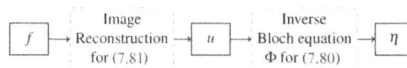

Figure 7.9. Flowchart of MRF reconstruction strategy by solving (7.81) and (7.80) in order.

for $i = 1, \ldots, n$. If the atoms d_l of D are not normalized, then l_i in (7.82) is obtained by minimizing $\|u_{i,:} - d_l\|^2 / \|d_l\|^2$ instead, where the numerator can be rewritten as $\|u_{i,:}\|^2 - 2\operatorname{Re}(\langle u_{i,:}, d_l\rangle) + \|d_l\|^2$, where $\|z\|^2 \triangleq \langle z, z\rangle$ is the standard norm of vector $z \in \mathbb{C}^n$, $\langle z, w\rangle \triangleq z^\top \bar{w} = \sum_{i=1}^n z_i \bar{w}_i$ is the standard inner product of two complex vectors $z, w \in \mathbb{C}^n$, and $\operatorname{Re}(\cdot)$ takes the real part of its complex-valued argument. Therefore, l_i in (7.82) is obtained by solving an equivalent problem

$$l_i = \underset{1 \leqslant l \leqslant |D|}{\arg\max} \frac{\operatorname{Re}(\langle u_{i,:}, d_l\rangle)}{\|d_l\|_2^2}, \tag{7.83}$$

which is how the dictionary matching is performed in the original MRF (Ma *et al* 2013).

MRF has several advantages over the standard MRI. First, MRF is an efficient data acquisition technique by varying excitation parameters α_t and TR$_t$, as well as the k-space undersampling mask $P_t \in \mathbb{R}^{m_t \times n}$, at a different time t, so the scan time can be significantly reduced. Second, by leveraging the temporal dynamics of Bloch equations featured by different magnetization parameters, it is possible to recover images faithfully with very limited amount of data. Third, MRF is capable of quantitatively inferring all magnetization parameters in η from k-space data in a single scan. For example, using a single-shot spiral trajectory with undersampling, one can acquire T1, T2, and PD maps of a whole-brain in less than 10 min (European-Society-of Radiology 2015).

However, MRF also comes with some challenging issues. For example, due to the short repetition times, the k-space data $f_t \in \mathbb{C}^{m_t}$ in each time frame must be highly undersampled $m_t \ll n$, from which it can be difficult to reconstruct $u_t \in \mathbb{C}^n$ with high accuracy. Furthermore, as we have seen above, the dictionary size $|D|$ depends on the number p of magnetization parameters in η. Therefore, $|D|$ increases exponentially fast in p as $|D|$ corresponds to all possible points of η in the discretization of space $\mathbb{R}^p$. A large MRF dictionary demands more hardware storage, and more seriously, a much longer computational time for finding the dictionary atom to match a temporal signal, which needs to be performed for all pixels (voxels). Therefore, the recent developments on MRF are dedicated to resolving these issues and make this technique more practical.

The main focus of the existing deep-learning-based MRF reconstruction methods is to replace the dictionary matching in (7.82) by a trained neural network. With a sufficient number of training data pairs of form (x, η), where $x = B(\eta)$, such a network can closely approximate Φ but requires may fewer computational resources for new data compared to the original dictionary matching approach (7.82) described in Ma *et al* (2013). More specifically, one can generate the training pairs by simulating the solutions $x = B(\eta) \in \mathbb{C}^T$ of the Bloch equations for various $\eta \in \mathbb{R}^p$. Then, a deep neural network is built and trained end-to-end to approximate the mapping from x to η. This is the common approach in the recent literature (Cohen *et al* 2018, Fang *et al* 2017, Golbabaee *et al* 2019, Hoppe *et al* 2017, Virtue *et al* 2017), which uses standard feedforward neural networks with fully connected layers but different choices of depth and width to mimic Φ in (7.82).

For example, in Golbabaee *et al* (2019), the MRF reconstruction network, termed MRF-Net, has five layers, including the input and output layers with matching widths T and p, respectively. The three hidden layers in between have sizes 10, 200, and 30, respectively. In Golbabaee *et al* (2019), the width of the first hidden layer is set to 10 as it is roughly the dimension of the manifold where the solutions of Bloch equations reside. However, often there are no particular reasons for the manually chosen width and depth of a neural network in the existing works. Nevertheless, the goal of the expensive offline training process is to encode the dictionary into the neural network Φ parameterized by Θ, so that the online storage of the dictionary D and the computational cost of matching in (7.82) can be significantly reduced, since only the storage of network parameters Θ in $\Phi(\cdot; \Theta)$ and the feedforward mapping are needed to identify the tissue types according to the new MRF signals. As the set-up and the training procedure are standard in the practice of deep neural networks, we omit the details here. Figure 7.10 shows an example reconstruction result of T_1 and T_2 maps by the MRF-Net on an *in vivo* MR brain image of 22.5×22.5 cm^2 FOV, 256×256 voxel spatial resolution with 5 mm slice thickness using a variable density spiral sampling with 89 interleaves. A baseline MRF dictionary with $|D| = 113\,781$ atoms is simulated on the fine grid of $T_1 \times T_2$ over $[100, 4000] \times [20\,600]$, each in milliseconds (msec). In this example, the MRF-Net reconstruction is over 60 times faster than the dictionary matching (DM) for this two-parameter (T_1 and T_2) MRF dataset. More details about the experimental set-up and result can be found in Golbabaee *et al* (2019).

The field of MRF reconstruction is still undergoing fast development. Deep learning is a promising tool and will play an important role in the advancement of MRF reconstruction techniques. So far, the image reconstruction (7.81) and the parameter inference (7.80) are still considered separately in the literature, and the main focus of the existing deep-learning-based MRF is on the latter. Leveraging the advanced image reconstruction techniques we introduced earlier can significantly reduce aliasing artifacts and improve the reconstruction quality for (7.81) compared to the zero-filling reconstruction or principal component analysis (PCA)-based approaches in the existing work. Furthermore, a more promising approach for MRF reconstruction is to integrate the image reconstruction (7.81) and the inference (7.80) into a unified reconstruction process to fully exploit the spatial–temporal information in these two coupled equations.

7.3.3 Synergized pulsing-imaging network

The synergized pulsing-imaging network (SPIN) (Lyu *et al* 2018) is a framework recently proposed to integrate the pulse sequence design for k-space data sampling and the image reconstruction into a streamlined and optimized workflow. In the SPIN scheme, the data sampling parameters and the parameters in the reconstruction network are learned jointly in the training process. The data sampling parameters include the angle-flip radiofrequency pulses α_t and the gradient signals $G_{x,t}$ and $G_{y,t}$ for each time interval $t = 1, \ldots, T$. These parameters determine how the k-space data are acquired. The parameters of the reconstruction network are

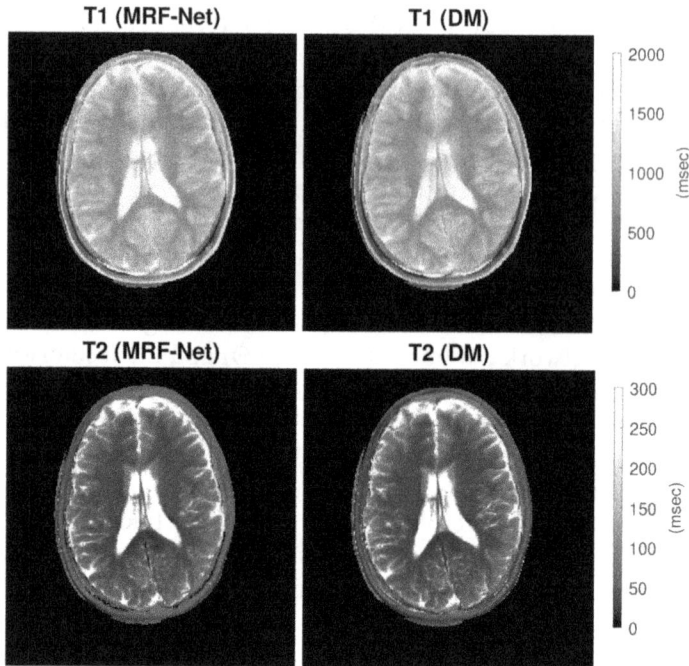

Figure 7.10. An example reconstruction of T_1 (top) and T_2 (bottom) maps using MRF-Net (left) and its comparison to dictionary matching (DM) baseline (right) Golbabaee *et al* (2019). Color bars indicate the quantitative values of the T_1 and T_2 maps. Reproduced with permission from (Golbabaee *et al* 2019). Copyright 2019 IEEE.

standard as in the deep-learning literature such that the feedforward mapping with the optimized parameters can generate high-quality images from significantly undersampled k-space data using the acquisition parameters mentioned above. The structure of the SPIN is illustrated in figure 7.11.

To optimize the two sets of parameters in SPIN, an alternating minimization scheme can be employed to optimize one set of the parameters while keeping the other set fixed (Lyu *et al* 2018). More specifically, with an initial set of acquisition parameters fixed, the parameters of the network for image reconstruction can be learned using the SGD algorithm to minimize a proper loss function, which can be, for example, set to a weighted sum of root mean squared errors (between the reconstructed image and the reference image from a full k-space data). We denote the final loss function value obtained in this update by L_0. With the optimal network parameters, the reconstruction network can directly map an undersampled k-space data (or its zero-filled reconstruction) to a reconstructed image of decent quality. Then, the set of k-space acquisition parameters can be in turn updated. To that end, each component in the first set of parameters is perturbed by a small quantity δ to estimate the partial derivative of the loss function with respect to that parameter. For example, one can perturb the first RF pulse flip angle α_1 to obtain a new one $\alpha_1' = \alpha_1 + \delta$, and the reconstruction network can be re-trained to obtain a new

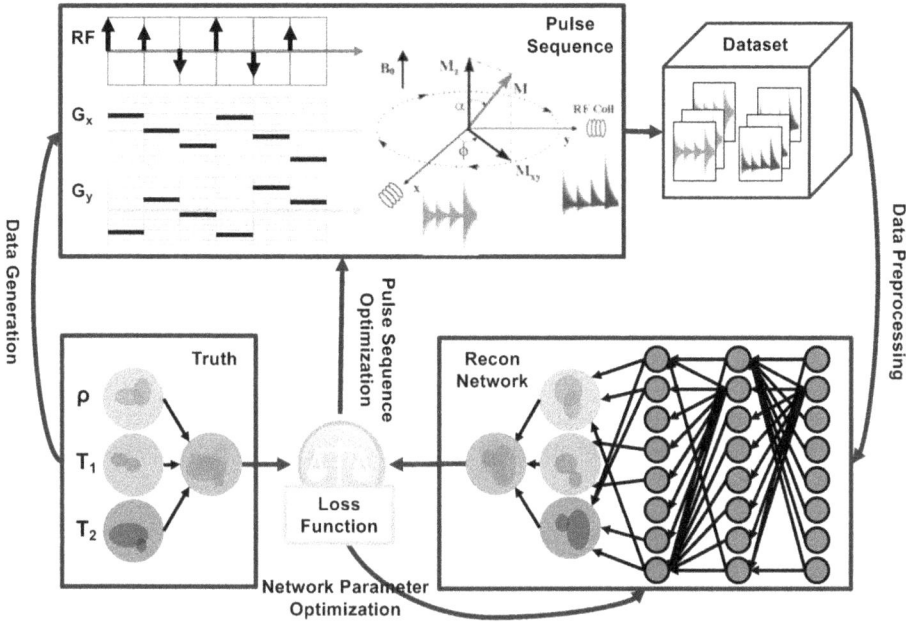

Figure 7.11. Flowchart of the SPIN (Lyu *et al* 2018). Given a training set of ground truth images ('Truth' box), one can generate partial *k*-spaced data ('Dataset' box) using specific acquisition parameters, such as flip-angle RF and excitation field gradients G_x and G_y ('Pulse Sequence' box), to be optimized. On the other hand, the partial *k*-space datasets are fed into the reconstruction network ('Recon Network' box) with network parameters to be learned to output reconstructed images. The loss function, set to measure the difference between the reconstructed images and the corresponding ground truth reference images using a proper metric such as root mean squared error, etc, contains both the acquisition parameters in the 'Pulse Sequence' box and the reconstruction parameters in the 'Recon Network' box. The two sets of parameters of the loss function can be optimized in an alternating fashion.

optimal loss function value L_1. Then, the partial derivative of the joint loss function with respect to α_1 can be approximated by

$$\frac{\partial F}{\partial \alpha_1} = \frac{L_1 - L_0}{\delta}. \tag{7.84}$$

The same process can be employed for other acquisition parameters $G_{x,t}$, $G_{y,t}$, etc. Finally, the acquisition parameters can be updated via gradient descent search:

$$\alpha_t \leftarrow \alpha_t - \eta \frac{\partial F}{\partial \alpha_t}, \quad G_{x,t} \leftarrow G_{x,t} - \eta \frac{\partial F}{\partial G_{x,t}}, \quad G_{y,t} \leftarrow G_{y,t} - \eta \frac{\partial F}{\partial G_{y,t}}, \tag{7.85}$$

for each time interval $t = 1, \dots, T$, where $\eta > 0$ is a step size also referred to as a learning rate, which can vary for different *k*-space acquisition parameters. These two steps are repeated until an overall convergence criterion is met.

A main intention behind the work (Lyu *et al* 2018) is to bypass the Fourier formulation, which is only approximate but has been the foundation for MRI reconstruction, and optimize the whole MRI process by jointly optimizing both acquisition parameters and reconstruction parameters in a unified machine learning framework. The overall computational complexity is rather high if a brute-force implementation is used, due to the repeated training, complicated computational tasks involved, and the slow convergence, in particular when the numbers of such parameters and the time intervals are large. However, the SPIN points to a promising direction that the most effective and accurate MRI may be in principle achieved when both acquisition and reconstruction are taken into consideration. To achieve this goal, there are many open questions, including the selection of acquisition parameters, the architecture of deep reconstruction networks, efficient numerical methods, and high-performance computing power such as quantum computing. Moreover, the idea of SPIN can be extended to other imaging modalities and their combinations, such as ultrasound imaging.

7.4 Miscellaneous topics*

7.4.1 Optimization with complex variables and Wirtinger calculus

A unique feature of MRI compared to other imaging modalities is that MRI data are acquired in the k-space or the Fourier domain. Although the magnetization parameters, such as T1, T2, PD, are all non-negative real numbers, intermediate and final MRI reconstructions are often complex-valued due to k-space undersampling, patient motion, data noise, system instability, computational errors, etc. Therefore, we often encounter optimization problems involving complex-valued variables throughout the algorithmic design and actual implementation for MRI image reconstruction.

The common practice in most existing deep MRI reconstruction methods is to treat every complex number $z = x + iy \in \mathbb{C}$ as a vector $(x, y) \in \mathbb{R}^2$. For example, as the basic building block of deep neural networks, each layer performs a nonlinear mapping in the form

$$g(z) = \sigma(w^{\mathsf{T}}z + b), \tag{7.86}$$

where $z = x + iy \in \mathbb{C}^n$, and $x, y \in \mathbb{R}^n$. The nonlinearity of g is is introduced by the activation σ, for which the standard choices, such as sigmoid, tanh, ReLU, etc, only accept real-valued inputs. To ensure the liner mapping $w^{\mathsf{T}}z + b$ supplies real-valued inputs for σ, the weight w is set to two channels w_x and w_y, both in $\mathbb{R}^n$, for the real and imaginary part of z, respectively, and $b \in \mathbb{R}$, such that

$$w^{\mathsf{T}}z + b = w_x^{\mathsf{T}}x + w_y^{\mathsf{T}}y + b. \tag{7.87}$$

In this case, the activation σ in (7.86) still receives real-valued inputs due to (7.87).

The separation of the real and imaginary parts of a complex-valued input $z \in \mathbb{C}^n$ is simple to implement, with a doubled number of neurons in the network which does not incur much difficulty in computation. However, the main issue with such a treatment is the loss of phase information, which can be critical to recover high-quality MRI images. To retain phase information, all the network parameters as well as the activation functions need to accommodate complex values. However, the standard calculus for real variables often does not apply to complex ones, and all the gradient related issues, such as partial derivatives, product rules, and chain rules, need to be rigorously carried out in the complex domain.

The exact definition and practice with complex calculus are studied in the field of complex analysis, an important branch of mathematical analysis. A significant difference between complex calculus and the commonly known real calculus is that a function $f: \mathbb{C} \to \mathbb{C}$, defined by $f(z) = u(x, y) + iv(x, y)$ for $z = x + iy$ with $x, y \in \mathbb{R}$ and $u, v: \mathbb{R}^2 \to \mathbb{R}$, is considered differentiable in the context of the classical complex analysis if and only if u and v satisfy the Cauchy–Riemann equation:

$$\frac{\partial u(x, y)}{\partial x} = \frac{\partial v(x, y)}{\partial y} \quad \text{and} \quad \frac{\partial v(x, y)}{\partial x} = -\frac{\partial u(x, y)}{\partial y}. \tag{7.88}$$

The Cauchy–Riemann equation (7.88) is a very strong condition. In fact, with (7.88), the function f is called holomorphic, or analytic, such that f can be differentiated infinitely many times, and the power series at every point is convergent. This is in sharp contrast to the derivative of a real-valued function with real variables, where the highest order of derivatives is defined.

However, the complex calculus mentioned above renders many functions in real-world applications (including those for CS-MRI) not differentiable in the context of the standard complex analysis. For example, it is easy to verify that (7.88) fails to hold for any real-valued function $f: \mathbb{C} \to \mathbb{R}$ with complex-valued input (unless $f(z) \equiv c$ for some constant $c \in \mathbb{R}$), and hence f is not differentiable. However, many functions, in particular the loss functions, in CS-MRI and deep-learning-based image reconstruction return values in $\mathbb{R}$ which have order ('<' relation between two real numbers) defined for any optimization to make sense. For example, consider the simple generic neural network with a single layer in the form of (7.86) but with (possibly complex-valued) activation σ, the least-square loss function for matching $g(z) = \sigma(w^\mathsf{T} z + b) \in \mathbb{C}$ in (7.86) and a given $u + iv \in \mathbb{C}$ is defined by

$$L(\Theta) = \frac{1}{2}|(u + iv) - g(z)|^2 = \frac{1}{2}|(u + iv) - \sigma(w^\mathsf{T} z + b)|^2, \tag{7.89}$$

where $u, v \in \mathbb{R}$ are given, $|z| \triangleq \sqrt{z\bar{z}} = \sqrt{x^2 + y^2}$ is the modulus of a complex number $z = x + iy \in \mathbb{C}$, and $\Theta \triangleq (w, b) \in \mathbb{C}^n \times \mathbb{C}$ is the parameter of this layer. Hence, $L(\Theta)$ does not satisfy the Cauchy–Riemann function (7.88) and is thus considered not differentiable in the context of complex analysis.

To overcome the issue with the strict requirement in complex analysis, we can adopt the so-called Wirtinger calculus which is between real calculus and complex analysis (Brandwood 1983, Kreutz-Delgado 2009, Wirtinger 1927). To this end, Wirtinger calculus considers a function $f(z) = u(x, y) + iv(x, y)$ differentiable as long as the partial derivatives of u and v with respect to x and y exist, and the derivative is defined by

$$\frac{\partial f}{\partial z} \triangleq \frac{1}{2}\left(\frac{\partial f}{\partial x} - i\frac{\partial f}{\partial y}\right) \quad \text{and} \quad \frac{\partial f}{\partial \bar{z}} \triangleq \frac{1}{2}\left(\frac{\partial f}{\partial x} + i\frac{\partial f}{\partial y}\right), \tag{7.90}$$

where $\partial f/\partial x \triangleq \partial u/\partial x + i\partial v/\partial x$ and $\partial f/\partial y \triangleq \partial u/\partial y + i\partial v/\partial y$. By the definitions in (7.90), we can derive that

$$\frac{\partial f}{\partial z} = \frac{1}{2}\left(\left(\frac{\partial u}{\partial x} + \frac{\partial v}{\partial y}\right) + i\left(\frac{\partial v}{\partial x} - \frac{\partial u}{\partial y}\right)\right) \quad \text{and}$$

$$\frac{\partial f}{\partial \bar{z}} = \frac{1}{2}\left(\left(\frac{\partial u}{\partial x} - \frac{\partial v}{\partial y}\right) + i\left(\frac{\partial v}{\partial x} + \frac{\partial u}{\partial y}\right)\right). \tag{7.91}$$

Note that if f is differentiable in the standard complex analysis, i.e. satisfies the Cauchy–Riemann equation (7.88), then (7.91) reduces to

$$\frac{\partial f}{\partial z} = \frac{\partial u}{\partial x} + i\frac{\partial v}{\partial x} \quad \text{and} \quad \frac{\partial f}{\partial \bar{z}} = 0, \tag{7.92}$$

where the former equation coincides with the one in the standard complex analysis, and the latter indicates that f, regarded as a function of $(z, \bar{z})$ instead of (x, y), is independent of $\bar{z}$. Therefore, the Wirtinger calculus can be considered as a relaxation of the standard complex calculus, and it can be applied to functions which are not considered differentiable in the classical sense.

To use Wirtinger calculus for CS-MRI and deep neural networks, we also need a series of identities including the product rule, the quotient rule, and the chain rule, etc. First of all, the Wirtinger calculus yields the following identities:

$$\frac{\partial \bar{f}}{\partial \bar{z}} = \overline{\frac{\partial f}{\partial z}} \quad \text{and} \quad \frac{\partial \bar{f}}{\partial z} = \overline{\frac{\partial f}{\partial \bar{z}}}. \tag{7.93}$$

In particular, if $f: \mathbb{C} \to \mathbb{R}$ is real-valued, then $f = \bar{f}$ and hence (7.93) implies

$$\frac{\partial f}{\partial z} = \overline{\frac{\partial f}{\partial \bar{z}}}. \tag{7.94}$$

Furthermore, the chain rule for the composition $g \circ h$ of two complex-valued functions $g, h: \mathbb{C} \to \mathbb{C}$ is

$$\frac{\partial g \circ h}{\partial z} = \frac{\partial g}{\partial h}\frac{\partial h}{\partial z} + \frac{\partial g}{\partial \bar{h}}\frac{\partial \bar{h}}{\partial z} \quad \text{and} \quad \frac{\partial g \circ h}{\partial \bar{z}} = \frac{\partial g}{\partial h}\frac{\partial h}{\partial \bar{z}} + \frac{\partial g}{\partial \bar{h}}\frac{\partial \bar{h}}{\partial \bar{z}}. \tag{7.95}$$

In addition, the product rule is the same as in real calculus, and the quotient rule can be obtained similarly based on the chain rule (7.95).

With the definition of Wirtinger calculus and all the necessary calculus rules ready, we can derive the gradient of the loss function L with respect to its arguments.

We take the simple loss function (7.89) as an example and show the optimality condition of a minimizer w. Let us use the following notations:

$$h(w) = z^T w + b \qquad (7.96a)$$

$$g(w) = (\sigma \circ h)(w) = \sigma(z^T w + b) \qquad (7.96b)$$

$$e(w) = (u + iv) - g(w) = (u + iv) - \sigma(z^T w + b) \qquad (7.96c)$$

$$L(w) = |e(w)|^2 / 2 = e(w)\overline{e(w)}/2. \qquad (7.96d)$$

First of all, due to the fact that L is real-valued, the optimality condition of w can be represented using either of the following two identities, which are equivalent due to (7.94):

$$\frac{\partial L}{\partial w} = 0 \quad \text{and} \quad \frac{\partial L}{\partial \overline{w}} = 0. \qquad (7.97)$$

The derivative $\partial L/\partial w$ is given by the chain rule (7.95) as

$$\frac{\partial L}{\partial w} = \frac{\partial L}{\partial e}\frac{\partial e}{\partial w} + \frac{\partial L}{\partial \overline{e}}\frac{\partial \overline{e}}{\partial w} = \frac{1}{2}\left(\overline{e(w)}\frac{\partial e}{\partial w} + e(w)\frac{\partial \overline{e}}{\partial w}\right) = -\frac{1}{2}\left(\overline{e(w)}\frac{\partial g}{\partial w} + e(w)\frac{\partial \overline{g}}{\partial w}\right), \qquad (7.98)$$

where we used (7.96d) to obtain the second equality and (7.96b) to obtain the last equality. Furthermore, we can compute $\partial g/\partial w$ and $\partial \overline{g}/\partial w$ as follows:

$$\frac{\partial g}{\partial w} = \frac{\partial g}{\partial h}\frac{\partial h}{\partial w} + \frac{\partial g}{\partial \overline{h}}\frac{\partial \overline{h}}{\partial w} = \frac{\partial g}{\partial h}z^T \quad \text{and} \quad \frac{\partial \overline{g}}{\partial w} = \frac{\partial \overline{g}}{\partial h}\frac{\partial h}{\partial w} + \frac{\partial \overline{g}}{\partial \overline{h}}\frac{\partial \overline{h}}{\partial w} = \frac{\overline{\partial g}}{\partial \overline{h}}z^T \qquad (7.99)$$

since h is linear and hence holomorphic to w, which implies $\partial \overline{h}/\partial w = \overline{\partial h/\partial \overline{w}} = 0$. We temporally leave $\partial g/\partial h$ as is since it depends on the specific choice of activation function σ, which we will instantiate in the next subsection. Substituting (7.99) into (7.98), we obtain

$$\frac{\partial L}{\partial w} = -\frac{1}{2}\left(\overline{e(w)}\frac{\partial g}{\partial h} + e(w)\frac{\overline{\partial g}}{\partial \overline{h}}\right)z^T. \qquad (7.100)$$

Using the gradient (7.100), we can perform a gradient descent search for the optimal w that minimizes the loss function L in (7.89). If the neural network is composed of multiple layers, then z in (7.100) is the output of the second-to-last layer, and we can compute all involved gradients step by step in the standard backpropagation manner.

7.4.2 Activation functions with complex variables

As we know, the activation functions in deep learning are critical as they introduce nonlinearity into the approximation of a complicated function. There are numerous candidates for a nonlinear activation function σ. For example, the sigmoid function, the hyperbolic tangent (tanh) function, and the rectified linear unit (ReLU) are a few common choices of activation functions:

$$\text{sigmoid}(x) = \frac{1}{1 + e^x}, \quad \tanh(x) = \frac{e^{2x} - 1}{e^{2x} + 1},$$
$$\text{ReLU}(x) = \max(0, x) = (x)_+$$

(7.101)

for $x \in \mathbb{R}$. However, these activation functions are defined for real-valued inputs, and hence are not suitable for deep MRI reconstruction when the phase information is considered and the complex calculus needs to be carried out. For example, the sigmoid function $\text{sigmoid}(z) = 1/(1 + e^z)$ is not bounded at $z = i\pi$ since $e^{i\pi} = -1$, and the ReLU function $\text{ReLU}(z) = \max(0, z)$ is not defined for complex number $z \in \mathbb{C}$ which, unlike $\mathbb{R}$, lacks ordering between its elements.

To overcome this issue, several activation functions for complex-valued inputs were proposed (Amin and Murase 2009, Georgiou and Koutsougeras 1992, Nitta 1997, Virtue *et al* 2017). Now we introduce two of these activation functions which preserve the phase of the input. The first one is the sigmoid of the negative logarithm (siglog) function (Georgiou and Koutsougeras 1992), which is a modification of the standard sigmoid function such that the magnitude of an input is mapped according to the standard sigmoid function in (7.101) and the phase is unaltered. That is, the siglog function is given by

$$\text{siglog}(z) = \sigma(-\log r)e^{i\alpha} = \frac{re^{i\alpha}}{1 + r} = \frac{z}{1 + |z|},$$

(7.102)

where $z = re^{i\alpha}$ is a complex number with magnitude (modulus) $r \in \mathbb{R}_+$ and phase (angle) $\alpha \in [0, 2\pi)$. Note that the magnitude of z is scaled into $[0, 1]$ which is independent of the phase. The derivative of $\text{siglog}(z)$ with respect to z and $\bar{z}$ can be computed accordingly.

The second one is the the cardioid function (Virtue *et al* 2017). Unlike the siglog function which extends the sigmoid function and keeps the phase information of the input complex number untouched, the cardioid function tends to extend the ReLU but also takes its phase information into consideration. More specifically, the cardioid function is defined as follows:

$$\text{cardioid}(z) = \frac{1}{2}(1 + \cos\alpha)z = \frac{1}{2}\left(1 + \frac{z + \bar{z}}{2|z|}\right)z$$

(7.103)

for any $z = re^{i\alpha}$. One can easily check that $\text{cardioid}(z)$ reduces to ReLU in (7.101) for real-valued z since $\alpha = 0$ for positive z and $\alpha = \pi$ for negative z. The comparison of the activation functions (7.102) and (7.103) is illustrated in figure 7.12.

For each activation function, an important matter is to compute the derivative with respect to its input. If we use the cardioid function as activation σ to obtain

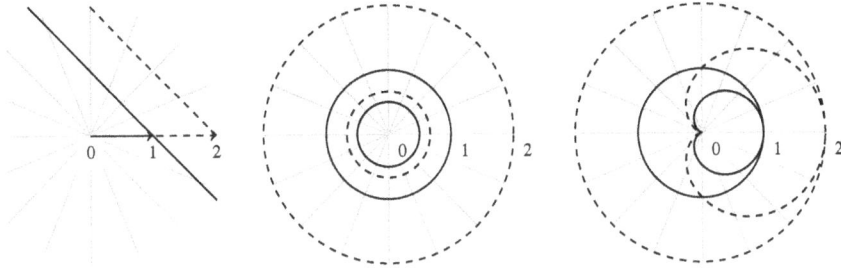

Figure 7.12. Comparison of activation functions in the complex plane $\mathbb{C}$. Gray curves indicate the value of the input complex number, and the corresponding black curves show the output value of the activation function. The solid and dashed gray-black pairs indicate two sets of complex numbers of different magnitudes. Left: ReLU($x + y$); middle: siglog(z); right: cardioid(z), for $z = x + \mathrm{i}y \in \mathbb{C}$. Note that the left function alters the phase of the input z, whereas the middle and the right ones do not.

$g(h) = \sigma(h)$ with complex-valued input h, then we can obtain the gradients of g with respect to $h = r\mathrm{e}^{\mathrm{i}\alpha}$ and its conjugate $\bar{h}$ as

$$\frac{\partial g}{\partial h} = \frac{1}{2}(1 + \cos\alpha) + \mathrm{i}\,\frac{\sin\alpha}{4} \quad \text{and} \quad \frac{\partial g}{\partial \bar{h}} = -\mathrm{i}\,\frac{\sin\alpha}{4}\frac{h}{\bar{h}}. \tag{7.104}$$

These are essentially the gradient $\partial g/\partial h$ and $\partial g/\partial \bar{h}$ we need in (7.99) and (7.100).

7.4.3 Optimal k-space sampling

One of the most important and practical problems in accelerated MR imaging is the k-space sampling strategy. In CS-MRI, the acceleration in imaging data acquisition is (nearly) the reduction factor of the sampling, i.e. the reciprocal of the under-sampling ratio of the k-space. For example, sampling 20% of the Fourier data yields 5 times faster data acquisition compared to the full k-space sampling. Under certain k-space data acquisition constraints, for a prescribed undersampling ratio r, it is important to design the most efficient sampling strategy in the k-space, i.e. the rn Fourier coefficients to acquire for an image with n pixels (also the number of Fourier coefficients in the k-space) such that the image can be recovered with highest faithfulness to the reference image reconstructed from full k-space data.

In the literature, there are a number of methods proposed to use specific shapes of undersampling patterns, such as the simple k_y direction subsampling in GRAPPA and SENSE, the spiral and radial trajectories, and Poisson density masks. With machine learning methods emerging in medical imaging research, we can consider to 'learn' an optimal sampling strategy for a specific imaging task provided that a sufficiently large training dataset is available. If the learning result suggests a specific sampling pattern (i.e. k-space mask), it is likely to work well for all images of the same type.

For this purpose, we need to specify the criterion of 'optimal sampling strategy'. The optimal strategy should depend on features of interest in the training set, as well as scan parameters, scanner characteristics, and the reconstruction method to be employed. Last but not least, it also depends on the measure or criterion for image quality evaluation. Therefore, it is reasonable and rigorous to define the optimal sampling strategy problem as follows: if a training set of images $\{u_{\text{true}}^{(i)} \in \mathbb{R}^n: 1 \leqslant i \leqslant N_{\text{train}}\}$ contains independent and identically distributed (i.i.d.) samples from the latent manifold of images to be reconstructed in $\mathbb{R}^n$, and one has decided to employ a specific reconstruction method φ and an image quality measure d, find the optimal k-space sampling pattern P with at most m (or r such that $m = rn$; either way r or m is prescribed) sampling locations, then the following minimization problem is to be solved:

$$\underset{|P| \leqslant m}{\text{minimize}} \, L(P), \quad \text{where} \; L(P) \triangleq \frac{1}{N_{\text{train}}} \sum_{i=1}^{N_{\text{train}}} d\left(\varphi\left(P\mathcal{F}u_{\text{true}}^{(i)}, P\right), u_{\text{true}}^{(i)}\right), \quad (7.105)$$

and $\varphi(f, P)$ is the reconstruction method (e.g. ISTA-Net, ADMM-Net, or Var-Net) with P and the corresponding partial k-space data f as an input and a reconstructed image u as an output, d is a pre-specified image quality measure (e.g. the squared error $d(u, u_{\text{true}}) = \|u - u_{\text{true}}\|^2$). Note that the physical constraint on the k-space sampling trajectory is not considered here, and P is only required to have at most m Fourier coefficients in $\mathbb{R}^n$.

The optimal sampling problem (7.105) is a combinatoric optimization problem, which is very challenging to solve in general. One of the commonly used solutions to this kind of problems is the greedy approach (Gözcü *et al* 2018). That is, one can start with an empty P, search for the most effective k-space location $p_1 \in \{1, \dots, n\}$ such that the objective function $L(P)$ in (7.105) is minimized with P determined by $\{p_1\}$ (in this case $|P| = 1$), and then add the second k-space location point $p_2 \in \{1, \dots, n\}$ that minimizes $L(P)$ for P determined by $\{p_1, p_2\}$ (hence $|P| = 2$), and so on, until m points $\{p_1, \dots, p_m\}$ are found. By this design, the search will terminate in m steps, and the k-space sampling pattern P determined by $\{p_1, \dots, p_m\}$ is the so-called greedy solution to (7.105). Although this greedy approach may not be able to find the globally optimal solution of (7.105), the practical performance is often satisfactory.

7.5 Further readings

Deep-learning-based MRI reconstruction is still an active research field. An overview of recent results in this field can be found in Lundervold and Lundervold (2019). Deep-learning techniques have been used in many other MR related imaging tasks beyond what has been described in this chapter. For example, image denoising (Benou *et al* 2017, Phophalia and Mitra 2017, Zhang *et al* 2015), parameter estimation (Dikaios *et al* 2014, Golkov *et al* 2016, Guo *et al* 2017, Gurbani *et al* 2018), end-to-end reconstruction (Hyun *et al* 2018, Jin *et al* 2017, Kwon *et al* 2017), artifact reduction (Küstner *et al* 2018, Kyathanahally *et al* 2018), classification (Amin and Murase 2009), quantitative susceptibility mapping (QSM) (Deistung *et al* 2013, 2017, Liu *et al* 2009, Rasmussen *et al* 2019, Yoon *et al* 2018), domain adaptation (Han *et al* 2018), domain transformation (Zhu *et al* 2018), super-resolution (Bahrami *et al* 2017, Chaudhari *et al* 2018, Liu *et al* 2018, Plenge *et al* 2012, Ropele *et al* 2010, Shi *et al* 2018, Shilling *et al* 2009, Van Steenkiste *et al* 2017,

Zeng *et al* 2018), segmentation (AlBadawy *et al* 2018, Bobo *et al* 2018, Cheng *et al* 2017, Cui *et al* 2018, Hoseini *et al* 2019, Le *et al* 2017, Perkuhn *et al* 2018, Wang *et al* 2017, Yang *et al* 2017, Yoo *et al* 2018), and data synthesis (Duchateau *et al* 2018, Kitchen and Seah 2017, Mok and Chung 2018, Nie *et al* 2017, Zhou *et al* 2018).

References

AlBadawy E A, Saha A and Mazurowski M A 2018 Deep learning for segmentation of brain tumors: impact of cross-institutional training and testing *Med. Phys.* **45** 1150–8

Amin M F and Murase K 2009 Single-layered complex-valued neural network for real-valued classification problems *Neurocomputing* **72** 945–55

Bahrami K, Shi F, Rekik I, Gao Y and Shen D 2017 7 T-guided super-resolution of 3 T MRI *Med. Phys.* **44** 1661–77

Benou A, Veksler R, Friedman A and Raviv T R 2017 Ensemble of expert deep neural networks for spatio-temporal denoising of contrast-enhanced MRI sequences *Med. Image Anal.* **42** 145–59

Bobo M F *et al* 2018 Fully convolutional neural networks improve abdominal organ segmentation *Medical Imaging 2018: Image Processing* **10574** 205724V

Brandwood D 1983 A complex gradient operator and its application in adaptive array theory *Proc. SPIE* **130** 11–16

Chaudhari A S, Fang Z, Kogan F, Wood J, Stevens K J, Gibbons E K, Lee J H, Gold G E and Hargreaves B A 2018 Super-resolution musculoskeletal MRI using deep learning *Magn. Reson. Med.* **80** 2139–54

Cheng R *et al* 2017 Automatic magnetic resonance prostate segmentation by deep learning with holistically nested networks *J. Med. Imaging* **4** 1

Cohen O, Zhu B and Rosen M S 2018 MR fingerprinting deep reconstruction network (DRONE) *Magn. Reson. Med.* **80** 885–94

Cui S, Mao L, Jiang J, Liu C and Xiong S 2018 Automatic semantic segmentation of brain gliomas from MRI images using a deep cascaded neural network *J. Healthcare Eng.* **2018** 1–14

Deistung A, Schäfer A, Schweser F, Biedermann U, Turner R and Reichenbach J R 2013 Toward *in vivo* histology: a comparison of quantitative susceptibility mapping (QSM) with magnitude-, phase-, and R_2*-imaging at ultra-high magnetic field strength *Neuroimage* **65** 299–314

Deistung A, Schweser F and Reichenbach J R 2017 Overview of quantitative susceptibility mapping *NMR Biomed.* **30** e3569

Dikaios N, Arridge S, Hamy V, Punwani S and Atkinson D 2014 Direct parametric reconstruction from undersampled (k, t)-space data in dynamic contrast enhanced MRI *Medical Image Anal.* **18** 989–1001

Duchateau N, Sermesant M, Delingette H and Ayache N 2018 Model-based generation of large databases of cardiac images: synthesis of pathological cine MR sequences from real healthy cases *IEEE Trans. Med. Imaging* **37** 755–66

European Society of Radiology 2015 Magnetic resonance fingerprinting—a promising new approach to obtain standardized imaging biomarkers from MRI *Insights Imaging* **6** 163–5

Fang Z, Chen Y, Lin W and Shen D 2017 Quantification of relaxation times in MR fingerprinting using deep learning *Proc. of the Int. Society for Magnetic Resonance in Medicine Scientific Meeting and Exhibition* https://www.ncbi.nlm.nih.gov/pmc/articles/PMC5909960/

Georgiou G M and Koutsougeras C 1992 Complex domain backpropagation *IEEE Trans. Circuits Syst. II* **39** 330–4

Golbabaee M, Chen D, Gómez P A, Menzel M I and Davies M E 2019 Geometry of deep learning for magnetic resonance fingerprinting *2019 IEEE Int. Conf. on Acoustics, Speech and Signal Processing (ICASSP 2019)* (Piscataway, NJ: IEEE) pp 7825–9

Golkov V, Dosovitskiy A, Sperl J I, Menzel M I, Czisch M, Sämann P, Brox T and Cremers D 2016 Q-space deep learning: twelve-fold shorter and model-free diffusion MRI scans *IEEE Trans. Med. Imaging* **35** 1344–51

Goodfellow I *et al* 2014 Generative adversarial nets *Advances in Neural Information Processing Systems* pp 2672–80

Gözcü B, Mahabadi R K, Li Y-H, Ilıcak E, Cukur T, Scarlett J and Cevher V 2018 Learning-based compressive MRI *IEEE Trans. Med. Imaging* **37** 1394–406

Guo Y, Lingala S G, Zhu Y, Lebel R M and Nayak K S 2017 Direct estimation of tracer-kinetic parameter maps from highly undersampled brain dynamic contrast enhanced MRI *Magn. Reson. Med.* **78** 1566–78

Gurbani S S *et al* 2018 A convolutional neural network to filter artifacts in spectroscopic MRI *Magn. Reson. Med.* **80** 1765–75

Hammernik K, Klatzer T, Kobler E, Recht M P, Sodickson D K, Pock T and Knoll F 2018 Learning a variational network for reconstruction of accelerated MRI data *Magn. Reson. Med.* **79** 3055–71

Han Y, Yoo J, Kim H H, Shin H J, Sung K and Ye J C 2018 Deep learning with domain adaptation for accelerated projection-reconstruction MR *Magn. Reson. Med.* **80** 1189–205

Hoppe E, Körzdörfer G, Würfl T, Wetzl J, Lugauer F, Pfeuffer J and Maier A 2017 Deep learning for magnetic resonance fingerprinting: a new approach for predicting quantitative parameter values from time series *Stud. Health Technol. Inform.* **243** 202–6 https://www.ncbi.nlm.nih.gov/pubmed/28883201

Hoseini F, Shahbahrami A and Bayat P 2019 AdaptAhead optimization algorithm for learning deep CNN applied to MRI segmentation *J. Dig. Imaging* **32** 105–11

Hyun C M, Kim H P, Lee S M, Lee S and Seo J K 2018 Deep learning for undersampled MRI reconstruction *Phys. Med. Biol.* **63** 135007

Jin K H, McCann M T, Froustey E and Unser M 2017 Deep convolutional neural network for inverse problems in imaging *IEEE Trans. Image Process.* **26** 4509–22

Kitchen A and Seah J 2017 Deep generative adversarial neural networks for realistic prostate lesion MRI synthesis, arXiv:1708.00129

Kreutz-Delgado K 2009 The complex gradient operator and the CR-calculus, arXiv:0906.4835

Küstner T, Liebgott A, Mauch L, Martirosian P, Bamberg F, Nikolaou K, Yang B, Schick F and Gatidis S 2018 Automated reference-free detection of motion artifacts in magnetic resonance images *Magn. Reson. Mater. Phys. Biol. Med.* **31** 243–56

Kwon K, Kim D and Park H 2017 A parallel MR imaging method using multilayer perceptron *Med. Phys.* **44** 6209–24

Kyathanahally S P, Döring A and Kreis R 2018 Deep learning approaches for detection and removal of ghosting artifacts in MR spectroscopy *Magn. Reson. Med.* **80** 851–63

Le M H, Chen J, Wang L, Wang Z, Liu W, Cheng K-T T and Yang X 2017 Automated diagnosis of prostate cancer in multi-parametric MRI based on multimodal convolutional neural networks *Phys. Med. Biol.* **62** 6497–514

Liu C, Wu X, Yu X, Tang Y, Zhang J and Zhou J 2018 Fusing multi-scale information in convolution network for MR image super-resolution reconstruction *Biomed. Eng. Online* **17** 114

Liu T, Spincemaille P, De Rochefort L, Kressler B and Wang Y 2009 Calculation of susceptibility through multiple orientation sampling (COSMOS): a method for conditioning the inverse problem from measured magnetic field map to susceptibility source image in MRI *Magn. Reson. Med.* **61** 196–204

Lundervold A S and Lundervold A 2019 An overview of deep learning in medical imaging focusing on MRI *Z. Med. Phys.* **29** 102–27

Lyu Q, Xu T, Shan H and Wang G 2018 A synergized pulsing-imaging network (SPIN), arXiv:1805.12006.

Ma D, Gulani V, Seiberlich N, Liu K, Sunshine J L, Duerk J L and Griswold M A 2013 Magnetic resonance fingerprinting *Nature* **495** 187

Mardani M, Gong E, Cheng J Y, Vasanawala S S, Zaharchuk G, Xing L and Pauly J M 2019 Deep generative adversarial neural networks for compressive sensing MRI *IEEE Trans. Med. Imaging* **38** 167–79

Mok T C and Chung A C 2018 Learning data augmentation for brain tumor segmentation with coarse-to-fine generative adversarial networks, arXiv:1805.11291

Nie D, Trullo R, Lian J, Petitjean C, Ruan S, Wang Q and Shen D 2017 Medical image synthesis with context-aware generative adversarial networks *Int. Conf. on Medical Image Computing and Computer-Assisted Intervention* (Berlin: Springer) pp 417–25

Nitta T 1997 An extension of the back-propagation algorithm to complex numbers *Neural Netw.* **10** 1391–415

Perkuhn M, Stavrinou P, Thiele F, Shakirin G, Mohan M, Garmpis D, Kabbasch C and Borggrefe J 2018 Clinical evaluation of a multiparametric deep learning model for glioblastoma segmentation using heterogeneous magnetic resonance imaging data from clinical routine *Invest. Radiol.* **53** 647–54

Phophalia A and Mitra S K 2017 3D MR image denoising using rough set and kernel PCA method *Magn. Reson. Imaging* **36** 135–45

Plenge E, Poot D H, Bernsen M, Kotek G, Houston G, Wielopolski P, van der Weerd L, Niessen W J and Meijering E 2012 Super-resolution methods in MRI: can they improve the trade-off between resolution, signal-to-noise ratio, and acquisition time? *Magn. Reson. Med.* **68** 1983–93

Qin C, Schlemper J, Caballero J, Price A N, Hajnal J V and Rueckert D 2019 Convolutional recurrent neural networks for dynamic MR image reconstruction *IEEE Trans. Med. Imaging* **38** 280–90

Rasmussen K G B *et al* 2019 DeepQSM-using deep learning to solve the dipole inversion for MRI susceptibility mapping *NeuroImage* **195** 373–83

Ropele S, Ebner F, Fazekas F and Reishofer G 2010 Super-resolution MRI using microscopic spatial modulation of magnetization *Magn. Reson. Med.* **64** 1671–5

Schlemper J, Caballero J, Hajnal J V, Price A N and Rueckert D 2018 A deep cascade of convolutional neural networks for dynamic MR image reconstruction *IEEE Trans. Med. Imaging* **37** 491–503

Shi J, Liu Q, Wang C, Zhang Q, Ying S and Xu H 2018 Super-resolution reconstruction of MR image with a novel residual learning network algorithm *Phys. Med. Biol.* **63** 085011

Shilling R Z, Robbie T Q, Bailloeul T, Mewes K, Mersereau R M and Brummer M E 2009 A super-resolution framework for 3-D high-resolution and high-contrast imaging using 2-D multislice MRI *IEEE Trans. Med. Imaging* **28** 633–44

Simonyan K and Zisserman A 2014 Very deep convolutional networks for large-scale image recognition, arXiv:1409.1556

Sun J *et al* 2016 Deep ADMN-NET for compressive sensing MRI *Advances in Neural Information Processing Systems* pp 10–18.

Van Steenkiste G, Poot D H, Jeurissen B, Den Dekker A J, Vanhevel F, Parizel P M and Sijbers J 2017 Super-resolution T_1 estimation: quantitative high resolution T_1 mapping from a set of low resolution T_1-weighted images with different slice orientations *Magn. Reson. Med.* **77** 1818–30

Virtue P, Stella X Y and Lustig M 2017 Better than real: complex-valued neural nets for MRI fingerprinting *2017 IEEE Int. Conf. on Image Processing (ICIP)* (Piscataway, NJ: IEEE) pp 3953–7

Wang X *et al* 2017 Searching for prostate cancer by fully automated magnetic resonance imaging classification: deep learning versus non-deep learning *Sci. Rep.* **7** 15415

Wirtinger W 1927 Zur formalen Theorie der Funktionen von mehr komplexen Veränderlichen *Math. Ann.* **97** 357–75

Yang G *et al* 2018 DAGAN: deep de-aliasing generative adversarial networks for fast compressed sensing MRI reconstruction *IEEE Trans. Med. Imaging* **37** 1310–21

Yang X, Liu C, Wang Z, Yang J, Min H L, Wang L and Cheng K-T T 2017 Co-trained convolutional neural networks for automated detection of prostate cancer in multi-parametric MRI *Med. Image Anal.* **42** 212–27

Yoo Y *et al* 2018 Deep learning of joint myelin and T1w MRI features in normal-appearing brain tissue to distinguish between multiple sclerosis patients and healthy controls *NeuroImage: Clinical* **17** 169–78

Yoon J *et al* 2018 Quantitative susceptibility mapping using deep neural network: QSMNET *NeuroImage* **179** 199–206

Zeng K, Zheng H, Cai C, Yang Y, Zhang K and Chen Z 2018 Simultaneous single-and multi-contrast super-resolution for brain MRI images based on a convolutional neural network *Comput. Biol. Med.* **99** 133–41

Zhang J and Ghanem B 2018 ISTA-Net: interpretable optimization-inspired deep network for image compressive sensing *Proc. of the IEEE Conf. on Computer Vision and Pattern Recognition* pp 1828–37

Zhang X, Xu Z, Jia N, Yang W, Feng Q, Chen W and Feng Y 2015 Denoising of 3D magnetic resonance images by using higher-order singular value decomposition *Med. Image Anal.* **19** 75–86

Zhou Y, Giffard-Roisin S, De Craene M, Camarasu-Pop S, D'Hooge J, Alessandrini M, Friboulet D, Sermesant M and Bernard O 2018 A framework for the generation of realistic synthetic cardiac ultrasound and magnetic resonance imaging sequences from the same virtual patients *IEEE Trans. Med. Imaging* **37** 741–54

Zhu B, Liu J Z, Cauley S F, Rosen B R and Rosen M S 2018 Image reconstruction by domain-transform manifold learning *Nature* **555** 487

Part IV

Others

IOP Publishing

Machine Learning for Tomographic Imaging

Ge Wang, Yi Zhang, Xiaojing Ye and Xuanqin Mou

Chapter 8

Modalities and integration

In the article entitled 'Perspective on deep imaging' (Wang 2016), we outlined a roadmap for the new area of deep-learning-based tomographic imaging for smart diagnosis and intervention, which is reproduced here as figure 8.1. This figure is quite simple but generally applicable to all imaging modalities and their combinations, including traditional image analysis as the red arrows, the new area of deep-learning-based tomographic reconstruction as the green arrows, and the potential frontiers to perform data-/image-driven intelligent medical interventions such as radiation therapy and robotic surgery as the blue arrow. The dashed arrows clearly suggest the end-to-end procedures in the machine learning framework, which are becoming hot topics in medical machine learning.

In chapters 4–7, two important imaging modalities, CT and MRI, have been covered in detail. In this chapter, we will describe the imaging principles and deep learning examples for several other important imaging modalities from nuclear emission tomography to ultrasound and optical imaging. Although these modalities are fundamentally different from CT and MRI, the deep-learning-based imaging

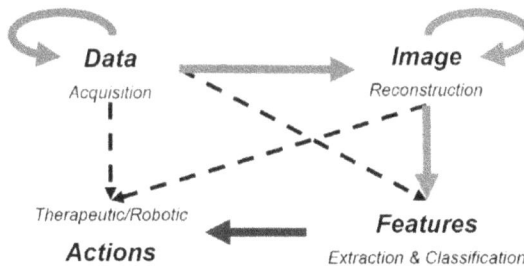

Figure 8.1. Big picture of deep imaging toward the full fusion of medical imaging and deep learning. There is a high likelihood that the direct paths from data to features and actions may need intermediate neural layers essentially equivalent to a reconstructed/processed image. Reproduced with permission from Wang (2016). Copyright 2016 IEEE.

ideas are very similar. Hence, we will be relatively brief in terms of their working principles and network-based implementations.

8.1 Nuclear emission tomography

Nuclear emission tomography works in two modes: positron emission tomography (PET) and single photon emission computed tomography (SPECT). CT and nuclear emission tomography have important similarities and differences, as listed in table 8.1. With a solid knowledge of CT imaging, the PET and SPECT imaging processes are not difficult to understand, and will be briefly explained in the following subsection. Then, recently developed neural networks for nuclear emission tomography will be described.

8.1.1 Emission data models

While the radiation source for CT is an x-ray tube, the radiation sources for nuclear emission tomography are radioactive tracers that are typically injected into a patient. These radioactive tracers are subject to radioactive decays, emitting gamma ray photons at energy levels similar to that of x-ray photons.

Two mechanisms are important for nuclear emission tomography, as shown in figure 8.2. In the first mechanism, a radionuclide emits a positron that will be quickly combined with a nearby electron, leading to a pair of gamma ray photons. This process is called annihilation. Fluorine-18 is a common example, producing gamma ray photons in pairs at 511 keV. For conservation of momentum, the paired gamma ray photons must travel in opposite directions. The biological tracer molecule for PET is fludeoxyglucose (FDG), an analog of glucose that moves with the

Table 8.1. Comparison between CT and PET/SPECT (more details in the following subsection).

Similarities	Differences (CT versus PET/SPECT)
Photon energy levels	External versus internal source
Line integral data	High versus low flux/resolution
Imaging geometry	Low versus high sensitivity
Reconstruction methods	Anatomical versus metabolic information

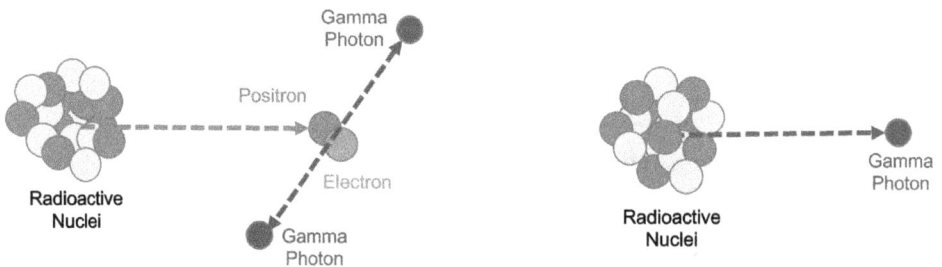

Figure 8.2. Gamma ray photons can be generated either in pairs (left) or singly (right) through positron emission and single gamma ray photon emission, respectively.

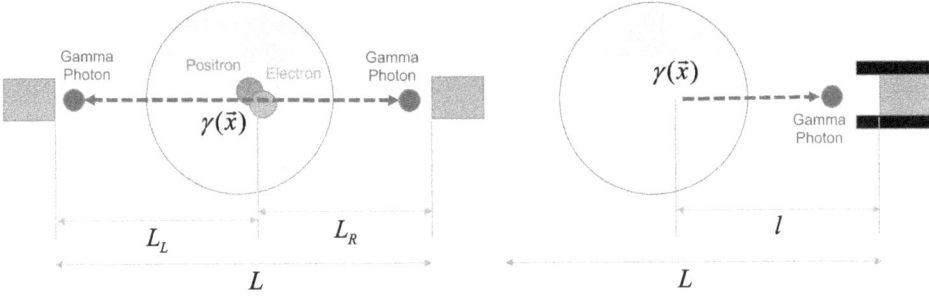

Figure 8.3. Data models for paired and single gamma ray photon emission are not the same, because the corresponding detector collimation schemes are different.

bloodstream and reveals cancer metastasis through metabolic trapping of FDG. In the second mechanism, instead of paired gamma ray photon emission, a single gamma ray photon is emitted at a time. The radioactive tracer Tc-99 m is the main example in this case, which is a pure gamma-emitter with a half-life of about 6 hours at a radiation energy level of 140 keV, and ideal for medical imaging.

The data acquisition processes in the two nuclear emission imaging modes are different, as shown in figure 8.3. In the first mode, the electronic collimation is performed to make sure that when and only when a pair of photons is simultaneously captured (within a very narrow time window) by two detectors, an event is reported along a line of response correcting the two detector cells. For a unit tracer at some point along a path L, the probability for one of the paired photons reaching the left (L) detector is expressed as

$$p_{L_L}(x) = e^{-\int_{L_L} \mu(s)ds}, \tag{8.1}$$

where $\mu(\cdot)$ is the linear attenuation coefficient at the relevant radiation energy level, and similarly the probability for the other of the paired photons to teach the right (R) detector is

$$p_{L_R}(x) = e^{-\int_{L_R} \mu(s)ds}. \tag{8.2}$$

Clearly, the joint probability for this pair of gamma ray photons to be simultaneously detected should be the product of these two probabilities, and the signal formed along the path L from the trace distribution $\gamma(\cdot)$ inside a field of view can be formulated as

$$d_L = \int_{L=L_L+L_R} \gamma(x)e^{-\int \mu(s)ds}dx = e^{-\int \mu(s)ds} \int_{L=L_L+L_R} \gamma(x)dx. \tag{8.3}$$

It can be seen from the above equation, if we know the attenuation background, that the signal along a line of response is nothing but a line integral, which is the same as what we have for CT. The tomographic image formation based on the paired gamma ray photon emission is called positron emission tomography (PET).

In the second mode, each gamma ray photon that escaped from the attenuating background can reach a detector cell of a gamma ray camera only through a mechanically collimated aperture. Hence, the number of detectable gamma ray photons from a location x in a tracer concentration $\gamma(\cdot)$ should be proportional to the following:

$$d(x) = \gamma(x)e^{-\int_{l(x)} \mu(s)ds}, \tag{8.4}$$

where $\mu(\cdot)$ is again the linear attenuation coefficient but at an energy level different from that in the case of paired gamma ray photon emission, and $l(x)$ is the path from the location x to the detector cell. Consequently, the total signal formed along the path L from the trace distribution $\gamma(\cdot)$ inside a field of view becomes

$$d_L = \int_L \gamma(x)e^{-\int_{l(x)} \mu(s)ds}dx. \tag{8.5}$$

The tomographic image formation based on single gamma ray photon emission is called single photon emission computed tomography (SPECT).

Although the PET and SPECT data models are both linear given a known attenuation background, the PET data are standard line integrals, while the SPECT data are attenuated linear integrals. This type of non-standard line integrals cannot be directly inverted using a CT algorithm, since the location-dependent attenuation factor $e^{-\int_{l(x)} \mu(s)ds}$ cannot be factorized out of the overall line integral $\int_L \gamma(x)e^{-\int_{l(x)} \mu(s)ds}dx$.

As mentioned earlier, both PET and SPECT data are highly noisy, because the number of gamma photons used for emission imaging is orders of magnitude smaller than the number of x-ray photons used for CT. The Poisson distribution is the physical model to statistically describe the gamma ray photon emission, which is the same for x-ray photon generation. Thanks to the elegant properties of independent Poisson noise, the likelihood of the underlying radioactive tracer distribution can easily be formulated, and then maximized for the so-called maximum likelihood image reconstruction applicable to PET and SPECT imaging.

Since PET and SPECT image reconstruction needs precise knowledge of the attenuation background, modern PET and SPECT scanners are commonly integrated with a CT scanner so that anatomical imaging can be seamlessly combined with functional/metabolic imaging. The PET-CT scanner shown in figure 8.4 has been widely used, and actually there are no standalone PET scanners in modern hospitals and clinics. The PET-CT scanner has been particularly instrumental for cancer diagnosis and treatment. The SPECT-CT scanner is similarly useful but the imaging performance of PET is generally better than that of SPECT, while PET-CT comes at a significantly higher cost than SPECT-CT (the tracers and gamma ray detectors are cheaper for SPECT than the PET counterparts).

8.1.2 Network-based emission tomography

Similar to what was described for CT, neural networks can be designed to perform emission tomographic imaging. There are several good examples of this type. It is

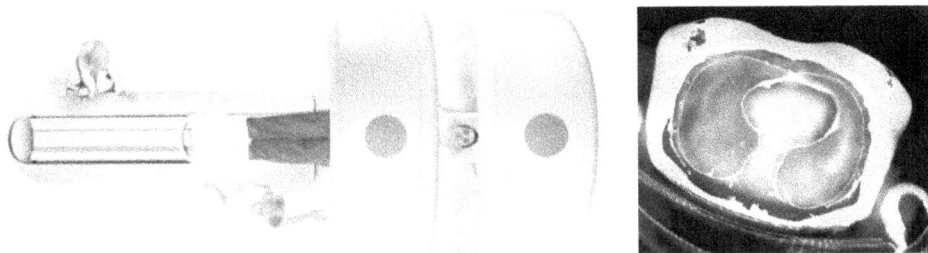

Figure 8.4. PET-CT scanner (left) for dual-modality imaging seamlessly integrating an anatomic CT image with a co-registered functional/metabolic PET map (right). Left panel reproduced with permission from Philips. Right panel adapted from figure 8 in Kim *et al* (2007). Copyright 2007 IEEE.

known that dynamic PET demands a high imaging speed and acquires low flux data. As a result, PET data are affected by strong Poisson noise. To address the Poisson noise, the maximum likelihood image reconstruction was formulated as the expectation maximization (EM) iteration (Shepp and Vardi 1982). Nevertheless, the image quality of dynamic PET reconstruction has been problematic, due to a low SNR and various model mismatches, such as the checkerboard effect from detector modularization. In 2017, the stacked sparse auto-encoder in figure 8.5 was proposed for dynamic PET imaging (Cui *et al* 2017). The dynamic PET reconstruction was formulated as a deep yet sparse auto-encoder for post-processing maximum likelihood reconstruction results, with the encoding layers extracting the features on multi-scales and the decoding layers recombining these multi-resolution features. Their results in Monte Carlo simulation and on clinical patient scans are promising, as shown in figure 8.5.

In 2018, a PET reconstruction algorithm was developed, regularized by a deep-learning-based prior (Kim *et al* 2018), as summarized in figure 8.6. In this algorithm, a denoising convolutional neural network was trained using full-dose images as the label and low-dose counterparts as the input which was formed via down-sampling the full-dose data so that the co-registration is assured. Tests at various noise levels showed that some image bias could be induced by the network due to the discrepancy between the real noise level of input data and the noise level assumed in the network training. Then, a local linear fitting function was utilized to minimize the bias. In numerical simulation and clinical imaging studies, the proposed network-based PET reconstruction outperformed the conventional reconstructions regularized by either total variation or nonlocal means.

To perform attenuation correction for PET without a transmission scan, the deep learning approach was used to generate CT images from uncorrected 18F-FDG PET images (Liu *et al* 2018), assuming that the attenuation effect in the PET images contains sufficient clues on the attenuation background. A convolutional encoder–decoder network was designed, trained with uncorrected volumetric head PET images, and co-registered CT counterparts, and tested on additional patient scans with encouraging results, as illustrated in figure 8.7.

In another interesting study, a deep residual convolutional neural network was proposed for PET image deblurring so that images acquired with large pixelated

Training:

Dynamic PET
reconstruction
images $x_1, x_2, \cdots, x_N$

SAE
model

Encoder 1 — input $x_1, x_2, \cdots, x_N$
hidden layer h_1

Encoder 2 — hidden layer h_1
hidden layer h_2

Decoder s+1 — hidden layer h_s
label layer

After training, get
weights W and bias b

PET reconstruction:

Dynamic PET sinogram
$\bar{y}_1, \bar{y}_2, \cdots, \bar{y}_N$

MLEM ⬇

Dynamic PET
reconstruction
images $\bar{x}_1, \bar{x}_2, \cdots, \bar{x}_N$

SAE model with initial
weights W and bias b

Output layer

Take Gaussian weighted
average of output

Reconstruction images $\bar{o}_i$

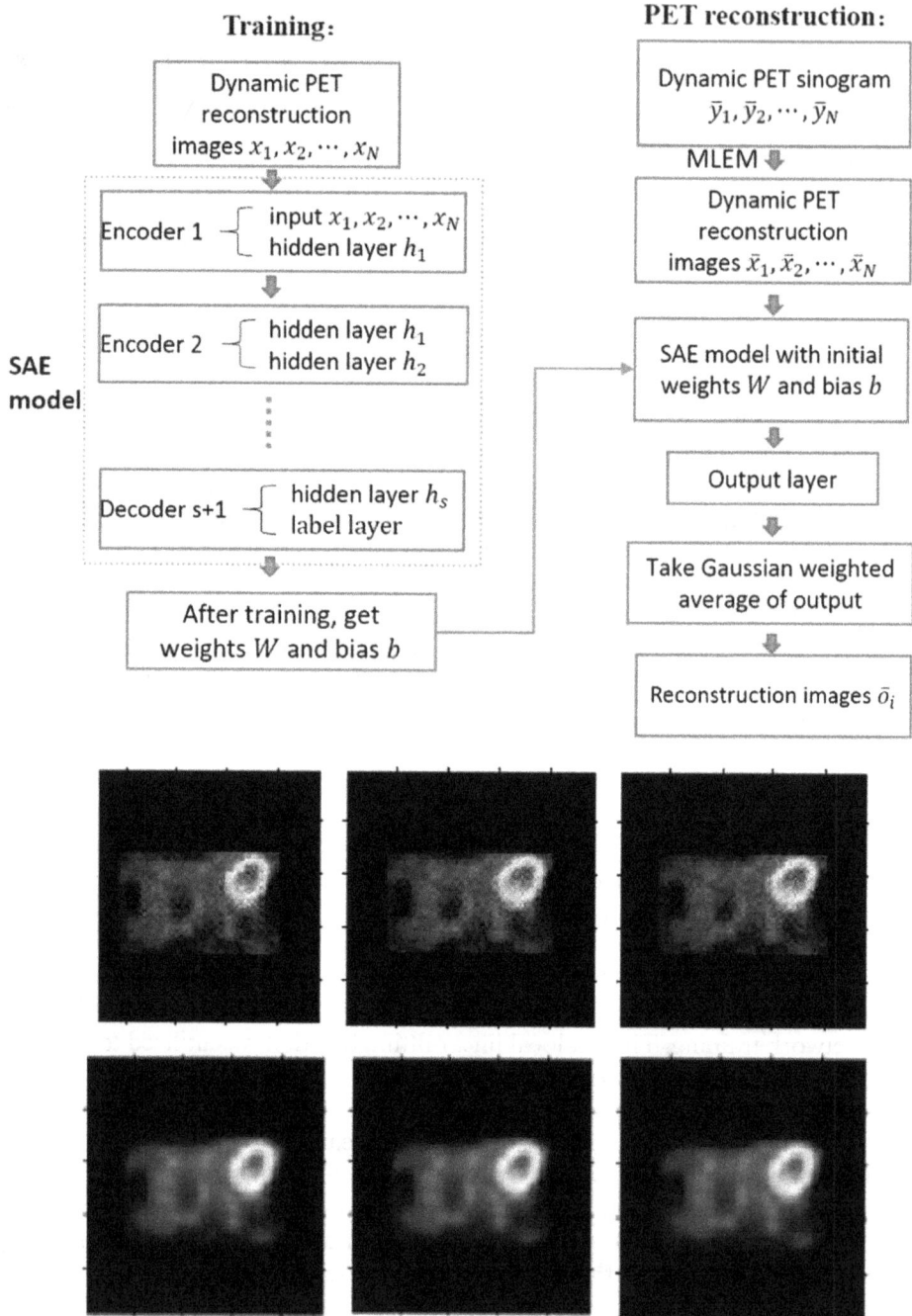

Figure 8.5. Sparse auto-encoder-based dynamic PET framework (top) and real heart maximum likelihood reconstructions for the second, third, and fourth frames (middle row) and the corresponding network-based reconstructions (bottom row). Reproduced from Cui *et al* (2017) under a CC BY 4.0 license.

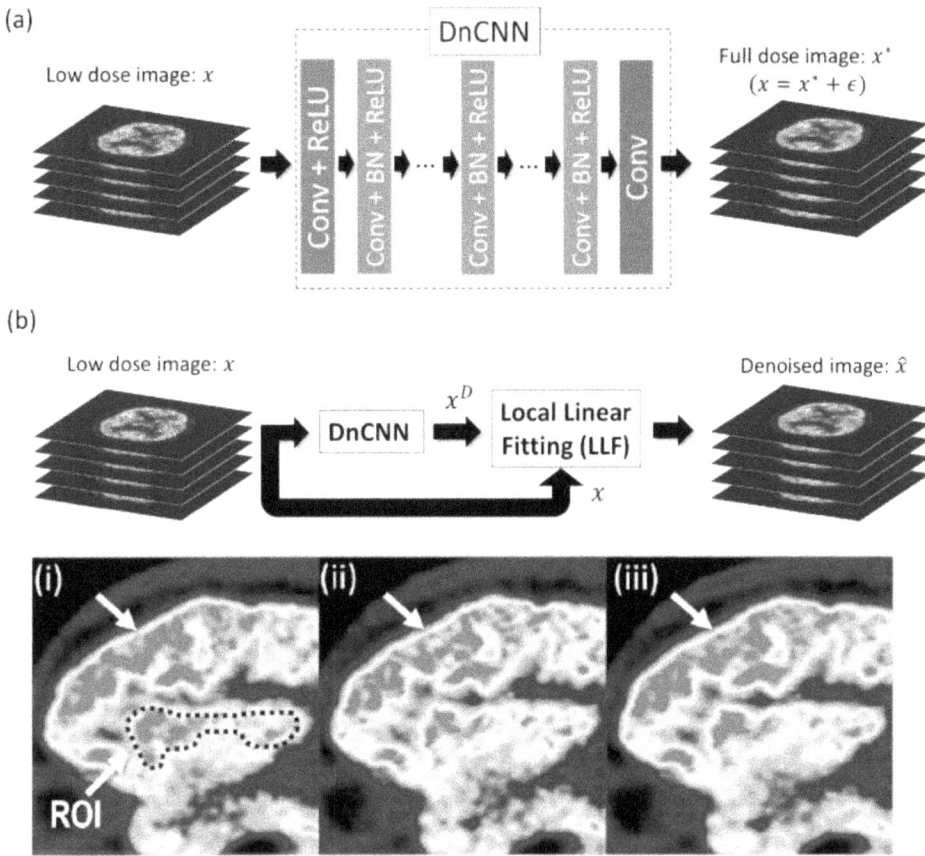

Figure 8.6. Denoising CNN (DnCNN) (top, a and b) trained with pairs of full-dose images and the corresponding down-sampled counterparts obtained via Poisson thinning at a pre-defined noise level coupled with local linear fitting, and representative magnified ROIs (bottom) reconstructed using maximum likelihood (i, full-dose data), nonlocal mean (ii, 1/10 dose data), and deep learning (iii, 1/10 dose data) methods. Reproduced with permission from Kim *et al* (2018). Copyright 2018 IEEE.

crystals can be improved toward the counterparts acquired with small pixelated crystals (Hong *et al* 2018). With the proposed network, the coarse blurry sinograms associated with large crystals were computationally mapped to refined sinograms in a data-driven manner. Unlike deblurring natural images, sinogram data contain periodic padding artifacts, and must be specifically addressed. The network was validated in simulation and preclinical imaging tests with impressive results, suggesting the feasibility of designing low-cost high-performance PET systems.

8.2 Ultrasound imaging

Unlike CT, MRI, PET, and SPECT scanners which are huge and heavy, ultrasound imaging scanners are compact and cost-effective, which are also called ultrasound

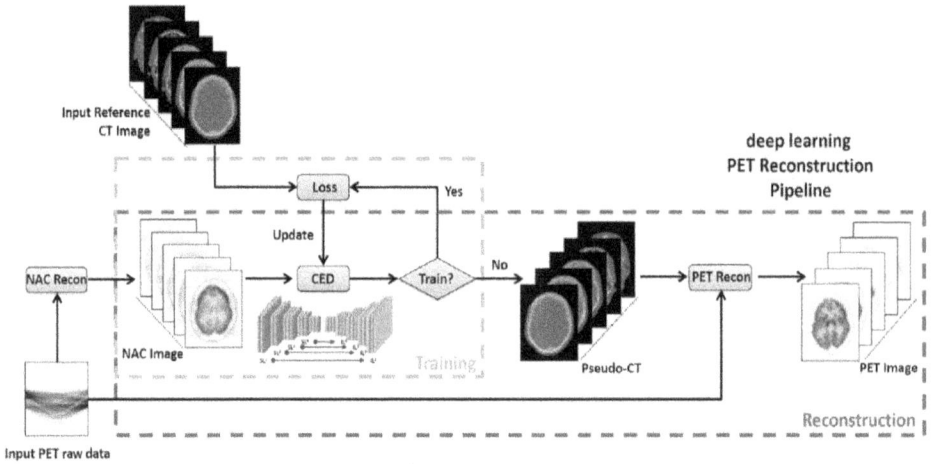

Figure 8.7. Deep learning process for estimating the attenuation background from an uncorrected PET reconstruction. Reproduced from Liu *et al* (2018) under a CC BY 4.0 License.

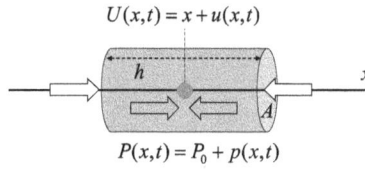

Figure 8.8. Differential element (the red cylindrical voxel with its mass center at the red dot) at a location x in a mechanical wave field, described by its relative displacement $u(x, t)$ and perturbing pressure $p(x, t)$.

probes or ultrasound transducers. Just like an **MRI** scanner which transmits and receives radio-frequency waves in the imaging process, an ultrasound scanner sends mechanical waves in the range of ultrasound frequencies into a patient's body and takes reflected, scattered, or transmitted ultrasound signals. The governing equation for ultrasound imaging is the mechanical wave equation which is in the same form as the electromagnetic wave equation for **MRI**. In this subsection, we briefly review the ultrasound imaging principles, and then describe some deep neural networks for ultrasound imaging.

8.2.1 Ultrasound scans

To explain how the ultrasound imaging data are generated, it is critical to have a basic understanding of the mechanical wave equation. Various interactions between ultrasound waves and biological tissues can be studied based on the wave equation. In the case of a homogeneous biological tissue background, 1D propagation of a mechanical wave can be modeled in terms of a differential relationship, illustrated in figure 8.8 under idealized conditions.

In reference to figure 8.8, let us analyze the dynamics of a voxel element in terms of its displacement $u(x, t)$ around its reference location x, driven by a perturbing pressure wave field $p(x, t)$. First, the acceleration of the voxel can be computed according to Newton's second law:

$$-\Delta p A = (\rho A h) \cdot \frac{\partial^2 u}{\partial t^2}, \tag{8.6}$$

where A is the cross-sectional area of the cylindrical (actually the voxel shape does not matter) voxel and $-\Delta p$ is the pressure across the voxel along the wave propagation direction, h is the height of the voxel along the x-axis, and ρ is the density of the voxel. Note that by the definition of the pressure wave field we have

$$\Delta p = \frac{\partial p}{\partial x} h. \tag{8.7}$$

On the other hand, the perturbing pressure is proportional to the relative volumetric change of the voxel under idealized conditions (the same temperature, etc). Hence, we have

$$p = -\kappa \frac{\Delta V}{V} = -\kappa \frac{A(\Delta u)}{A h} = -\kappa \frac{A\left(\frac{\partial u}{\partial x} h\right)}{A h}, \tag{8.8}$$

where κ is a coefficient (also known as the volumetric modulus of elasticity), and V is the reference volume of the voxel. In deriving the above equation, the fact has been utilized that $\Delta u = \frac{\partial u}{\partial x} h$, which means that it is the difference between the displacements of the left and right surfaces of the cylindrical voxel that gives the relative volumetric change of the voxel. Now, we combine the above three equations (first combining the last two to express Δp in terms of $\frac{\partial^2 u}{\partial x^2}$, and and then plugging this into the first equation), and the 1D mechanical wave equation is immediately obtained as follows:

$$\kappa \frac{\partial^2 u}{\partial x^2} = \rho \frac{\partial^2 u}{\partial \partial t^2} \tag{8.9}$$

or equivalently

$$\frac{\partial^2 u}{\partial x^2} = c^2 \frac{\partial^2 u}{\partial \partial t^2}, \tag{8.10}$$

where $c = \kappa/\rho$ is the speed of the mechanical wave. If the frequency of the wave is in the ultrasonic range, we have an ultrasound wave. It can be easily verified that the general solution to the wave equation is $u(x, t) = A u_1(x - ct) + B u_2(x + ct)$, and a specific exemplary solution is $u(x, t) = u_0 \sin\left(\frac{2\pi}{\lambda}(x - ct)\right)$, where λ is the wavelength.

Given the 1D wave equation in the homogeneous medium, it is fairly straightforward to extend it to the 3D case and even account for an inhomogeneous background. In any voxel, a 3D wave field along the wave propagation direction can be modeled as a 1D problem. Also, even if an acoustic background is inhomogeneous, we can treat a sufficiently small voxel and its neighborhood as homogeneous, and the differential wave equation for the homogeneous background can be locally applied. With these considerations and tedious steps, interactions between ultrasound waves and biological tissues can be analyzed, at least computationally.

Despite the conceptual simplicity of the wave equation, its solution characteristics are interesting and often complicated. Solutions can be obtained to reflect various types of interactions between ultrasound waves and biological tissues, including transmission, refraction, absorption, scattering, and reflection (quite similar to those between x-rays and biological tissues; for example, just as backscattering imaging can be performed with x-rays, ultrasound echoes can be received for imaging).

These wave-based interactions can be conveniently perceived according to the Huygens principle. Roughly speaking, the Huygens principle claims that in a wave field every element on a wavefront (the frontier of a propagating wave) can be regarded as a small source of waves. All such sourcelets on the wavefront generate wavelets moving forward at the same speed as determined by the wave equation and forming the future wavefront as the envelope tangent to all the forward wavelets, as illustrated in figure 8.9.

The ultrasound wave can be elegantly produced and sensed using a piezoelectric device, which is called an ultrasound transducer. The ultrasound transducer is made of the 'magical' matter called piezoelectric material that couples mechanical and electrical vibrations bi-directionally, as shown in figure 8.10.

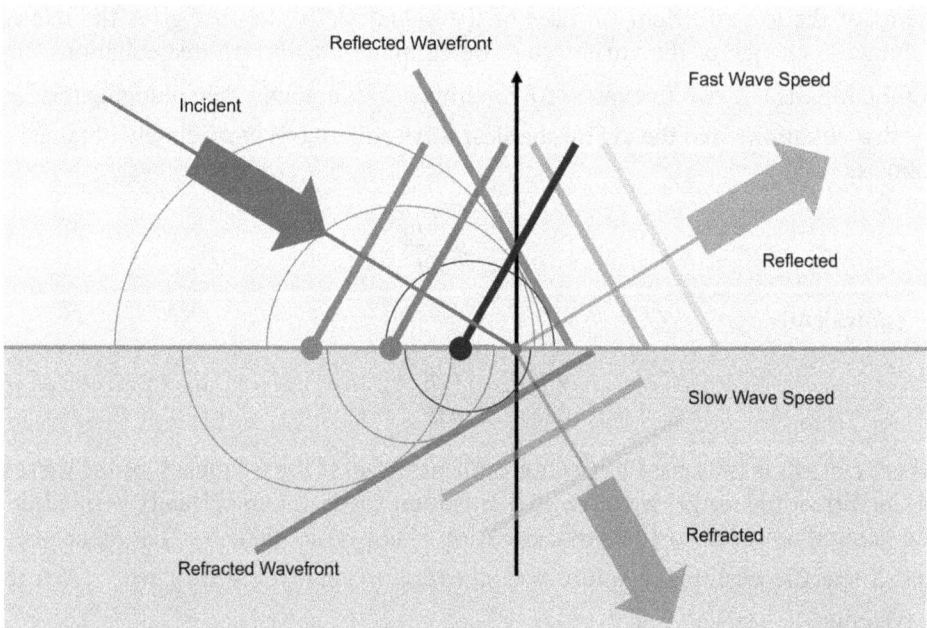

Figure 8.9. Huygens principle predicting the next wavefront from the current one. The incident wavefront (the solid red bar in the top medium) first perturbs the red dot, and then similarly the green and blue ones. These dots emit wavelets at fast and slow speeds in the top and bottom media, respectively. The envelopes of the wavelets in the top and bottom media form the wavefronts of the reflected and refracted wavefronts, respectively.

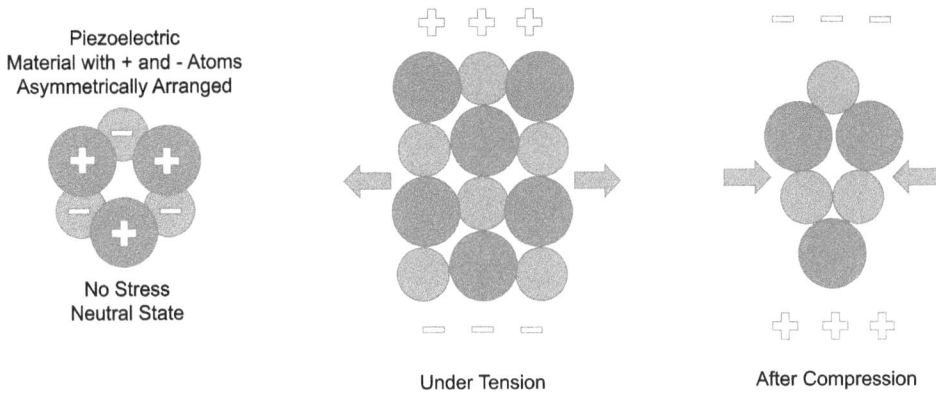

Figure 8.10. Piezoelectric effect is due to the asymmetric electrostatic structure in the piezoelectric material. An alternating electrical potential at an ultrasound frequency across a piece of piezoelectric material will compel it into a periodic mechanical vibration to generate an ultrasound wave. On the other hand, alternating compression/relaxation of the piezoelectric material will make the corresponding electrical potential change across the piezoelectric material to signify the mechanical/ultrasound wave.

An ultrasound transducer can work in various modes. The A-mode, also referred to as the amplitude mode, is the simplest mode. In the A-mode, the transducer sends a pencil beam into a patient, detects a series of echoes, and plots the amplitudes of these echoes as a function of the interaction depth computed with the echo time and the sound speed. The B-mode, also known as the brightness mode, is a repeated application of the transducer in the A-mode. In the B-mode, the transducer scans along a line or curve on a patient's body surface to form a 2D image through the patient. Then, the C-mode forms a 2D image via echo-time gating to select a plane (i.e., a coronal plane) orthogonal to B-mode images. Furthermore, the M-mode, or the motion mode, ultrasound imaging is used to monitor moving features by performing either A-mode or B-mode imaging multiple times. In each time frame, we have a snapshot of internal structures, and over time an ultrasound video is obtained to reveal dynamic features such as cardiac functions. Note that ultrasound transducers can be miniaturized and assembled into 1D or 2D arrays. Each array element can be electronically and independently controlled to serve as an ultrasound sourcelet. Collectively, these sourcelets can emulate any wavefront to focus and scan freely; for example, we can focus at different depths or perform a linear scan electronically, as illustrated in figure 8.11.

Finally, let us mention another important ultrasound imaging mode—Doppler imaging. The Doppler effect is well known, which is the perceived/detected frequency change when a sound source is moving relative to a listener. When a sound source such as a police car is moving towards you, you hear the siren sound of a higher frequency. This is because the sound waves hit your ear drum more frequently as the distance between you and the police car becomes shorter and shorter, while the frequency of the sound source is constant. Mathematically, the Doppler frequency shift can be understood in reference to figure 8.12.

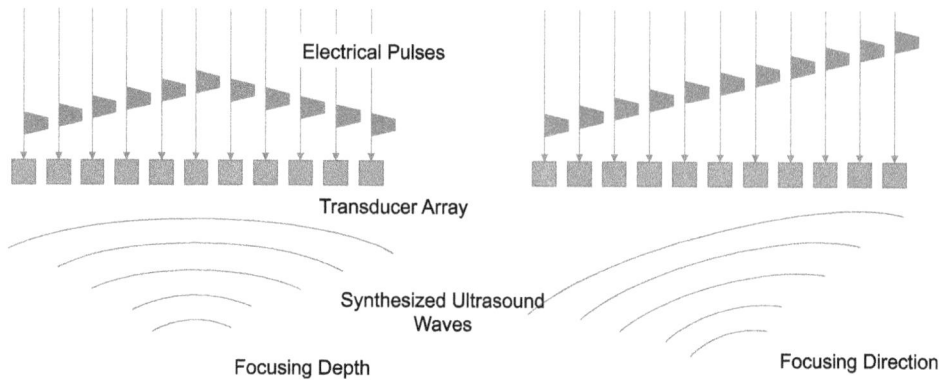

Figure 8.11. Ultrasound transducers in a linear array can be activated in a synchronized fashion to focus at different depths (left) or scan over a plane in the B-mode (right). In either case, the beam forming process can be appreciated according to the Huygens principle in which each elemental transducer is a sourcelet, and all the sourcelets are coordinated to form a desired wavefront. Note that this idea can be used in the reverse order to detect signals at various depths or from different orientations.

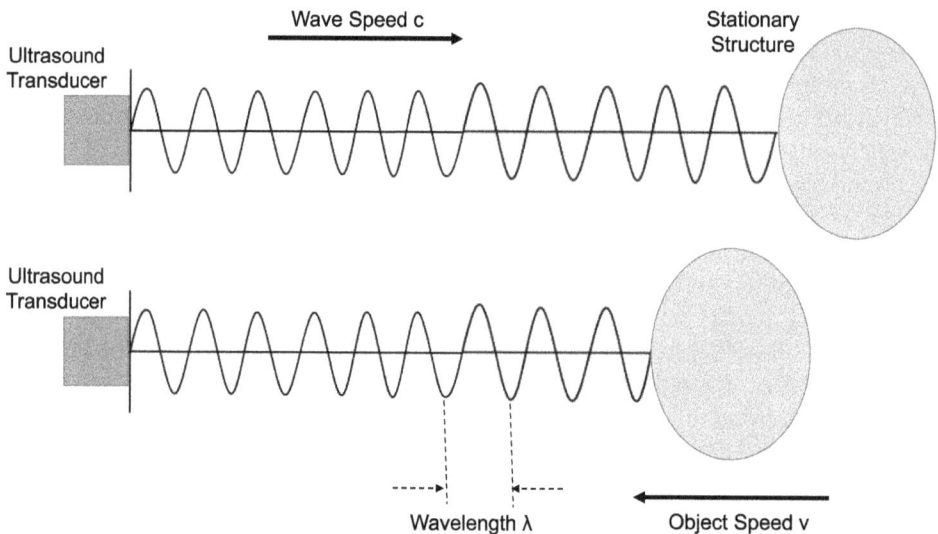

Figure 8.12. Doppler frequency shift due to relative motion between a transducer and an object. If no relative motion exists between the transducer and the object (top), the reflected ultrasound wave is received by the transducer at the same frequency as that of the incident wave; otherwise (bottom), a Doppler frequency shift is induced (a positive shift for a positive v, and vice versa).

Suppose that we send an ultrasound wave as a series of sinusoidal pulses at frequency $f = c/\lambda$ and speed c in a homogeneous medium and have an object to be probed by the ultrasound beam. If the object is stationary, it senses and reflects the wave at the same frequency f, so the train of echoes back to the transducer has the same frequency f. Without loss of generality, if the object is moved toward the transducer at a speed v, then the number of pulses the object encounters will be more than that in the stationary case. Specifically, the number of pulses hitting the structure within a unit time will be the number of pulses emitted by the transducer in the stationary case, which is $f = c/\lambda$, plus the number of additional pulses covered by the moving object, which is v/λ. Hence, the apparent frequency perceived by the moving object is $f' = (c + v)/\lambda$, and the difference between the frequency of pulses the moving object reflects to the transducer and the ultrasound frequency in the stationary medium can be computed as $\Delta f = f' - f = v/\lambda = fv/c$. Since the reflected sound wave travels at the same speed c in a stationary medium, the wavelength of the reflected sound wave is $\lambda' = c/f' = \lambda c/(c + v) \approx \lambda$, since c is practically much greater than v. Eventually, the reflected pulses will be received by the transducer that is in relative motion towards the object of interest. Hence, when the reflected pulses are received by the transducer, the second frequency change is induced, which is approximately the same Δf. Therefore, the Doppler frequency change the transducer sees will be $f_D = 2\Delta f = 2fv/c$. Generally speaking, the velocity of a moving structure may make an angle θ with the direction of the ultrasound wave propagation, instead of being parallel to the wave direction. In this general case, the Doppler frequency shift is expressed as

$$f_D = \frac{2fv\cos\theta}{c}, \tag{8.11}$$

where $v\cos\theta$ is the relevant velocity component which is parallel to the wave direction.

8.2.2 Network-based ultrasound imaging

Ultrasound scattering in heterogeneous biological tissues is a major problem in ultrasound imaging. Speckles are formed in ultrasound images that are due to constructive interferences of scattered ultrasound waves, instead of real structures. A classic method for ultrasound imaging with suppressed speckles performs nonlocal means (NLM) filtration, yielding substantial despeckling results but still suffering significantly from residual artifacts. For better despeckling, a two-stage convolutional neural network was proposed for ultrasound imaging (Yu *et al* 2018), modified from the so-called PCANet (Chan *et al* 2015). The proposed network extracts features of image patches for similarity computation in figure 8.13, in which the parametric rectified linear unit (PReLU, also known as leaky ReLU) is used as the activation function to extract deep features. The PCANet consists of three components: cascaded PCA, binary hashing, and block-wise histogram analysis. In PCANet, multi-stage convolution kernels are first learnt with PCA, and output features are obtained via binary hashing and histogram analysis. PCANet can be trained more efficiently than CNN, because no numerical optimization is involved. This model extracts and concatenates features together to compute the refined structural similarity measures between image patches, form the weighted mean

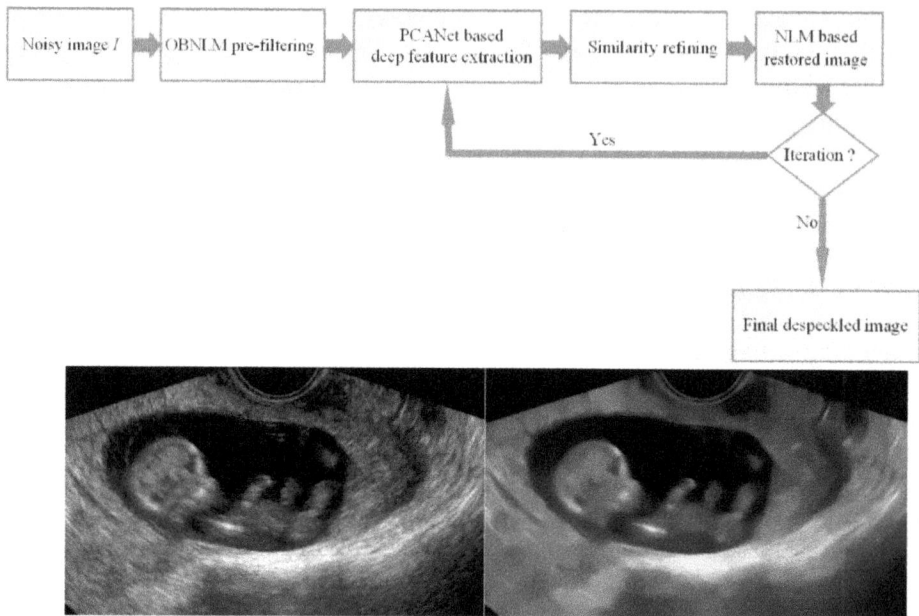

Figure 8.13. Ultrasound despeckling network (top) to compute structural similarity measures, along with an original fetal ultrasound image (bottom left) and the network-based despeckling result (bottom right), where OBNLM means optimized Bayesian nonlocal means. Reproduced from Yu *et al* (2018) under a CC BY 4.0 license.

according to these similarity measures, and produce a final despeckled image. As compared to several competing methods, the proposed network-based ultrasound imaging despeckling method produced the best structural similarity (SSIM) values for synthetic images, and the highest equivalent number of looks (ENL, which is a parameter originally for synthetic aperture radar (SAR) images; it reflects the degree of averaging operation during imaging and/or post-processing for despeckling) for clinical ultrasound images.

8.3 Optical imaging

Optical imaging also has a significant role in clinical and preclinical applications. Depending on whether the involved light can be treated as waves or particles, we have optical tomographic imaging in the interferometric or diffusive mode. Optical coherence tomography (OCT) is a great optical imaging modality for ophthalmology widely used in eye clinics. An OCT imager is an optical interferometer constructed to take advantage of light interference, which will be explained in the next subsection. On the other hand, diffuse optical tomography (DOT) and optical molecular tomography, including fluorescence molecular tomography (FMT) and bioluminescence tomography (BLT), assume a light beam consisting of photons that propagate in biological tissues diffusively, which will be also covered in the next subsection.

8.3.1 Interferometric and diffusive imaging

The Michelson interferometer, invented by Albert Michelson, was used to test if the speed of light is constant or not in 1887. In the experiment, the speed of light was demonstrated to be constant in the supposed 'aether' as the medium for light propagation, which directly motivated the development of Einstein's special relativity and is considered the beginning of modern physics. In 2015, a huge version of the Michelson interferometer, the Laser Interferometer Gravitational-wave Observatory (LIGO) at CalTech, confirmed gravitational waves for the first time. Let us explain the working principle of the Michelson interferometer, as shown in figure 8.14. The device consists of a light source, a beam splitter, two optical mirrors, and a light detector. In the device, a light beam from the source is divided by the beam splitter into two optical paths. Each of the divided light beams is reflected back by the corresponding mirror and directed toward the splitter. Then, the two beams through the splitter are recombined at and detected by the detector. The resulting intensity measurement is modulated by the phase difference between the two beams, thereby sensitively reflecting the optical path difference up to an accuracy in an order of a wavelength of the light wave.

Let us denote the amplitudes of the two split monochromatic laser beams along the first and second optical paths as $E_1(t) = A\sin(\omega t)$ and $E_2(t) = A\sin(\omega t + \phi)$, where $\phi = 2\pi\Delta L/\lambda = 2\pi \cdot (2(L_1 - L_2))/\lambda$, noting that the optical path difference is really $\Delta L = 2(L_1 - L_2)$ because of the round trip along each optical arm, ω and λ are the angular frequency and wavelength of the laser beam, respectively. The

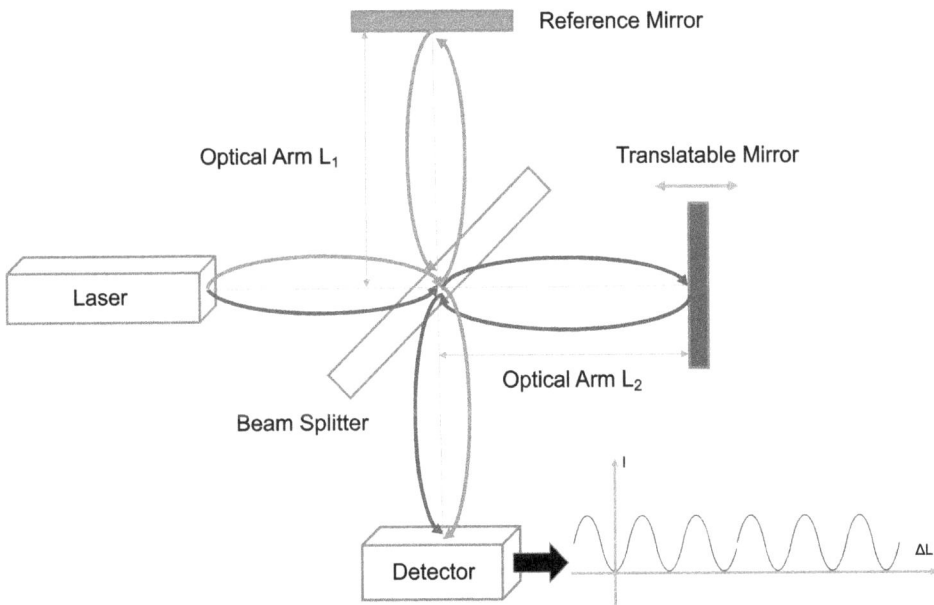

Figure 8.14. Michelson optical interferometer to sense the optical path difference between the first (red) and second (blue) arms.

instantaneous intensity of the recombined signal from the two split laser beams can be computed as follows:

$$
\begin{aligned}
E^2(t) &= A^2[\sin(\omega t) + \sin(\omega t + \phi)]^2 \\
&= A^2[\sin^2(\omega t) + \cos(\phi) - \cos(2\omega t + \phi) + \sin^2(\omega t + \phi)],
\end{aligned}
\tag{8.12}
$$

where we have used the identity $\sin(x)\sin(y) = \frac{1}{2}(\cos(x - y) - \cos(x + y))$. Then, the average intensity over one period $T \in [0, 2\pi/\omega]$ can be obtained as follows:

$$
I = \int_0^T E^2(t)\mathrm{d}t = 2I + 2I\cos(\phi) = 2I[1 + \cos(\phi)],
\tag{8.13}
$$

where we have used $\int_0^T \sin^2(\omega t)\mathrm{d}t = 1/2$ and denoted $I = A^2/2$. Hence, the intensity measured by the detector of the interferometer in figure 8.14 can be expressed as a sinusoidal function oscillating between the doubled intensity of either split beam (constructive interference between the two split beams) and zero (destructive interference between them) as a function of the optical path difference between the two optical arms:

$$
I = 2I[1 + \cos(2\pi\Delta L/\lambda)] \quad \text{or} \quad I = 2I[1 + \cos(2\pi\Delta L f/c)],
\tag{8.14}
$$

where f is the frequency, and c is the speed of light.

For simplicity, let us select the distance unit so that $c = 1$, and only focus on the oscillating term. Now, let us extend the monochromatic case to the polychromatic case. With a light source of a polychromatic spectrum, we have the recorded recombined light intensity at the detector as a function of the frequency of the laser light:

$$
S(\Delta L, f) = 2I(f)\cos(2\pi\Delta L f),
\tag{8.15}
$$

where the average intensity of each laser component depends on the frequency of that component. Then, the total signal intensity detected by the interferometer is as follows:

$$
S(\Delta L) = 2\int_0^\infty S(\Delta L, f)\cos(2\pi\Delta f)\mathrm{d}f,
\tag{8.16}
$$

which is just the well-known cosine transform as shown in figure 8.15, a special case of the Fourier transform. If we like, we can similarly formulate this relationship in terms of a sine transform, or more generally a Fourier transform.

In addition to interferometric imaging, diffuse optical imaging is also popular in the literature. In this context, light is treated as photons propagating through biological tissues and reaching optical detectors as signals, with which images of optical properties or light sources are reconstructed. To describe this radiative transfer process, the concept of radiance should be first established, as shown in

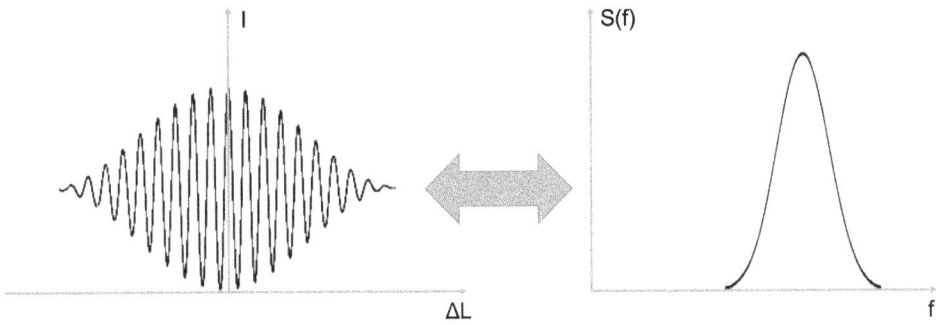

Figure 8.15. In the optical interferometer, the detected intensity of the recombined laser beams and the energy spectrum of a broad band laser source form a Fourier transform pair (the cosine transfer in this case).

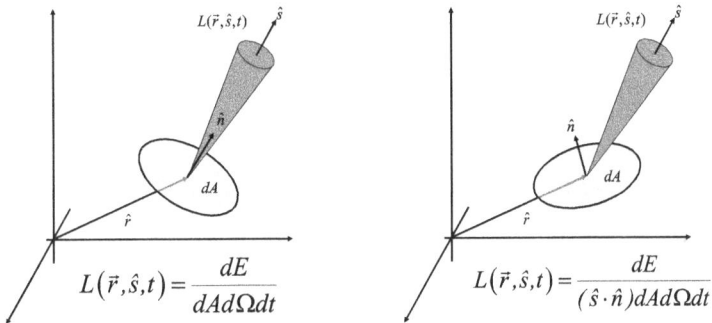

Figure 8.16. Radiance measures radiative energy flow from any area element into any solid angle, completely describing a radiative field in a 3D space. The normal direction of the area element can be in parallel to the direction of an energy flow (left) or point in another direction (right). In the latter case, a cosine correction factor is needed to normalize the energy intensity.

figure 8.16. First, let us imagine a small area element with its normal direction defined in a 3D space, and radiative energy in the form of photons happens to flow along this normal direction into a certain solid angle. Evidently, this energy flow can be described as the total energy flown through a unit area into a unit solid angle per a unit time. If the normal direction of the area element is not in parallel to the direction of the energy flow, a cosine correction factor must be applied to find a projected area element whose normal is in parallel to the direction of the energy flow. Mathematically, the radiance L is reasonably defined as follows:

$$L(\vec{r}, \hat{s}, t) = \frac{dE}{(\hat{s} \cdot \hat{n}) dA d\Omega dt}, \tag{8.17}$$

where $\vec{r}$, $\hat{s}$, $\hat{n}$, A, and Ω denote the spatial location, the direction of a radiative energy flow, the normal vector of an area element, the elemental area, and the solid angle, and t stands for time. The unit of radiance is $W\ m^{-2}\ sr^{-1}$.

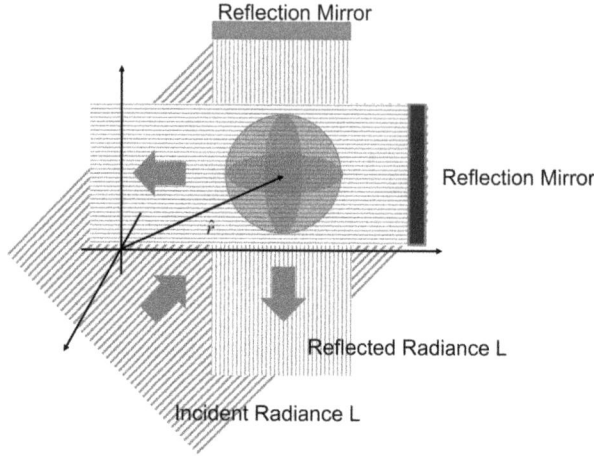

Figure 8.17. Fluence measured at a location is interpreted as the integrated radiance over a sphere. In this illustration, the fluence is measured as $3L$, which is three times the incident radiance L due to the contributions from the primary radiance and two reflected components.

Based on the radiance, we can define the fluence rate as follows:

$$\Phi(\vec{r}, t) = \int_{4\pi} L(\vec{r}, \hat{s}, t)\mathrm{d}\Omega, \tag{8.18}$$

which can be understood as the total energy an imaginary black sphere centralized at a spatial location would intercept, as shown in figure 8.17. The unit of radiance is W m^{-2}.

Furthermore, we define energy density u (J m^{-3}), fluence F (J m^{-2}), and current density (W m^{-2}) as follows:

$$u = \frac{\mathrm{d}E}{\mathrm{d}V} = \int_{4\pi} \frac{L(\vec{r}, \hat{s}, t)\mathrm{d}A\mathrm{d}\Omega\mathrm{d}t}{\mathrm{d}A(c\mathrm{d}t)} = \frac{\Phi(\vec{r}, t)}{c} \tag{8.19}$$

$$F(\vec{r}) = \int_{-\infty}^{+\infty} \Phi(\vec{r}, t)\mathrm{d}t \tag{8.20}$$

$$\vec{J}(\vec{r}, t) = \int_{4\pi} \hat{s}L(\vec{r}, \hat{s}, t)\mathrm{d}\Omega. \tag{8.21}$$

Now, let us derive an integro-differential equation to describe the dynamics of a radiative field according to the energy preservation principle. As shown in figure 8.18, along a given direction of radiative transfer (say, the horizontal direction without loss of generality), the radiance at a voxel can be increased by a radiative source in the voxel and a positive scattering component from incident radiance components towards the voxel from all directions. On the other hand, the radiance at that voxel can also be decreased by a positive divergence component (i.e. energy flows more out of the voxel than comes into it) and the attenuation mechanism. Hence, the overall energy balance explains the change of the energy content in the

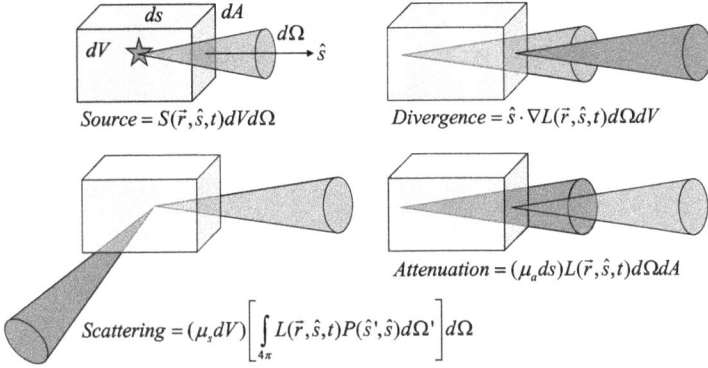

Figure 8.18. Dynamics of the radiative transfer process described in terms of radiance according to the energy conservation principle.

voxel in the following formula (detailed steps are omitted, which can be figured out based on the idea described above) which is referred to as the radiative transfer/transport equation (RTE) or Boltzmann equation:

$$
\frac{\partial L(\vec{r}, \hat{s}, t)/c}{\partial t} = -\hat{s} \cdot \nabla L(\vec{r}, \hat{s}, t) - \mu_a L(\vec{r}, \hat{s}, t)
$$

$$
+ \mu_s \int_{4\pi} L(\vec{r}, \hat{s}', t) P(\hat{s}', \hat{s}) d\Omega' + S(\vec{r}, \hat{s}, t), \quad (8.22)
$$

where $P(\hat{s}', \hat{s})$ is the phase function, often approximated as the Henyey–Greenstein function which only relies on the angle between the incident and scattered directions of a photon. For more details, see the excellent textbook (Wang and Wu 2007).

RTE involves six variables (three coordinates for a spatial location, two angular variables for the direction of radiance, and time), and is computationally complicated to solve. In practice, RTE can be simplified into the so-called diffusion equation under the conditions that the medium is weakly attenuating, strongly diffusive, and a photon experiences a sufficient number of scattering events, which is called the diffusion approximation. The diffusion equation describes the fluence field as follows:

$$
\frac{\partial \Phi(\vec{r}, t)}{c\partial t} + \mu_a \Phi(\vec{r}, t) - \nabla \cdot [D\nabla\Phi(\vec{r}, t)] = S(\vec{r}, t), \quad (8.23)
$$

where $D = (1/3) \cdot (\mu_a + \mu_s')$, $\mu_s' = \mu_s(1 - g)$, μ_a, μ_s, and g are the absorption, scattering, and anisotropy factor of the Henyey–Greenstein function, respectively. The physical meaning of g is the averaged scattering angle of photons. Ideal backscattering, isotropic scattering, and forward scattering are characterized by $g = -1$, 0, and 1, respectively. Typically, g is between 0 and 1. Again, see the textbook by Wang and Wu (2007) for more details.

RTE can be solved via Monte Carlo simulation as shown in figure 8.19, which has now become practical thanks to the rapid development of high-performance

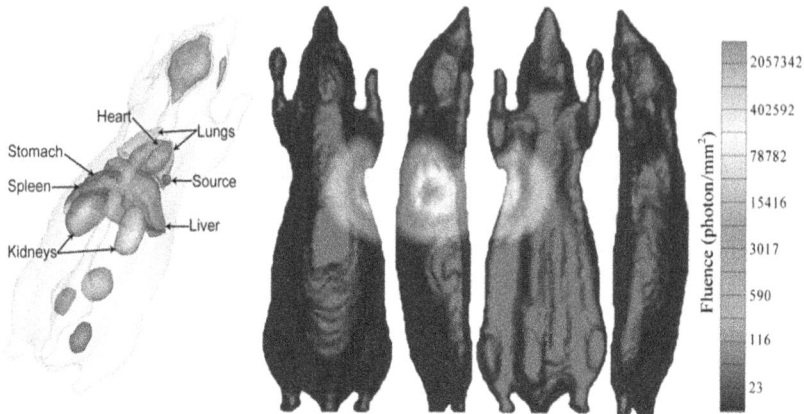

Figure 8.19. Monte Carlo simulation for bioluminescence tomography of a heterogeneous mouse. The geometry of the mouse with major organs near a bioluminescence source (left), and the fluence on the mouse body surface as viewed from four orthogonal directions. Reproduced with permission from Shen and Wang (2010). Copyright 2010 Institute of Physics Publishing.

computing techniques. Interestingly, Monte Carlo simulation can be accelerated using deep learning techniques as well (Huang and Wang 2017, Shen *et al* 2018).

8.3.2 Network-based optical imaging

In several ways, optical coherence tomography (OCT) works like ultrasound imaging. As mentioned earlier, speckle noise is troublesome for ultrasound imaging. Similarly, despeckling is important for OCT to improve visualization and analysis of optical interferometric images. In a recent study illustrated in figure 8.20, an end-to-end conditional generative adversarial network (cGAN) was proposed for both speckle noise reduction in the case of retinal OCT imaging (Ma *et al* 2018). In particular, the edge loss was utilized to preserve high-frequency information. After training the network with commercial OCT images, excellent denoising results were obtained, outperforming existing methods and having a good generalizability over other OCT images.

In addition to image noise reduction, image resolution improvement or deblurring is also highly desirable. Image resolution degradation is due to multiple physical factors such as a limited optical aperture and underlying physiological motion. Classic deblurring algorithms are model-based, and model mismatches are unavoidable. Since deblurring is a highly ill-posed inverse problem, any model mismatch could produce strong artifacts, which are further worsened by low-quality data. A deep neural network is a data-driven non-model framework, and promises to produce superior and stable results in many applications. In a recent study, traditional model-based deblurring and CNN-based learning were combined to improve retinal OCT images for super-resolution (Lian *et al* 2018), outperforming the model-based OCT deblurring approach. To address speckle noise and down-sampling induced quality loss, a generative adversarial network-based approach was developed for simultaneously denoising and deblurring OCT images (Huang

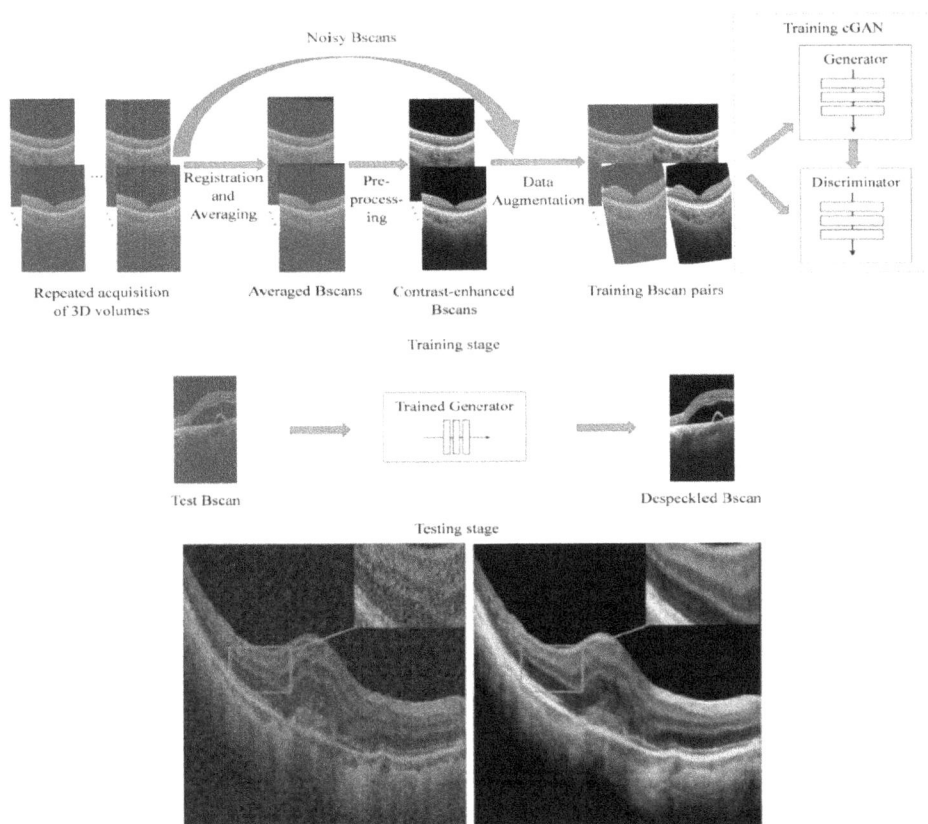

Figure 8.20. Proposed cGAN for OCT denoising (top), along with a noisy retinal OCT image (bottom left), and the despeckled counterpart using the cGAN (bottom right). Reprinted with permission from Ma *et al* (2018). Copyright 2018 The Optical Society of America.

et al 2019). The resultant network is called SDSR-OCT. It was trained into three super-resolution models with gain factors of 2, 4, and 8 compensating for the corresponding down-sampling rates, respectively. This approach was quantitatively and qualitatively evaluated against popular algorithms with promising results, as shown in figure 8.21.

Bioluminescence tomography (BLT) is an optical molecular imaging mode to reconstruct a bioluminescence source distribution inside an organism, in particular a small animal, from optical flux measured on the body surface (Wang *et al* 2004, 2006, Klose *et al* 2010, Mollard *et al* 2016, Gao *et al* 2017), as illustrated in figure 8.22. While OCT relies on interferences of light waves, the optical signals involved in BLT are incoherent and diffusive. Since BLT performs image reconstruction only based on passive, diffusive, and surface-based observation, it is perhaps the most challenging problem among all tomographic modes.

BLT is invaluable for cancer research in live mouse models of almost all human diseases. To realize the huge potential of BLT, the accuracy and reliability of BLT image reconstruction must be optimized by overcoming workflow complexity,

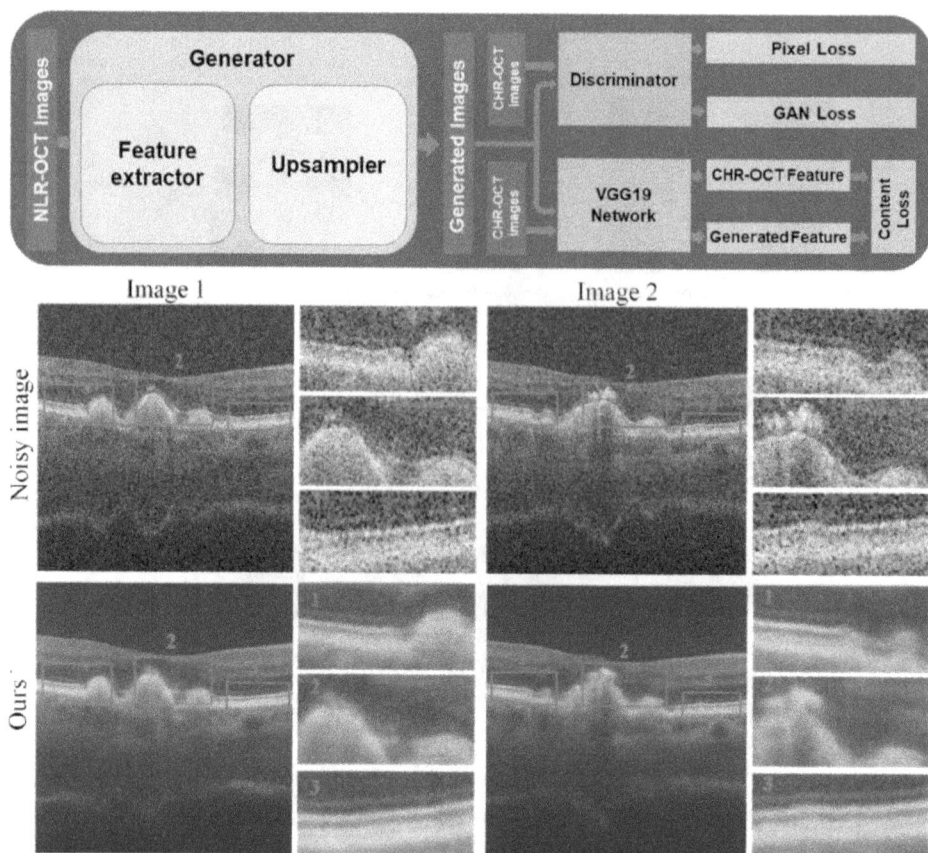

Figure 8.21. GAN-based OCT denoising and deblurring. The overall network architecture (top), and deep-learning-based denoising and deblurring results from two retinal OCT images (bottom), quantitatively and qualitatively outperforming the corresponding results obtained using the BM3D + BICUBIC, NWSR, and SRCNN methods (not included). Reproduced with permission from Huang *et al* (2019). Copyright 2019 The Optical Society of America.

model mismatches, data noise, and ill-posedness. In a recent project, an inverse problem simulation (IPS) approach was proposed to improve the quality of *in vivo* tumor BLT with a multilayer-perceptron network (Gao *et al* 2018). The IPS network consists of four hidden layers, one input layer, and one output layer. The input is the surface flux, and the output is the underlying bioluminescent light source distribution. The concept of the permissible source region is utilized to reduce the number of unknowns in the brain region, and the number of neurons in each layer is equal to the number of nodes in the permissible region. In contrast to the common practice of solving the BLT problem based on a linear system with several significant approximations, the IPS network learned a data-driven relationship from the surface measurement to internal source parameters. In numerical and orthotopic glioma

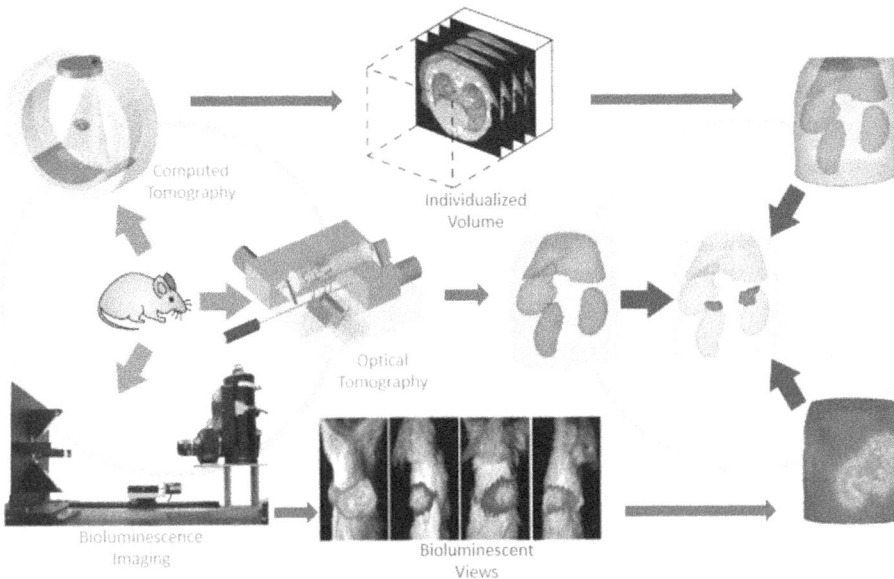

Figure 8.22. Bioluminescence-coded mouse is scanned in a BLT system chamber to measure external bioluminescent views around the mouse, coupled with a tomographic anatomical scan mode (such as micro-CT) and an optical scan for tissue properties to construct an accurate forward imaging model. This forward process is then inverted to reconstruct a 3D distribution of bioluminescent sources for analysis of bioluminescently labeled molecular and cellular features.

experiments, the IPS network produced satisfactory BLT reconstruction results over the conventional approach, as shown in figure 8.23.

8.4 Integrated imaging

The PET-CT scanner in figure 8.24 is a great example of multimodality imaging. The PET-MRI scanner in figure 8.24 is another powerful hybrid imaging technology, in which a PET-MRI scan performs both a PET scan and an MRI scan simultaneously or successively to harness the strengths of PET and MRI. MRI produces excellent contrast, resolution, and functions of biological tissues and organs, PET reveals metabolic abnormalities with radioactive molecular targeting. Clinically, PET-MRI brings the benefits of rich soft tissue contrasts and functional info from MRI to the dual-modality images and is better than PET-CT in terms of MRI merits. On the other hand, PET-MRI loses the advantage of PET-CT in terms of the CT-based attenuation background correction necessary for quantitative PET imaging. As shown in figure 8.25, a recent study suggests the feasibility of mapping LAVA Flex T1 and T2 MRI images with a 3D deep neural network to an attenuation background for quantitative PET of the pelvis (Bradshaw *et al* 2018). The trained network is capable of material decomposition into air, water, fat, and bone, and finally a CT-like continuous attenuation map. Their results demonstrate that this network-based approach enables PET attenuation correction in the pelvis.

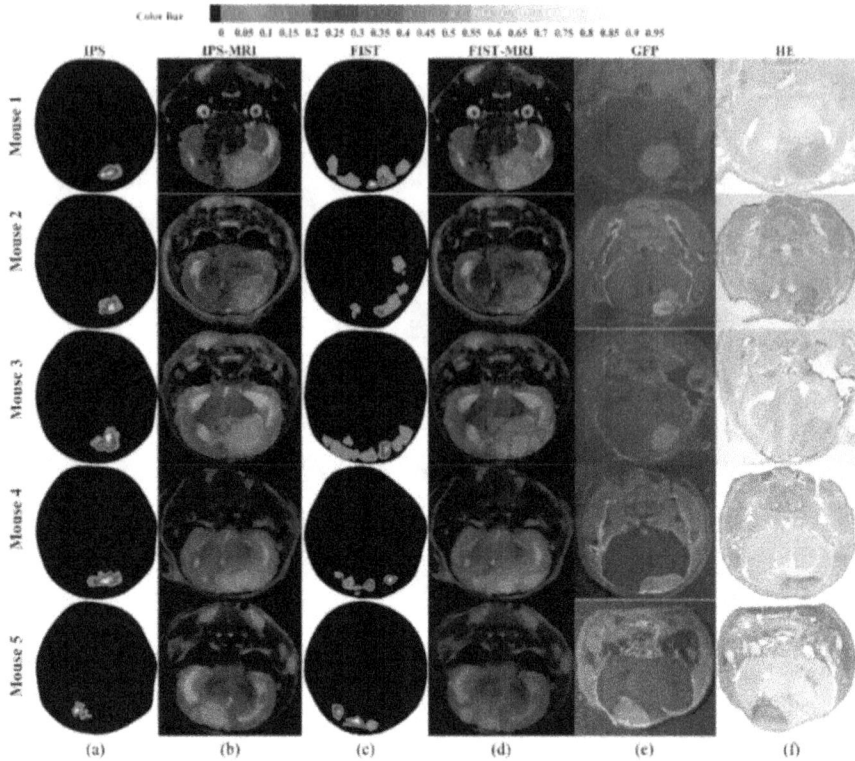

Figure 8.23. BLT of orthotopic glioma in five live mouse models. The first two columns are the transverse slices showing bioluminescence source distributions reconstructed by IPS and the corresponding BLT-MRI fusions, the third and fourth columns are the counterparts reconstructed with the fast iterative shrinkage threshold (FIST) algorithm, and the last two columns present the GFP imaging and H&E staining results of the cryo-slices, respectively. Reproduced with permission from Gao *et al* (2018). Copyright 2018 The Optical Society of America.

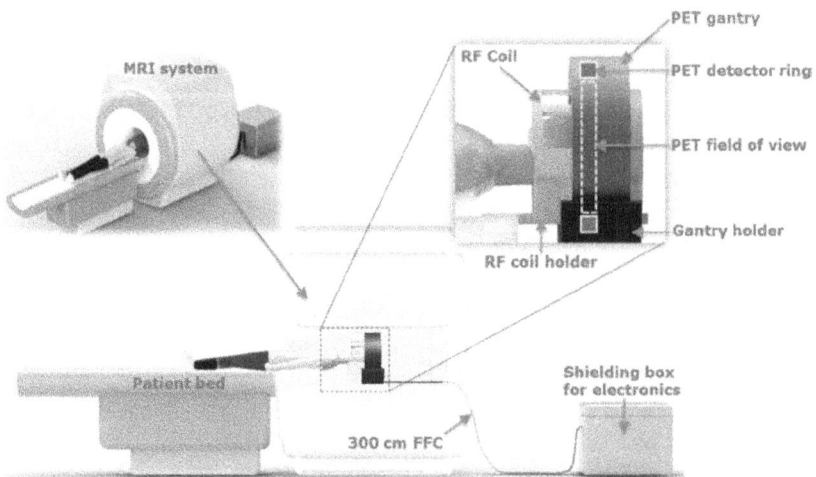

Figure 8.24. PET-MRI scanner allowing simultaneous PET and MRI scans. Reproduced with permission from Hong *et al* (2013). Copyright 2013 American Association of Physicists in Medicine.

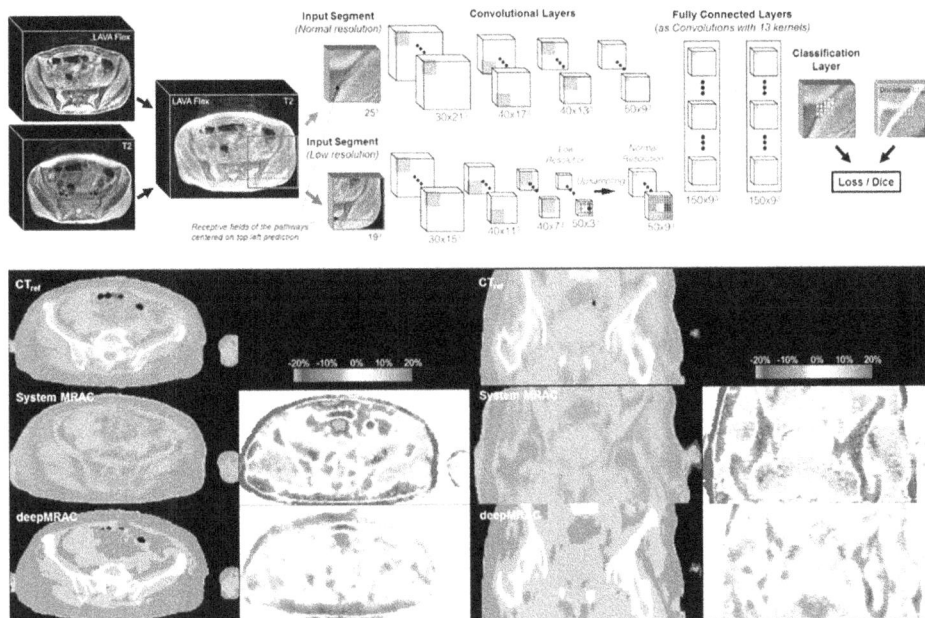

Figure 8.25. DeepMedic network (originally for brain MR segmentation) adapted for pelvis MRI (top). The network consists of two 3D CNNs with different receptive fields and is combined with fully connected layers to predict four output classes after training with reference CT images. Comparison of the reference axial and coronal CT images (bottom, first row) to the current system MRI-based attenuation estimation and PET correction (bottom, second row), and the proposed learned counterparts (bottom, third row), along with the PET error maps, relative to the results of the reference CT scan Bradshaw *et al* (2018).

Also of great interest is the work by Shen's group (Xiang *et al* 2017) to obtain high-quality PET images from a low-dose radioactive tracer injection. As shown in figure 8.26, a deep neural network was proposed to predict a standard-dose PET image from the combination of a low-dose PET scan and the associated T1-weighted MRI scan. With their network, an end-to-end mapping was obtained between the inputs and the high-quality PET output. Also, the CNN modules were integrated with an auto-context strategy for iterative refinement. The network was successfully validated with real brain PET/MRI scans.

A future opportunity is to construct a simultaneous CT-MRI scanner. Towards this direction, Wang's group and collaborators performed a top-level design along with feasibility and utility analyses in (Wang *et al* 2015, Xi *et al* 2016), as illustrated in figure 8.27. Integrated multimodality imaging systems such as PET-CT and MRI-PET gained acceptance as valuable clinical and research tools after initial skepticism, but CT-MRI has not been attempted, largely due to technical challenges, despite its great promise. CT offers a nearly ideal map of morphology at fine resolution and high speed. MRI captures functional, flow-sensitive, and tissue-specific signals. If a simultaneous/contemporaneous CT-MRI scanner is built, we would be uniquely equipped to image the intrinsic complexity and dynamic

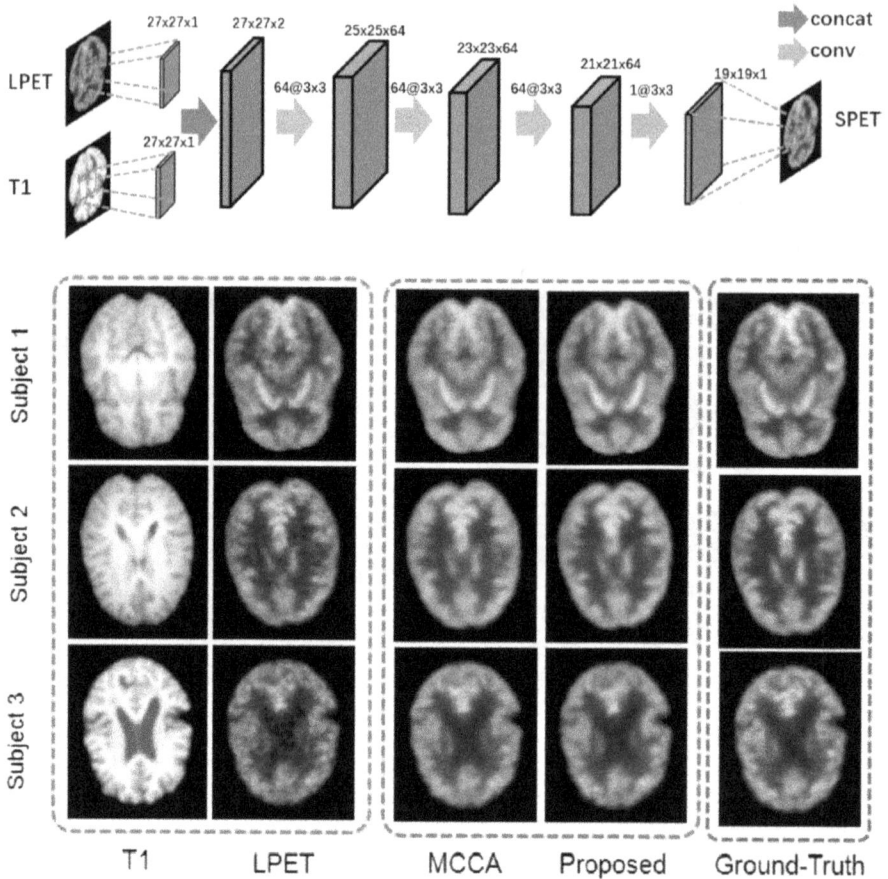

Figure 8.26. Four-layer CNN (top) designed to estimate a standard-dose PET (SPET) scan from a low-dose PET (LPET) scan and a T1-weighted MRI scan, and the visual comparison (bottom) of the proposed method and the MCCA method (Xiang *et al* 2017). Reproduced with permission from Xiang *et al* (2017). Copyright 2017 Elsevier.

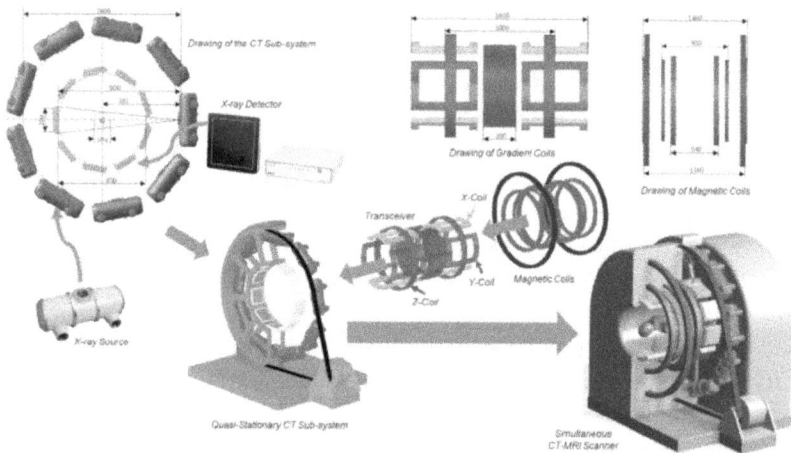

Figure 8.27. First top-level design of a CT-MRI scanner that delivers clinically relevant imaging performance in a local region of interest. Reproduced with permission from Wang *et al* (2015) under a CC BY 3.0 license.

character of biological and pathological processes, particularly in non-contrast/ contrast-enhanced cardiovascular and oncological applications as well as in the context of radiation therapy (Jia *et al* 2017).

Although a physical prototype scanner has not been built yet for simultaneous CT-MRI imaging, a number of papers were published to convert MRI images to CT images using deep learning techniques. MRI-based synthesis of CT images is a new area of interest for applications in attenuation correction for MRI-guided radio-therapy planning. In 2017, a deep convolutional neural network (DCNN) was designed for MRI-based CT image generation (Han 2017). The proposed network consists of 27 convolutional layers interleaved with pooling and un-pooling layers. The network was trained via transfer learning and refined with limited CT and T1-weighted MR images. The proposed network produced the best mean absolute error for all subjects compared to the atlas-based method. In a follow-up study, the generative adversarial network was adapted to synthesize CT images (Emami *et al* 2018). The generator is a residual network (ResNet) taking T1-weighted MRI images as the input. The discriminator is a CNN. The image quality metrics include the mean absolute error, structural similarity index, and peak signal-to-noise ratio. The results suggest a decent performance, as shown in figure 8.28.

In another relevant study, the selection of MRI inputs was investigated with respect to the resulting output using a fixed neural network (Leynes and Larson 2018). It was found that Dixon MRI images may be sufficient for generation of CT images, and ZTE MRI images help capture air content. Independently, several synthetic CT algorithms were compared, including segmentation, atlas-based, and deep learning techniques (Arabi *et al* 2018). The data indicate that the deep CNN approach produced the best synthetic CT values within organs, with accurate organ segmentation and dosimetric distribution. However, the CNN approach suffered from vulnerability to anatomy, evidenced by performance outliers (Arabi *et al* 2018).

In 2012, Wang's group and collaborators published a conceptual paper on omni-tomography for a grand fusion of all relevant tomographic modalities ('all-in-one') to acquire different datasets simultaneously ('all-at-once') and capture multi-physics interactions (Wang *et al* 2012). While the initial concept of omni-tomography is based on interior tomography for localized imaging, the recent development on MR-based polarized tracer imaging (Zheng *et al* 2016) suggests a possibility of omni-tomography for global imaging (Gjesteby *et al* 2018). The feasibility of global omni-tomography relies on the availability of polarized radioactive tracers. Clinically attractive polarizable tracers are not available yet for this purpose but their feasibility is not against physics. Such a tracer can be flexibly manipulated in the MRI framework in any way we like (anywhere and anytime); see figure 8.29 for a general idea on how to steer a volume of interest to emit gamma rays into a specified direction, see figure 8.30 for an artistic rendering of a hybrid CT-SPECT-MRI scanner, and read (Gjesteby *et al* 2018) for more details.

There is perhaps no argument that the ultimate integration of all imaging modalities in a single gantry is an ideal hardware framework. It is underlined here that deep imaging should be a unified computational framework for image

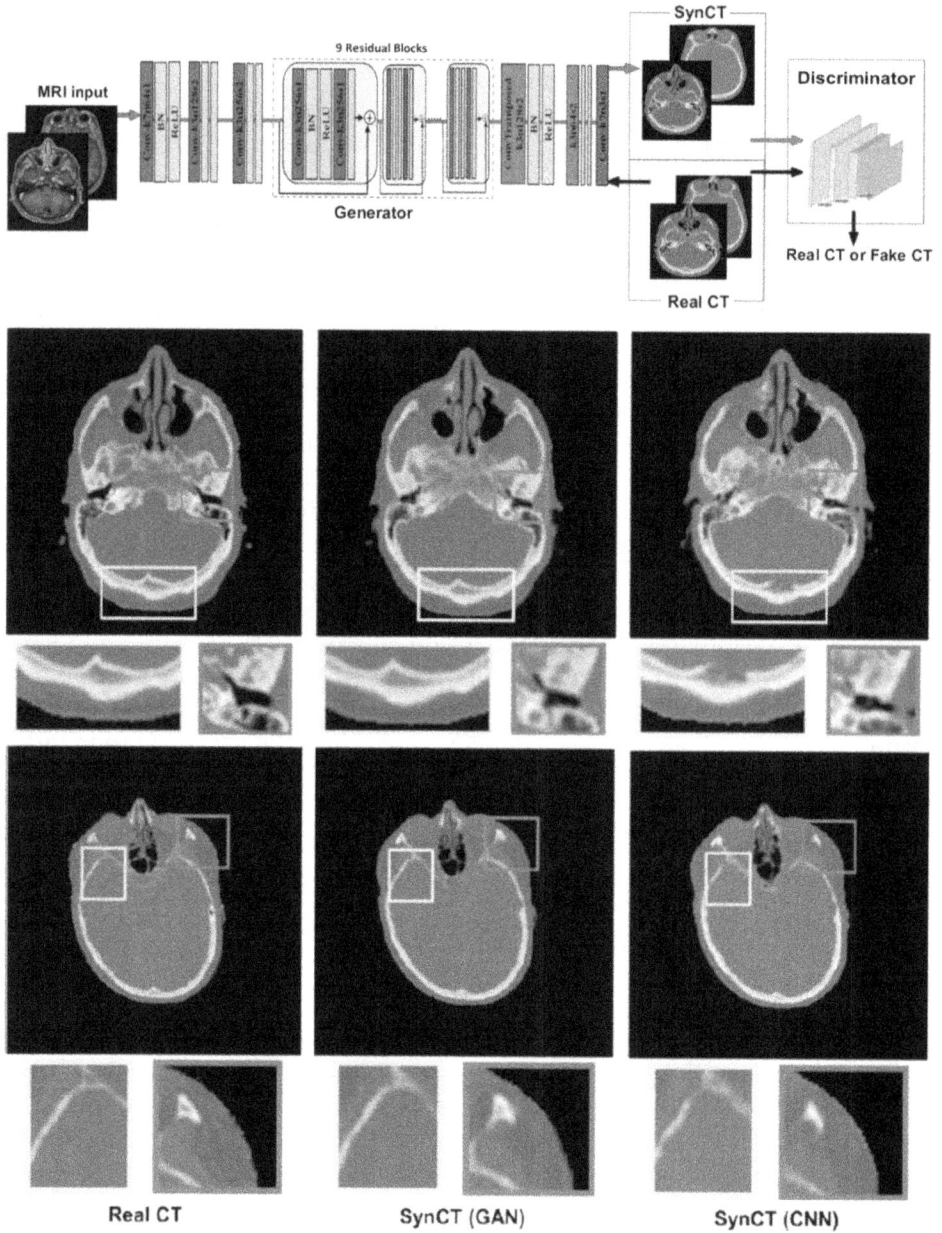

Figure 8.28. GAN architecture (top) for MRI-based synthesis of CT images, and comparison with MRI-based CT images synthesized with a relatively simpler U-net. Reprinted with permission from Emami *et al* (2018). Copyright American Association of Physicists in Medicine.

reconstruction and analysis (Wang 2016). In contrast to the traditional reconstruction algorithms which are regularized with various generic constraints from non-negativity to sparsity and low dimensional manifold, the current neural networks for image reconstruction are regularized with big data. In the future, we believe that

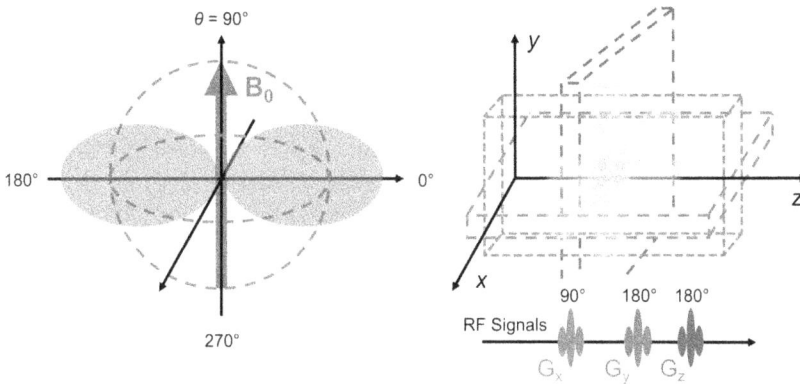

Figure 8.29. Gamma ray photons from a polarized radioactive tracer have a preferred direction (left, in the purple region), which can be manipulated at any place in the field of view to any specified direction when a background magnetic field B_0 is modulated by appropriate gradient fields generated by a suitable pulse sequence (right). See Zheng *et al* (2016) and Gjesteby *et al* (2018) for more details.

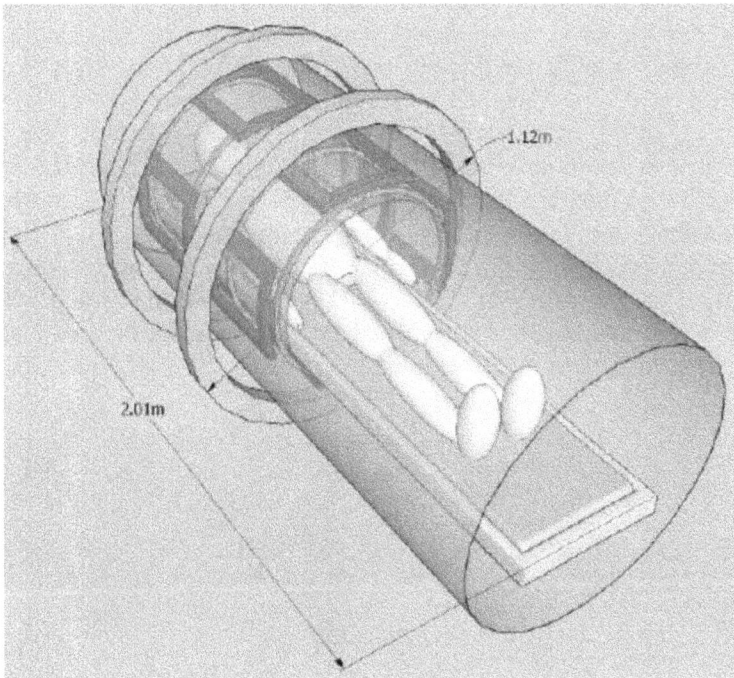

Figure 8.30. Omni-tomography based on manipulating polarized radiotracers in the MR framework allowing simultaneous emission, transmission, and MR imaging (the imaging system works in principle but it needs clinically desirable polarizable tracers to be developed). Reproduced with permission from Gjesteby *et al* (2018). Copyright 2018 IEEE.

multimodality data can be utilized to regularize individual modality image reconstruction via machine learning/deep learning.

Once we have medical images from single or multiple modalities, experts or machines need to analyze the images to extract diagnostic information. Such analyses will eventually be aided, to an increasingly large degree, by deep learning. This transition is a hot area of research known as 'radiomics'. In the traditional sense, radiomics is pattern recognition dealing with a huge space of handcrafted and/or network generated features from images; i.e. from images to features. Now, both image reconstruction and image analysis can be performed by neural networks, and we have an opportunity to combine and optimize these networks to go from tomographic raw data (and other information such as genetic profiling data) to the final diagnostic decision, which we call 'rawdiomics' (raw data/info+omics) (Kalra *et al* 2018).

Taking CT-based radiomics as an example, currently only a single CT image volume is used for radiomics. Since we do not have a perfect reconstruction algorithm when data are not perfect, there is necessarily information loss in the reconstruction process. With multiple representative reconstruction algorithms working together, the information loss will be minimized in differently reconstructed images which contain complementary information. Furthermore, if the neural network is developed to be sufficiently powerful, all the information in raw tomographic data should be directly extractable without any loss from image reconstruction that was intended for human eyes. This concept is shown in figure 8.31.

The progress in rawdiomics research and translation underlines the importance of raw data. Unfortunately, many imaging vendors, such as CT companies, are protective of their raw data and have little incentive to open data formats to the research community at large. In special cases, the problem can be solved with an academic–industrial partnership. Our own experience with imaging vendors has been positive and successful. Alternatively or complementarily, raw data can be generated using realistic simulation aided by machine learning. This will also solve

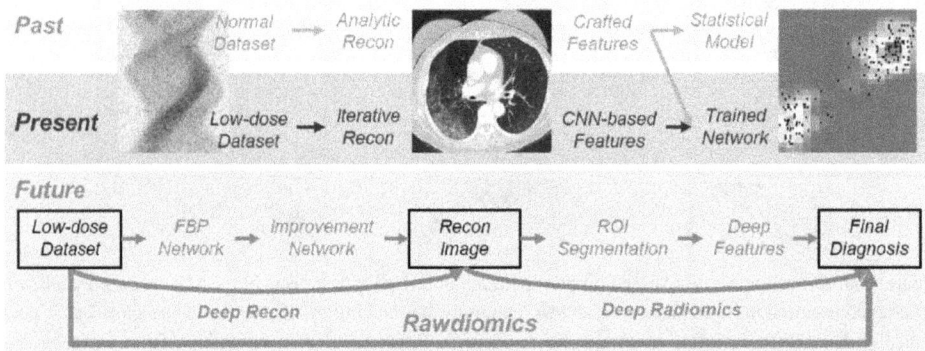

Figure 8.31. Rawdiomics for an end-to-end mapping from tomographic raw data to diagnostic reporting (Kalra *et al* 2018), with a conceptual example in the CT context.

the patient privacy issue, and avoid other troubles in obtaining big data. For example, we recently proposed a generative adversarial network (GAN) for low-dose CT simulation, which is an inverse process of network-based low-dose CT denoising. In the GAN framework, the proposed generator is an encoder–decoder network, a shortcut connection from input to output facilitates the generation of a noise distribution. Also, a conditional batch normalization layer is plugged between the encoder and the decoder. After the model is trained, a Gaussian noise generator serves as the latent variable controlling the noise in generated CT images. Trained with the Mayo Low-dose CT Challenge datasets, the proposed network produced low-dose CT images visually comparable to the ground-truth low-dose CT images. The simulated low-dose CT images can be used to verify the robustness of the current low-dose CT denoising models and also help perform other imaging tasks.

It is our belief that the convergence of multiple imaging modalities and various image reconstruction and image analysis methods is becoming critical to enable smart medicine with hybrid-/integrated radiomics/rawdiomics in the artificial intelligence framework.

8.5 Final remarks

There are two scientific approaches to reasoning, which are deduction and induction. Accordingly, we have two associated schools of AI/ML. Deduction goes from 'general' to 'specific', and is a 'top-down' approach. Decades ago, rule-based expert systems were popular, and fifth-generation computer research was a hot topic, initiated by Japanese scholars. In these settings, we hope to reason from general rules to specific claims/instances. On the other hand, induction works from data toward knowledge or information. This is 'bottom-up' or data-driven. This is how Newton formulated $F = ma$ based on experimental observations. We believe that the future of AI/ML will be an integration of deduction and induction approaches. An idea how to generate semantic knowledge and rules in a data driven fashion was mentioned in a presentation given in University of Minnesota: https://ima.umn.edu/2019-2020/SW10.14-18.19/28288.

Currently, the mainstream approach of machine learning is inductive or data-driven, and hence big data is of paramount importance. A medical AI/ML application is a whole workflow. The prerequisite of such an application is the availability of big data that is diverse, representative, and sufficient to train neural networks, in particular deep ones, so that these trained networks can generalize well into new cases.

There are many medical datasets that are publicly available, and more are emerging worldwide. For example, several MRI datasets were used for the TMI special issue on machine-learning-based image reconstruction published in 2018 (Wang *et al* 2018). A good example is the MRI BrainWeb Data. Specifically, Simulated Brain Database (SBD) images were generated using an MRI simulator, developed at the McConnell Brain Imaging Centre. The simulator computes according to the Bloch equations for NMR signal simulation with noise and partial volume effects incorporated. The tissue MR parameters are under control. As another example, in the MRI IXI (Information eXtraction from Images) there are 600 MR images from normal, healthy subjects scanned at hospitals in London. The MRI data for each subject allow reconstruction of T1, T2, and PD-weighted images, MRA images, and diffusion-weighted images in 15 directions. The most popular CT dataset for machine learning is the AAPM-Mayo Low-dose CT Challenge Database. The image dataset contains 6K 1mm thickness full-dose CT images from ten patients, with low-dose counterparts simulated by noise insertion. Furthermore, ADNI stands for the Alzheimer's Disease Neuroimaging Initiative, a longitudinal multicenter study to develop clinical, imaging, genetic, and biochemical biomarkers. BraTS means the Multimodal Brain Tumor Segmentation Challenge, recently updated with more clinically acquired 3T multimodal MRI scans and the ground truth manually labeled. TCIA is a service which hosts medical images of cancer. The data are organized as 'collections', related by a disease (e.g. lung cancer), modality (MRI, CT, etc), or research focus. In some cases, supporting data are available, such as patient genomics and pathology.

For machine-learning-based medical imaging, data can be generated in several modes, including (i) real scanning, (ii) numerical simulation, (iii) physical emulation, and (iv) dataset generation with adversarial learning. The learning-based data generation is far from being fully explored but holds great potential.

Given the rapid progress being made in theoretical, technical, and FDA regulatory aspects, we believe that machine learning algorithms will become the mainstream for medical imaging. Eventually, we hope that AI/ML will be able to assist or even replace radiologists, pathologists, and even oncologists and surgeons. There seem much more exciting opportunities ahead to synergize imaging and therapy in the AI/ML framework. In other words, we do not worry at all about any near future risk of entering another AI winter.

References

Arabi H, Dowling J A, Burgos N, Han X, Greer P B, Koutsouvelis N and Zaidi H 2018 Comparative study of algorithms for synthetic CT generation from MRI: consequences for MRI-guided radiation planning in the pelvic region *Med. Phys.* **45** 5218–33

Bradshaw T J, Zhao G Y, Jang H, Liu F and McMillan A B 2018 Feasibility of deep learning-based PET/MR attenuation correction in the pelvis using only diagnostic MR images *Tomography* **4** 138–47

Chan T-H, Jia K, Gao S, Lu J, Zeng Z and Ma Y 2015 PCANet: a simple deep learning baseline for image classification? *IEEE Trans. Image Process.* **24** 5017–32

Cui J A, Liu X, Wang Y L and Liu H F 2017 Deep reconstruction model for dynamic PET images *Plos One* **12**

Emami H, Dong M, Nejad-Davarani S P and Glide-Hurst C K 2018 Generating synthetic CTS from magnetic resonance images using generative adversarial networks *Med. Phys.* **45** 3627–36

Gao Y, Wang K, An Y S X, Meng H J and Tian J 2018 Nonmodel-based bioluminescence tomography using a machine-learning reconstruction strategy *Optica* **11** 1451–4

Gao Y, Wang K, Jiang S, Liu Y, Ai T and Tian J 2017 Bioluminescence tomography based on Gaussian weighted Laplace prior regularization for *in vivo* morphological imaging of glioma *IEEE Trans. Med. Imaging* **36** 2343–54

Gjesteby L, Cong W, Yang Q, Qian C and Wang G 2018 Simultaneous emission-transmission tomography in an MRI hardware framework *IEEE Trans. Radiat. Plasma Med. Sci.* **2** 326–36

Han X 2017 MR-based synthetic CT generation using a deep convolutional neural network method *Med. Phys.* **44** 1408–19

Hong K J *et al* 2013 A prototype MR insertable brain PET using tileable GAPD arrays *Med. Phys.* **40** 042503

Hong X, Zan Y L, Weng F H, Tao W J, Peng Q Y and Huang Q 2018 Enhancing the image quality via transferred deep residual learning of coarse PET sinograms *IEEE Trans. Med. Imaging* **37** 2322–32

Huang L and Wang L 2017 Accelerated Monte Carlo simulations with restricted Boltzmann machines *Phys. Rev.* B **95** 035105

Huang Y, Lu Z, Shao Z, Ran M, Zhou J, Fang L and Zhang Y 2019 Simultaneous denoising and super-resolution of optical coherence tomography images based on generative adversarial network *Opt. Express* **27** 12289–307

Jia X, Tian Z, Xi Y, Jiang S B and Wang G 2017 New concept on an integrated interior magnetic resonance imaging and medical linear accelerator system for radiation therapy *J. Med. Imaging* **4** 015004

Kalra M, Wang G and Orton C G 2018 Radiomics in lung cancer: its time is here *Med. Phys.* **45** 997–1000

Kim J, Cai W, Eberl S and Feng D 2007 Real-time volume rendering visualization of dual-modality PET/CT images with interactive fuzzy thresholding segmentation *IEEE Trans. Inf. Tech. Biomed.* **11** 161–9

Kim K, Wu D F, Gong K, Dutta J, Kim J H, Son Y D, Kim H K, El Fakhri G and Li Q Z 2018 Penalized PET reconstruction using deep learning prior and local linear fitting *IEEE Trans. Med. Imaging* **37** 1478–87

Klose A D, Beattie B J, Dehghani H, Vider L, Le C, Ponomarev V and Blasberg R 2010 *In vivo* bioluminescence tomography with a blocking-off finite-difference SP$_3$ method and MRI/CT coregistration *Med. Phys.* **37** 329–38

Leynes A P and Larson P E Z 2018 Synthetic CT generation using MRI with deep learning: how does the selection of input images affect the resulting synthetic CT? *2018 IEEE Int. Conf. on Acoustics Speech and Signal Processing (ICASSP)* pp 6692–6

Lian J, Hou S J, Sui X D, Xu F Z and Zheng Y J 2018 Deblurring retinal optical coherence tomography via a convolutional neural network with anisotropic and double convolution layer *IET Comput. Vis.* **12** 900–7

Liu F, Jang H, Kijowski R, Bradshaw T and McMillan A B 2018 Deep learning MR imaging-based attenuation correction for PET/MR imaging *Radiology* **286** 676–84

Ma Y H, Chen X J, Zhu W F, Cheng X N, Xiang D H and Shi F 2018 Speckle noise reduction in optical coherence tomography images based on edge-sensitive CGAN *Biomed. Opt. Express* **9** 5129–46

Mollard S, Fanciullino R, Giacometti S, Serdjebi C, Benzekry S and Ciccolini J 2016 *In vivo* bioluminescence tomography for monitoring breast tumor growth and metastatic spreading: comparative study and mathematical modeling *Sci. Rep.* **6** 36173

Shen H, Liu J and Fu L 2018 Self-learning Monte Carlo with deep neural networks *Phys. Rev. B* **97** 205140

Shen H and Wang G 2010 A tetrahedron-based inhomogeneous Monte Carlo optical simulator *Phys. Med. Biol.* **55** 947–62

Shepp L A and Vardi Y 1982 Maximum likelihood reconstruction for emission tomography *IEEE Trans. Med. Imaging* **1** 113–22

Wang G 2016 A perspective on deep imaging *IEEE Access* **4** 8914–24

Wang G, Cong W, Durairaj K, Qian X, Shen H, Sinn P, Hoffman E, McLennan G and Henry M 2006 *In vivo* mouse studies with bioluminescence tomography *Opt. Express* **14** 7801–9

Wang G, Kalra M, Murugan V, Xi Y, Gjesteby L, Getzin M, Yang Q, Cong W and Vannier M 2015 Vision 20/20: simultaneous CT-MRI-next chapter of multimodality imaging *Med. Phys.* **42** 5879–89

Wang G, Li Y and Jiang M 2004 Uniqueness theorems in bioluminescence tomography *Med. Phys.* **31** 2289–99

Wang G, Ye J C, Mueller K and Fessler J A 2018 Image reconstruction is a new frontier of machine learning *IEEE Trans. Med. Imaging* **37** 1289–96

Wang G *et al* 2012 Towards omni-tomography-grand fusion of multiple modalities for simultaneous interior tomography *PLoS One* **7** e39700

Wang L V and Wu H-i 2007 *Biomedical Optics: Principles and Imaging* (Hoboken, NJ: Wiley)

Xi Y, Zhao J, Bennett J R, Stacy M R, Sinusas A J and Wang G 2016 Simultaneous CT-MRI reconstruction for constrained imaging geometries using structural coupling and compressive sensing *IEEE Trans. Biomed. Eng.* **63** 1301–9

Xiang L, Qiao Y, Nie D, An L, Lin W L, Wang Q and Shen D G 2017 Deep auto-context convolutional neural networks for standard-dose PET image estimation from low-dose PET/MRI *Neurocomputing* **267** 406–16

Yu H Q, Ding M Y, Zhang X M and Wu J B 2018 PCANET based nonlocal means method for speckle noise removal in ultrasound images *PLoS One* **13**

Zheng Y, Miller G W, Tobias W A and Cates G D 2016 A method for imaging and spectroscopy using gamma-rays and magnetic resonance *Nature* **537** 652–5

Chapter 9

Image quality assessment

This chapter is relevant to all image reconstruction chapters because any reconstruction algorithm or tomographic neural network that produces images as the final outcome should be evaluated in a meaningful and systematic way for optimization and characterization. In this sense, all the materials here are supplemental (which can be read as needed), foundational (which use basic knowledge on mathematical/signal analysis that you are probably already familiar with), and yet innovative (some of which on neural-network-based image quality assessment are quite recent). That is to say, this chapter is practically useful and fairly independent, and can be read earlier or later (except for the network-based assessment part that should be read after reading the general knowledge on neural networks; especially chapter 3).

Since any imaging modality is intended to produce images that are supposed to be faithful, detailed, and informative, image quality assessment (IQA) is an important aspect of imaging, in particular medical imaging. Defining quality objectively is not straightforward, since quality is a context-sensitive concept. That being said, we have certain ways of defining image quality that are widely used. While some ways are applicable to all types of images, others are more system-specific or task-specific. These will be explained in the following sections.

9.1 General measures

Generally speaking, how to measure discrepancies between two signals or images is a fundamental question and has been extensively studied. In this section, we describe the three most influential approaches. In the first subsection, we focus on classic distances. In the second subsection, we present structural similarity. In the third subsection, we explain several key concepts of information theory.

9.1.1 Classical distances

The simplest way to assess image quality is to compute a distance, such as the Euclidean distance, between a medical image and its ground truth or gold-standard

doi:10.1088/978-0-7503-2216-4ch9   9-1

that can be somehow measured or specified. The common measure between an image in question x and a gold-standard reference y is the mean squared error (MSE). This is defined as

$$\text{MSE}(x, y) = \frac{1}{n} \sum_{i=1}^{n} (x_i - y_i)^2, \tag{9.1}$$

where i is the index through n pixels or voxels. MSE measures how different an image is pixel/voxel-wise from the reference. A lower MSE is expected to indicate better image quality. MSE has several variants such as the root mean squared error (RMSE),

$$\text{RMSE}(x, y) = \sqrt{\text{MSE}(x, y)} = \sqrt{\frac{1}{n} \sum_{i=1}^{n} (x_i - y_i)^2}, \tag{9.2}$$

the mean absolute error (MAE),

$$\text{MAE}(x, y) = \frac{1}{n} \sum_{i=1}^{n} |x_i - y_i|, \tag{9.3}$$

and the mean absolute percentage error (MAPE),

$$\text{MAPE}(x, y) = \frac{1}{n} \sum_{i=1}^{n} \left| \frac{x_i - y_i}{y_i} \right|. \tag{9.4}$$

These variations are desirable in various situations, since the squared difference can be too large when the error is large. Note that the MSE and MAE are special cases of the general p-norm distance $D_p(x, y) = \left((1/n) \sum_{i=1}^{n} |x_i - y_i|^p \right)^{1/p}$.

The MSE is quite reasonable, as it is essentially a measure of the area between two curves, or more relevantly the measure of the volume between the two surfaces, which are a 2D image of interest and the reference image, respectively. This measure is compatible with error analysis in a transformed/channelized space using coefficients of an orthonormal linear transform. Recall that Parseval's identity in Fourier analysis states that the area under a squared function is the same as the area of its Fourier transform after being squared: $\int_{-\infty}^{\infty} |f(t)|^2 dt = \int_{-\infty}^{\infty} |\hat{f}(s)|^2 ds$.

Although all the above errors are useful, they often fail to agree with our visual perception, and cannot perform well in many practical applications. As shown in figure 9.1, relative to the ground truth, all the compromised images suffer from the same amount of MSE but they give very different visual impressions; please review chapter 2 on the human vision system (HVS).

9.1.2 Structural similarity

It can be seen in figure 9.1 how all the sample images differ from the ground truth in different ways. For example, the top two images are visually pleasing with minor variations. On the other hand, the bottom three images have poorer quality due to

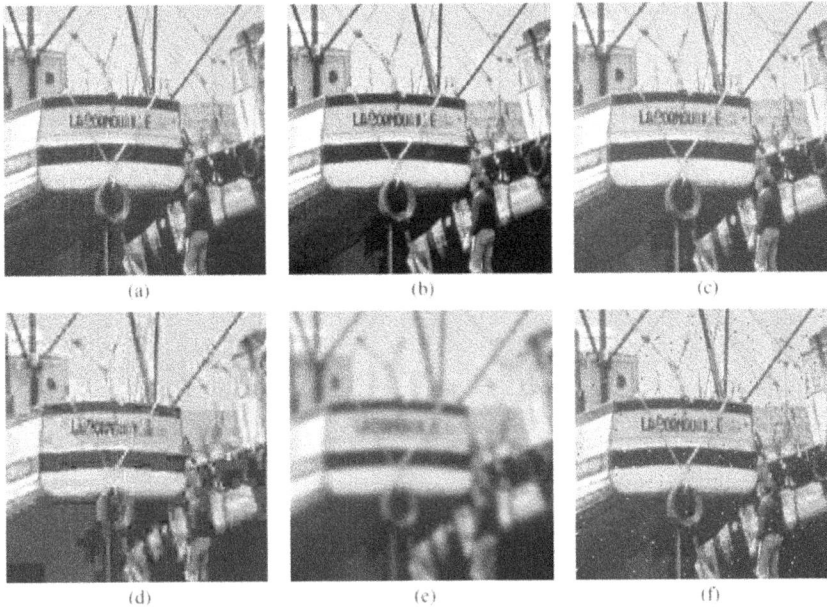

Figure 9.1. Relative to the standard image (a), the five compromised images (b–f) all have the same MSE of 225 but look very different. On the other hand, the SSIM measure (to be explained below) effectively reflects the visually perceived image quality. Reproduced with permission from Wang *et al* (2004). Copyright 2004 IEEE.

their blocky, blurring, or noisy appearance, respectively. Therefore, we cannot rely on MSE alone to assess image quality in a way consistent with the HVS.

This brings us to a biomimetic perspective of image quality assessment based on the HVS that is an invaluable gift to us. As described in chapter 2, the HVS extracts structural information and is adapted for contextual changes. This is why all the compromised images look so different despite the same MSE measure for each of them. This structurally oriented assessment uses the top-down approach in contrast to the pixel-wise comparison and overall error-pooling, required by the bottom-up approach.

Then, how to assess structural distortion in images? A classic result is the structural similarity (SSIM) (Wang *et al* 2004). This measure computes a structural distortion in an image that is consistent with the HVS. Figure 9.2 shows the mechanism by which the SSIM measurement works. SSIM incorporates three aspects of the difference between two images (in practice, image patches of equal size are used, and the SSIM between two images is the average of SSIM between image patch pairs by sliding through the entire images) are used in terms of luminance, contrast, and structure, respectively. To be clear, luminance is the measure of the average value of an image, contrast is the variability within the image. Once we normalize two images to have the same luminance and contrast levels, we are ready to compare their structures. Formally, we wish to incorporate

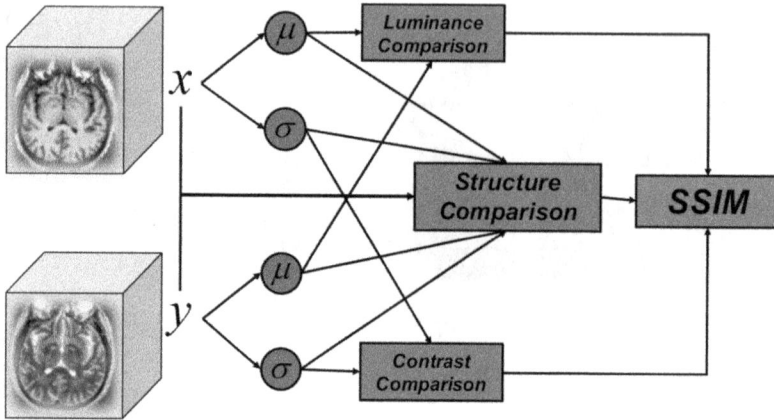

Figure 9.2. Structural similarity (SSIM) integrates differences in luminance, contrast, and structure, and forms a composite number upper-bounded by 1 to reflect an overall visual impression.

these three aspects into a single measure summarizing the similarity between two image patches x and y both with N pixels,

$$S(x, y) = f(l(x, y), c(x, y), s(x, y)), \qquad (9.5)$$

where l is the difference in luminance, c is the difference in contrast, and s is the similarity in structure. When formulating such a similarity measure, we reasonably request the measure to satisfy the following postulates: symmetry, boundedness, and unique maximum:

$$\text{Symmetry:} \quad S(x, y) = S(y, x) \qquad (9.6)$$

$$\text{Boundedness:} \quad S(x, y) \leqslant 1 \qquad (9.7)$$

$$\text{Unique maximum:} \quad S(x, y) = 1 \\ \text{if and only if the two images are the same.} \qquad (9.8)$$

To find an SSIM value, first we need to compute the mean and deviation of each image. The mean is

$$\mu_x \triangleq \mu(x) = \frac{1}{N} \sum_{i=1}^{N} x_i. \qquad (9.9)$$

Note that if we remove the mean from an image, $x - \mu_x$, we have a luminance-normalized image. The standard deviation is computed as

$$\sigma_x(x) = \left(\frac{1}{N-1} \sum_{i=1}^{N} (x_i - \mu_x)^2 \right)^{1/2}, \qquad (9.10)$$

where the denominator $N - 1$ (instead of N) is for an unbiased estimate. After we scale a zero-mean image with its deviation, $(x - \mu_x)/\sigma_x$, we have a mean-contrast-normalized image.

Now, let us perform the luminance comparison in the following heuristically formulated scheme:

$$l(x, y) = \frac{2\mu_x\mu_y + C_1}{\mu_x^2 + \mu_y^2 + C_1}, \tag{9.11}$$

where the constant C_1 is included to avoid instability when $\mu_x^2 + \mu_y^2$ is close to zero. Specifically, we could set $C_1 = (K_1 L)^2$, where K_1 is a very small constant, and L is the range of pixel/voxel values (for example, $L = 2^8 - 1 = 255$ for 8 bit grayscale images). To see how the above expression satisfies the postulates we have previously listed, we give the following simple computation assuming $\mu_x = x > 0$, $\mu_y = a > 0$ and a zero C_1. Letting $f(x) \triangleq l(x, y)$:

$$l(x, y) = f(x) = \frac{2ax}{a^2 + x^2}, \tag{9.12}$$

we have

$$f'(x) = \frac{2a(a^2 + x^2) - 2x(2ax)}{(a^2 + x^2)^2} = \frac{2a(a^2 - x^2)}{(a^2 + x^2)^2} \tag{9.13}$$

and therefore $f'(x) = 0$ if and only if $x = a$. Then, we can use a similar expression for the contrast comparison:

$$c(x, y) = \frac{2\sigma_x\sigma_y + C_2}{\sigma_x^2 + \sigma_y^2 + C_2}, \tag{9.14}$$

where $C_2 = (K_2 L)^2$, K_2 being another small constant like K_1.

Note also that for the same amount of mean or contrast difference ($\Delta\mu = \mu_y - \mu_x$ or $\Delta\sigma = \sigma_y - \sigma_x$), the corresponding comparison will basically depend on the ratio between the difference and the background. This point can be appreciated by looking at the following estimation after ignoring the offset constant C:

$$\begin{aligned} f(x, x + \Delta x) &= \frac{2x(x + \Delta x)}{x^2 + (x + \Delta x)^2} \\ &= \frac{2x^2 + 2x\Delta x}{2x^2 + 2x\Delta x + (\Delta x)^2} \approx \frac{1}{1 + \frac{(\Delta x)^2}{2x^2}} \approx 1 - \frac{(\Delta x)^2}{2x^2}. \end{aligned} \tag{9.15}$$

In other words, a change in stimulus is measured in terms of this ratio, being consistent with the HVS, which is commonly referred to as Weber's law, as illustrated in figure 9.3. Now, we are ready to compare the structural similarity

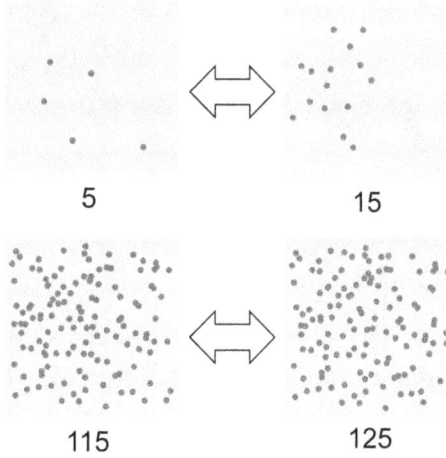

Figure 9.3. Example showing how the HVS interprets a difference within a background. The two images in the top row are in better contrast than the image pair in the bottom row even though the number of dots differs by the same amount (i.e., 10) in the two rows.

between two normalized images $(x - \mu_x)/\sigma_x$ and $(y - \mu_y)/\sigma_y$. The correlation between the two images is an effective measure for this purpose, based on the Cauchy–Schwarz inequality. Since the correlation between the normalized images is the same as the correlation coefficient between the original un-normalized images, it can be used for structural comparison as follows:

$$s(x, y) = \frac{\sigma_{xy} + C_3}{\sigma_x \sigma_y + C_3} \tag{9.16}$$

with the constant C_3 for the same reason as previously mentioned, and

$$\sigma_{xy} = \frac{1}{N - 1} \sum_{i=1}^{N} (x_i - \mu_x)(y_i - \mu_y). \tag{9.17}$$

Note that the denominator is again $N - 1$, and the value of $S(x, y)$ will be no more than 1 due to the Cauchy–Schwarz inequality and reach 1 if and only if $x = ky$ where k is a constant.

Combining (9.14), (9.16), and (9.17), we obtain the SSIM as follows:

$$\text{SSIM}(x, y) = [l(x, y)]^\alpha [c(x, y)]^\beta [s(x, y)]^\gamma, \tag{9.18}$$

where α, β, and γ are parameters to adjust the relative importance of the three components. This expression can be simplified by setting $\alpha = \beta = \gamma = 1$ and $C_3 = C_2/2$, yielding the following widely used form of SSIM:

$$\text{SSIM}(x, y) = \frac{(2\mu_x \mu_y + C_1)(2\sigma_{xy} + C_2)}{\left(\mu_x^2 + \mu_y^2 + C_1\right)\left(\sigma_x^2 + \sigma_y^2 + C_2\right)}. \tag{9.19}$$

The concept of SSIM has been extended in various ways, such as in the cases of color images, time-varying signals, and multi-scale analysis.

9.1.3 Information measures

From the information theoretic perspective, when an imaging system produces an image, a measurement is performed on an underlying object, and inherent information in the image needs to be extracted by a radiologist, a pathologist, or a computer program. Given the rapid development of machine learning techniques, computerized image readers are being actively developed, and will help and eventually replace human experts in many cases. This whole workflow includes multiple steps from data acquisition, through image reconstruction, processing, and analysis, to the final diagnostic report (even including prognostic information). Each of the steps should be optimized to maximize the information content or minimize the information loss.

Information theory is an important branch of modern science, which was pioneered by Claude E Shannon to find the fundamental capability of signal processing and communication (Shannon 1997). In information theory, random variables or processes are used as mathematical models. Among important concepts in this field, entropy and mutual information are the most important and particularly relevant to imaging. In the following, we describe the essential idea of information theory, which demands some abstract thinking and substantial effort to digest the material well; for more details, please read appendix A.

The entropy for a discrete random variable is defined as

$$H(X) = \sum_{x \in X} p(x) \log_2 \frac{1}{p(x)}, \tag{9.20}$$

where the discrete random variable x is defined in the space X with a probability $p(x)$. The meaning of the second factor $-\log_2 p(x)$ on the right-hand side of (9.20) is the number of bits necessary to represent permissible values of the variable x. If you toss a coin, you have two possible events, and you only need 1 bit to record your outcome (for example, 0 for heads, and 1 for tails). If you play a regular octahedral dice, you have eight equally likely results, and you need 3 bits to record the outcome. Of course, if you play a regular dice which has six faces, you still need 3 bits to represent an outcome, since the bit is the minimum unit of the digital memory. Actually, you can play this regular dice many times and group the outcomes for the most efficient utilization of bits so that $\sum_{x \in X} p(x) \log_2 1/p(x)$ is the lower bound of the average code length. In other words, with the probability $p(x)$ as the weighting factor for $-\log_2 p(x)$, the entropy is nothing but the statistical expectation of minimum code length. Clearly, the larger the entropy, the more the uncertainty is associated with the random variable, and the larger the number of bits needed for encoding experimental outcomes; for more details, see standard textbooks on information theory such as (Cover and Thomas 2006).

For a rigorous justification of the entropy, we can use an axiomatic approach (Cover and Thomas 2006). Specifically, we request that the information gain $I(x)$ from the observation of an event x with probability $p(x)$ satisfies the following properties: (i) $I(x)$ is monotonically decreasing in $p(x)$ (an increased likelihood of an event decreases the information from the observed x); (ii) $I(x)$ is non-negative (which is the nature of information gain); (iii) $I(1) = 0$ (an event that happens with certainty carries no information); and (iv) $I(x, y) = I(x) + I(y)$ (the information gain from a collection of independent events is the sum of individual information gains each of which is associated with an event in the set). It can be proved that the definition of entropy $\sum_{x \in X} p(x) \log_2 1/p(x)$ meets all these requirements and gives the minimum average coding length.

For two discrete random variables x and y in their domains X and Y with probabilities $p(x)$ and $p(y)$ as well as joint probability $p(x, y)$, we have the conditional probability distributions as follows:

$$p(x|y) = \frac{p(x, y)}{p(y)} \text{ and } p(y|x) = \frac{p(x, y)}{p(x)}. \tag{9.21}$$

If the two random variables are independent, each conditional probability will be independent of the conditioned variable. If the variables are dependent, how should we measure their dependence?

From the information theoretic viewpoint, we should compute how much information we obtain on one variable after we know the value of the other variable. This consideration leads to the concepts of conditional entropy, total entropy, mutual information, and Kullback–Leibler distance.

The conditional entropy $H(Y|X)$ and total entropy $H(X, Y)$ are naturally defined as

$$H(Y|X) = -\sum_{x \in X} p(x) \sum_{y \in Y} p(y|x) \log_2 p(y|x) \tag{9.22}$$

$$H(X, Y) = -\sum_{x \in X, y \in Y} p(x, y) \log_2 p(x, y). \tag{9.23}$$

For one who is insightful enough, it should be immediately clear that the total entropy should be the sum of the entropy of one variable and the entropy conditioned on that variable; that is,

$$H(X, Y) = H(X) + H(Y|X). \tag{9.24}$$

If we have a composite system Z that consists of two independent subsystems X and Y, with three corresponding random variables x, y, and z with probabilities $p(x)$, $p(y)$, and $p(z)$, then $p(z) = p(x)p(y)$. Then, the total entropy must be the sum of the component entropies (i.e. entropy is additive):

$$\begin{aligned}
H(Z) &= -\sum_{z \in Z} p(z) \log_2 p(z) \\
&= -\sum_{x \in X, y \in Y} p(x)p(y) \log_2[p(x)p(y)] \\
&= -\sum_{x \in X} p(x) \log_2 p(x) - \sum_{y \in Y} p(y) \log_2 p(y) \\
&= H(X) + H(Y).
\end{aligned}$$

To prove (9.24) for a general case that x and y are not independent, we only need to point out that $p(x, y) = p(x)p(y|x)$ from (9.21). In this case, we have $H(Y|X) \leqslant H(Y)$ (i.e. we would have some information about y once we know x if the two variables are statistically relevant). Hence, we can define the mutual information as

$$I_{X,Y} = H(Y) - H(Y|X) = H(X) - H(X|Y). \tag{9.25}$$

Equation (9.25) can be reformulated as follows:

$$
\begin{aligned}
I_{X,Y} &= H(Y) - H(Y|X) \\
&= H(Y) + H(X) - H(X, Y) \\
&= -\sum_{y \in Y} p(y) \log_2 p(y) - \sum_{x \in X} p(x) \log_2 p(x) + \sum_{x \in X, y \in Y} p(x, y) \log_2 p(x, y) \\
&= -\sum_{x \in X, y \in Y} p(x, y) \log_2 p(y) - \sum_{x \in X, y \in Y} p(x, y) \log_2 p(x) + \sum_{x \in X, y \in Y} p(x, y) \\
&\quad \log_2 p(x, y) \\
&= \sum_{x \in X, y \in Y} p(x, y) \log_2 \frac{p(x, y)}{p(x)p(y)}.
\end{aligned}
$$

That is

$$I_{X,Y} = \sum_{x \in X, y \in Y} p(x, y) \log_2 \frac{p(x, y)}{p(x)p(y)}. \tag{9.26}$$

Geometrically, the mutual information measures how far a joint probability distribution is from an independent distribution given by the product of the marginal distributions.

Motivated by the concept of mutual information, the Kullback–Leibler distance between two probability distributions $p(x)$ and $q(x)$ can be defined in the same spirit of (9.26) as follows:

$$D(q\|p) = \sum_{x \in X} q(x) \log_2 \left(\frac{q(x)}{p(x)} \right), \tag{9.27}$$

with the convention that $0 \log_2 0 = 0$ (motivated by $\lim_{x \to 0^+} x \log_2(x) = 0$, i.e. an event with zero probability carries no information) and $0 \log_2(0/0) = 0$ (an interpretation is that zero components are considered as the same). It can be shown that $D(p\|q)$ is convex with respect to $q(x)$, non-negative, and equal to zero if and only if $p(x)$ and $q(x)$ are indistinguishable. Hence, $D(p\|q)$ is qualified as a distance measure, although it is not symmetric, i.e. $D(p\|q) \neq D(q\|p)$ in general.

In subsection 9.1.1, we have described the least squared measure as the Euclidean distance. There are actually many types of distances, and by any measure the Kullback–Leibler distance is among the most important ones in many applications; for example, in image processing (registration) and machine learning. Also, in subsection 9.1.2, we have introduced SSIM, which involves a correlation coefficient (9.17) to measure the structural similarity. While the correlation coefficient only reflects the linear correlation, the mutual information (9.26) effectively captures the nonlinear dependence in the sense of information theory.

9.2 System-specific indices

Since tomographic scanners produce images, we are interested in system specifications on various aspects of the image quality. These specifications are nominal, and often provided by manufactures and monitored by medical physicists for quality control. Technological advancements lead to constantly improved system-specific image quality indices.

First, data noise is what we do not want but it is unavoidable with any medical imaging measurement. The noise will demonstrate itself as random fluctuations on top of presumably true signals, because of the probabilistic nature of the underlying physics. Given data noise and imperfectness of imaging system components, images computed from noisy data necessarily contain noise as well. The signal-to-noise ratio (SNR) is typically defined as the ratio of the magnitude of a signal and the standard deviation of noise,

$$\text{SNR} = \frac{A_{\text{Signal}}}{\sigma_{\text{Noise}}}, \tag{9.28}$$

where A denotes the signal magnitude, and σ is noise deviation. Sometimes, the SNR is also defined as the ratio between the power of the signal and the variance of the noise. Then, the peak SNR (PSNR) can be defined as

$$\text{PSNR} = 10\log\left(\frac{A_{\text{Signal-Max}}}{\sigma_{\text{Noise}}}\right)^2, \tag{9.29}$$

where $A_{\text{Signal-Max}}$ is the maximum signal in an image. Clearly, information extraction will be more difficult with a lower SNR than in the case of a higher SNR. A noise-interfered image of 'Deep Recon' is shown in figure 9.4. The noise in the above PSNR definition can be replaced by MSE, leading to an alternative PSNR measure as follows:

$$\text{PSNR} = 10\log\frac{L^2}{\text{MSE}}, \tag{9.30}$$

where L is the range of pixel/voxel values. If a CT image pixel value is coded in 12 bits, L is equal to $2^{12} - 1 = 4095$. This PSNR is convenient to compare images. A closely relevant measure of SNR is the contrast-to-noise ratio (CNR) defined as

$$\text{CNR} = \left|\frac{A_{\text{Signal-1}} - A_{\text{Signal-2}}}{\sigma_{\text{Noise}}}\right|, \tag{9.31}$$

which is focused on the difference between two signals, normalized by the noise deviation.

Figure 9.4. Signal embedded in different noisy backgrounds from weak (top), moderate (middle), and strong (bottom) noise respectively.

Second, various kinds of image resolution are used to quantify the resolving power of an imaging system. The target to be resolved by the imaging system can be structural details, shading differences, temporal changes, or spectral contents, which correspond to spatial resolution (or simply, resolution), contrast resolution (sometimes, contrast), temporal resolution (i.e., time resolution), and spectral resolution (also, energy resolution) respectively. Quantitatively, resolution of any kind is a measure of the smallest difference between two components of interest with which they can be still differentiated.

For a point object modeled as a delta function, an imaging system will not be capable of capturing it perfectly. The resultant image will be a blurred version of this point, which is called the impulse response or point spread function (PSF) of the imaging system. Then, spatial resolution is the distance between two bright spots at which they can be told apart and within which they become indistinguishable. Suppose that the spots are of bell shape, which is also referred to as Gaussian blurring, when they overlap too much they will be blurred together. Hence, it is heuristic to define spatial resolution as the full width at half maximum (FWHM), as shown in figure 9.5. When the separation is less than the FWHM, the two spots are fused into one cluster and we cannot resolve them visually.

FWHM is only one of many spatial resolution measures. Another good resolution measure is the modulation transfer function (MTF). With an ideal point/line object as an input, the output of the imaging system will be a blurred version of the original structure. The Fourier transform is then performed for the input and the output, respectively. Their resultant Fourier spectra are $I(f)$ and $O(f)$, respectively. MTF is defined as

$$\text{MTF}(f) = \frac{|O(f)|}{|I(f)|}, \tag{9.32}$$

where f is the spatial frequency. A typical MTF curve is shown in figure 9.6. Note that noise has significant components from low through high frequencies, and the so-called white noise has a uniform Fourier spectrum. When the value of MTF goes below the noise level at a sufficiently high frequency, image details and noise

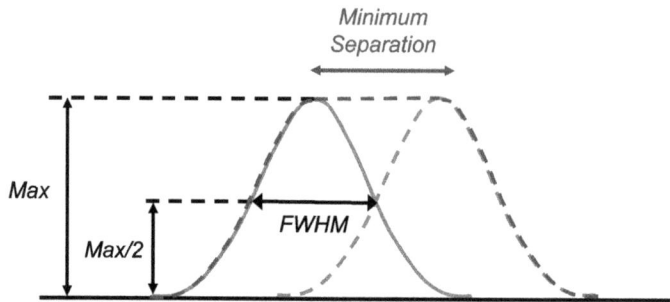

Figure 9.5. Image resolution defined as the FWHM of a point spread function, which is often in a Gaussian form for CT, PET, and SPECT scanners. When two points (red solid and dashed curves) are separated by a distance less than the FWHM, they will be merged into a single cluster and cannot be visually recognized as two peaks (blue dashed curve).

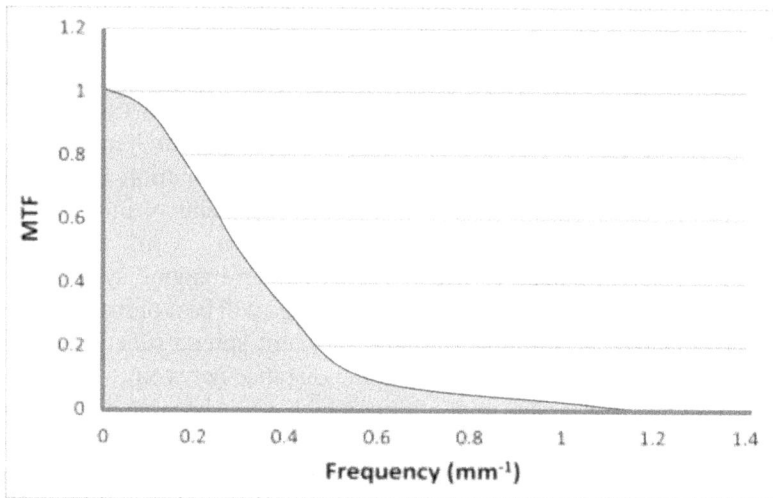

Figure 9.6. Normalized MTF curve.

fluctuations are no longer separable, and it is reasonable to associate the spatial resolution to that critical frequency. In a similar spirit, the spatial resolution can be specified in terms of certain line pairs per millimeter or centimeter that can be barely resolved, as shown in figure 9.7 (with any smaller separation the lines cannot be reliably recognized). A recent image example from an ultrahigh resolution CT scanner is shown in figure 9.8.

Another important type of resolution is contrast resolution, which targets subtle shading differences. For example, nodules may look very similar to normal biological soft tissues in CT images, and they do not present good contrast when image noise is strong or the imaging protocol is not optimized, such as when the x-ray tube voltage is too high, making x-ray photons too energetic and penetrating a patient too easily. Figure 9.9 shows significant differences between two dental radiograms with different contrasts.

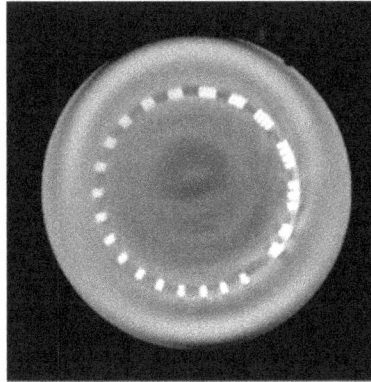

Figure 9.7. CT performance phantom Catphan 528 (left) used to measure spatial resolution, indicating a resolution measure of ten line pairs per centimeter (right) on a universal CT benchtop system at the AI-based X-ray Imaging System (AXIS) Lab at RPI.

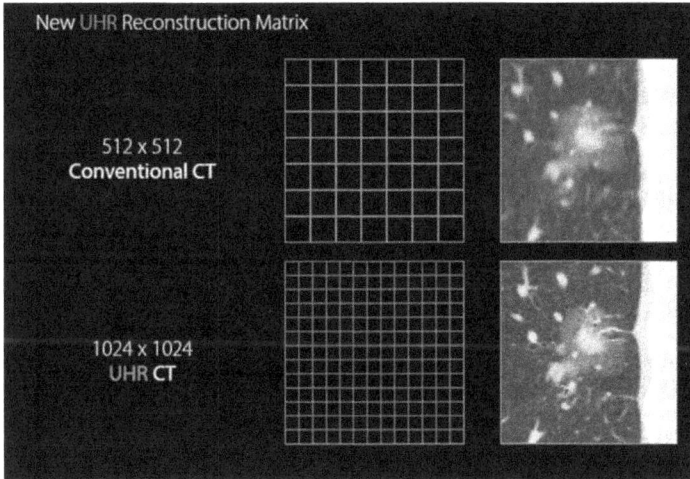

Figure 9.8. Canon medical CT scanner equipped with 0.25 mm detector technology to improve the standard spatial resolution of multirow-detector CT (MDCT) (top) significantly for ultrahigh resolution CT (UHRCT) (bottom) (courtesy of Fujita Health University Hospital, Japan).

Temporal resolution measures how quickly an imaging system can take a snapshot so that a moving structure, such as a beating heart, can be 'frozen' to avoid motion blurring. For example, ultrasound imaging forms an image quickly, and is suitable to determine cardiac functions. In contrast, magnetic resonance imaging (MRI) cannot have both fine spatial resolution and fast temporal resolution at the same time. As shown in figure 9.10, the temporal resolution of cardiac CT can be improved by not only a higher gantry rotation speed but also more advanced image reconstruction algorithms for motion correction. Deep learning based cardiac CT and cardiac MRI are two related good topics.

Spectral resolution is also highly important. For optical imaging, we can use various fluorescent probes to label biological biomarkers in different colors. Thus,

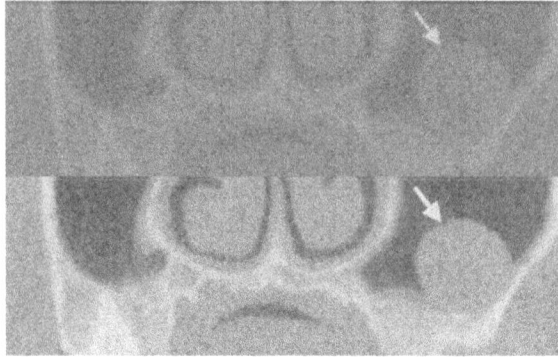

Figure 9.9. Two dental x-ray radiographic images from a patient in low (top) and high (bottom) contrast, respectively, where the yellow arrow indicates a pseudo-cyst (mucus retention) in the left maxillary sinus.

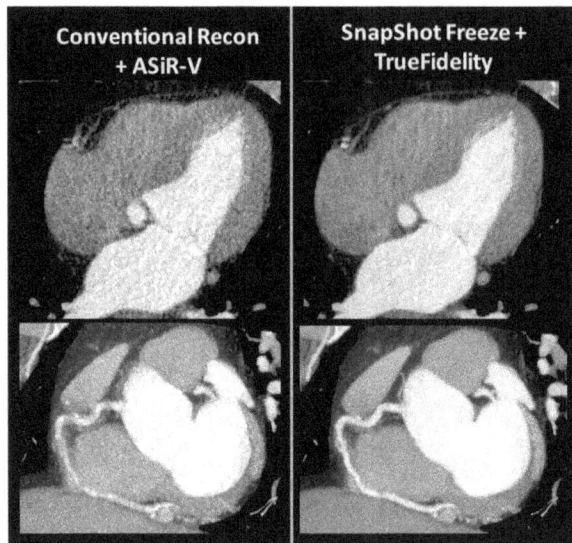

Figure 9.10. Cardiac CT motion artifacts at a high heart rate (left column), and motion correction (right column) using the smart reconstruction technique Snapshot Freeze delivering a temporal resolution of 30 ms. Reproduced with permission from General Electric.

we can produce colorful images. As a further example, x-ray images used to be in grayscale but now a micro-CT or CT scanner can be equipped with the so-called photon-counting detector to measure x-ray photons individually and energy specifically. However, the response of a photon-counting detector to a monochromatic radiation source such as Ba for 80 keV will spread over an energy range around 80 keV and blur the energy resolution, as illustrated in figure 9.11. Note that the spectral blurring depends on x-ray energy, detector materials, and design details. As a result, the detected x-ray energy profile will be a blurred version of the ideal profile. With x-ray energy sensitive data, the K-edge of contrast tracers can be resolved for chemically specific material decomposition.

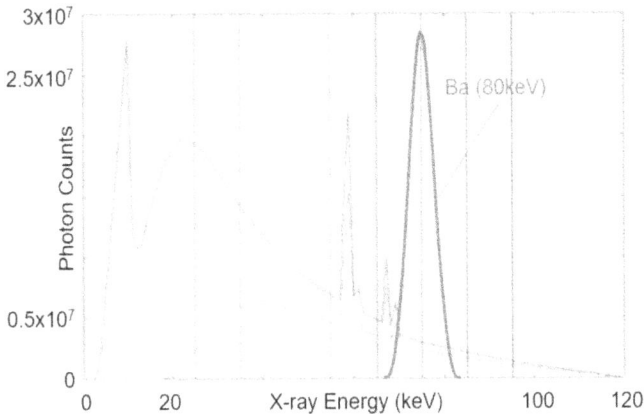

Figure 9.11. Original x-ray energy spectrum (light blue curve), attenuated spectrum (orange curve), and photon-counting detector response (red curve) to a monochromatic radiation source Ba for 80 keV. Given the finite energy resolution of about 10 keV, a limited number of photon-counting energy bins are used, delimited by user-specified energy thresholds (vertical colored lines). Reproduced with permission from Mathew Getzin's PhD dissertation advised by Dr Ge Wang (Getzin 2019)

In addition to various resolution measures, the concept of image artifacts must be explained. Image artifacts are those structures in images that are not real. Hence, the artifacts are ghosts in the image domain. The sources and types of imaging modality-dependent image artifacts are multiple, and can be compensated for or eliminated using dedicated techniques.

For example, when metallic implants are present inside a patient, x-rays cannot penetrate them effectively, leading to poor SNR data or missing data. When we reconstruct a CT image from compromised x-ray data, streaking and darkening artifacts will appear around the metal parts. These artifacts are not real but they block anatomical and pathological features. Figure 9.12 shows two cases, in which there are the true, uncorrected images, and images corrected using metal artifact reduction (MAR) algorithms including a state-of-the-art method called normalized MAR (NMAR) and our in-house convolutional neural network (CNN) method.

Image blurring is another example of image artifacts. Modern medicine heavily relies on CT and MRI images. Since pathological features are small at an early stage, we want to have CT and MRI images that are as clear as possible, revealing small features and even subtle details. However, the higher the requirements for image clarity, the more complicated an imaging scanner needs to be, and the more expensive the healthcare costs. Making an image clearer overall and closer to the ground truth is a special topic of image processing, called deblurring or super-resolution imaging, meaning something we could do via image processing so that an image from an imaging system can be improved to a resolution better than the original resolution of the system (which sounds like a kind of free lunch).

Recently, machine learning, in particular deep learning, has become an extremely hot topic, given its tremendous successes in image processing and computer vision. To a great degree, deep neural networks can simulate the HVS. The HVS is amazing. When you see a horse blocked by trees in a forest, you can perceive the

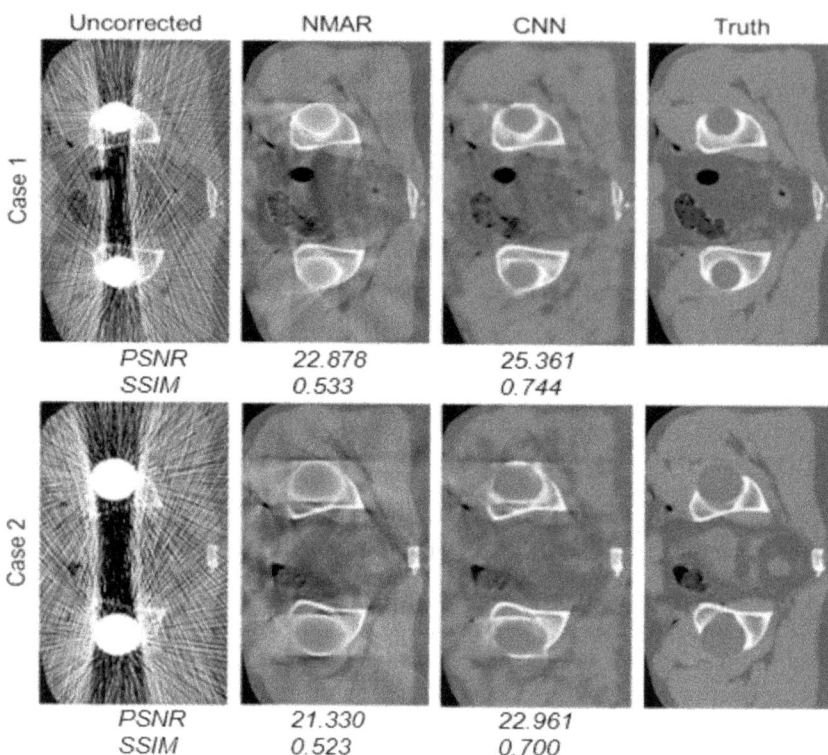

	Uncorrected	NMAR	CNN	Truth

Figure 9.12. Metal artifact affected, corrected and true CT images. Reproduced with permission from Lars Gjesteby's PhD dissertation advised by Dr Ge Wang (Gjesteby 2018).

horse in its entirety. In that case, you can do so because you have extensive knowledge of trees and horses as well as visual reasoning power in the context, and you can infer confidently what the blocked features of a horse should be. This visual perception can now be achieved via image processing using deep learning techniques. Specifically, a deep neural network can be designed to mimic our neurological network responsible for vision, and trained on a big dataset consisting of numerous pairs of low-quality (blocked, blurred, or otherwise compromised) and corresponding high-quality images (say, unblocked horses). After training the deep artificial neural network, it becomes a human-like expert in this application domain: whenever a compromised image is presented to the network, it will process the image, and finally deliver a more informative counterpart.

In 2016, we wrote a perspective on deep-learning-based tomographic imaging (Wang 2016), which is the basis for an *IEEE Transactions on Medical Imaging* special issue on this theme (and also for our machine-learning-based tomographic imaging patent applications) (Wang *et al* 2018). In that perspective, we presented a machine-learning-based super-resolution imaging example. Recently, we posted two arXiv papers to improve medical CT and MRI images towards super-resolution (Lyu *et al* 2018, You *et al* 2019). While traditional deblurring methods could only improve CT or MRI image resolution by a modest fraction, deep learning methods

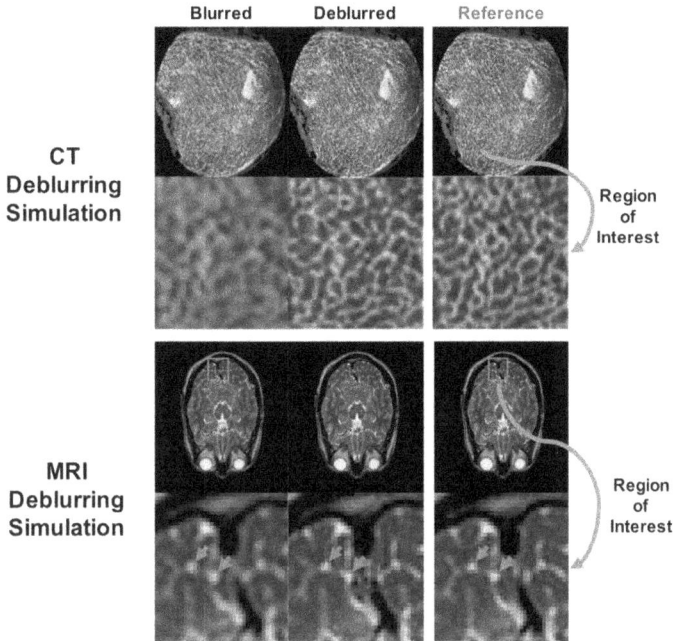

Figure 9.13. Our simulated CT and MRI super-resolution results, which show resolution improvements of 100% or more.

can do much better, with 100% or even 200% resolution improvement in our simulation studies, as shown in figure 9.13. The major implication is that super-resolution deep learning could computationally turn a cheap imaging device into a decent one by making existing images reliably and faithfully clearer.

9.3 Task-specific performance

The specifications of an imaging system include multiple system-specific parameters/indices, and not all of them are equally important for a given diagnostic/interventional task. Most imaging studies are for a single disease/procedure, and image quality assessment should be task-based to optimize the clinical performance. In this section, we will introduce a few basic measures that are of direct clinical interest.

After reading an image, an interpretation must be given, which can be correct or incorrect, depending on inherent image quality and also subsequent image analysis by an expert, who may be outstanding or on a learning curve. As an example, let us say that we wish to detect a lung nodule in a patient. Only four mutually exclusive scenarios are possible: (i) there is a nodule, and it is detected, known as a true positive (TP); (ii) there is no nodule, and it is confirmed as no nodule, known as a true negative (TN); (iii) a nodule exists but it is not detected and falsely reported as no nodule, known as a false negative (FN); and (iv) no nodule exists, but a tumor is falsely reported, known as a false positive (FP). Depending on the task, false negatives and false positives can have different degrees of adverse effects. For example, in the case of lung screening, a false negative is usually far worse than a false positive.

Two fundamental concepts in medicine are sensitivity and specificity. Sensitivity is defined as $TP/(TP + FN)$, where the numerator is the number of true positive reports, and the denominator is the total number of all positive cases. Sensitivity is the likelihood of detecting a pathological feature positively when the case is indeed positive; in other words, it is a measure of how surely we can say 'yes'. On the other hand, specificity is defined as $TN/(TN + FP)$, where the numerator is the number of true negative reports, and the denominator is the total number of all negative cases. Specificity is the likelihood of confirming a case as negative when the case is indeed negative; in other words, it is a measure of how surely we can say 'no'. Two related concepts are positive predictive value (PPV) and negative predictive value (NPV). PPV is defined by $TP/(TP + FP)$, where the denominator is the total number of all positively reported cases. PPV is the fraction of patients who have positive results that are actually positive. On the other hand, NPV is defined as $TN/(TN + FN)$, where the denominator is the total number of all negatively reported cases. NPV is the fraction of patients who have negative results that are actually negative. Furthermore, diagnostic accuracy (DA) is the ratio between the number of correctly reported cases and the number of patients, prevalence (PR) is the ratio between the number of positive cases and the number of involved subjects.

To highlight how PPV and NPV are different from sensitivity and specificity, see table 9.1. In this lung CT screening study, sensitivity is 20/30 (66.7%), specificity is 1700/1750 (97.1%), PPV is 20/70 (28.6%), NPV is 1700/1710 (99.4%), DA is (20+ 1700)/1780 (96.6%), and PR is 30/1780 (1.7%).

Now, we are ready to discuss a practically important but slightly confusing tool: receiver operating characteristic (ROC) analysis. This tool relies on a curve in terms of specificity and sensitivity to depict the diagnostic performance of an imaging study. The plot focuses on sensitivity as a function of '1-specificity' which is the rate of false positives or false alarms, as shown in figure 9.14.

To understand figure 9.14 better, let us consider two extreme cases. First, what if we blindly claim that all the subjects have diseases? In this case, all healthy patients are called diseased, and the rate of false positives will be 100% (specificity will be 0%). On the other hand, sensitivity will be 100%, since no patient with disease will be called healthy. This case is represented by the diagnostic performance point, or the characteristic point, at the top-right corner. Second, what if we blindly claim all

Table 9.1. Diagnostic analysis on lung CT screening results from 1780 subjects.

CT reading	Lung nodule existence/non-existence		
	Yes	No	Total
Positive	20	50	70
Negative	10	1700	1710
Total	30	1750	1780

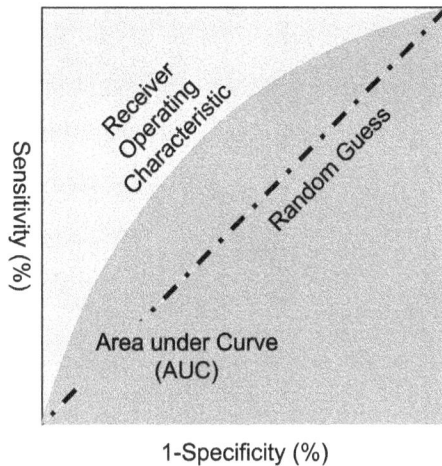

Figure 9.14. Receiver operating characteristic (ROC) curve with the horizontal and vertical axes for '1-specificity' (false positive or false alarm rate) and sensitivity, respectively.

patients are healthy? In this case, all diseased patients are called healthy, and the rate of false positives will be 0% (specificity will be 100%), but sensitivity will be 0%, since no patient is called diseased. This case is the point at the bottom-left corner. Third, we can just flip a fair coin to decide if we call a subject healthy or diseased. Then, both sensitivity and specificity will be 50%. If the coin is not fair, the characteristic point will be somewhere on the diagonal.

Clinically, a population is categorized into two groups: healthy and diseased. In a common case, there are differences between image features from the two groups, potentially making a good distinction between those who are healthy and those who are diseased. However, often the distribution of these features has a significant overlap between healthy and diseased cases, resulting in no clear division between the two classes. Due to this ambiguity, different thresholds can be set to achieve varying levels of sensitivity and specificity. In other words, many possible combinations of sensitivity and specificity are possible, with the characteristic points above the diagonal (otherwise, we can just switch the classification results 'healthy' and 'diseased').

As shown in figure 9.15, with different thresholds the corresponding diagnostic decisions will lead to various selectivity and specificity values. Each decision rule (threshold) will define a unique characteristic point in the sensitivity–specificity plane. All such points form an ROC curve as a representation of the diagnostic performance. Better ROC curves will be those bulging farther to the upper left corner, having greater areas under the ROC curve.

It should be underlined that ROC analysis can be applied not only to assess image quality but also to evaluate doctors' expertise or the capability of image analysis software in extracting diagnostic information from images. In this way, we can measure an individual doctor's performance with regard to a diagnostic task, since each doctor has his/her own interpretation and therefore his/her own ROC curve.

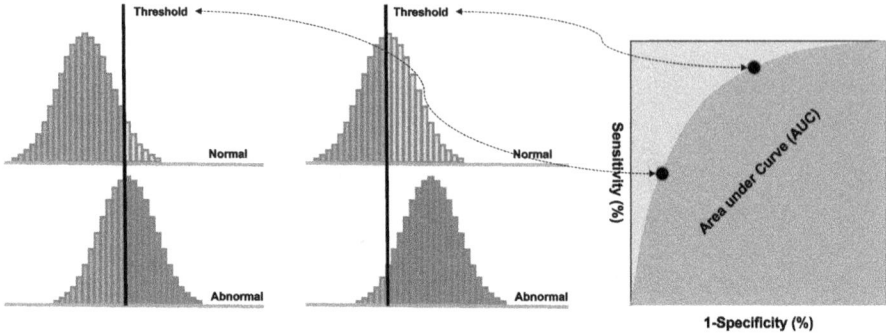

Figure 9.15. Area under the ROC curve as a good measure of the diagnostic performance.

Similarly, we can compare computer-aided diagnostic programs with trained radiologists. In all these cases, better doctors or higher quality software packages will follow an ROC curve closer to the upper left corner. It is widely believed that the day is not far away that artificial intelligence will outperform average radiologists in most diagnostic tasks (Wang *et al* 2017)!

To assess an imaging system, a protocol, or a computer aided diagnostic (CAD) software package for a specific diagnostic task, a human reader study can be performed in terms of ROC curves such as areas under these curves. However, radiologists are rather expensive, and also a human reader study is time-consuming. Therefore, it is highly desirable to have a cost-effective system that automatically performs a diagnostic task as reasonably well as a radiologist. This will greatly facilitate the evaluation and translation of new imaging techniques.

To appreciate how a machine can perform a radiologist's job, let us first describe the algorithmic mechanism called the model observer, including the ideal observer, Hotelling observer, and channelized Hotelling observer (He and Park 2013). Let us start with a linear imaging system model, which is a good approximation for a majority of imaging systems and typically in a discrete form:

$$\mathbf{g} = \mathbf{M}\mathbf{f} + \mathbf{n}, \tag{9.33}$$

where $\mathbf{f}$ is a vector for the patient ground truth being imaged, $\mathbf{M}$ is an operator that represents an imaging system, $\mathbf{n}$ is a noise background in the imaging process, and $\mathbf{g}$ is an image vector. Suppose that there are two cases to be reported based on an image produced by the imaging system: positive or negative (for example, there is a nodule or not). This binary classification task can be formulated as a statistical hypothesis test:

$$\begin{aligned} &\mathbf{H}_0\colon \mathbf{g} = \mathbf{M}\mathbf{b} + \mathbf{n} \\ &\mathbf{H}_1\colon \mathbf{g} = \mathbf{M}(\mathbf{b} + \mathbf{s}) + \mathbf{n}, \end{aligned} \tag{9.34}$$

where $\mathbf{H}_0$ is the outcome from a negative case, $\mathbf{H}_1$ is for a positive case, $\mathbf{b}$ is the background image (or the features of an image that do not correspond to any pathological signature), and $\mathbf{s}$ is a signal (or the features of interest that indicate a positive result, such as a nodule).

In an ideal case, the observer utilizes all statistical information available on the diagnostic task to maximize the diagnostic performance. The ideal observer should extract all relevant features, have all prior knowledge, and achieve the best diagnostic performance as measured by the area under the ROC curve among all the observers, based on the same image g. It is conceptually clear that the ideal observer should make its decision according to the likelihood ratio:

$$LR(\mathbf{g}) = \frac{p(\mathbf{g}|H_1)}{p(\mathbf{g}|H_0)}, \tag{9.35}$$

where $p(\mathbf{g}|H_0)$ and $p(\mathbf{g}|H_1)$ are conditional probability density functions, which are generally too complicated to be available. Recognizing that it is practically infeasible to compute the likelihood ratio, we can take a linear system assumption that an image can be linearly mapped to an informative feature $\mathbf{h}$ (that is, $\mathbf{h}$ is the inner product of the image vector and a template vector), and then a decision can be made according to this feature $\mathbf{h}$. If this feature is closer to the distribution of feature values from a healthy population, we infer that the image should come from a healthy individual. On the other hand, if the feature is closer to the distribution of diseased features, the image suggests an abnormality. To formulate a reasonable closeness to either of the distributions, we can define the following SNR:

$$SNR = \frac{\mu_1 - \mu_0}{\sqrt{\frac{\sigma_0^2 + \sigma_1^2}{2}}}, \tag{9.36}$$

where μ_0 and μ_1 are the means of the normal and abnormal distributions, and σ_0 and σ_1 are associated standard deviations respectively. The resultant optimal observer is called the Hotelling observer. The Hotelling observer and the ideal observer are the same if the involved images are from multivariate Gaussian distributions with any hypothesis-related difference only reflected in the mean vectors under the hypotheses H_0 and H_1. As an extension of the Hotelling observer, the channelized Hotelling observer will utilize a number of features of an image in preselected channels such as different regions in the Fourier domain. There are various other kinds of model observers, most of which were designed according to classic pattern recognition theory.

Very recently, Zeng et al (2019) at the FDA characterized the two earliest deep-learning-based denoising algorithms (Chen et al 2017a, 2017b), and compared them with filtered backprojection (FBP) and model-based iterative reconstruction (MBIR) via total variation minimization in terms of standard and task-specific metrics including contrast-to-noise ratio, modulation transfer function, and noise power spectrum, as well as low-contrast detectability. The training data were from the AAPM Low-dose CT Grand Challenge including full-dose and simulated quarter-dose patient CT FBP images. The testing data are some standard phantoms that are piece-wise uniform and rather different from the training data. The result is shown in figure 9.16. It was observed that similar to model-based iterative reconstruction, these earliest deep learning networks have nonlinear behavior, and

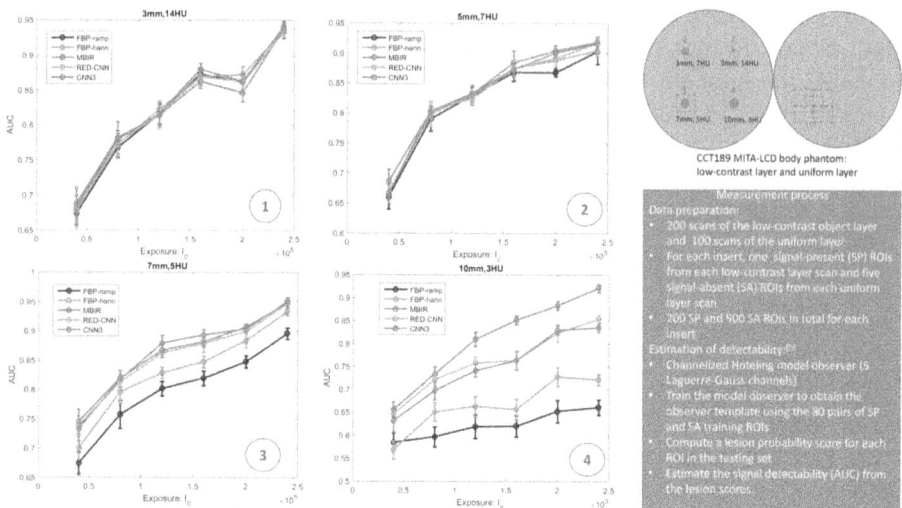

Figure 9.16. Comparative study between two of the earliest deep learning methods, filtered backprojection (FBP) and model-based iterative reconstruction (MBIR) via total variation minimization on the low-contrast detectability using a piece-wise uniform phantom. Reproduced with permission from Zeng *et al* (2019). Copyright 2019 The American Association for Physicists in Medicine.

shifted the noise frequency peak toward the DC frequency. With the piece-wise uniform phantom data, these earliest deep learning networks performed inferiorly to MBIR in the contrast–noise ratio and modulation transfer function, but delivered similar task-based performance when the sizes of simulated lesions are small ($\leqslant 7$ mm) in a channelized Hotelling model observer study. Clearly, more efforts are needed for both network development and systematic evaluation.

9.4 Network-based observers*

Although the use of the channelized Hotelling observer is popular and successful in the medical imaging field, machine learning, in particular deep learning, has opened a door to new opportunities for image quality assessment, aided by neural networks driven by big data. In principle, a neural-network-based observer should be able to do a better job than a classic observer when the underlying probability density functions are high-dimensional, too complicated, and the optimal decision processes are nonlinear and nontrivial.

As mentioned earlier, to avoid time-consuming human reader studies, numerical model observers were developed, and used to mimic the performance of human observers. Since the ideal observer is impractical, the Hotelling observer was extensively studied, but it is subject to major approximations such as linearity. Deep learning holds a great promise in modeling human observers for several reasons. Most importantly, a deep neural network is highly nonlinear, data-driven, and consistently demonstrates superior performance as compared to classic data analysis methods. In several recent studies, machine-learning-based model observers were designed and produced promising results.

In 2017, a convolutional neural network (CNN) was used as a model observer. This CNN model observer was compared to alternative model observers including the relevance vector machine (RVM)-based observer and channelized Hotelling observer (CHO) for SPECT-based target detection (Massanes and Brankov 2017). In this study, the data from a previous study were re-used, and the task was for a human observer to detect a small Gaussian object in the presence of correlated noise whose spectrum was defined by a power law. For this task, different CNN nodes, configurations and layers were evaluated from a single hidden layer network to a more complex network consisting of multiple convolutional layers. 80% of the data were used for training, and the rest for testing. Then, the CNN setting was selected of the highest agreement with the human observer on the testing dataset. It was demonstrated that the CNN-based model observer achieved decent agreement with the human readers (figure 9.17) (Massanes and Brankov 2017).

In 2018, a neural-network-based observer was constructed for CT (Kopp *et al* 2018). In this study, a softmax regression model was trained as the model observer, and compared with the channelized Hotelling observer (CHO). Labeled CT images of the Catphan phantom along with confidence ratings were obtained in a reader study for a task of detecting different signals (disks of specified sizes and elevated intensities) at various noise levels reconstructed using representative reconstruction algorithms. It was reported that the neural-network-based model observer produced excellent agreement with an experienced radiologist (figure 9.18) (Kopp *et al* 2018).

The above two early network-based observers are exemplary but simplistic, since they only applied a brute-force fitting scheme to generic neural networks. Along this direction, there is great space for imagination, suggesting numerous research opportunities. While traditional observers were designed based on classic statistical models, and often in light of linear system theory, neural-network-based observers can be much more powerful. An inspiring fact is that in many image classification tasks, deep neural networks have achieved performance metrics better than or on par with what a human can do on the ImageNet dataset. The ImageNet dataset is well known, and organized in the WordNet hierarchy. Every concept in WordNet,

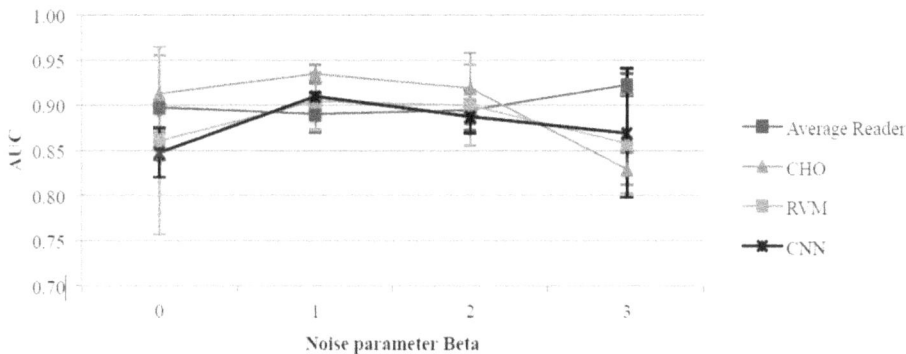

Figure 9.17. Comparative results in the experimental setting where the training was performed on 3600 samples. The error bars represent 95% of the confidence interval. Reproduced with permission from Massanes and Brankov (2017). Copyright 2017 SPIE.

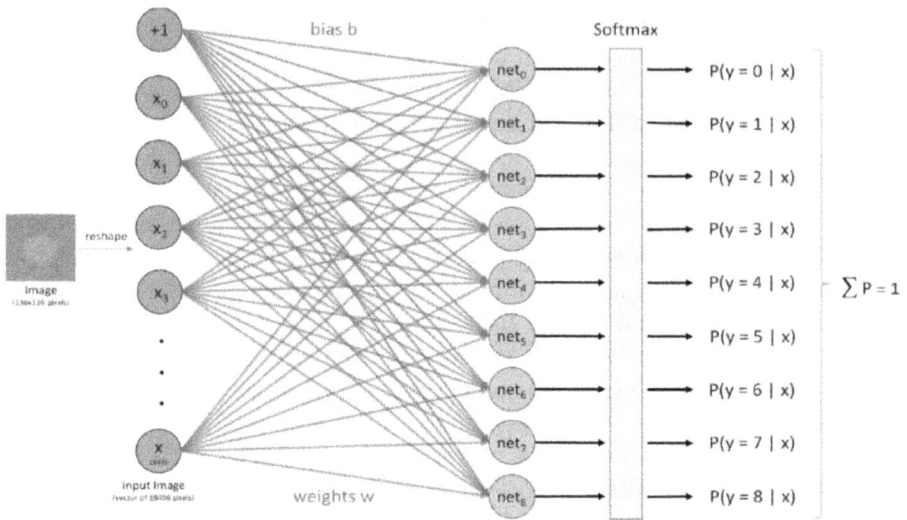

Figure 9.18. Neural network for softmax regression. For an input image x, the neural network gives the probability P that a radiologist's confidence rating is y. Reproduced with permission from Kopp *et al* (2018). Copyright 2018 SPIE.

described by multiple words or phrases, is called a 'synset'. There are over 100 000 synsets in WordNet, each of which is illustrated with about 1000 images. In terms of the so-called top-five classification error rate on the ImageNet dataset (top-five error means the percentage of those test samples that are not covered by the top-five suggested answers), human error is about 5.1%, and the error with a state-of-the-art neural network is now only 3.57% (figure 9.19) (Dodge and Karam 2017). It can be hypothesized that these neural networks for general image classification tasks may have well reflected or at least effectively simulated the HVS, and shed light on the optimal design of next-generation numerical observers.

To harvest perceptual information extracted by the deep neural networks designed for general image analysis tasks, we proposed using a generative adversarial network (GAN) with a perceptual loss that senses the difference in an appropriate feature space for two main reasons (Yang *et al* 2018). First, human perception is not pixel-wise formed, rather it extracts semantic features (even topological features). Thus, we can use a pre-trained deep CNN, such as the famous VGG, for feature extraction at low and intermediate levels where natural images in the ImageNet dataset and medical images should have very similar characteristics. Second, medical images are not uniformly distributed in a high-dimensional Euclidean space, instead they are on a low-dimensional manifold. By computing intrinsic image features, we can compare images meaningfully on a relevant manifold by calculating the geodesic distance. Therefore, the use of the perceptual loss allows structural comparison in a more advanced fashion than what SSIM can do. Our deep neural network with the perceptual loss is shown in figure 9.20, and produced excellent low-dose CT denoising results, shown in figure 9.21.

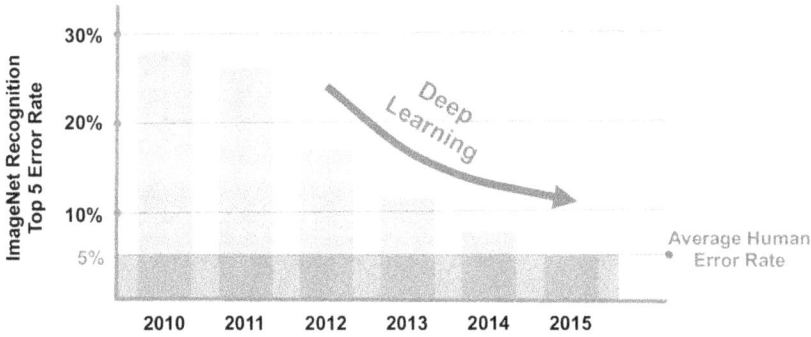

Figure 9.19. Deep neural networks with significant performance improvements over recent years, being competitive with the human-level performance on the ImageNet dataset.

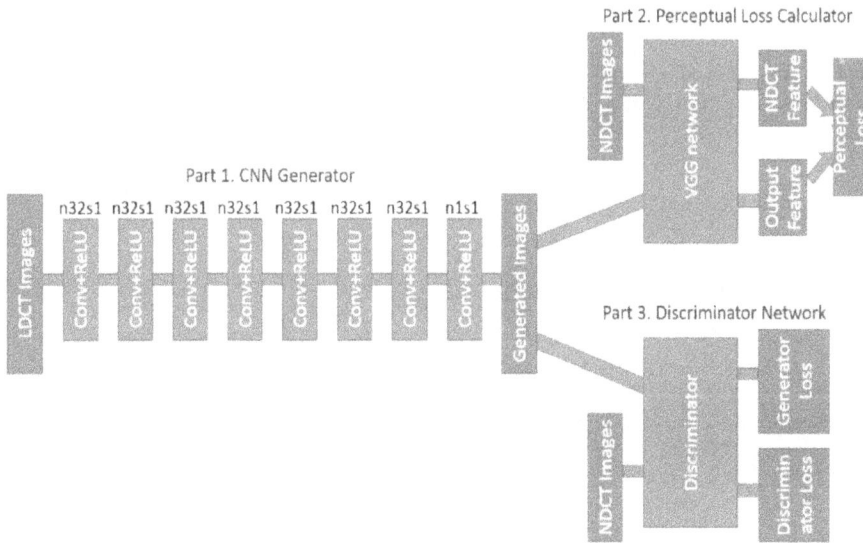

Figure 9.20. Our proposed deep neural network with the perceptual loss for low-dose CT denoising. Reproduced with permission from Yang *et al* (2018). Copyright 2018 IEEE.

For those who are not familiar with VGG, it refers to a deep convolutional network developed by the Visual Geometry Group (VGG) at University of Oxford, who have achieved decent performance on the ImageNet dataset (Simonyan and Zisserman 2014). VGG is very famous, since it has a publicly transparent structure and weights, and is directly downloadable from the website http://www.robots.ox.ac. uk/vgg/research/very_deep. The overall architecture of VGG is shown in figure 9.22.

In 2018, a deep-learning-based model observer was developed at Mayo Clinic that operates on patient images and was shown to simulate the human observer well in the context of low-contrast lesion detection (Gong *et al* 2018). Their network-based observer integrates a pre-trained convolutional residual network (ResNet)

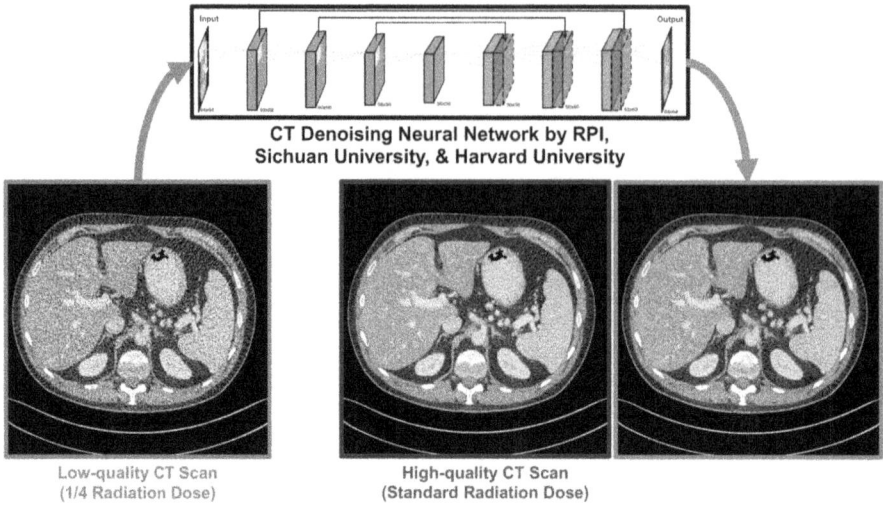

**CT Denoising Neural Network by RPI,
Sichuan University, & Harvard University**

Low-quality CT Scan
(1/4 Radiation Dose)

High-quality CT Scan
(Standard Radiation Dose)

Figure 9.21. Low-dose CT denoising example showing that a deep neural network equipped with an adversarial/perceptual loss produced an excellent denoising result. For more details on the denoising algorithm, see Shan *et al* (2018). Reproduced with permission from Shan *et al* (2018). Copyright 2018 IEEE.

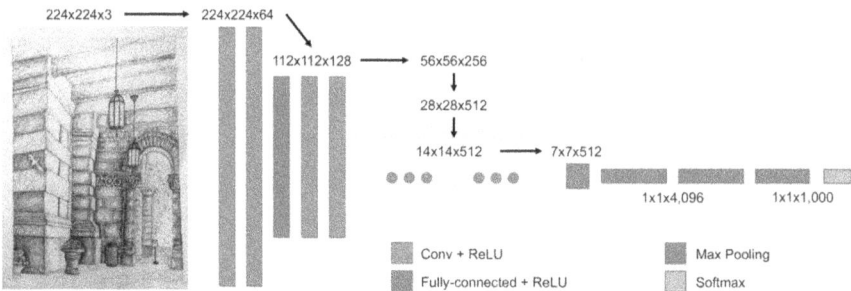

Figure 9.22. Overall architecture of VGG whose trained parameters are publicly downloadable. Image credit: Hannah Wang.

with partial least squares regression (PLSR) and an internal-noise generator, as shown in figure 9.23.

ResNet performs so-called residual learning, which is a breakthrough in the deep learning field (He *et al* 2016). For deep neural networks, residual learning facilitates the training process and improves the overall performance of the network. Also, assuming similarity between the deep network and the HVS, early layers of ResNet were used to extract low-level features. Then, PLSR was trained over extracted features to generate model observer statistics for testing. PLSR is a statistical method, related to principal component regression (Lin *et al* 2012, Panchuk *et al* 2018). Different from the principal component analysis that only focuses on the input domain, PLSR maximizes the covariance between the output/response and input/independent variables by projecting the input and output variables to new

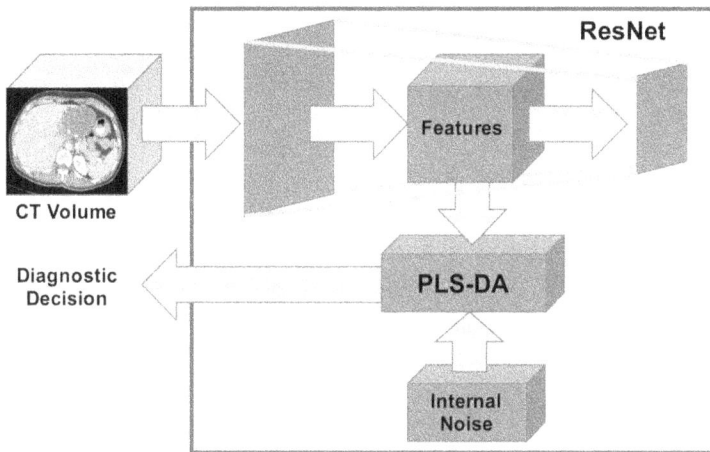

Figure 9.23. Network-based model observer developed at Mayo (Gong *et al* 2018).

spaces of reduced dimensionalities and establishing the relationship between the new spaces. If the output/response is categorical, we call the techniques the partial least squares discriminant analysis (PLS-DA). PLS regression is a good method of choice when the number of input variables is huge and correlated. For a geometric explanation of PLS techniques, see Dunn (2019).

To train and test the ResNet/PLSR-based model observer, seven abdominal CT scans were collected and reconstructed. Then, CT images of liver lesions were modified into lesion models of 5, 7, 9, and 11 mm in size at contrast levels of 15, 20, and 25 in Hounsfield units (HU) respectively, and digitally inserted into patient liver images that were produced or simulated at normal, half, and quarter radiation dose levels respectively. Four experts were asked to make a two-alternative-forced-choice (2AFC) in the detection task under 12 experimental conditions, while the network-based observer performed the same task after an appropriate internal noise was embedded in the model observer under the corresponding experimental condition. It was found that the network-based observer and the human experts performed very similarly, with the Pearson correlation coefficient up to 0.968 (95% confidence interval [0.888, 0.991]) and a negative Bland–Altman agreement result indicating no statistical difference between the network and human observers.

9.5 Final remarks*

We have discussed four aspects of image quality assessment in the above four sections respectively. Various measures are defined to quantify how well an imaging system performs in different aspects. For example, the least squared measure tells how close the gap is between an acquired image and the reference, SSIM shows structural similarity as the name SSIM indicates, and the area under the ROC curve is specific to a clinical task, which can be measured by various observers.

It is underlined that precision medicine is based on optimal individualized task-based medical imaging. While eventually machine-learning-based medical imaging will replace radiologists in many cases, the current consensus is that in the near future machine-learning-based medical imaging and image analysis should only assist radiologists in challenging cases. Therefore, there are two directions for us to pursue: a human-mimicking model observer and an intelligent image analyzer. As a model observer, the goal is to behave like a human, including making human-like mistakes. Such a model observer can be used as a human substitute in human reader studies. On the other hand, an intelligent image analyzer is to produce image analysis results as close to the truth as possible in whatsoever ways that work. In the machine learning framework, both the observer and the analyzer can be implemented as deep neural networks. For the same imaging task, the network architectures can be similar. Depending on the purpose of the network, the training process may target either the human interpretation if the network is intended as an observer or the ground truth if the network is for standalone image analysis.

Intended for either an observer or an image analyzer, advances in neural networks can be directly useful. It is a comforting goal to develop imaging networks that can first simulate or duplicate what a radiologist can do. In a good sense, letting a network mimic a radiologist is letting a network learn from a human. With such a network, various reconstruction networks can be rapidly evaluated, optimized, and accelerated, minimizing the validation efforts by radiologists. Also, if such an observer is validated to be as good as or better than a radiologist, then its use in place of a radiologist will become mature.

Because network-based model observers need big data of good quality, there is a critical need to have publicly shared and professionally labeled datasets for different diseases. There are multiple well-known datasets already, as mentioned in the last section of chapter 8. For example, a set of clinical data and images was prepared and

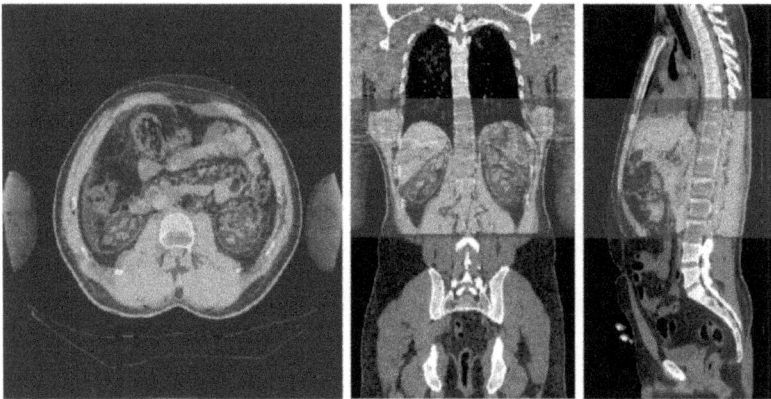

Figure 9.24. NIH visible human CT volume turned into a corresponding spectral CT counterpart by deformable mapping of a spectral CT patient scan onto a visible human CT scan (Meng B, Yang J, Ai D N, Fu T Y, and Wang G with the Beijing Institute of Technology (Beijing, China) and Rensselaer Polytechnic Institute (Troy, NY, USA)). The spectral CT images were produced via dual-energy CT-based material decomposition on a GE CT scanner in Beijing, China. Such a colorized CT volume can be deformed into various CT volumes of other human bodies, potentially with added pathological features. Reproduced with permission from the Beijing Institute of Technology and Rensselaer Polytechnic Institute.

authorized by Mayo Clinics for the 2016 NIH-AAPM-Mayo Clinic Low Dose CT Grand Challenge (http://www.aapm.org/grandchallenge/lowdosect). Given the tremendous cost involved with expert-labeled big data, a desirable way to generate big data for medical imaging should be realistic simulation. In this regard, a pilot study shows how to synthesize spectral CT images from the NIH visible human project data via dual-energy CT and deformable registration, as shown in figure 9.24. This remains a major challenge in the field and a great direction for research.

References

Chen H, Zhang Y, Kalra M K, Lin F, Chen Y, Liao P, Zhou J and Wang G 2017a Low-dose CT with a residual encoder–decoder convolutional neural network *IEEE Trans. Med. Imaging* **36** 2524–35

Chen H, Zhang Y, Zhang W, Liao P, Li K, Zhou J and Wang G 2017b Low-dose CT via convolutional neural network *Biomed. Opt. Express* **8** 679–94

Cover T M and Thomas J A 2006 *Elements of Information Theory* 2nd edn (Hoboken, NJ: Wiley)

Dodge S and Karam L 2017 A study and comparison of human and deep learning recognition performance under visual distortions *2017 26th Int. Conf. on Computer Communication and Networks (ICCCN) (IEEE)* pp 1–7

Dunn K 2019 Latent variable modelling *Process Improvement Using Data* (https://learnche.org/pid/PID.pdf?497-02b3)

He K, Zhang X, Ren S and Sun J 2016 Deep residual learning for image recognition *Proc. of the IEEE Conf. on Computer Vision and Pattern Recognition* pp 770–8

He X and Park S 2013 Model observers in medical imaging research *Theranostics* **3** 774–86

Getzin M 2019 System-wide advances in photon-counting CT: Corrections, similations, and image analysis *PhD Dissertation* Rensselaer Polytechnic Institute

Glesteby L 2018 CT metal artifact reduction with machine learning and photon-counting techniques *PhD Dissertation* Rensselaer Polytechnic Institute

Gong H, Yu L, Leng S and McCollough C 2018 A deep learning based model observer for low-contrast object detection task in x-ray computed tomography *60th Annual Meeting of the American Association of Physicists in Medicine, APM ePoster Library. Gong H.* (Jul 31, 2018) 217515; TU-K-202-5

Kopp F K, Catalano M, Pfeiffer D, Rummeny E J and Noël P B 2018 Evaluation of a machine learning-based model observer for x-ray CT *Proc. SPIE* **10577** 105770S

Lin M I, Groves W A, Freivalds A, Lee E G and Harper M 2012 Comparison of artificial neural network (ANN) and partial least squares (PLS) regression models for predicting respiratory ventilation: an exploratory study *Eur. J. Appl. Physiol.* **112** 1603–11

Lyu Q, You C, Shan H and Wang G 2019 Super-resolution MRI through deep learning *Proc. SPIE 11113, Developments in X-Ray Tomography* **XII** 111130X

Massanes F and Brankov J G 2017 Evaluation of CNN as anthropomorphic model observer *Medical Imaging 2017: Image Perception, Observer Performance, and Technology Assessment* vol 10136 (International Society for Optics and Photonics) p 101360Q

Panchuk V, Semenov V, Legin A and Kirsanov D 2018 Signal smoothing with PLS regression *Anal. Chem.* **90** 5959–64

Shan H, Zhang Y, Yang Q, Kruger U, Kalra M K, Sun L, Cong W and Wang G 2018 3-D convolutional encoder-decoder network for low-dose CT via transfer learning from a 2-D trained network *IEEE Trans. Med. Imaging* **37** 1522–34

Shannon C E 1997 The mathematical theory of communication *MD Comput.* **14** 306–17

Simonyan K and Zisserman A 2014 Very deep convolutional networks for large-scale image recognition, arXiv:1409.1556

Wang G 2016 A perspective on deep imaging *IEEE Access* **4** 8914–24

Wang G, Kalra M and Orton C G 2017 Machine learning will transform radiology significantly within the next 5 years *Med. Phys.* **44** 2041–4

Wang G, Ye J C, Mueller K and Fessler J A 2018 Image reconstruction is a new frontier of machine learning *IEEE Trans. Med. Imaging* **37** 1289–96

Wang Z, Bovik A C, Sheikh H R and Simoncelli E P 2004 Image quality assessment: from error visibility to structural similarity *IEEE Trans. Image Process.* **13** 600–12

Yang Q *et al* 2018 Low-dose CT image denoising using a generative adversarial network with Wasserstein distance and perceptual loss *IEEE Trans. Med. Imaging* **37** 1348–57

You C *et al* 2019 CT super-resolution GAN constrained by the identical, residual, and cycle learning ensemble (GAN-circle) *IEEE Trans. Med. Imaging*

Zeng R, Divel S, Li Q and Myers K 2019 Performance evaluation of deep learning methods applied to CT image reconstruction *Med. Phys.* **46** E162

Chapter 10

Quantum computing*

While deep learning remains the mainstream of the machine learning/artificial intelligence field, it will not dominate forever, like any other technologies. It is anticipated that more advanced forms of machine learning/artificial intelligence will combine data-driven/bottom-up/inductive learning and rule-based/top-down/deductive learning to be more intelligent like humans, and unsupervised learning from a limited number of samples will become feasible and play a more important role. Furthermore, the discovery of rules and even the construction of axioms will be generally possible in the future.

To invent the future of machine learning/artificial intelligence, algorithms will be more complicated, since the phenomena and processes to be dealt with will be more subtle and overwhelming. Hence, the computing power must be accordingly improved. However, Moore's law (the observation that the number of digital logic components is roughly doubled every year) has been close to saturation when the size of electronic components is below 10 nm, as shown in figure 10.1. The boundary between classical and quantum physics is arguably about 1 nm. As the size of digital logic components becomes sufficiently small, quantum effects must be taken into account to make classical digital logic work well. This will be technically rather challenging.

Modern computers are all built according to Boolean algebra, in which binary codes are stored and processed in a circuit consisting of logic gates that are operated in a deterministic fashion. In sharp contrast to the deterministic nature of Boolean algebra, quantum computers will work according to quantum mechanics, in which data are stored and processed as superposition of states in a parallel and probabilistic fashion.

Richard Feynman was a pioneer of quantum computing and discussed simulating physics with quantum computers in 1982 (Feynman 1982). In a nutshell, when the objective is to study a complicated quantum system, to describe the state of the system the computational cost will grow exponentially with the complexity of the system for a classical computer due to the probabilistic superposition of the

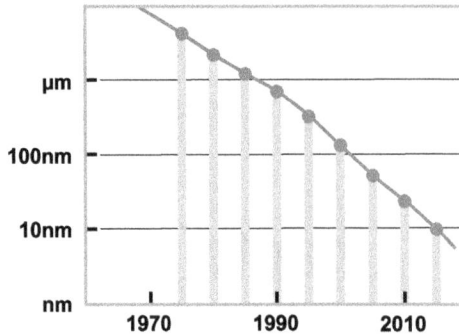

Figure 10.1. Challenges to Moore's law, as physics below 10 nm is entering the quantum world. Image designed in reference to Anthony (2015).

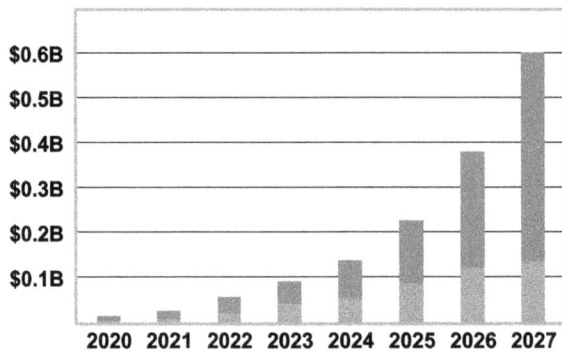

Figure 10.2. Seven year forecasts of quantum computing expenditure (green for research and red for application). Image designed based on CIR (2018).

involved observable variables. To appreciate this point a little further, recall that when you perform a Monte Carlo simulation on a classical computer, random events are traced branch by branch at a tremendous computational cost. Naturally, it is more desirable to use a quantum computer to simulate a quantum system, since a quantum computer computes in parallel, which will be detailed in this chapter.

Over the past few decades, the field of quantum computing has been greatly advanced. Despite debates on how useful the quantum computing techniques will be and if the answer is positive how soon they will be practical (Dyakonov 2018), the momentum of quantum computing seems too significant to be ignored, as shown in figure 10.2, and we should be prepared. More importantly, our own understanding of this technology also suggests its great potential which will be realized sooner or later, and all depend on how rapidly the technology could evolve, of which we are optimistic.

To learn quantum computing, first we must become familiar with quantum mechanics to a good degree. A good sign of familiarity with quantum mechanics is that we feel truly confused about quantum interplays. As Richard Feynman said (Feynman 1985): 'No, you're not going to be able to understand it.... You see, my physics students don't understand it either. That is because I don't understand it.

Nobody does…. The theory of quantum electrodynamics describes Nature as absurd from the point of view of common sense. And it agrees fully with an experiment. So I hope that you can accept Nature as She is—absurd.' Hence, in the following section, we will look at a few aspects of quantum mechanics to get confused and thus prepared. 'Cogito, ergo sum' is a philosophical claim in Latin by Descartes, meaning that 'I think, therefore I am'. As an analogy, in the current context, if we get confused by it, we start understanding it.

10.1 Wave–particle duality

Without much disagreement, quantum mechanics can be introduced with the double-slit experiment. Originally, the double-slit experiment was used to demonstrate that light is a wave. As shown in figure 10.3, if a monochromatic wave field from the left side propagates towards an opaque screen with two open slits, the wave will be blocked everywhere except for the two narrow openings. Through both small apertures, the wave field becomes two coherent small wave sources according to the Huygens principle. Then, the net effect of the two light wavelets at any spot on the detection screen depends on the difference between the distance from the top wavelet to the spot location and that from the bottom wavelet to the same location. Let us denote the distance between the two slits and the distance between the two screens as d and D respectively. At any point on the detection screen such as A or B in figure 10.3, the path difference Δ from the two slits to the point of interest can easily be computed as

$$\Delta = d \sin \theta \quad \text{when } D \gg d. \tag{10.1}$$

Clearly, we will have constructive and destructive interferences when $d \sin \theta = m\lambda$ and $d \sin \theta = (1/2)m\lambda$, respectively, where m and λ are integers and the wavelength

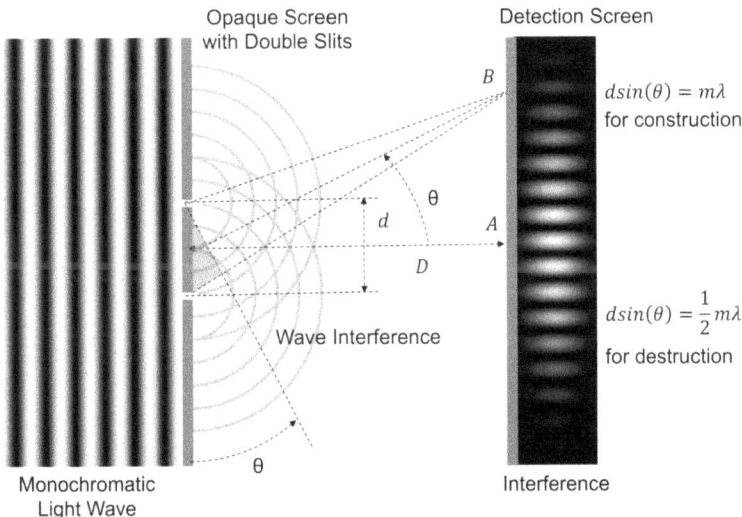

Figure 10.3. Double-slit experiment showing that light is a wave, as evidenced by the interference pattern.

of the light wave respectively. This experiment was initially performed with light, and yielded the interference pattern as explained above, in support of the concept that light is a wave.

On the other hand, light is emitted as single photons, each at a discretized energy level. Einstein won his Nobel Prize for his work in 1905 on quantum physics showing that light was composed of particles, which were later called photons. Insightfully, he explained the photoelectric effect. He pointed out that the current induced by light on a metal surface should vary according to the color of the light, and if the color is not right the current will not be induced regardless of the light intensity.

The photoelectric effect is commonly utilized in x-ray imaging, not only demonstrated inside a patient as a major type of interaction between x-ray photons and biological tissues, but also utilized in modern x-ray photon-counting detectors. When an x-ray tube is operated to generate an open beam, x-ray photons in the beam can be individually counted in an energy-discriminative fashion. Figure 10.4 shows an exemplary distribution of x-ray photon-counting data collected during multiple relatively weak exposures by a detector element in the Medipix detector assembly on the MARS micro-CT scanner at Rensselaer Polytechnic Institute (Troy, NY, USA) (https://www.marsbioimaging.com/mars/).

It is now well known that light, x-rays, and any other matter, all have wave–particle duality. For example, x-ray imaging can now be done to form images that show both particle and wave properties, aided by x-ray gratings which are extensions of the double-slit screen experiment; see figure 10.5.

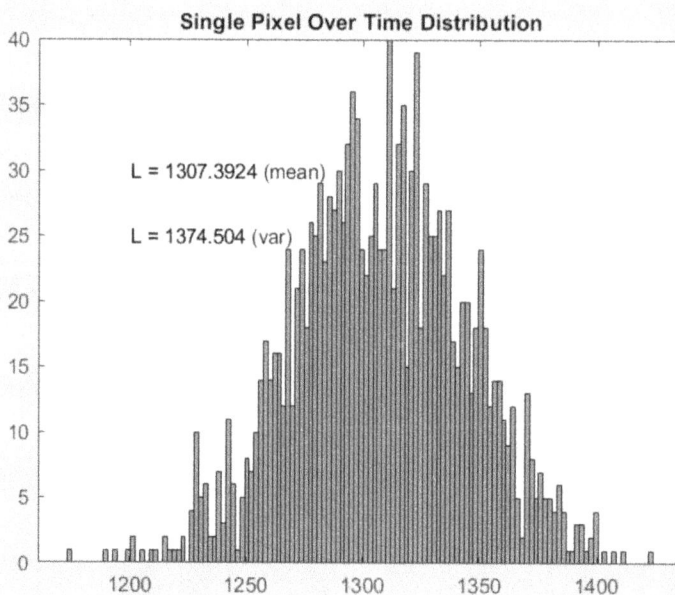

Figure 10.4. The numbers of x-ray photons recorded by a detector element in a state-of-the-art photon-counting detector during multiple x-ray exposures, performed at the AI-based X-ray Imaging System (AXIS) Lab, Rensselaer Polytechnic Institute.

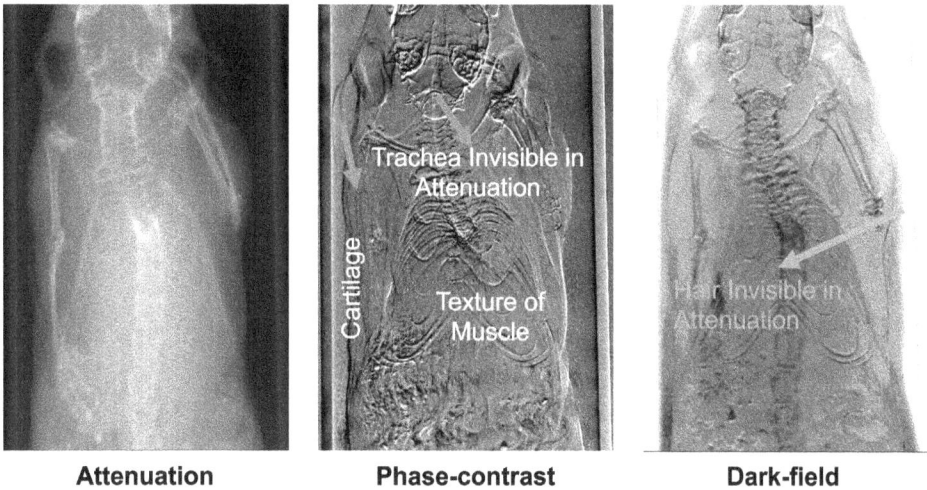

Attenuation   **Phase-contrast**   **Dark-field**

Figure 10.5. X-ray images in multiple contrasts, including attenuation (left), phase-contrast (middle), and small angle scattering (right), acquired in the AI-based X-ray Imaging System (AXIS) Lab, Rensselaer Polytechnic Institute. While attenuation and small angle scatter images have a natural particle-based interpretation, phase-contract imaging must be formulated in terms of wave propagation.

An interesting question is: when does a photon behave like a particle or a wave? Alternatively: is a micro-focus x-ray tube a source of x-ray waves or x-ray photons? As clearly illustrated in figure 10.4, an open beam of x-rays contains many individual photons. What will happen if we insert an opaque double-slit screen into the open x-ray beam? Intuitively, these individual x-ray photons could only go through the slits, should project onto the detection screen, and form two bright lines. However, the reality is that although the through-slit beam can be made so weak that photons are individually detected at each time interval, collectively they still form the same interference pattern as that in figure 10.3, despite the random noise arising from the photon-counting process, as shown in figure 10.6.

The condition is critical that the photons are singly emitted towards the double-slit screen and individually detected on the detection screen. The observed interference in figure 10.6 leads to the conclusion that an individual photon must go through the two slits simultaneously so that the interference can be explained! On the other hand, on the detection screen, each photon is individually detected one at a time, and we wonder which slit it came through. In figure 10.7, through which slit a photon goes is monitored to satisfy our curiosity, and then the interference pattern disappears, which means that the photon behaves as a classical particle!

By now we have already become amazed by the wave–particle duality. In the following, let us summarize a number of famous quantum puzzles in terms of A, B, C, D, and E.

Ambiguity/uncertainty. Just like classic physics, quantum physics needs measurement. However, in principle, not all physical quantities can be precisely measured at the same time. This is expressed by Heisenberg's uncertainty principle on the

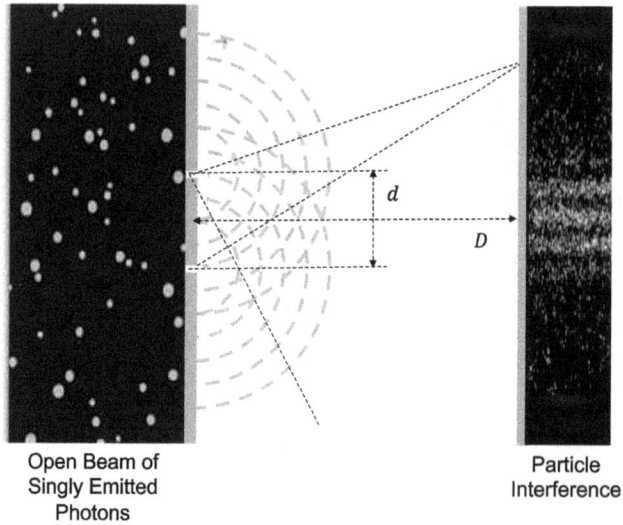

Figure 10.6. Double-slit experiment showing that an open weak beam of individual photons still interfere, even if they are singly emitted.

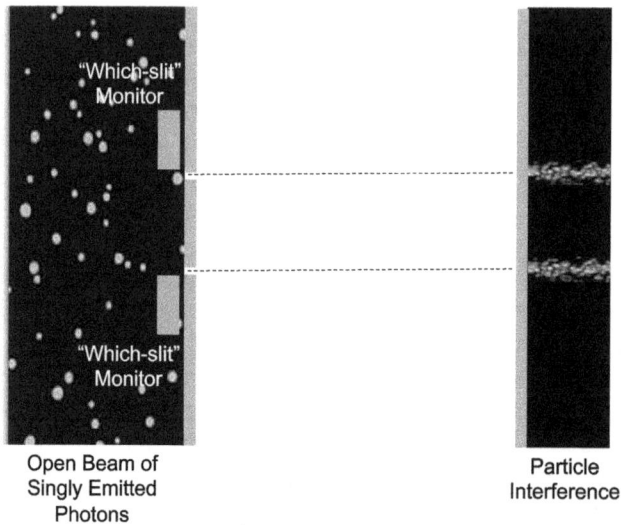

Figure 10.7. Double-slit experiment showing that individual photons in an open beam behave classically if they are monitored before the double-slit screen; otherwise, they propagate as waves.

intrinsic limit to the precision of the measurement on some paired physical quantities, such as the position x and momentum p of an x-ray photon:

$$\Delta x \Delta p \geqslant \frac{h}{2}, \tag{10.2}$$

where the errors of the position and momentum are measured as the standard deviation, and h is the Planck constant. While the speed of light sets the upper bound of physical object speed, the Planck constant sets the lower bound of coupled physical measurements. According to Heisenberg's principle, if we put a particle into a large region, we can define its momentum precisely. Alternatively, we can measure its momentum accurately but we cannot accurately localize it at the same time. That is, the position and momentum of the particle cannot be known well at the same time.

For this physical uncertainty principle, we have a mathematical prototype. For a square-integrable zero-mean function, its dispersion is defined as the variance of its squared version. Under some moderate conditions, the product of the dispersion of the original function and the dispersion of the corresponding Fourier transform cannot be less than $1/(16\pi^2)$.

Bang/collapse. Based on the double-slit experimental results, we have to admit that before measurement a photon propagates as a wave, and once it is measured all of sudden it is collapsed into a particle according to a probabilistic distribution that is obtained as the squared amplitude of the wave formed through interference. Therefore, the wave is probabilistic, and according to its probabilistic structure a particle occurs during the measurement as an interaction between the particle and a detecting material. This interaction is a 'small bang', in contrast to the 'Big Bang'.

Before a photon is observed, it or its probabilistic distribution goes into two slits (actually into the whole space, known as a wave). When the photon is detected on the detection screen, it becomes (collapses into) a single particle and will not appear anywhere else. Hence, after a monitoring device is set-up, photons become particles, as shown in in figure 10.7. It is a mystery how the wave/field at another location knows that the probability density should immediately vanish there right after a collapsing event happens in the detection screen.

Cat/co-existence/superposition. The fact that a single photon can go through two distinct slits at the same time is explained as it goes through the slits simultaneously in the probability sense. We can measure to determine which slit the photon goes through, as done in figure 10.7. Before the measurement, the photon goes through both slits so that we can explain the interference pattern in figure 10.7. In the metaphor known as 'Schrödinger's cat', quantum and classical results are coupled. Before an observation, we have superposition of two states: alive and dead (corresponding to a superposed status; for example, the propagation through two slits simultaneously, or the combination of decayed and not decayed emission states of an alpha particle). Thus, the cat will be in a superposition of dead (the alpha particle emitted to kill the cat) and alive (the alpha particle not emitted) cats, which is rather counter-intuitive!

Delayed choice. The weirdness of the wave–particle duality is further illustrated in Wheeler's delayed-choice experiment. In his experiment, the two-path interferometer is configured after a single photon has entered it. The device can either combine the two paths for interference or measure along which path the photon has come. It is the after-the-fact choice (that is, after the time taken by the photon having

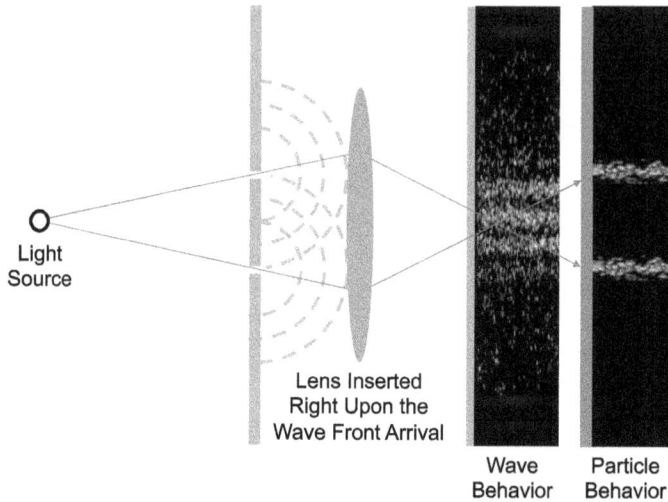

Figure 10.8. The delayed-choice experiment in which after a photon had gone through one or both the slits the choice was made to use either the light blue detection screen or the red detection screen coupled with the green lens focusing the slits onto the corresponding locations on the red screen.

followed either one or both of the paths) that determines how the photon really behaves (Jacques *et al* 2007)!

The idea behind this intriguing experiment can be visualized in figure 10.8. First, a point light source of low intensity emits a spherical wave or random photons isotropically so that at most only one photon could reach one or both slits within a narrow time window. After the wave or photon reaches a neighborhood of the principle plane of the green lens, a decision is rapidly made to use one of the two measuring set-ups: (i) the light blue detection screen or (ii) the green lens associated with the red detection screen to focus the slits onto the respective locations on the red screen. With the selection of the first measuring device, the interference pattern is observed; but with the selection of the second measuring device, the particle behavior is demonstrated as two projected stripes on the red screen, without any interference. Intuitively, if the interference pattern is seen, the photon must have gone through both the slits; otherwise, no interference means that the photon went through one of the slits. Either traveling one way or both, the fact had been fixed before the time instant when the decision was made as to which measuring device would be used. However, the experimental results show that the after-the-fact random decision could retrospectively affect the past behavior of the photon, which is strikingly counter-intuitive.

Entanglement. To demonstrate the incompleteness of quantum mechanics, in 1935 Einstein, Podolsky, and Rosen (EPR) published a thought experiment leading to the so-called EPR paradox. They imagined that a pion at rest decayed into a pair of photons. These photons together have a conserved zero momentum; that is, they must travel in opposite directions with opposite spins; i.e. if the spin vector of one photon is $S_1 = (x_1, y_1, z_1)$, then the spin vector of the other photon must be

$S_2 = (x_2, y_2, z_2) = (-x_1, -y_1, -z_1)$. If one photon is measured to be spin-up along an axis (say, $x_1 > 0$), then the corresponding component of the other photon must be spin-down (i.e. $x_2 < 0$), no matter how far they are separated. This is called quantum entanglement, in which the quantum states of two (or more) objects are described with reference to each other in a deterministic way.

Einstein's viewpoint is that the reason for quantum entanglement is the existence of hidden variables that can explain the instantaneous communication between the entangled photons in terms of their quantum states. In other words, Einstein believed that the states were determined before the measurement, or even before the entanglement. This is in sharp contrast to the Copenhagen interpretation advocated by Bohr.

Interestingly, in 1964 Bell made a breakthrough with his inequality allowing a critical test on whether or not the hidden variables exist. In the aforementioned case of two entangled photons, specified by hidden variables, each photon has spin values in eight cases with different probabilities, as shown in figure 10.9 and table 10.1. For a given photon, we can only perform one measurement on the spin value along a single axis, i.e. the x-, y-, or z-axis. For the two entangled photons, we can perform such a measurement twice on each of them, which is equivalent to measuring the spin values of either photon twice, and obtain the correlation functions also listed in table 10.1. Similar correlation functions can be also derived from quantum mechanics. It is underlined that the correlation functions derived from classical physics (figure 10.9 and table 10.1), which are referred to as Bell's inequalities (for example, it can be proved that $|p_{xz} - p_{zy}| \leqslant 1 + p_{xy}$), are significantly different from those based on quantum physics (not shown here), and the differences between the classical and quantum predictions can be experimentally tested. Therefore, aided by Bell's inequality, whether or not hidden local variables exist can be experimentally answered. Up to now, all experimental data indeed exclude the existence of hidden variables, opposite to what Einstein thought. Further exploration is still going on and will offer more insights in the future.

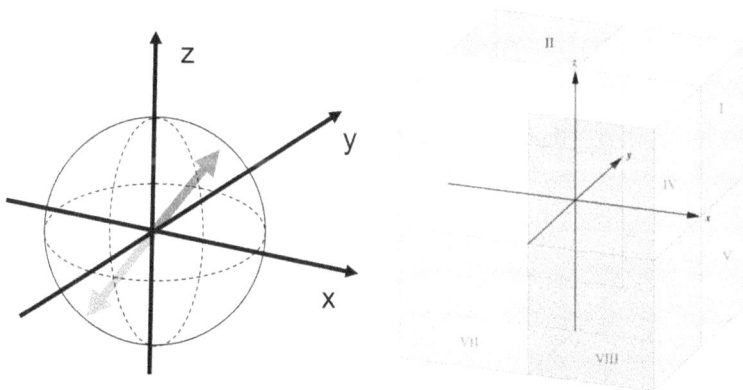

Figure 10.9. Two entangled particles in opposite quadrants with zero momentum in total. If one (red) is measured to be clockwise on an axis, the other (green) must be counterclockwise.

Table 10.1. States of two entangled photons and their correlations between selected corresponding components can be computed assuming the existence of 'hidden variables'.

	1st spin			2nd spin			Prob.	Correlation			
	x	y	z	x	y	z		p_{xx}	p_{xy}	p_{xz}	p_{zy}
I	+	+	+	−	−	−	p_1	−1	−1	−1	−1
II	−	+	+	+	−	−	p_2	−1	+1	+1	−1
III	−	−	+	+	+	−	p_3	−1	−1	+1	+1
IV	+	−	+	−	+	−	p_4	−1	+1	−1	+1
V	+	+	−	−	−	+	p_5	−1	−1	+1	+1
VI	−	+	−	+	−	+	p_6	−1	+1	−1	+1
VII	−	−	−	+	+	+	p_7	−1	−1	−1	−1
VIII	+	−	−	−	+	+	p_8	−1	+1	+1	−1

10.2 Quantum gates

It is assumed that you are fairly familiar with digital computers, both their architectural design and programming; otherwise, you will not have read until this point (if this is not the case, you can search the Internet to fill in your knowledge gap). Simply speaking, digital computers work with binary codes, which are strings of zeros and ones. A basic memory unit can be made to hold the minimum 'bit' of information, being either '0' or '1'. Such a single bit is not very useful, but 8 bits together can represent 2^8 integers from 0 to 255. Given a number of bits as an input, a digital electronic logic gate can output a result as a Boolean function of the binary input. There are many kinds of logic gates, some of which are called universal logic gates. As indicated by their name, universal gates can be combined to implement any Boolean function without involving any other gate types. NAND and NOR gates are examples of universal gates. These electronic components not only store information in the binary format and process data through digital logic operations, but also coordinate the executions of the logic operations in terms of programming, since such a coordination can be viewed as Boolean logic operations implemented as a function of time under binary control.

Quantum computers share certain similarities with digital computers but at the same time differ from them dramatically. As far as information coding is concerned, while the minimum information unit for digital signal processing is a 'bit', the basic unit for quantum computing is a quantum bit, known as a 'qubit', implemented with a quantum device of two states such as spin-up and spin-down or vertical and horizontal polarization, which can be measured only once along one axis. The quantum device for a qubit is quite different from the device for a classic bit, and remains an active area of research. Candidate technologies for holding a qubit are shown in figure 10.10.

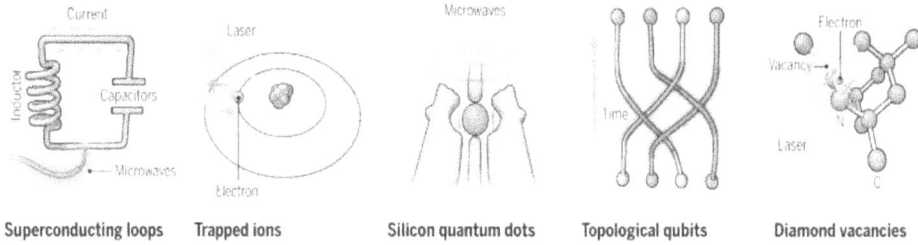

Figure 10.10. Superconducting loops, trapped ions, silicon quantum dots, topological qubits, and diamond vacancies as candidates for holding qubits. Reproduced with permission from Popkin (2016). Copyright 2016 the American Association for the Advancement of Science.

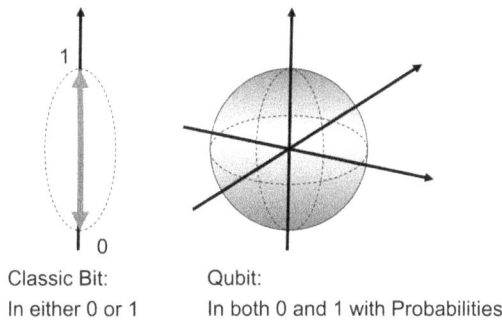

Figure 10.11. In contrast to a classical bit (left) which can only be either '0' or '1', a qubit (right) represents a superposition of '0' and '1' which could be any point on the unit sphere.

As shown in figure 10.11, in a classic bit, we can put either '0' or '1' but not both at the same time. However, in a qubit, the states of '0' and '1' can 'co-exist' at the same time. It is this superposition of two states that allows parallel information storage and processing. Furthermore, not all superpositions of '0' and '1' are the same. For a given qubit at a specific time, the probabilities for the qubit to show its state as '0' or '1' are generally different. When this qubit is immediately measured, the superposition will turn out to be either '0' or '1' according to the corresponding probability. The superposition is a wave function (also simply referred to as a wavefunction). The measurement forces the wave function to select or collaps into either '0' or '1'. Hence, quantum computing, starting from qubits, is probabilistic or statistical, instead of deterministic like classical computation.

In the so-called 'bra and ket' notation, a qubit $|\Psi\rangle$ is expressed as a linear combination of two quantum states $|0\rangle$ and $|1\rangle$:

$$|\Psi\rangle = \alpha|0\rangle + \beta|1\rangle, \tag{10.3}$$

where α and β are complex numbers, also called complex amplitudes, giving probabilities $|\alpha|^2$ and $|\beta|^2$ for the qubit to show $|0\rangle$ and $|1\rangle$, respectively, during a measurement. Thus, $|\alpha|^2 + |\beta|^2 = 1$. Since a complex number has two degrees of

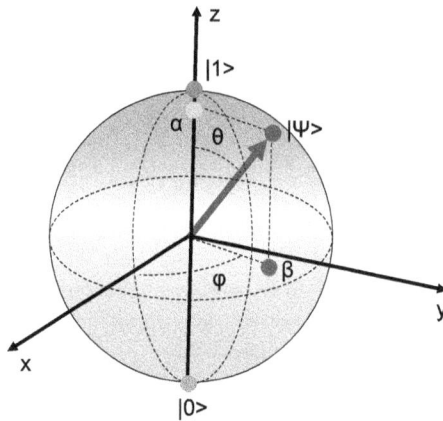

Figure 10.12. Bloch sphere visualizing a qubit superposition in terms of a linear combination of two quantum states in the complex domain.

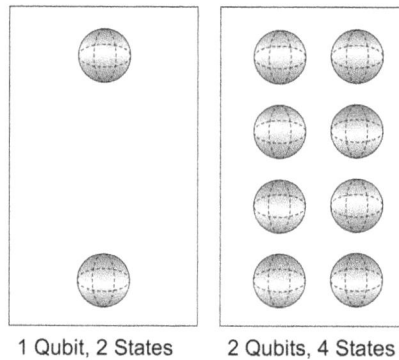

1 Qubit, 2 States   2 Qubits, 4 States

Figure 10.13. Encoding with 1 and 2 qubits, respectively. Each of the codes is associated with a unique probability according to which the state will show up during a measurement.

freedom and the quantum states are constrained to have a unit probability, there are three degrees of freedom in the superposition $|\Psi\rangle$:

$$\begin{cases} \alpha = e^{i\phi} \cos\dfrac{\theta}{2} \\ \beta = e^{i(\phi+\psi)} \sin\dfrac{\theta}{2} \end{cases}, \qquad (10.4)$$

where the phase factors do not affect the measurement probabilities but are important in forming interference patterns and implementing quantum computing. Sometimes, for convenience the phase factor of α can be set to zero. Generally, the qubit $|\Psi\rangle$ can be visualized in the Bloch sphere in figure 10.12.

Again, just like bits are used collectively, qubits must be multiple for meaningful computations. As shown in figure 10.13, when we only use one qubit, we use two complex numbers to specify the linear combination and the probabilities at which the two quantum states may be measured. When we have two qubits, four quantum states may co-exist, and during a measurement the system will collapse into one of the four states according to the corresponding probability. Furthermore, when we have n qubits, $N = 2^n$ quantum states can co-exist with various collapsing probabilities, and be expressed as

$$|\Psi\rangle = \alpha_1|1\rangle + \alpha_2|2\rangle + \cdots + \alpha_N|N\rangle \tag{10.5}$$

with the probability $|\alpha_n|^2$ for the nth state to show up in a measurement.

Instead of measuring $|\Psi\rangle$ to let it turn into one of the quantum states, we can transform it into another superposition:

$$|Z\rangle = \beta_1|1\rangle + \beta_2|2\rangle + \cdots + \beta_N|N\rangle. \tag{10.6}$$

Many quantum transformations are possible under the condition that these transforms must be linear, since physically meaningful operations on qubits are linear. In other words, the complex coefficients of the superposition before and after the transformation can be linked as follows:

$$U\begin{bmatrix} \alpha_1 \\ \alpha_2 \\ \vdots \\ \alpha_N \end{bmatrix} = \begin{bmatrix} \beta_1 \\ \beta_2 \\ \vdots \\ \beta_N \end{bmatrix}, \tag{10.7}$$

where U is a complex-valued matrix. Furthermore, from the perspective of measurement, the U matrix must be unitary so that $\sum_{n=1}^{N}|\alpha_n|^2 = 1$ means $\sum_{n=1}^{N}|\beta_n|^2 = 1$; i.e., if we measure the quantum superposition either before or after the transformation, then one and only one of the co-existent quantum states must be observed according to the corresponding probability $|\alpha_n|^2$ or $|\beta_n|^2$.

The fact that a quantum superposition $|\Psi\rangle$ can be changed through a unitary transform to another superposition $|Z\rangle$ suggests that $|\Psi\rangle$ can be measured in terms of not only its computational basis $|n\rangle$ but also its projections via mutually orthogonal projectors $P_1, P_2, \ldots, P_M$, each of which projects onto a subspace S_i of a Hilbert space S and all of which allow a unique representation of $|\Psi\rangle$ in S.

Either classic bits or qubits need to be processed for computational purposes. Classic bits are processed in terms of digital logic gates and their combinations. Not surprisingly, qubits must be processed with quantum gates. These quantum gates implement various unitary transformations as permitted by quantum mechanics. Digital logic gates are specialized circuits governed by the Maxwell equations under proper initial and boundary constraints. Correspondingly, quantum gates are quantum devices governed by the Schrödinger equation, also constrained by appropriate conditions.

The simplest quantum gates only process one qubit. Four popular one-qubit gates are X, Z, H, and T gates. Let the input qubit be $|\Psi\rangle = \alpha|0\rangle + \beta|1\rangle$, then the output of the X gate is given by

$$X(|\Psi\rangle) = \begin{bmatrix} 0 & 1 \\ 1 & 0 \end{bmatrix}\begin{bmatrix} \alpha \\ \beta \end{bmatrix} = \beta|0\rangle + \alpha|1\rangle \tag{10.8}$$

and you can think of the term X as standing for exchange.

The outputs of the Z, H, and T gates are, respectively, given by

$$Z(|\Psi\rangle) = \begin{bmatrix} 1 & 0 \\ 0 & -1 \end{bmatrix} \begin{bmatrix} \alpha \\ \beta \end{bmatrix} = \alpha|0\rangle - \beta|1\rangle \tag{10.9}$$

$$H(|\Psi\rangle) = \frac{1}{\sqrt{2}} \begin{bmatrix} 1 & 1 \\ 1 & -1 \end{bmatrix} \begin{bmatrix} \alpha \\ \beta \end{bmatrix} = \frac{1}{\sqrt{2}}[(\alpha + \beta)|0\rangle + (\alpha - \beta)|1\rangle] \tag{10.10}$$

$$T(|\Psi\rangle) = \begin{bmatrix} 1 & 0 \\ 0 & e^{i(\pi/4)} \end{bmatrix} \begin{bmatrix} \alpha \\ \beta \end{bmatrix} = \alpha|0\rangle + e^{i(\pi/4)}\beta|1\rangle. \tag{10.11}$$

The symbols and matrix definitions of these four gates are presented in figure 10.14.

A more interesting quantum gate takes more than one qubit as its input and has multiple output components as well. The most important quantum gate is probably the so-called 'Controlled NOT gate', which is referred to as the CNOT gate, that transfers two input qubits into two output qubits. Specifically, the quantum state of the two input qubits $\alpha_1|00\rangle + \alpha_2|01\rangle + \alpha_3|10\rangle + \alpha_4|11\rangle$ will be transformed by the CNOT gate into $\alpha_1|00\rangle + \alpha_2|01\rangle + \alpha_4|10\rangle + \alpha_3|11\rangle$. Mathematically, we have

$$\text{CNOT}(|\Psi_1\rangle, |\Psi_2\rangle) = \begin{bmatrix} 1 & 0 & 0 & 0 \\ 0 & 1 & 0 & 0 \\ 0 & 0 & 0 & 1 \\ 0 & 0 & 1 & 0 \end{bmatrix} \begin{bmatrix} \alpha_1 \\ \alpha_2 \\ \alpha_3 \\ \alpha_4 \end{bmatrix} = \alpha_1|00\rangle + \alpha_2|01\rangle + \alpha_4|10\rangle + \alpha_3|11\rangle, \tag{10.12}$$

which flips the second qubit (the right qubit) when the first qubit (the left qubit) is 1, and otherwise (when the first qubit is 0) it does nothing to the second qubit. The CNOT gate is a universal gate in the sense that any quantum circuit can be implemented using CNOT gates and single-qubit rotations. A classic analog of the CNOT gate is the XOR gate for digital logic, since the second output qubit (right qubit) is the XOR of the two input qubits (see figure 10.15).

As an example, let us see how two qubits can be entangled together with two quantum gates H and CNOT. As shown in figure 10.16, at time t_0 the two input qubits are both set to $|0\rangle$. Then, the H gate turns the top qubit into a superposition of

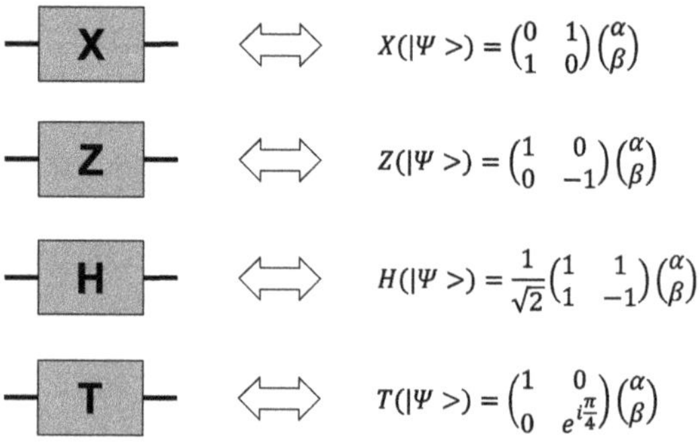

Figure 10.14. Symbols (left) and the matrix definitions (right) of the four popular quantum gates X, Z, H, and T.

two equally likely states, while the bottom qubit remains at $|0\rangle$. Finally, under the control of the top qubit the CNOT gate entangles the two qubits to take the same state, either $|0\rangle$ or $|1\rangle$.

Up to this point, we have described the concepts of qubits and the functionalities of popular quantum gates. If you have followed the text so far, you have understood the essence of quantum computing: quantum information is coded in qubits as a superposition of multiple (exponentially growing with the number of qubits) quantum states, processed through quantum gates in the form of unitary transformation, and eventually read out in a classical measurement. Once a measurement is performed, the superposition is totally destroyed to yield one of many quantum states in a computational basis or through a projective process. These points are all we need to perform quantum computing, just like we only need a chess set and a limited number of rules to play the game. In other words, as a quantum computing programmer, we have now obtained sufficient knowledge on the computational model, and the rest is up to our imagination and capabilities so that quantum algorithms/circuits can be designed to do wonderful things.

10.3 Quantum algorithms

As far as the intrinsic computing capabilities are concerned, quantum computers are at least as powerful as classic computers. This can be easily appreciated by comparing the bits and qubits, universal digital logic gates, and universal quantum gates, as well as topological similarities between digital circuits and quantum circuits. Indeed, qubits can be set as classic bits (say, a 100% chance to be in one of the two states), quantum gates can be made to perform digital logic operations, and a quantum computer can be made to emulate a classical computer, which is awkward but we are convinced that quantum computing is a valid concept.

Quantum computing is implemented with quantum algorithms. Studies have shown that quantum computers can perform arithmetic operations, and evaluate Boolean functions. However, we are not satisfied to find a quantum way to do what classical computers can do. The main motivation for quantum computing is to increase our computing capabilities dramatically; for example, can we use a quantum computer to perform certain computational tasks exponentially faster than a classical computer?

To demonstrate a fundamental advantage of quantum computing over classical computing, let us explain the simplest yet very important Deutsch problem and solution.

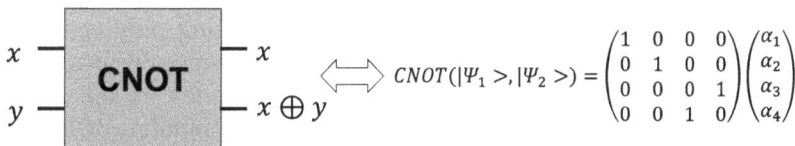

$$CNOT(|\Psi_1\rangle, |\Psi_2\rangle) = \begin{pmatrix} 1 & 0 & 0 & 0 \\ 0 & 1 & 0 & 0 \\ 0 & 0 & 0 & 1 \\ 0 & 0 & 1 & 0 \end{pmatrix} \begin{pmatrix} \alpha_1 \\ \alpha_2 \\ \alpha_3 \\ \alpha_4 \end{pmatrix}$$

Figure 10.15. Symbol (left) and definition (right) of the quantum gate CNOT, where we call the top input qubit the control qubit, and the bottom input qubit the target qubit. The top and bottom qubits are also called the first and second qubits.

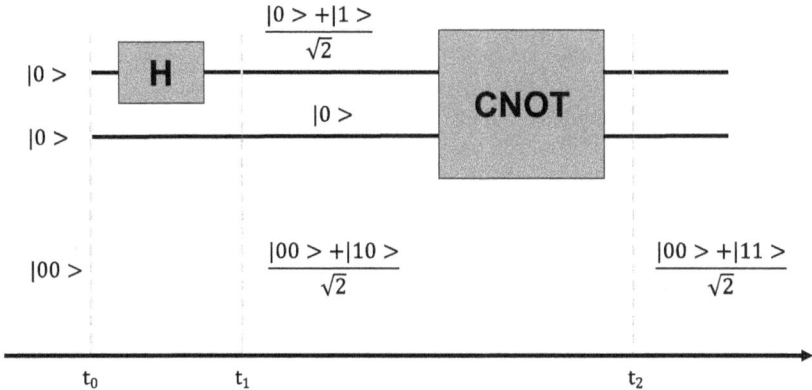

Figure 10.16. Two qubits both at $|0\rangle$ are entangled so that they must take the same state, $|0\rangle$ or $|1\rangle$ with the same chance upon measurement.

Table 10.2. Definition of the black box function F.

Input	Output			
	i	ii	iii	iv
0	0	1	0	1
1	0	0	1	1

The purpose is to appreciate how the superposition in qubits can be translated into a much higher computational efficiency.

To define the Deutsch problem, we have a black box function F that maps $\{0, 1\}$ to $\{0, 1\}$. In terms of the input–output relationship, the function F can be characterized as in table 10.2. We can evaluate the given black box function by feeding it with '0' and '1', respectively, and observing its output. The function F must be in one of the four possible cases. In other words, the function F can be characterized after two evaluations; i.e., we evaluate $F(0)$ and $F(1)$ respectively. For our current interest, we are only interested in whether F is 'constant' or 'balanced'. In cases (i) and (iv), we call F constant since its output is uniformly 0 (in case (i)) or 1 (in case (iv)). Otherwise, F is called 'balanced', since its output has both 0 and 1 in cases (ii) and (iii).

Next, let us see how the function F can be characterized with a doubled computational efficiency; i.e., how to characterize F using only one evaluation instead of two. This appears impossible at the first glance since one evaluation $F(x)$ is insufficient to know $F(1 - x)$, and after only one evaluation we still do not know if F is constant or balanced. Actually, with quantum computing it is feasible to characterize F with only one evaluation, enabled by the superposition nature of qubits. It is the superposition that makes a fundamental difference between quantum and classical computation, allowing an exponential gain with quantum computing in some computational tasks such as solving the Deutsch problem.

To solve the Deutsch problem, let us first transform the function F into an equivalent two-qubit quantum gate D in figure 10.17. The gate D maps $|xy\rangle$ to $|x(y \oplus F(x))\rangle$. The binary output of the function F is a special case of the qubit state, and can be stored in a qubit. The XOR operation can be directly implemented using a CNOT gate. Then, given any y value the knowledge on F is equivalent to the knowledge on D (we just need to look at $y \oplus F(x)$ and should be able to recover $F(x)$ based on the value of y).

It can be easily verified that for any specified F a corresponding unitary matrix representation of gate D can be always found. For example, case (ii) in table 10.2, $F(0) = 1$ and $F(1) = 0$, implies the specific map from $|xy\rangle$ to $|x(y \oplus F(x))\rangle$, and we can easily obtain the mapping table and the corresponding matrix in table 10.3.

Now, we are ready to determine if the function F (equivalently, gate D) is constant or balanced after only one evaluation, instead of the two evaluations required from the classical perspective. The algorithm to solve this functional classification problem is called the Deutsch algorithm, as shown in figure 10.18. Let us first give you a general idea of the Deutsch algorithm. With the constant input $|01\rangle$, the measured output of the H gate after the D gate will always be $F(0) \oplus F(1)$ (further explanation below), and as such the output 0 will indicate that F is constant (i.e. $F(0)$ and $F(1)$ must be the same); otherwise F is balanced ($F(0)$ differs from $F(1)$). It is underlined that the function F is characterized with only one evaluation, thanks to the superposition of qubits implemented by the two H gates before the D gate so that all four possible quantum states are presented to the D gate simultaneously, enabling parallel computing. Although this algorithm is among the simplest quantum algorithms, it does reveal the essential secret why quantum computing can be highly efficient and revolutionary.

To understand the inner-workings of the very clever Deutsch algorithm illustrated in figure 10.18, it is worth examining the information flow step by step. First, the initial input $|01\rangle$ is processed by the two H gates before the D gate to produce the following results as the input to the D gate:

$$H(|0\rangle) = \frac{1}{\sqrt{2}}\begin{bmatrix} 1 & 1 \\ 1 & -1 \end{bmatrix}\begin{bmatrix} 1 \\ 0 \end{bmatrix} = \frac{1}{\sqrt{2}}[|0\rangle + |1\rangle] \tag{10.13}$$

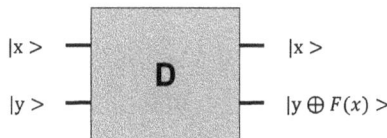

Figure 10.17. Function F is transformed to an equivalent quantum gate D.

Table 10.3. Function of the quantum gate D for case (ii) in table 10.2, i.e. $F(0) = 1$ and $F(1) = 0$, along with the matrix representation for the quantum gate D, where $\oplus$ denotes the XOR logic operation.

$|xy\rangle$	$|x(y \oplus F(x))\rangle$ (in case (ii) in table 10.2)	Matrix representation for gate D			
00	$0(0 \oplus F(0)) = 0(0 \oplus 1) = 01$	0	1	0	0
01	$0(1 \oplus F(0)) = 0(1 \oplus 1) = 00$	1	0	0	0
10	$1(0 \oplus F(1)) = 1(0 \oplus 0) = 10$	0	0	1	0
11	$1(1 \oplus F(1)) = 1(1 \oplus 0) = 11$	0	0	0	1

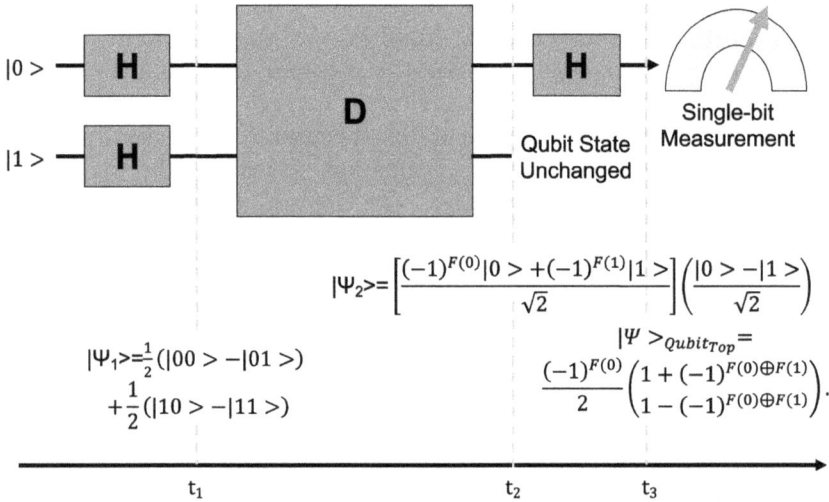

$$|\Psi_2\rangle = \left[\frac{(-1)^{F(0)}|0\rangle + (-1)^{F(1)}|1\rangle}{\sqrt{2}}\right]\left(\frac{|0\rangle - |1\rangle}{\sqrt{2}}\right)$$

$$|\Psi_1\rangle = \frac{1}{2}(|00\rangle - |01\rangle) + \frac{1}{2}(|10\rangle - |11\rangle)$$

$$|\Psi\rangle_{Qubit_{Top}} = \frac{(-1)^{F(0)}}{2}\begin{pmatrix} 1 + (-1)^{F(0)\oplus F(1)} \\ 1 - (-1)^{F(0)\oplus F(1)} \end{pmatrix}.$$

Figure 10.18. Quantum circuit implementing the Deutsch algorithm.

$$H(|1\rangle) = \frac{1}{\sqrt{2}}\begin{bmatrix} 1 & 1 \\ 1 & -1 \end{bmatrix}\begin{bmatrix} 0 \\ 1 \end{bmatrix} = \frac{1}{\sqrt{2}}[|0\rangle - |1\rangle], \tag{10.14}$$

which can be regrouped into the following quantum superposition:

$$|\Psi_1\rangle = \frac{1}{\sqrt{2}}[|0\rangle + |1\rangle] \cdot \frac{1}{\sqrt{2}}[|0\rangle - |1\rangle] = \frac{1}{2}[|00\rangle - |01\rangle] + \frac{1}{2}[|10\rangle - |11\rangle], \tag{10.15}$$

where the purple factor (keep this factor in mind, it will be referenced below) is the quantum state of the bottom qubit input to gate D in figure 10.18.

Then, the above superposition is processed by the D gate. To formulate the output of the D gate, we need the following relationship that can be directly verified for a binary variable x:

$$|0 \oplus x\rangle - |1 \oplus x\rangle = (-1)^x(|0\rangle - |1\rangle). \tag{10.16}$$

To make sure that the above equation is correct, when $x = 0$ we have

$$|0 \oplus x\rangle - |1 \oplus x\rangle = |0 \oplus 0\rangle - |1 \oplus 0\rangle - |0\rangle - |1\rangle = (-1)^0(|0\rangle - |1\rangle) \tag{10.17}$$

and when $x = 1$, we have

$$|0 \oplus x\rangle - |1 \oplus x\rangle|0 \oplus 1\rangle - |1 \oplus 1\rangle|1\rangle - |0\rangle = (-1)^1(|0\rangle - |1\rangle). \qquad (10.18)$$

Then, the output of gate D is as follows:

$$|\Psi_2\rangle = \frac{1}{2}(|0(0 \oplus F(0))\rangle - |0(1 \oplus F(0))\rangle) + \frac{1}{2}(|1(0 \oplus F(1))\rangle - |1(1 \oplus F(1))\rangle)$$

$$= \frac{1}{2}|0\rangle(|0 \oplus F(0)\rangle - |1 \oplus F(0)\rangle) + \frac{1}{2}|1\rangle(|0 \oplus F(1)\rangle - |1 \oplus F(1)\rangle)$$

$$= \frac{1}{2}(-1)^{F(0)}|0\rangle(|0\rangle - |1\rangle) + \frac{1}{2}(-1)^{F(1)}|1\rangle(|0\rangle - |1\rangle)$$

$$= \left[\frac{(-1)^{F(0)}|0\rangle + (-1)^{F(1)}|1\rangle}{\sqrt{2}}\right]\left(\frac{|0\rangle - |1\rangle}{\sqrt{2}}\right),$$

where the purple factor is exactly the same as the quantum state of the bottom qubit input to gate D in figure 10.18, while the first factor is the quantum state of the top qubit output of gate D in figure 10.18, in which two simultaneous evaluations $F(0)$ and $F(1)$ become involved. A direct measurement on the top qubit output of gate D is not informative at all, since regardless of the values of $F(0)$ and $F(1)$, $|0\rangle$ and $|1\rangle$ will be observed with equal chance.

The magic happens when the top qubit output is further processed by another H gate, which yields the following outcome:

$$|\Psi\rangle_{\text{Qubit}_{\text{Top}}} = \frac{1}{\sqrt{2}}\begin{pmatrix} 1 & 1 \\ 1 & -1 \end{pmatrix}\begin{pmatrix} \frac{(-1)^{F(0)}}{\sqrt{2}} \\ \frac{(-1)^{F(1)}}{\sqrt{2}} \end{pmatrix}$$

$$= \frac{1}{2}\begin{pmatrix} 1 & 1 \\ 1 & -1 \end{pmatrix}\begin{pmatrix} (-1)^{F(0)} \\ (-1)^{F(1)} \end{pmatrix}$$

$$= \frac{1}{2}\begin{pmatrix} (-1)^{F(0)} + (-1)^{F(1)} \\ (-1)^{F(0)} - (-1)^{F(1)} \end{pmatrix}$$

$$= \frac{(-1)^{F(0)}}{2}\begin{pmatrix} 1 + (-1)^{F(0) \oplus F(1)} \\ 1 - (-1)^{F(0) \oplus F(1)} \end{pmatrix}.$$

Hence, when $F(0) \oplus F(1) = 0$, we have $|\Psi\rangle_{\text{Qubit}_{\text{Top}}} = (-1)^{F(0)}\begin{bmatrix} 1 \\ 0 \end{bmatrix}$; otherwise $F(0) \oplus F(1) = 1$, and $|\Psi\rangle_{\text{Qubit}_{\text{Top}}} = (-1)^{F(0)}\begin{bmatrix} 0 \\ 1 \end{bmatrix}$. That is, if the measurement on the output of the last H gate reveals $|0\rangle$, the function F is constant; otherwise, the measure outcome $|1\rangle$ indicates that the function F is balanced.

The quantum computing workflow to solve the Deutsch problem is rather inspiring. The two H gates before the D gate demonstrates how to utilize the

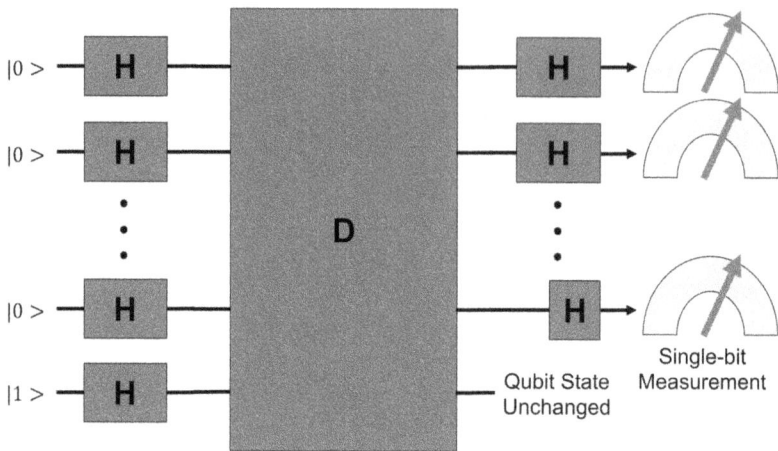

Figure 10.19. Quantum circuit implementing the Deutsch–Jozsa algorithm. As you may correctly guess, if all the measurements on the outputs of the final layer of the H gates are zeros, the function F is constant; otherwise, it is balanced.

superposition nature of qubits to present all possible input combinations to the D gate. The H gate after the D gate shows the information on the solution can be extracted through a constructive inference so that the output of the last H gate will directly provide the answer with 100% certainty.

Not surprisingly, the Deutsch algorithm was extended to the Deutsch–Jozsa algorithm to classify an n bit function into the constant or balanced classes with only a single evaluation (Guide *et al* 2003 and https://en.wikipedia.org/wiki/Deutsch-Jozsa_algorithm). Specifically, the Deutsch–Jozsa algorithm solves the following problem. Suppose that a function F maps from $\{0, 1\}^n$ to $\{0, 1\}$ for a positive integer n, and can only be constant, meaning that $F(x) = 0$ or 1 for all inputs, or balanced, meaning that $F(x) = 0$ for and only for half of the inputs whose values are arbitrary. The problem is to decide if F is constant or balanced. While we have not found any major practical applications of the Deutsch–Jozsa algorithm, this prototype, as shown in figure 10.19, is very enlightening in the sense of establishing the feasibility that a quantum algorithm can be deterministic and yet exponentially faster than any classical algorithm.

To date, a number of excellent quantum algorithms have been reported that are exponentially more efficient on a quantum computer than any algorithm on a classical computer. For example, the Simon algorithm is to find s in the input domain for a black box function F from $\{0, 1\}^n$ to $\{0, 1\}^n$ satisfying the condition that $F(x) = F(y)$ if and only if $x \oplus y \in \{0^n, s\}$. This algorithm inspired the Shor algorithm for integer factorization, which means to find all prime factors of an integer N. Integer factorization is a central problem of cryptography, and has important financial, military, and other applications. For more quantum algorithms, the reader is referred to a number of excellent textbooks and lecture notes such as https://home-pages.cwi.nl/~rdewolf/, in which the quantum Fourier transform is also covered. In the same spirit of the algorithms we have discussed above, the quantum Fourier transform can be easily configured as a unitary transform characterized by an $N \times N$ matrix, where $N = 2^n$, which can be implemented using $O(n^2)$ quantum gates (gate complexity) in $O(n)$ stages (time complexity). Recall that the computational complexity of the fast Fourier transform is $O(N \log N) = O(n2^n)$ steps, which is much slower than the quantum counterpart. Finally, quantum algorithms can be deterministic, such as the Deutsch algorithm, or, most generally, stochastic since the measurement on quantum computing results is intrinsically probabilistic.

10.4 Quantum machine learning

Quantum machine learning is an emerging field of machine learning/artificial intelligence. Roughly speaking, anything at the intersection of quantum computing and machine learning can be called quantum machine learning; in particular quantum algorithms designed to perform machine learning tasks with either quantum or classical data. Since quantum computers are still premature, most quantum machine learning algorithms are mathematically studied, and at most tested on small (up to 100 qubits) or special (~1000 qubits) quantum prototypes (Biamonte *et al* 2017).

Among quantum machine learning algorithms, those for solving a system of linear equations were well studied as building blocks of large-scale learning software. A good example is the quantum algorithm for matrix inversion, which is computationally more efficient than classical algorithms, and needed for machine learning such as least-squares fitting (Schuld *et al* 2016). The quantum matrix can be inverted using the quantum eigenvalue (or phase) estimation algorithm. For quantum phase estimation, given a unitary matrix U and a quantum state $|\Psi\rangle$, what is sought after is the eigenvalue (or phase) $e^{i2\pi\theta}$ such that $U|\Psi\rangle = e^{i2\pi\theta}|\Psi\rangle$. The quantum eigenvalue estimation algorithm can find all the eigenvalues with a high probability and a low error rate inversely proportional to the number of quantum operations (Cleve *et al* 1998).

In addition to linear algebraic tasks, the quantum algorithms are also computationally advantageous for searching. The most famous algorithm in this category is the Grover algorithm, which can find with a high probability the input to a black box function for a specified output (Grover 2001). The Grover algorithm improves unstructured searches, and is very useful for clustering and other learning tasks.

Since classical neural networks are currently quite a hot topic, quantum neural networks will naturally be of great interest. Quantum neural networks can be built in different ways, such as based on the Deutsch quantum model, which is the quantum counterpart of the Turning machine (Gupta and Zia 2001). A number of schemes were proposed to design a quantum equivalent of the perceptron as the building block of a quantum neural network. A key issue is how to implement the activation function. An exemplary implementation of the activation function is the quantum circuit operated via the aforementioned quantum phase estimation process (Schuld *et al* 2015). Specifically, the quantum perceptron has a normalized inner product ϕ between the input vector and the weight vector in the range [0, 1] encoded into the phase of a quantum state $|x_1, x_2, \ldots, x_n\rangle$, and outputs a measurement outcome 1 or 0 with a high probability depending on whether ϕ is greater than a pre-specified threshold (typically, 1/2) or not. The phase estimation can be performed up to a desirable precision as allowed by a sufficiently long string of τ qubits, and we only need the first qubit to know if the estimated phase is greater than 1/2 or not, which implements the activation function for the quantum perceptron.

To simulate a classic perceptron, several ingredients are necessary for a quantum perception, which are an input vector, an inner product, and a nonlinear activation. These can be achieved in several ways. In the recent work on a multidimensional input quantum perceptron (MDIQP) (Yamamoto *et al* 2018), a classical input is assumed to be first mapped into a quantum vector of length N. Then, ancillary qubits are used to steer the input to evolve in phase. Finally, the ancillary qubits are measured as the output.

Let us first look at a single-qubit quantum perception as shown in figure 10.20 (Yamamoto *et al* 2018). Although it is quite simplistic, it shows the idea behind a quantum perceptron nicely. In this single-qubit case, the input qubit $|x_i\rangle = \alpha|0\rangle + \beta|1\rangle$ is paired with a single ancillary qubit $|0\rangle$. After the initial H gate, the ancillary qubit $|0\rangle$ is tilted onto the Bloch sphere equator so that the tilted qubit has equal likelihood to be measured as $|0\rangle$ or $|1\rangle$. Then, the output of the initial H gate and the original input qubit are fed as the input to the W gate whose unitary matrix is defined as follows:

$$
W = \begin{bmatrix} 1 & 0 & 0 & 0 \\ 0 & 1 & 0 & 0 \\ 0 & 0 & e^{i\phi} & 0 \\ 0 & 0 & 0 & e^{i\theta} \end{bmatrix},
\tag{10.19}
$$

which plays the same role as the weighting factor in a classical single input perceptron. It can be shown (the steps omitted) that when $\phi = 0$ we have the

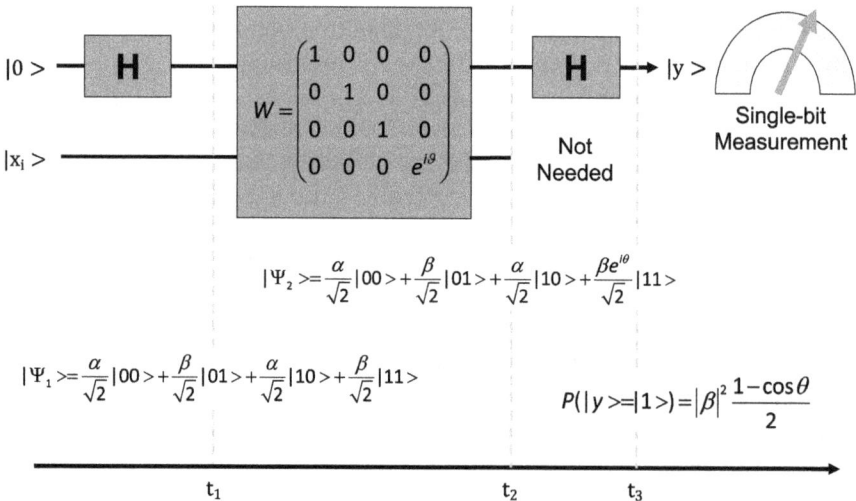

$$|\Psi_2\rangle = \frac{\alpha}{\sqrt{2}}|00\rangle + \frac{\beta}{\sqrt{2}}|01\rangle + \frac{\alpha}{\sqrt{2}}|10\rangle + \frac{\beta e^{i\theta}}{\sqrt{2}}|11\rangle$$

$$|\Psi_1\rangle = \frac{\alpha}{\sqrt{2}}|00\rangle + \frac{\beta}{\sqrt{2}}|01\rangle + \frac{\alpha}{\sqrt{2}}|10\rangle + \frac{\beta}{\sqrt{2}}|11\rangle$$

$$P(|y\rangle = |1\rangle) = |\beta|^2 \frac{1 - \cos\theta}{2}$$

Figure 10.20. Single-qubit quantum perceptron (Yamamoto *et al* 2018), in which a phase estimation is performed so that the output qubit measurement has a nonlinearly (sinusoidally) modulated probability.

following probability of obtaining $|1\rangle$ through measuring the output of the H gate after the matrix W:

$$P(|y\rangle = |1\rangle) = |\beta|^2 \frac{1 - \cos\theta}{2}, \qquad (10.20)$$

which implements a cosine-modulated nonlinear activation operation. Note that the unitary transform is invertible, so quantum computing is invertible, which is generally not the case with classical computation.

Now, we can extend the single-qubit perceptron to a multiple qubit counterpart. Specifically, we prepare the input of $2n$ qubits consisting of n zero states $|0\rangle$ and n states $|x_1, x_2, ..., x_n\rangle$ to the perceptron, denoted as $|0, 0, ..., 0\rangle|x_1, x_2, ..., x_n\rangle = |0, 0, ..., 0\rangle|\Psi_0\rangle$. Then, n H gates transform n zero states into 2^n states with equal chance, encoded as $|J\rangle = |J_1, J_2, ..., J_n\rangle$. Furthermore, a unitary transformation involving the weight parameters of the perceptron is needed to write the inner product ϕ into the phase of the quantum state. Finally, the resultant superposition is evolved via the inverse quantum Fourier transform. Figure 10.21 illustrates a two-qubit quantum perceptron (Yamamoto *et al* 2018).

Once we have a quantum perceptron/neuron, a quantum neural network, even a deep quantum neural network, can be constructed. This is such a new area with too many fresh ideas and major challenges (Biamonte *et al* 2017) to be fully covered in this chapter. However, it is hoped that the introductory materials provided here will

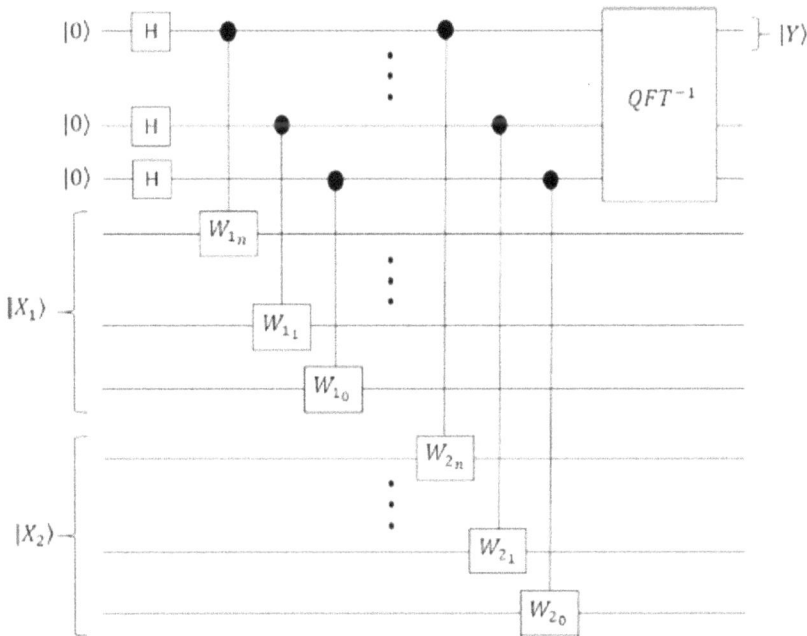

Figure 10.21. A two-qubit quantum perceptron performing the activation via positive interference of wave functions using the inverse quantum Fourier transform (QFT) (Yamamoto *et al* 2018). Reprinted with permission from Yamamoto *et al* (2018). Copyright 2018 Springer Nature.

lay a good foundation for imaging researchers to get prepared for this potentially revolutionary perspective and explore further if interested.

10.5 Final remarks

Not all opinions on quantum computing are positive. In November 2018, a negative opinion on quantum computing was expressed by Dyakonov in *IEEE Spectrum* (Dyakonov 2018). Despite the huge amount of research literature and frequent news exposure, no practical results have been obtained after numerous efforts over decades. Thus, Dyakonov asked a natural question: 'When will useful quantum computers be constructed?' He recognized that the answers range from 5 to 10 years, to 20 to 30 years, and gave his pessimistic answer 'Not in the foreseeable future', based on his understanding of the technical challenges involved.

First of all, the superposition of quantum states in qubits is argued to be not only the origin of the power of quantum computing but also the reason for its fragility. It was estimated that the number of qubits should be at least between 1000 and 100 000 to be practically useful, representing a number of variables larger than the number of particles in the Universe. Since qubits store continuous variables, whose values cannot be exactly set, errors are unavoidable. To make error-corrections for quantum information, each qubit may have to be coupled with 1000 redundant qubits for reliability, which can be mathematically proved but might be beyond our near-future engineering capability. Then, the qubits need to be processed by quantum gates/circuits for computational tasks. In the roadmap by the Advanced Research and Development Activity, a funding agency of the US intelligence community, a 2012 goal 'requires on the order of 50 physical qubits' and 'multiple logical qubits through the full range of operations required for fault-tolerant' quantum computing, which has not been achieved yet. The number of qubits for current experimental studies is typically below 10, and much less than 50, needless to say 1000.

Dyakonov considers that the most promising quantum computing technology is the *D*-wave systems, based on interconnected Josephson junctions cooled to very low temperatures. Recently, 49-, 50-, and 72-qubit chips have been prototyped by Intel, IBM, and Google, respectively. However, he is skeptical that a practical quantum computer will be made soon, and furthermore, 'the quantum computing fervor is nearing its end. That's because a few decades is the maximum lifetime of any big bubble in technology or science.'

Although Dyakonov's arguments made important points and are highly respected, his viewpoints are not of the mainstream. For example, a rebuttal was published by *Quantum Computing Report* (2019), advocating 'quantum computing to become viable in the coming years' for several strong reasons.

The first reason is based on an analogy with the invention of the light bulb in 1879, for which Edison tried 6000 materials until his success. Now, 58 organizations are developing quantum hardware in eight different technology areas, and billions of dollars are being spent yearly by the USA, Europe, China, and others. It is not unreasonable to expect that at least one of the groups will make a breakthrough for scalable quantum devices and then enjoy rapid growth. Also, quantum computers can be used to perform approximate or probabilistic computations for suboptimal answers, so there is no need to be very accurate. For example, quantum computing results in various simulations may be already very useful. Furthermore, hybrid classic/quantum optimization algorithms can be designed, such as quantum approximate optimization algorithms (QAOA) and variational quantum eigensolver (VQE), to tolerate quantum errors and make massive quantum error correction unnecessary.

With quantum computing, the authors of this book are on the proactive side because of the confidence we have gained from the history of technological developments, and that is why we have written this chapter. This proactive attitude is encouraged by multiple types of emerging evidence, such as 3D quantum Hall efforts, room-temperature superconductors, and the great progress in quantum algorithm research.

As the final remark, given the weirdness and utilities of quantum mechanics, quantum computing, and quantum machine learning, there is an increasing interest in the theory and technologies of quantum mechanics, and naturally the interpretations to quantum mechanics (Bunge 2005 and https://en.wikipedia.org/wiki/Interpretations_of_quantum_mechanics) is an ever-fascinating topic for leisure time. Earlier in this chapter we have presented quantum puzzles as A, B, C, D, and E. Let us examine them again from a dual-domain perspective to have a somehow unified understanding and a somewhat new interpretation. This interpretation is motivated by the similarity between the uncertainty principles in quantum mechanics and the counterpart in Fourier analysis. Here, we would like to extrapolate the mathematical structure underlying the Fourier transform to the Heisenberg uncertainty principle. Specifically, just like a function and its Fourier transform are related through a global transform, we suggest that a particle and its wave function are similarly related through a global transform (on this suggested linkage, a student commented as follows: it is better to say that the wave function in the position space and that in the momentum space are related by a global transform). Each uncertainty principle is thus attributed to its underlying global transformation. If this is true, then the seemingly local interaction of particles is also a global interaction of the associated wave functions. In this view, a conventional solid object and a probabilistic wave function are equally real and two sides of the same thing. While the propagation of a wave function is governed by the Schrödinger equation and limited by the speed of light, the particle behavior as a collapse of the wave function or a realization of the global transform is instantaneous.

References

Anthony S 2015 Intel forges ahead to 10 nm, will move away from silicon at 7 nm *Ars Technica* https://arstechnica.com/gadgets/2015/02/intel-forges-ahead-to-10nm-will-move-away-from-silicon-at-7nm/

Biamonte J, Wittek P, Pancotti N, Rebentrost P, Wiebe N and Lloyd S 2017 Quantum machine learning *Nature* **549** 195–202

Bunge M 2005 Survey of the interpretations of quantum mechanics *Am. J. Phys.* **24** 272

CIR 2018 Who is using quantum computers and why? *CIR* https://cir-inc.com/blog/using-quantum-computers/

Cleve R, Ekert A, Macchiavello C and Mosca M 1998 Quantum algorithms revisited *Proc. R. Soc.* A **454** 339–54

Dyakonov M 2018 The case against quantum computing *IEEE Spectrum* https://spectrum.ieee.org/computing/hardware/the-case-against-quantum-computing

Feynman R P 1982 Simulating physics with computers *Int. J. Theor. Phys.* **21** 467–88

Feynman R P 1985 *QED: The Strange Theory of Light and Matter Alix G Mautner Memorial Lectures* (Princeton, NJ: Princeton University)

Grover L K 2001 From Schrodinger's equation to the quantum search algorithm *Am. J. Phys.* **69** 769–77

Guide S, Riebe M, Lancaster G P T, Becher C, Eschner J, Häffner H, Schmidt-Kaler F, Chuang I L and Blatt R 2003 Implementation of the Deutsch–Jozsa algorithm on an ion-trap quantum computer *Nature* **421** 48–50

Gupta S and Zia R K P 2001 Quantum neural networks *J. Comp. Syst. Sci.* **63** 355–83

Jacques V, Wu E, Grosshans F, Treussart F, Grangier P, Aspect A and Roch J-F 2007 Experimental realization of Wheeler's delayed-choice Gedanken experiment *Science* **315** 966–8

Popkin G 2016 Quest for qubits *Science* **354** 1091–3

Quantum Computing Report 2019 The case against quantum computing—a rebuttal *Quantum Computing Report* https://quantumcomputingreport.com/our-take/the-case-against-quantum-computing-a-rebuttal/

Schuld M, Sinayskiy I and Petruccione F 2015 Simulating a perceptron on a quantum computer *Phys. Lett.* A **379** 660–3

Schuld M, Sinayskiy I and Petruccione F 2016 Prediction by linear regression on a quantum computer *Phys. Rev.* A **94** 022342

Yamamoto A Y *et al* 2018 Simulation of a multidimensional input quantum perceptron *Quantum Inf. Process.* **17** 128

Appendix A

Math and statistics basics

A.1 Numerical optimization

A.1.1 Basics in optimization

Numerical optimization is a branch of applied mathematics aimed at finding the solution of an optimization problem, often presented as a minimization, maximization, or saddle-point problem, subject to constraints on the solution. For example, a standard minimization problem is to find an $x \in X$ that minimizes a given function f:

$$\underset{x \in X}{\text{minimize}}\, f(x), \tag{A.1}$$

where $X \subset \mathbb{R}^n$ is the set that characterizes additional constraints on x, and $f(x)$ is called the objective function (also called the cost function or loss function). If X is explicitly given or easy to specify, such as $X = \overline{B(x^*, r)} \triangleq \{x \in \mathbb{R}^n \,|\|x - x^*\| \leqslant r\}$, the closed ball centered at x with radius $r > 0$ in $\mathbb{R}^n$, then the problem (A.1) is called the set constrained minimization. If X is given implicitly, such as $X = \{x \in \mathbb{R}^n \,|g(x) = 0, h(x) \leqslant 0\}$, then (A.1) is called the function-constrained minimization, where $g(x) = 0$ and $h(x) \leqslant 0$ are called the equality and inequality constraint functions, respectively. The maximization problem, $\text{maximize}_{x \in X}\, f(x)$, can also be considered, but mathematically it is equivalent to (A.1) with $-f(x)$ as the objective function. In this appendix, we mainly consider an optimization problem in the form of (A.1).

The goal of (A.1) is to find an optimal solution x that attains the minimum value of f over X. We call x^* a global minimizer of (A.1) if $f(x) \geqslant f(x^*)$ for all $x \in X$. In contrast, we call x^* a local minimizer if $\exists \varepsilon > 0$ such that $f(x) \geqslant f(x^*)$ for all $x \in B(x^*, \varepsilon) \cap X$, i.e. x^* is optimal in its local neighborhood. We call a minimizer x^*, global or local, 'strict' if the equality holds only if $x = x^*$. In most scenarios, we want to find a global minimizer of (A.1). If the admissible set X is convex, i.e. $\lambda x + (1 - \lambda)y \in X$, and f is a convex function, i.e. $f(\lambda x + (1 - \lambda)y) \leqslant \lambda f(x) + (1 - \lambda)f(y)$, for all $x, y \in X$ and $\lambda \in [0, 1]$, then one can easily show that any local minimizer is also global for (A.1). Such a problem is called the convex optimization problem, which has been well studied with numerous theoretical and algorithmic results. However, there may exist a large number of local minimizers for a non-convex optimization problem, i.e. either X or f is non-convex. In the non-convex case, we may be content with a local minimizer, since finding a global one can be very difficult.

In this appendix, we mostly focus on the case where the objective and constraint functions are differentiable. In particular, we assume the objective function f has a Lipschitz continuous gradient ∇f, i.e. there exists $L > 0$ such that $\|\nabla f(x) - \nabla f(y)\| \leqslant L\|x - y\|$ for all $x, y \in X$.

Since most optimization algorithms are designed as iterative procedures that gradually update an existing estimate x to a new (hopefully better) estimate, it is important to determine in which direction x should go in each iteration. To this end, we call $d \in \mathbb{R}^n$ a feasible direction at $x \in X$ if $\exists \varepsilon > 0$ such that $x + \alpha d \in X$ for all $\alpha \in [0, \varepsilon]$, i.e. one remains inside X by moving not too far along direction d. From calculus, we know the directional derivative of f in the direction d is defined by

$$\lim_{\alpha \to 0} \frac{f(x + \alpha d) - f(x)}{\alpha} = d^{\mathsf{T}} \nabla f(x). \tag{A.2}$$

Hence, it is natural to use a descent direction d such that the directional derivative in (A.2) is negative, then the new estimate $x + \alpha d$ satisfies $f(x + \alpha d) < f(x)$ for a properly chosen step size $\alpha > 0$.

It is straightforward to verify that x^* is a local minimizer only if it satisfies the first-order necessary condition: for any feasible direction d at x^*, there is $d^{\mathsf{T}} \nabla f(x^*) \geqslant 0$. If x^* is an interior point of X, i.e. $\exists \delta > 0$ such that $B(x^*, \delta) \subset X$, then x^* is a local minimizer only if $\nabla f(x^*) = 0$. Therefore, given a minimization problem (A.1), we can narrow down the set of local minimizers by first finding the critical points $\{x \in X | \nabla f(x) = 0\}$ and those on the boundary ∂X of X, and then investigating these points further using a second-order condition or similar to determine their optimality.

In the remainder of this section, we list several commonly used optimization algorithms and present the basic theory of constrained optimization. For more comprehensive treatment of numerical optimization, we refer the reader to several excellent textbooks, such as (Bertsekas *et al* 2003, Boyd and Vandenberghe 2004, Chong and Zak 2013, Nesterov 2004, Nocedal and Wright 2000).

A.1.2 Unconstrained optimization algorithms

As we have seen above, an intuitive approach to solving the minimization problem (A.1) is to iterate the following procedure:

$$x^{(k+1)} = x^{(k)} + \alpha_k d^{(k)}, \tag{A.3}$$

where $d^{(k)} \in \mathbb{R}^n$ is a descent direction at the kth iterate $x^{(k)}$. For properly chosen step size $\alpha_k > 0$, we hope that the next iterate $x^{(k+1)}$ further reduces the objective function. The descent direction should make the right-hand side of (A.2) negative, and hence an obvious choice is $d^{(k)} = -\nabla f(x^{(k)})$. This choice yields the basic gradient descent algorithm. It can be shown that, if the minimization problem (A.1) is unconstrained (i.e. $X = \mathbb{R}^n$) and f is convex, then the gradient method with constant step size is convergent for any $\alpha \in (0, 2/L)$ from any initial $x^{(0)}$. However, a constant step size policy is often inefficient in practice: a large step size may cause the iterates $\{x^{(k)}\}$ to diverge, whereas a small step size results in many iterations and slow progress. There are various types of line search strategies to choose proper step sizes α_k automatically, given $x^{(k)}$ and $d^{(k)}$. We refer the interested reader to (Nocedal and Wright 2000) for more discussion regarding this topic.

A main issue with the gradient descent method using $d^{(k)} = -\nabla f(x^{(k)})$ is that the negative gradient direction may not be an effective descent direction, in particular when the problem f is ill-conditioned where $-\nabla f(x^{(k)})$ is not pointing to the minimizer. Therefore, there are a variety of much more effective algorithms that modify $g^{(k)} \triangleq \nabla f(x^{(k)})$ to obtain $d^{(k)}$. For example, Newton's method sets $d^{(k)} = -H(x^{(k)})^{-1}g^{(k)}$, where $H(x^{(k)}) = \nabla^2 f(x^{(k)})$ is the Hessian matrix of f at $x^{(k)}$. However, inverting the Hessian matrix can be very expensive for larger problem size n unless the problem has a very special structure, and sometimes impossible if $H(x^{(k)})$ is singular. Therefore, we mostly just use the first-order methods which only require computations of the gradient ∇f in real-world problems.

Among various first-order methods, nonlinear conjugate gradient and quasi-Newton methods are most commonly used for unconstrained minimization problems. Given an initial $x^{(0)}$, the conjugate gradient method computes $g^{(0)} = \nabla f(x^{(0)})$ and $d^{(0)} = -g^{(0)}$. Then, for iteration $k = 0, 1, 2, \ldots$, the conjugate gradient method iterates the following steps:

$$\alpha_k = \arg \min_{\alpha > 0} f(x^{(k)} + \alpha_k d^{(k)}) \qquad (A.4a)$$

$$x^{(k+1)} = x^{(k)} + \alpha_k d^{(k)} \qquad (A.4b)$$

$$d^{(k+1)} = -g^{(k+1)} + \beta_k d^{(k)}, \qquad (A.4c)$$

where the univariate optimization (A.4a) is often (approximately) solved by the secant method, a simple univariate optimization method (Chong and Zak 2013). The gradient $g^{(k+1)} = \nabla f(x^{(k+1)})$ is computed based on the new iterate $x^{(k+1)}$ in (A.4b). The coefficient β_k in (A.4b) has several common choices (Chong and Zak 2013), including

Hestenes–Stiefel: $\beta_k = \langle g^{(k+1)}, g^{(k+1)} - g^{(k)} \rangle / \langle H_{k+1}d^{(k)}, g^{(k+1)} - g^{(k)} \rangle$

Polak–Ribière: $\beta_k = \langle g^{(k+1)}, g^{(k+1)} - g^{(k)} \rangle \|g^{(k)}\|^2$

Fletcher–Reeves: $\beta_k = \|g^{(k+1)}\|^2 / \|g^{(k)}\|^2$.

The quasi-Newton method is a class of numerical algorithms that employ simple and effective updates of H_k to approximate the computationally expensive inverse of Hessian matrices $\nabla^2 f(x^{(k)})$ in Newton's method. Given an initial $x^{(0)}$ and a symmetric positive definite matrix H_0, the quasi-Newton method computes $g^{(0)} = \nabla f x^{(0)}$ and then iterates the following steps:

$$d^{(k)} = -H_k g^{(k)}, \tag{A.5a}$$

$$\alpha_k = \arg \min_{\alpha \geq 0} f(x^{(k)} + \alpha d^{(k)}), \tag{A.5b}$$

$$x^{(k+1)} = x^{(k)} + \alpha_k d^{(k)}. \tag{A.5c}$$

There are different ways to compute the matrix H_k in the literature. The most commonly used one was developed by Broyden, Fletcher, Goldfarb, and Shannon (BFGS):

$$
\begin{aligned}
H_{k+1} = H_k &+ \left(1 + \frac{\Delta g^{(k)\mathsf{T}} H_k \Delta g^{(k)}}{\Delta g^{(k)\mathsf{T}} \Delta x^{(k)}} \right) \frac{\Delta x^{(k)} \Delta x^{(k)\mathsf{T}}}{\Delta x^{(k)\mathsf{T}} \Delta g^{(k)}} \\
&- \frac{H_k \Delta g^{(k)} \Delta x^{(k)\mathsf{T}} + (H_k \Delta g^{(k)} \Delta x^{(k)\mathsf{T}})^{\mathsf{T}}}{\Delta g^{(k)\mathsf{T}} \Delta x^{(k)}},
\end{aligned}
\tag{A.6}
$$

where $\Delta x^{(k)} = x^{(k)} - x^{(k-1)}$ and $\Delta g^{(k)} = g^{(k)} - g^{(k-1)}$. Note that the denominators are all inner products and hence are scalars, and the numerators such as $\Delta x^{(k)} \Delta x^{(k)\mathsf{T}}$ are rank-one matrices.

The iteration steps (A.3) must be terminated in finitely many steps in practice. It is usually better to let the algorithm terminate itself rather than setting the number of iterations manually in advance. One common choice for automatic termination is the use of a proper stopping criterion, such as $\|x^{(k+1)} - x^{(k)}\| / \|x^{(k+1)}\| \leq \varepsilon_{\text{tol}}$ for some prescribed tolerance $\varepsilon_{\text{tol}} > 0$, indicating that more iterations cannot further make sufficient improvements to $x^{(k)}$, and hence may stop for the sake of computational efficiency.

A.1.3 Stochastic gradient descent methods

The classical gradient descent methods introduced above require the computation of the gradient ∇f of the objective function f in every iteration. In many problems, such as those in deep learning, computing the full gradient ∇f is expensive since f may consist of many terms (e.g. the loss function $f(x) = \sum_{j=1}^{m} f_j(x)$ in deep learning is the sum of m squared errors over all training data pairs where m is of the order of thousands to even millions). Moreover, computing the full gradient of f can be wasteful since some $\nabla f_j(x)$ and $\nabla f_{j'}(x)$ (or more terms) may cancel each other. Hence, it is practically more efficient to compute the gradient of a mini-batch of f_j in each iteration. More precisely, one can select (or sample) a mini-batch $\{f_j | j \in D^{(k)}\}$ where $D^{(k)} = \{j_1, \ldots, j_{m'}\} \subset [m]$ is of size $m' \ll m$, and use $\sum_{l=1}^{m'} \nabla f_{j_l}$ in place of ∇f in iteration k.

The modifications above are within the class of stochastic gradient descent (SGD) methods, which are still undergoing fast developments in the field of optimization research. Here we introduce several SGD methods commonly used in the deep learning community. For simplicity, we denote the sum over a mini-batch $\{f_j | j \in D^{(k)}\}$ in iteration k by $F_k(x) \triangleq \sum_{j \in D^{(k)}} f_j(x)$. The first SGD method is called an adaptive gradient, or AdaGrad, whose iteration steps are given by

$$r^{(k+1)} = r^{(k)} + \|\nabla F_k(x^{(k)})\|^2 \tag{A.7a}$$

$$x^{(k+1)} = x^{(k)} - \alpha \nabla F_k(x^{(k)})/\sqrt{r^{(k+1)}}, \tag{A.7b}$$

where $r^{(0)} = 0$. AdaGrad (A.7) puts equal weights on all past (squared norm of) gradients. To place more weight on the recent gradients, the root mean square propagation, or RMSProp, generalizes the scheme (A.7) to

$$r^{(k+1)} = \lambda r^{(k)} + (1 - \lambda)\|\nabla F_k(x^{(k)})\|^2 \tag{A.8a}$$

$$x^{(k+1)} = x^{(k)} - \alpha \nabla F_k(x^{(k)})/\sqrt{r^{(k+1)}}, \tag{A.8b}$$

where $\lambda \in [0, 1)$. RMSProp renders a greater weight on the recent gradient by using a smaller λ, and otherwise by selecting a larger one. The adaptive moment estimate method, or ADAM, is a variation and further improvement of the two methods above:

$$m^{(k+1)} = \lambda_1 m^{(k)} + (1 - \lambda_1)\nabla F_k(x^{(k)}) \tag{A.9a}$$

$$r^{(k+1)} = \lambda_2 r^{(k)} + (1 - \lambda_2)\|\nabla F_k(x^{(k)})\|^2 \tag{A.9b}$$

$$\hat{m}^{(k+1)} = m^{(k+1)}/(1 - \lambda_1^k) \tag{A.9c}$$

$$\hat{r}^{(k+1)} = r^{(k+1)}/(1 - \lambda_2^k) \tag{A.9d}$$

$$x^{(k+1)} = x^{(k)} - \alpha \hat{m}^{(k+1)}/\sqrt{r^{(k+1)} + \varepsilon} \tag{A.9e}$$

with $m^{(0)} = 0 \in \mathbb{R}^n$ and $r^{(0)} = 0$, where typical choices of the parameters above are $\alpha = 0.001$, $\lambda_1 = 0.9$, $\lambda_2 = 0.999$, and $\varepsilon = 10^{-8}$. In ADAM (A.9), $m^{(k)}$ and $r^{(k)}$ represent the estimate of the expectation (first moment) and (norm of) second moment of the gradient, respectively. The scaling by $1/(1 - \lambda_i^k)$ for $i = 1, 2$ is for bias correction. The offset ε is to avoid division by 0 in case $\hat{r}^{(k)}$ is too small. It is easy to verify that ADAM reduces to several existing SGD methods (asymptotically), including AdaGrad and RMSProp above, with specific choices of λ_i and α.

A.1.4 Theory of constrained optimization

To conclude this section, we briefly introduce the theory of constrained optimization. In general, constrained optimization is more challenging to solve compared to their unconstrained counterparts since not all search directions are feasible in the constrained case. A common practice in this case is to modify the gradient descent method (A.3) to the projected gradient descent:

$$x^{(k+1)} = \Pi_X(x^{(k)} + \alpha_k d^{(k)}), \tag{A.10}$$

where $\Pi_X(b) = \arg\min_{x \in X} \|x - b\|$ is called the projection of b onto X. Such an approach is useful only if the projection can be performed at a low computational cost, which may not always be the case.

The fundamental theory of constrained optimization is based on the Lagrangian of the problem and related optimality conditions. Consider a constrained optimization problem with multiple equality and inequality constraints as follows:

$$\underset{x \in \mathbb{R}^d}{\text{minimize}} \ f(x), \tag{A.11a}$$

$$\text{subject to} \ g_i(x) = 0, \quad i = 1, \dots, n, \tag{A.11b}$$

$$h_j(x) \leqslant 0, \quad j = 1, \dots, m, \tag{A.11c}$$

where $g_i(x)$ and $h_j(x)$ are the ith equality constraint function and jth inequality constraint functions, respectively. To derive the optimality condition of a solution x^* to (A.11), we first form its Lagrangian:

$$L(x; \lambda, \mu) = f(x) + \lambda^\top g(x) + \mu^\top h(x), \tag{A.12}$$

where $g(x) \triangleq (g_1(x), \dots, g_n(x)) \in \mathbb{R}^n$, $h(x) \triangleq (h_1(x), \dots, h_m(x)) \in \mathbb{R}^m$. In (A.12), $\lambda \in \mathbb{R}^n$ and $\mu \in \mathbb{R}^m_+$ are the Lagrangian multipliers for the equality and inequality constraints, respectively. The constrained optimization problem (A.11) is equivalent to the following min–max problem:

$$\min_x \max_{\lambda, \mu \geqslant 0} L(x; \lambda, \mu). \tag{A.13}$$

In fact, it can be easily verified that

$$\tilde{f}(x) \triangleq \max_{\lambda, \mu \geqslant 0} L(x; \lambda, \mu) = \begin{cases} f(x) & \text{if } g(x) = 0 \text{ and } h(x) \leqslant 0, \\ +\infty & \text{otherwise}. \end{cases} \tag{A.14}$$

Therefore, (A.13) is equivalent to minimize$_x$ $\tilde{f}(x)$, which is identical to the original constrained minimization (A.11).

To find a solution of (A.11), we first determine the necessary optimality condition, called the Karush–Kuhn–Tucker (KKT) condition:

$$\nabla_x L(x; \lambda, \mu) = \nabla f(x) + \lambda^\top \nabla g(x) + \mu^\top \nabla h(x) = 0, \tag{A.15a}$$

$$g(x) = 0, \quad h(x) \leqslant 0, \quad \mu \geqslant 0, \tag{A.15b}$$

$$\mu_j h_j(x) = 0 \text{ for all } j = 1, \dots, m. \tag{A.15c}$$

The last equation (A.15c) is called the complementary slackness condition, which implies that either μ_j or $h_j(x)$ must be 0. If $h_j(x) = 0$, then the constraint is called active, and otherwise it is inactive. In practice, we first search for points that satisfy the KKT condition (A.15), and then determine their optimality by further checking the second-order conditions (Chong and Zak 2013). There are numerous methods developed to solve the KKT system (A.15), in particular in the convex optimization case, and they are often based on different principles and vary for different application problems. We refer the interested reader to Bertsekas *et al* (2003), Boyd and Vandenberghe (2004), Nesterov (2004), Nocedal and Wright (2000) for more discussion on this topic.

A.2 Statistical inferences

Machine learning is closely related to statistics. One common task in both machine learning and statistics is to build a suitable probabilistic model with unknown model parameters to describe the random outcomes of a process or experiment. These parameters are then inferred when the experiment is repeated and a sufficient amount of data are collected. There are two closely related approaches for parameter inference: maximum likelihood estimate (MLE) and Bayesian estimate (or maximum *a posteriori*, MAP) (Casella and Berger 1990).

In MLE, we first assume that the observable data are samples $\{x^{(i)}|i = 1, \ldots, N\}$ (often assumed to be independently and identically distributed or i.i.d.) from a distribution $p(x; \theta)$ with a known expression of the distribution function p but an unknown parameter θ. This is often denoted by $x^{(i)} \sim p(x; \theta)$ i.i.d. Given the observations $\{x^{(i)}\}$, we would infer that the parameter is θ^*—the one under which such observations are most likely. That is,

$$\theta^* = \arg \max_{\theta} L(\theta; \{x^{(i)}\}) \text{ where } L(\theta; \{x^{(i)}\}) \triangleq \prod_{i=1}^{N} p(x^{(i)}; \theta). \qquad (A.16)$$

In (A.16), $L(\theta; \{x^{(i)}\})$ is called the likelihood function of θ given the samples $\{x^{(i)}\}$. Due to the exponential form of most probability distribution functions, it is often more convenient to find θ^* by minimizing $-\log L(\theta; \{x^{(i)}\}) = -\sum_{i=1}^{N} \log p(x^{(i)}; \theta)$.

The Bayesian estimate is based on a fundamental identity in probability theory, called Bayes' rule. Bayes' rule exploits the joint and conditional probabilities of two random variables X and Y:

$$p(y|x) = \frac{p(x, y)}{p(x)} = \frac{p(x|y)p(y)}{p(x)}, \qquad (A.17)$$

where $p(x)$ is the marginal distribution of X, $p(x, y)$ is the joint distribution of X and Y, and $p(x|y)$ is the conditional distribution of X given Y. In the Bayesian estimate, we treat the unknown parameter θ as a 'random variable' jointly with the observations x. We then provide a guess of its distribution $p(\theta)$, called the prior, and find θ^* that maximize the posterior distribution $p(\theta|x)$ given a sample x. More precisely, with the samples $x^{(i)} \sim p(x|\theta)$ i.i.d., we apply Bayes' rule to obtain

$$p(\theta|\{x^{(i)}\}) = \frac{p(\{x^{(i)}\}|\theta)p(\theta)}{p(\{x^{(i)}\})} \propto p(\{x^{(i)}\}|\theta)p(\theta) = p(\theta)\prod_{i=1}^{N} p(x^{(i)}|\theta). \quad (A.18)$$

To find θ^* that maximizes the posterior distribution $p(\theta|\{x^{(i)}\})$, we can just minimize the negative logarithm of the right-hand side (since the denominator $p(\{x^{(i)}\})$ does not involve θ):

$$\theta^* = \arg\min_{\theta} \left(-\log p(\theta) - \sum_{i=1}^{N} \log p(x^{(i)}|\theta) \right). \quad (A.19)$$

Note that the last term in the objective function is the same as the negative logarithm of the likelihood function (A.16). Through the additional prior term, we can impose our preference or experience on θ in the search for these optimal model parameters.

A.3 Information theory

Information theory is a branch of applied mathematics and statistics to study the information and uncertainties in random variables for signal processing, analysis, and communications (Gray 2011, MacKay and MacKay 2003, Martin and England 2011).

A.3.1 Entropy

In information theory, the uncertainty (randomness) of a random variable $X \sim p(x)$ is quantified by the differential entropy, or information entropy, or simply entropy for short:

$$H(X) \triangleq \mathbb{E}[-\log(X)] = -\int p(x) \log p(x) \mathrm{d}x, \quad (A.20)$$

where X is a continuous random variable with probability density function p. Differential entropy is a generalization of the Shannon entropy of a discrete random variable X with probability mass function $p \in \Delta^n \triangleq \{p \in \mathbb{R}^n \mid \sum_{i=1}^{n} p_i = 1, 0 \leqslant p_i \leqslant 1\}$, which is defined as $H(X) = -\sum_{i=1}^{n} p_i \log p_i$. Here, the logarithm is taken with natural base e, and there is a constant scaling change if a logarithm with other base is used.

Briefly speaking, the entropy $H(X)$ is a measure to quantify the amount of uncertainty in X: the greater amount of randomness X has, the larger $H(X)$ is. Note that $H(X) \geqslant 0$ for any discrete random variable X, but $H(X)$ can be negative for continuous ones: the former is easy to verify since $0 \leqslant p_i \leqslant 1$ and hence $p_i \log p_i \leqslant 0$ for all i (we take the convention that $r \log r = 0$ for $r = 0$ due to $\lim_{r \to 0^+} r \log r = 0$ by the l'Hôpital's rule); the latter has examples such as $H(X) = \log \alpha$ for the uniform random variable $X \sim \text{Uniform}(0, \alpha)$ with probability density function $p(x) = 1/\alpha$ over interval $(0, \alpha)$, for which $H(X) \to -\infty$ as $\alpha \to 0$. Another example is $H(X) = (1/2) \cdot \log(2\pi e\sigma^2)$ for the Gaussian random variable $X \sim N(0, \sigma^2)$ with probability density function $p(x) = (2\pi\sigma^2)^{-1/2}e^{-x^2/(2\sigma^2)}$, for which $H(X) \to -\infty$ as

$\sigma \to 0$. Without loss of generality, we consider continuous random variables and use integral notations of entropy hereafter.

We can also define the joint entropy of two random variables X and Y with joint distribution $p(x, y)$ by

$$H(X, Y) = \mathbb{E}_{(X,Y)}[-\log p(X, Y)] = -\iint p(x, y) \log p(x, y) \mathrm{d}x \mathrm{d}y. \quad (A.21)$$

Moreover, the conditional entropy is defined by

$$H(X|Y) = H(X, Y) - H(Y) = \mathbb{E}_{(X,Y)}[-\log p(X|Y)]$$
$$= -\iint p(x, y) \log p(x|y) \mathrm{d}x \mathrm{d}y. \quad (A.22)$$

If $H(Y)$ is the amount of randomness in Y, and $H(X, Y)$ is the total amount of randomness in (X, Y), then the conditional entropy $H(X|Y)$ measures the amount of randomness left in X when Y is observed. In fact, it is straightforward to verify that $H(X|Y) = 0$ if and only if X can be completely determined by Y (e.g. X is a function of Y); and $H(X|Y) = H(X)$ if and only if X and Y are independent, in which case $H(X, Y) = H(X) + H(Y)$. In general, there is always $H(X|Y) \leqslant H(X)$, which can be verified as follows: let $p_X(x)$ and $p_Y(y)$ be the marginal distributions of X and Y, respectively, then there is

$$H(X|Y) = -\iint p(x, y) \log \frac{p(x, y)}{p_Y(y)} \mathrm{d}x \mathrm{d}y$$
$$= -\iint p_X(x) \frac{p(x, y)}{p_X(x)} \log \frac{p(x, y)}{p_Y(y)} \mathrm{d}x \mathrm{d}y$$
$$= \int p_X(x) \left[\int \frac{p(x, y)}{p_X(x)} \log \frac{p_Y(y)}{p(x, y)} \mathrm{d}y \right] \mathrm{d}x$$
$$\leqslant \int p_X(x) \log \left[\int \frac{p(x, y)}{p_X(x)} \cdot \frac{p_Y(y)}{p(x, y)} \mathrm{d}y \right] \mathrm{d}x$$
$$= \int p_X(x) \log \frac{1}{p_X(x)} \mathrm{d}x = H(X),$$

where the inequality is due to $\mathbb{E}[\log(g(X, Y))] \leqslant \log(\mathbb{E}[g(X, Y)])$, Jensen's inequality applied to $g(x, y) \triangleq p_Y(y)/p(x, y)$, and expectation taken over $Y|X$ with density $p(y|x) = p(x, y)/p_X(x)$, and that the logarithm is concave. One can also verify that $H(X|Y, Z) \leqslant H(X|Z)$:

$$H(X|Y, Z) = \mathbb{E}_{(Y,Z)}[H(X|Y, Z)] = \mathbb{E}_Y[\mathbb{E}_{Z|Y}[H(X|Y, Z)]]$$
$$= \mathbb{E}_Y[\mathbb{E}_{Z|Y}[H((X|Y)|Z)]]$$
$$\leqslant \mathbb{E}_Y[\mathbb{E}_{Z|Y}[H(X|Y)]] = \mathbb{E}_Y[H(X|Y)] = H(X|Y).$$

Moreover, using the definition (A.22), it is easy to verify that $H(X, Y|Z)=H(Y|Z) + H(X|Y, Z)$, which can be generalized to the so-called chain rule for conditional entropy:

$$H(X_1, \ldots, X_k|Z) = H(X_1|Z) + H(X_2|X_1, Z) + \cdots + H(X_k|X_1, \ldots, X_{k-1}, Z). \quad \text{(A.23)}$$

The chain rule (A.23) also implies that $H(X_1, \ldots, H_k|Z) \leqslant \sum_{i=1}^{k} H(X_i|Z)$.

A.3.2 Mutual information

In addition to conditional entropy, mutual information is commonly used to measure the dependence between two random variables X and Y, or the information 'shared' by these two random variables:

$$I(X, Y) \triangleq H(X) + H(Y) - H(X, Y) = \iint p(x, y) \log \frac{p(x, y)}{p(x)p(y)} dx dy. \quad \text{(A.24)}$$

Unlike the conditional entropy, the mutual information $I(X, Y)$ is symmetric in X and Y. Moreover, $I(X, Y) = H(X) - H(X|Y) \geqslant 0$ due to the definition of conditional entropy (A.22). In fact, $I(X, Y) = 0$ if X and Y are independent; and the greater the dependence between X and Y, the larger $I(X, Y)$. Note that $I(X, X) = H(X)$. Using the definition of $I(X, Y)$ and that $H(X|Y, Z) \leqslant H(X|Y)$, we can easily verify that $I(X, (Y, Z)) \geqslant I(X, Y)$ for any random variables X, Y, and Z.

A.3.3 Kullback–Leibler divergence

In information theory, it is a common problem to measure the similarity of two probability distributions $p(x)$ and $q(x)$. One of the common choices is the Kullback–Leibler (KL) divergence, defined as

$$\text{KL}(p||q) = \mathbb{E}_{X \sim p}\left[\log \frac{p(X)}{q(X)}\right] = \int p(x) \log \frac{p(x)}{q(x)} dx. \quad \text{(A.25)}$$

KL divergence is always nonnegative, and it becomes 0 if only if the two distributions p and q are identical. The larger $\text{KL}(p||q)$ is, the less similar that p and q are. Although frequently used to measure the distance between two probability distributions, strictly speaking, the KL divergence is not a distance since it is not symmetric in its two arguments, and it does not satisfy the triangle inequality in general.

It can also be verified that the mutual information $I(X, Y)$ between X and Y defined in (A.24) is in fact the KL divergence between the joint distribution $p(x, y)$ and the product of their marginal distributions $p(x)p(y)$. If X and Y are independent, then $p(x, y) = p(x)p(y)$, which minimizes the KL divergence between them to 0, and hence $I(X, Y) = 0$, a fact we have mentioned above.

References

Bertsekas D P, Nedić A and Ozdaglar A E 2003 *Convex Analysis and Optimization* (Belmont: Athena Scientific)

Boyd S and Vandenberghe L 2004 *Convex Optimization* (Cambridge: Cambridge University Press)

Casella G and Berger R L 1990 *Statistical Inference* (Pacific Grove, CA: Wadsworth and Brooks/ Cole)

Chong E K and Zak S H 2013 *An Introduction to Optimization* vol 76 (Hoboke, NJ: Wiley)

Gray R M 2011 *Entropy and Information Theory* (New York: Springer)

MacKay D J and MacKay D J 2003 *Information Theory, Inference and Learning Algorithms* (Cambridge: Cambridge University Press)

Martin N F and England J W 2011 *Mathematical Theory of Entropy* vol 12 (Cambridge: Cambridge University Press)

Nesterov Y 2004 *Introductory Lectures on Convex Optimization: A Basic Course* vol 87 (New York: Springer)

Nocedal J and Wright S 2000 *Numerical Optimization* (New York: Springer)

IOP Publishing

Machine Learning for Tomographic Imaging

Ge Wang, Yi Zhang, Xiaojing Ye and Xuanqin Mou

Appendix B

Hands-on networks

B.1 Open source toolkits for deep learning

With the impressive successes of deep learning methods in various fields, several well-known open source toolkits were developed by different groups. A number of popular and complicated operations, such as the backpropagation through different layers, were simplified by packaging these operations into high-level commands programmable through a user-friendly interface, which greatly facilitates the development and application of deep learning algorithms. Several representative toolkits of this type are described as follows.

(a) Tensorflow (http://www.tensorflow.org/) is an open source software library created by Google that is used to implement machine learning, in particular deep learning systems. It works efficiently with mathematical expressions involving multi-dimensional arrays and supports deep neural networks concepts by design. It enables GPU/CPU computing in the sense that the same code can be executed on both serial and parallel computing architectures.

(b) Pytorch (https://pytorch.org/) is a machine learning library for the programming language Python, based on the Torch library used for applications such as deep learning and natural language processing. It is primarily developed by Facebook's artificial-intelligence research group. Pytorch has two major features: (i) tensor computing (such as NumPy) that can be accelerated on graphics processing units (GPUs) and (ii) deep neural networks using an automatic differentiation method.

(c) Caffe (https://caffe.berkeleyvision.org/) was originally designed by the Berkeley Vision and Learning Center (BVLC), and has been improved by the group and other community contributors. Of all the open source tools for deep learning, it is probably the most widely used. Its framework is a BSD-licensed C++ library with Python and MATLAB bindings for constructing and training deep models.

doi:10.1088/978-0-7503-2216-4ch12